Maritime Transportation

The environmental and human costs of marine accidents are high, and risks are considerable. At the same time, expectations from society for the safety of maritime transportation, like most other activities, increase continuously. To meet these expectations, systematic methods for understanding and managing the risks in a cost-efficient manner are needed. This book provides readers with an understanding of how to approach this problem.

Firmly set within the context of the maritime industry, systematic methods for safety management and risk assessment are described. The legal framework and the risk picture within the maritime industry provide necessary context. Safety management is a continuous and wide-ranging process, with a set of methods and tools to support the process. The book provides guidance on how to approach safety management, with many examples from the maritime industry to illustrate practical use.

This extensively revised new edition addresses the needs of students and professionals working in shipping management, ship design and naval architecture, and transport management, as well as safety management, insurance, and accident investigation.

Maritime Transportation
Safety Management and
Risk Analysis

Second Edition

Stein Haugen and Svein Kristiansen

Routledge
Taylor & Francis Group

LONDON AND NEW YORK

First published 2023
by Routledge
4 Park Square, Milton Park, Abingdon, Oxon OX14 4RN

and by Routledge
605 Third Avenue, New York, NY 10158

Routledge is an imprint of the Taylor & Francis Group, an informa business

© 2023 Stein Haugen and Svein Kristiansen

British Library Cataloguing-in-Publication Data
A catalogue record for this book is available from the British Library

Library of Congress Cataloging-in-Publication Data
Names: Haugen, Stein, author. | Kristiansen, Svein, 1942- author.
Title: Maritime transportation : safety management and risk analysis / Stein Haugen and Svein Kristiansen.
Description: Second edition. | New York : Routledge, 2023. |
Includes bibliographical references and index.
Identifiers: LCCN 2022024536 | ISBN 9780367518561 (paperback) |
ISBN 9780367518578 (hardback) | ISBN 9781003055464 (ebook)
Subjects: LCSH: Navigation—Safety measures. | Navigation—Risk assessment. |
Shipping—Safety measures. | Shipping—Risk assessment. |
Ships—Safety measures. | Ships—Safety regulations.
Classification: LCC VK200 .H38 2023 | DDC 623.88/8—dc23/eng/20220831
LC record available at https://lccn.loc.gov/2022024536

ISBN: 978-0-367-51857-8 (hbk)
ISBN: 978-0-367-51856-1 (pbk)
ISBN: 978-1-003-05546-4 (ebk)

DOI: 10.4324/9781003055464

Typeset in Sabon
by codeMantra

Contents

Preface

Safety in maritime transportation has improved steadily over the past decades, as well as since the first edition of this book was published. This is evident from development in accident statistics, fatalities, and spills. Despite this positive trend, we still experience accidents and events that tell us that we still have a long way to go. The authors of this book have witnessed this on our own doorstep, off the coast of Norway.

In the fall of 2018, the Norwegian Royal Navy frigate *Helge Ingstad* collided with the oil tanker TS Sola on the west coast of Norway. The frigate was severely damaged, and after an unsuccessful beaching, it sank.

One of the key factors in the collision was that the bridge team on *Helge Ingstad* was unable to distinguish between the tanker and the oil terminal that the tanker had just left. It was clear that the navigation was flawed in many respects partly due to unexperienced personnel. This could be traced back to a naval policy of not having the right balance between investing in technology and spending on training and competence.

Another matter in this accident was also that soon after the accident, the naval command put the blame on the tanker and the shipyard. Their view was that the frigate sank due to wrong design, whereas subsequent investigations showed that the crew lacked a fundamental understanding of damage control. This points at an important lesson: Only proper, in-depth accident investigation can tell you what the real causes of an accident are, and without this knowledge, we cannot learn the right lessons to avoid future accidents. Learning is a key attribute of a safety-oriented organization.

Another recent example from Norway occurred in the spring of 2019, when the cruise vessel *Viking Sky* with 1,373 persons onboard lost power off the west coast of Norway under heavy weather and sea conditions. The vessel started to drift toward the coast. Drifting continued for nearly 1 hour before the anchors were engaged, and the vessel stopped only one ship-length from the shore. Control was regained with tug assistance, and no casualties were experienced. Analysis showed that if the engine loss had taken place a few minutes later or with slightly more southern weather, the distance to shore would have been half of the experienced drift distance and a disaster with loss of ship and many persons would most likely have occurred.

The heavy weather was forecast, but the ship still set off on its trip. The Norwegian Coastal Express, in practice also a cruise ship, crosses the same stretch of the coast twice a day and knows the conditions very well but had decided to remain in port. The case raises the question about assessing the situation and the consequences of setting out to sea. Given the hard environmental conditions, the vessel had in practice no own evacuation means as the lifeboats and rafts would have been impossible to operate. This raises the question of whether the risk was properly assessed before setting off to sea.

The cause of the loss of power was an automatic shutdown of the engines due to loss of lubricant. The automatic shutdown was there to protect damage to the engines. This illustrates the increasing complexity of modern technological systems and how important it is to do systematic analysis of the effect of failures and losses. Comprehensive risk analysis is a valuable tool in this process.

These two examples illustrate the importance of the topics that are covered in this book. To us, it is a reminder that the knowledge that we try to impart is needed in the industry. It is therefore our hope that maybe this book can contribute in a modest way to the improvement of safety in maritime transportation even further.

The writing of this book has been a close cooperation between the author of the first edition, Professor Svein Kristiansen, and Professor Stein Haugen. The present edition is to a large degree based on teaching experience and lecture notes for Master's and PhD students. We have shared the main responsibility for writing the chapters between us but have also commented on and corrected the chapters written by the other. We therefore regard the whole book as a joint project, not dividing authorship for the individual chapters between us.

Finally, we would like to express our thanks to former PhD students at the Department of Marine Technology at NTNU, Sheng Xu, who helped us with reference checking. We would also like to thank the publishers, who have patiently answered the questions we have had and have extended our deadline whenever we have run into problems with completing the work in time.

Stein Haugen
Svein Kristiansen

Authors

Professor Stein Haugen had been working at the Norwegian University of Science and Technology, in the Department of Marine Technology until 2021. He has extensive experience from industry and recently joined the consultancy company Safetec Nordic AS. His teaching and research has been in risk assessment and safety management, for the maritime, oil and gas, and process industry. He has extensive international experience, having worked in the UK for two years and in Italy for a year, as a visiting professor.

Professor Svein Kristiansen, PhD,has been teaching ship design and safety management at The Norwegian University of Science and Technology. He has also served as advisor to Sintef Ocean and Safetec Nordic AS and as External Examiner at Strathclyde University. His research interest is risk analysis and human factors. The research on accident analysis was linked to several EU-funded programs. He has been a visiting scholar at Scripps Institution of Oceanography (UC San Diego) and University of Valencia.

Chapter 1

Introduction

1.1 BACKGROUND

In the year 1120, the White Ship, a newly refitted vessel, was offered to King Henry I of England to take him from Normandy in France back to England (Wikipedia – The White ship story). He had already made other arrangements for his return, but many of his relatives and other nobles, including his heir, chose to use the ship. Upon departure, there were about 300 people on board. Shortly after leaving port, the ship struck a rock and capsized, causing many fatalities. The result of this was that King Henry was left with only one in-wedlock daughter that the nobles of England were reluctant to accept. When the king died in 1135, a civil war resulted that lasted until 1153, with a devastating effect for England.

Few maritime accidents have had such massive effects on a country as the sinking of the White Ship, but we can find many examples of maritime disasters that have caused large losses, many fatalities, and severe environmental effects. Accidents have also caused significant changes to legislation and regulations of maritime safety over the years.

The maritime industry has been and is very important for worldwide trade and the global economy. Waterborne transport of goods has for centuries been the basis for trade of basic materials and goods between nations and regions. Economic prosperity and development are to a large degree founded on division of labor and economic activity – in other words specialization. For this to be achievable, cheap and efficient transport is a necessity.

The cost of maritime transport is very competitive compared with land and air transport. The increase incurred to the total product cost by shipping only represents some few percent of the total cost. As an example, the cost of transporting a container from China to Europe cost around $2000 in 2020 (Financial Times 2021). The weak aspect of waterborne transport is longer transport time due to lower ship speed, congestion in harbors, and less efficient integration with other forms of transport and distribution.

DOI: 10.4324/9781003055464-1

Shipping has been under attack for unacceptable safety and environmental performance. This point will be discussed in Chapter 2. For now, it is however worth noting the paradox underlying the fact that shipping provides cheap and efficient transport but still seems to have low safety standards compared to other forms of transport (e.g. air and rail). Efficient transport should be able to pay for acceptable safety.

1.2 INTERNATIONAL TRADE AND SHIPPING

Seaborne transport of raw materials and manufactured goods is the backbone of international trade and shipping represents in the order of 70–80% of the total trade. In an increasingly globalized economy, the annual growth rate of shipping has been 2.5%. In broad terms the shipping market not only consists of tanker, general cargo, and dry bulk transport, but consists also of a range of specialized ships like container ships, ro-ro vesséls, ferries, and passenger vessels to mention a few. General cargo vessels are smaller than tankers and bulk carriers and may explain why they are greater in number than the specialized ships.

The economics of shipping is strongly influenced by the variation in international trade and thereby demand for transport services. This is especially true for tanker shipping where the charter rate (payment) may vary with a factor of almost 10. The transport market also influences cost of new ships and the price of scrapped vessels. A classic discussion is whether the volatility in the shipping market hinders a systematic effort to improve safety in this sector.

Seaborne transport is the cheapest mode of transport compared to road and air transport. On the other hand, the speed of sea transport is definitely slower than the other modes. This means that there is a division of labor where shipping stands for high-volume and low-cost goods (tank and bulk) compared to air transport that primarily carries high-cost and light cargoes. Container ships with relative higher speed may be placed between these markets as they transport manufactured good in great volumes.

The ship owning company or manager is at the outset responsible for safe operation and will be blamed after serious accidents and losses. It has been shown that the safety of shipping often has been questioned after accidents with human losses or serious environmental damage. Control of safety in shipping is said to be demanding due to the complicated organization or large numbers of actors like maritime administration, shipyards, classification, insurance, etc. One of the issues is the freedom of shipowners to flag a vessel under an administration with a lenient or minimal safety control regime. A more recent issue is the emerging realization that the charterer (cargo owner) may have significant influence on the operation without becoming liable in the case of accidents.

1.3 RISK AND SAFETY

Fundamentally, we may ask ourselves the question of why we should strive to improve safety? After all, there are both costs and practical implications of doing this.

The obvious answer is of course that we do this to reduce human suffering. As a general rule, people should not be injured or killed by activities that we as humans initiate ourselves. This is the ethical point of view underlying efforts to reduce risk. However, we soon move into the realm of economics, because we also see advantages of undertaking certain activities that involve risk. As already stated, shipping provides cheap and efficient transport, and this is a benefit to us that we are not necessarily willing to give up even if it means that people are killed and injured. Here, it is important to note that the benefits primarily are relevant for those who have a need for transport of goods (although in the end it will benefit all who buy those goods), while the risks have to be borne mainly by the crew of the ships. Different groups thus have different interests, and this destroys the balancing of risks vs benefits.

In addition, there are also financial incentives for reducing risk. Accidents are expensive, especially serious accidents where both ship and cargo may be lost. This is however not enough to reduce risk to a level that would be considered as acceptable (although this can vary very much between countries, cultures, and companies).

Therefore, we cannot leave it to the market to find the "right" level of risk. Instead, governments will normally take on the role of introducing and enforcing sufficiently strict requirements for safety. In the case of shipping, governments have however largely given this task over to IMO, the International Maritime Organization.

This may be one explanation why the risk level in the maritime industry is relatively high. In IMO, all countries have to agree and there will therefore be a tendency that the safety standards end up being lower than what, e.g., many developed countries would like to have.

On the other hand, air transport is also a very international business, regulated by international organizations. It can therefore be argued that the functioning of the regulatory regimes is quite similar. However, air transport is regarded as very safe and much safer than transport by sea. The fact that it is an international business is therefore probably not sufficient to explain the relatively low safety standards.

Safety improvements in shipping came historically as a response to serious and catastrophic accidents and very often with a narrow perspective on technical aspects or specific regulations.

The use of qualitative methods for risk assessment has a long history within the maritime industry. Quantitative studies were only used for very specific purposes, often as a result of requirements from other industries using ships for transportation and other purposes, e.g., the offshore oil and gas industry. The use of quantitative methods has also opened for an opportunity to compare broader technical concepts or safety measures on a common scale. Formal Safety Assessment is a designation used in a number of different contexts and industries (e.g., the nuclear industry) in order to describe a rational and systematic risk-based approach for safety assessment. Formal Safety Assessment also focuses on safety measures in a wider perspective involving technology, human resources, and organizational solutions. Another key difference is the shift from specific safety requirements to risk-based goals.

Bureau Veritas (BV) has published a guidance note that lists applicable risk analysis methods for marine and offshore applications (BV 2017). This provides a comprehensive overview of risk analysis in BV's rules, in IMO rules, in industry and for inland navigation vessels. The document lists both mandatory and recommended use of risk analysis. As far as the IMO rules go, regulations referring to risk analysis are included in five regulations/codes:

- SOLAS (International Convention for the Safety of Life at Sea) – Mainly for alternative designs of life-saving appliances, machinery, and safety systems (with regard to fire and explosion). Mostly, methods like HAZOP (Hazard and Operability Study), PHA (Preliminary Hazard Analysis), FMEA (Failure Modes and Effects Analysis), and What-if analyses are suggested. This applies to ships in general. For oil tankers, there is also a requirement to apply risk analysis to electrical systems that do not conform to standards and that are installed in hazardous location.
- IGC code (International Code for the Construction and Equipment of Ships carrying Liquefied Gases in Bulk, IGC 2016) – Regulations for liquified gas carriers, safety systems, electricity and automation, and structure. There are mandatory requirements for analysis of risk of fire propagation, FMEA of electrical systems and HAZID of automation system, integrated system, and membrane tanks. Alternative arrangements may also be accepted based on risk analysis.
- IGF code (International Code of Safety for Ship Using Gases or Other Low-flashpoint Fuels, IGF 2015a) – Requirements for mandatory analyses of machinery and structure (fuel system, fuel containment system, bunkering stations, tank connection space, airlock and ventilation inlets to accommodation, and machinery spaces). HAZID (Hazard Identification Study), HAZOP (Hazard and Operability Study), and FMECA (Failure Mode and Criticality Analysis), but also more detailed studies of fire and explosion and gas dispersion.
- HSC 2000 code (International Code of Safety for High-Speed Craft (2000), 2008 Edition) – Requirements for FMEA on control system, electrical system, gas turbine, machinery systems, and stabilization system.

- Polar code (International Code for Ships Operating in Polar Waters 2015b) – HAZID and FAS for ice operation.

Various international organizations also provide industry best practices that cover risk analysis. These include OCIMF (Oil Companies International Marine Forum), SIGTTO (Society of International Gas Tanker and Terminal Operators), and BIMCO (Baltic and International Maritime Council).

1.4 ANATOMY OF AN ACCIDENT

Risk reduction is based on an understanding of how and why an accident took place. Key elements in the description of an accident are events and causal factors. The "how" element is reflected by observable events like loss of power, wrong operation of a control function, or capsizing of the vessel. The "why" element is somewhat more complex in the sense that it requires deeper analysis of the so-called accident mechanisms. To understand the loss of power event, it may be necessary to look into the maintenance history and investigate degradation processes in the machinery system. A pilot may handle a rudder control system erroneously due to the fact that he was unfamiliar with it. His mental model of how the system functioned was wrong.

In order to reach a good understanding of the accident in question, it is important to find out how events and causal factors interact with each other. Over the time, different accident theories or models have been proposed. The general trend is that accident concept has become "richer" by introducing more and more elements:

- Relevant safety regulations and potential shortcomings
- Shipping company in terms of organization and management, and weaknesses
- Seaway in terms of infrastructure (traffic management and monitoring)
- Seaway in terms of weather, sea state, and exposed marine traffic
- The vessel status and its operational history
- Personnel involved and their competence
- System failures and related degradation processes
- Wrong operation by personnel and description of their misconceptions
- Vessel interaction with sea, weather, and seaway
- Loss of vessel integrity and related damage processes

The outcome of the accident analysis should be a set of recommendations that address both regulation, safety management, personnel training, and design modification. Accident analysis is primarily undertaken by national institutions independent of the maritime safety organizations. The accident reports are available to the public and are increasingly saved in databases and may also support ongoing investigations.

Accident analysis is in its nature a retrospective approach to safety management and has therefore been subject to criticism. The argument is that things change and therefore may make lessons from the past obsolete because technology is under continuous development and modern safety management is characterized by feedback and learning. A compromise view is that improved safety must be attained by combining both a retroactive and proactive approach. The latter means to have an open mind and ability to envision the conditions under which the shipping industry will operate in the future and thereby identify new risks.

1.5 MANAGING RISK

Safety management is a systematic and coordinated approach where accident analysis and risk analysis are key elements. The management of risk should address all life cycle phases of a system or an operation: planning, design, fabrication, operation, and maintenance. This lesson is sometimes forgotten and can be exemplified with the Exxon Valdez grounding and oil spill. The plan to export oil from Alaska to southern parts of the United States was subject to elaborate planning with focus on both safety and environmental protection. It was decided to choose ship transport from Alyeska rather than by pipe transport mainly for environmental reasons. In order to maintain a high safety level for the inshore sailing from the oil terminal, both a traffic separation scheme and a vessel traffic management system were established. Due to cost-cutting and increasing complacency, both the operation of vessels and the vessel traffic management system degraded and were important causal factors.

Every safety management process involves the following stages:

- Establish the management context
- Risk analysis
- Risk evaluation
- Propose measures to reduce risk
- Decide and implement measures

The starting point is to decide which risks to study and identify relevant regulations and clarify what approach to apply. The risk analysis process will result in a set of risk factors and associated risk level in terms of expected probability and loss potential. The risks may be related to fatalities, environmental damage, and economic losses. The risk evaluation is based on risk acceptance criteria. The selection of measures to reduce risk must balance risk reduction with the cost of implementing the measures. The comparison of risk and cost is somewhat controversial in the sense that our immediate feeling is that putting a monetary value on human life is unacceptable. However, in practice the society has accepted that such evaluations are done.

It is important to acknowledge that safety management is only one aspect of the management process. Any company has to a have broader view of the challenges to secure competitiveness and obtain economic profit in a long-range perspective. This means that risk control is also coupled with strategic objectives, organization, and personnel policy.

1.6 MOTIVATION FOR WRITING THE BOOK

The first edition of this book was published in 2005 and was then intended as a textbook for courses in universities and maritime academies. In the period since 2005, a lot has happened in the maritime industry and the field has also developed rapidly. We have therefore seen a need for updating the book to reflect these changes.

In general, safety management and methods for risk assessment have developed significantly in the period. Methods from this are accepted as useful and are commonly in use in a wide range of applications, including within the maritime industry.

Traditionally, safety regulations in the maritime industry have been driven by accidents. As is discussed later in the book, many of the important regulations have been introduced as a result of serious accidents. As long as technological development is slow, this may be an acceptable approach because we can learn from minor accidents and mishaps and improve technology and regulations based on this. However, with the increasing pace of development, more systematic methods for identifying what can go wrong and to act on this before accidents happen are necessary. This is where systematic safety management and risk assessment can be used.

Further, new ship concepts and new technology that can increase efficiency, reduce costs, and reduce risk are also being continuously developed. New developments can also challenge existing rules and regulations. The traditional rule-making process in IMO has difficulties keeping up with the pace of development. IMO therefore also allows a process for approval of alternative designs, based on risk assessment.

The maritime industry is conservative, and the value of risk assessment is often questioned, due to the uncertainties associated with risk assessment but also due to lack of experience with the methods, lack of data, and general lack of knowledge. We hope that this book can contribute to "demystify" risk assessment in some respects, and maybe make a small contribution to increase knowledge and recognition of safety management and risk assessment.

1.7 SCOPE

As the title states, the scope of the book is safety management, with particular emphasis on risk assessment since this is a crucial part of any safety management process.

IMO introduced the International Safety Management Code (ISM code) in 1994 (IMO 1994). This book covers safety management as specified by the code but has a wider perspective on the topic. After reading the book, the reader should be able to have a better understanding of the rationale for many of the requirements in the ISM code, including tools and methods that can help to satisfy both the letter and the intention of the code.

The book is aimed at the maritime industry, although the general framework for safety management and the methods for risk analysis, accident investigation, etc. are generic and could be applied to many other industries also. However, we have illustrated this with examples from the maritime industry and of course also other information specific for the industry.

1.8 HOW TO USE THE BOOK

The book is intended to give an introduction to the topics covered, meaning that no prior knowledge of the field is required. We see mainly three ways that this book can be used:

- It can be used as a textbook for courses taught at maritime academies and universities. This was the original purpose of the first edition, and we have retained a structure where the book can be used chapter by chapter in a course.
- It can also be used as an introduction to the topic for those who are interested to learn more about this, without attending a course. This may be professionals already working in the industry, those working as designated persons according to the ISM code, or others who need to know more about the topics in the book.
- Third, the chapters have, as far as possible, been written such that they can be read individually without having to read the whole book sequentially. For those who want to learn more about a specific topic, it is therefore also possible to select only the relevant parts and read those. Readers who have very limited knowledge of the topic beforehand may however find it useful to read Chapter 3 Terminology and perhaps also Chapter 5 Safety Management System first, to get an introduction to the topic.

REFERENCES

BV (2017). *Index on Applicable Risk Analysis for Marine and Offshore*, Guidance note NI 635 DT R00 E, Dec 2017. Bureau Veritas.

Financial Times (2021). *Shipping Costs Quadruple to Record Highs on China-Europe 'Bottleneck'*, https://www.ft.com/content/ad5e1a80-cecf-4b18-9035-ee50be9adfc6, accessed 11. May 2021.

IMO (1994). *ISM Code - International Safety Management Code*, 1994, IMO.

IMO (2008). *HSC 2000 Code - International Code of Safety for High-Speed Craft*. 2008 Edition, IMO.

IMO (2015a). *IGF Code 2015: Code of Safety for Ships Using Gases or Other Low-Flashpoint Fuels*. MSC.391(95), June 2015.

IMO (2015b). *Polar Code - International Code for Ships Operating in Polar Waters*. 2015 Edition, IMO.

IMO (2016). *IGC code (International Code for the Construction and Equipment of Ships Carrying Liquefied Gases in Bulk*. 2016 Edition, IMO.

Chapter 2

The risk picture

2.1 INTRODUCTION

Maritime transport has traditionally been a high-risk industry, with many serious accidents. In terms of loss of life, accidents with passenger ships are commonly the most serious events. However, extensive environmental damage is related to crude oil transport and numerous occupational accidents have resulted in fatalities or severe injuries over the centuries.

In this chapter an overview will be provided of the risk as it has been experienced within maritime transport. To do this, several types of information will be presented. Initially, an overview of the activity level in the industry will be given. This will be followed by an overview of the types of accidents that may occur and the losses that have occurred over the last few decades.

This forms an important backdrop for risk management. If we want to manage and reduce risk, we also need to have a thorough understanding of what the risk is, what types of accidents can occur and why and how often they occur. Without this, we do not know where to focus our efforts for reducing risk.

2.1.1 Major accidents

These are accidents that are characterized by a potential for large consequences. This is not necessarily realized in all cases, but the consequences typically have the potential to extend far beyond the location and people directly involved and being in the proximity of where the accident occurs. Examples are as follows:

- Fire on a cruise ship – this may start as a small fire, caused by careless use of electrical equipment, smoking, or some other activity, but can in the worst case escalate to the whole ship and cause many fatalities and extensive loss of life. The cost will also be very large.

DOI: 10.4324/9781003055464-2

- Collisions between two ships – This can cause extensive damage to both ships, escalate to fire and explosion, and leading to loss of one or both ships, many fatalities, and also environmental damage depending on the cargo.

All the accidents described in Section 2.3 may be characterized as major accidents.

Typical of these accidents is also that they are caused by a number of causal and contributing factors. Another name for this type of accident is therefore also organizational accidents (Reason, 1997), to underline the fact that contributing factors often can be traced back to the organizations involved. This type of accident is sometimes also called a concept accident. Major accidents are comparatively rare.

There is no standard terminology or grouping of major accidents. Some of the common accident categories used are as follows:

- Collision
  - Impact with other ship
  - Impact with icebergs
- Allision
  - Impact with various fixed objects (offshore platforms, bridges, quays, etc.)
- Grounding
  - Impact with submerged objects or seafloor
- Stranding
  - Impact with shore
- Fire and explosion
  - Cargo, fuel, other sources
- Loss of stability (capsize)
  - Cargo shifting, free liquid surface, wrong loading, flooding, icing
- Loss of buoyancy
  - Flooding, excessive heel
- Foundering
  - Ship sinking
- Besetting in ice
  - Stuck in ice

The lack of standard definitions means that different sources classify these accident categories differently and sometimes combine them. Grounding and stranding can be combined, and sometimes flooding can be defined as a separate category (above it is defined as possible causes of both loss of stability and loss of buoyancy). As an example, the Norwegian Maritime Authority uses the following accident categories in their database:

- Grounding
- Collision
- Fire and explosion
- Allision
- Capsizing

On the other hand, EMSA uses the following categories (EMSA, 2019):

- Grounding/stranding
- Collision
- Fire/Explosion
- Contact
- Capsizing/Listing
- Damage to ship equipment
- Flooding/Foundering
- Hull failure
- Loss of control

Differences in definitions also mean that different sources of data may be difficult to compare.

It should also be noted that the accident categories that are defined can be linked in some accident cases. As an example, a collision between two ships may lead to flooding and this may in turn lead to capsizing/listing. How this is recorded in accident databases may also vary. This underlines the need for understanding how accident categories are defined in different sources and publications.

2.1.2 Occupational accidents

These are accidents that may cause fatalities or serious injuries, but they will not have the potential to cause injury or damage outside the immediate vicinity of where the accident occurs or to others than those directly involved. Some examples are as follows:

- Fall from a ladder – this may seriously injure or even kill the person falling and injuries may also occur to persons standing below. Consequences outside this are however very unlikely.
- Crushing of fingers in machinery – Serious injury may occur, but again it is hard to envisage serious consequences outside this.

Occupational accidents are sometimes also called personal accidents. Relatively speaking, occupational accidents are far more frequent than major accidents. The causes of these types of accidents are normally considered to be much easier to identify and not as complex as for major accidents.

This is correct in the sense that often it is sufficient with one mistake or one failure for the accident to happen (person slips on the ladder or one rung on the ladder fails). However, the reasons why these mistakes and failures occur may well be more complex and involve the organization (Was the person in a hurry because of pressure from his supervisor to get the job done? Had there been a cutback on maintenance that led to failure of the ladder?).

Occupational accidents are mostly generic, in the sense that most types of accidents can occur in many different industries. Standards for industry in general, like ISO 12100 (2010), may therefore be relevant to consult also for the maritime industry.

2.1.3 Environmental factors

Accident statistics show that a relatively large portion of the impact accidents and especially collisions occur in poor visibility. This fact is not surprising considering that the navigation is dependent on radar and other electronic aids and without the support of direct visual observation of fairway and traffic. It has also been suggested that the technological development and for instance introduction of ARPA (Automatic Radar Plotting Aids) has led to reduced vigilance by the navigator. As a part of the traffic studies in the Dover Strait, the effect of visibility was also studied. It was concluded that the visibility factor was quite large and even greater than the effect of the particular encounter situation itself (Lewison, 1980). A traffic separation scheme (TSS) was implemented in Dover Strait in 1977. The effect of visibility was studied before and after implementation of the TSS. Visibility may be defined in various ways but in the present investigation three classes were applied as shown in Table 2.1.

The number of collisions per encounter before and after the implementation of TSS is shown in Figure 2.1. An encounter is defined as two vessels passing each other with less than 0.5 nautical mile separation. There was a reduction in collision probability in reduced visibility conditions but on the other hand there was an increase in clear weather.

Apart from the "before" and "after" effect of TSS, it could be concluded that the relative collision risk for the different ranges of visibility remained fairly constant. A visibility effect model was therefore developed, based on data from both before and after introduction of TSS. The Fog Collision Risk Index gives the number of collisions per encounter as a function of the relative incidence of the visibility ranges:

Table 2.1 Visibility range

Clear	Mist/Fog	Thick/Dense
Greater than 4 km	200 m–4 km	Less than 200 m

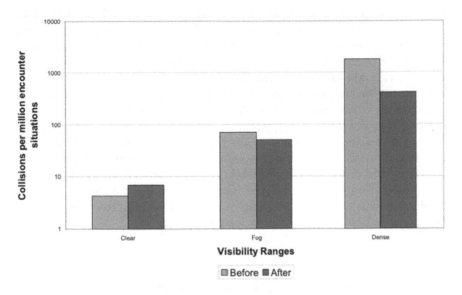

Figure 2.1 Collision probability before and after implementation of TSS for varying visibility.

Table 2.2 Relative visibility incidence vs. collision probability for the Dover Strait

Visibility (k)	Clear	Mist/Fog	Thick/Dense
Relative visibility incidence (VI)$_k$	0.9457	0.0446	0.0097
Collision probability (P$_k$) per million encounter	6	60	1800

$$FCRI = \sum_{k=1}^{3} P_k VI_k = \left(P_1 VI_1 + P_2 VI_2 + P_3 VI_3 \right) \qquad (2.1)$$

where

P_k = Probability of collision per million encounters
VI_k = Fraction of time that the visibility is in the range k
k = Visibility range: 1 = Clear, 2 = Fog, 3 = Dense

The estimated parameters of the model are shown in Table 2.2.

The data shows the dramatic effect of reduced visibility on the collision risk. Although the relative frequency of visibility "Thick/dense" is less than 1%, the probability increases with a factor of 300. The resulting value for Dover Strait was:

$$FCRI = 25.8 \cdot 10^{-6} \text{ collisions/encounters within } 0.5 \text{ nm} \qquad (2.2)$$

The contribution of "Thick/dense" to this value is 68% ($0.0097 \cdot 1800 = 17.5$). This means that without this contribution, Fog Collision Risk Index would have been approximately 8, or one-third of the calculated value. It can therefore be concluded that even though very low visibility in most areas is relatively rare, the effect on navigational safety is dramatic.

In a study by Heij and Knapp (2015), it was found that waves and wind also have an effect on risk levels, although they identified large variations in the effect between ship types and geographical areas. They looked at changes in the probability of an incident with changes in weather. Differences in weather during spring/summer and autumn/winter were identified, and differences in probability of weather-related incidents were determined from statistics. For the North Atlantic, it was found that the number of weather-related incidents was approximately two times higher during autumn/winter than during spring/summer.

2.1.4 Flag

Alderton and Winchester (2002) looked at casualty records for the period of 1997–1999 and calculated casualty rates for all flag states. The average rate for the world was found to be 1.6 per 100 ship years, with the highest rate for a flag state being 23.3 and the lowest 0.2. The objective of the study was to compare Flags of Convenience with more traditional flags, and some findings indicate that there are differences. If the flags are grouped into three, Flags of Convenience, traditional flags, and international registers established by traditional flags (like the Norwegian International Ship Register), some differences can be observed. Not surprisingly, Flags of Convenience have the highest average casualty rate, at 3.58 per 100 ship years while traditional flags are significantly lower at 1.36. This difference is also statistically significant. Interesting is also that the international registers have an average rate of 2.11, i.e., 50–60% higher than the traditional flags.

Another study looking at flags was performed by Li and Wonham (1999). This study looked at a much longer period (20 years) and in their results, they ended up dividing the world fleet into three groups that they called the best, the middle, and the worst:

- The best (total loss rates less than 0.2%): Russia, China, Brazil, Sweden, Hong Kong, Poland, the Netherlands, and Australia
- The middle group (loss rates between 0.2% and 0.6%): Germany, France, Japan, Canada, the USA, the UK, Mexico, Liberia, India, Malaysia, Indonesia, Peru, Singapore, Italy, Bahamas, Norway, Spain, Turkey, Denmark, and the Philippines
- The worst (loss rate above 0.75%): South Korea, Panama, Greece, Malta, Saint Vincent, Taiwan, Cyprus, and Honduras

Table 2.3 White/Gray/Black list of flags from Paris MoU, 2020–2021

White list (top 16 flags)	Gray list	Black list
United Kingdom	Morocco	Tunisia
Norway	Saudi Arabia	Cook Islands
Bahamas	Poland	St Kitts and Nevis
Netherlands	Algeria	Mongolia
Denmark	Curacao	Sierra Leone
Marshall Islands	Thailand	Belize
Singapore	Kazakhstan	Palau
Hong Kong	Lebanon	Ukraine
Japan	Azerbaijan	Tanzania
Bermuda	Iran	Moldova
Germany	Switzerland	Togo
Cayman Islands	India	Albania
Liberia	Vanuatu	Comoros
Sweden	Egypt	
France	St Vincent and Grenadines	
Isle of Man	Tuvalu	

Paris MoU (2020).

It may be noted that many traditional maritime nations are in the middle group in this study.

Both of these studies are fairly old and things may have changed significantly since then. Paris MoU uses flag to prioritize inspection of vessels, and they divide the flags into three groups, namely, Black, Gray, and White (Table 2.3). This is based on the rate of detentions of ships. Detention is not a direct measure of risk, but it is still regarded as a fairly good indicator since a detention means that serious deficiencies relevant for safety are identified in an inspection. The lists are updated annually and the most recent list at the time of writing was applicable from 1. July 2020 (parismou. org). In total, there are 70 flags on this list, of which 41 are on the white list.

Only the top 16 from the white list are shown, while the complete gray and black lists are included. The white list is topped by a number of traditional shipping nations, like the UK and Norway, but we can also see many flags that traditionally are regarded as Flags of Convenience, e.g., Bahamas, Marshall Islands, and Cayman Islands. There is therefore no clear pattern to distinguish traditional flags and Flags of Convenience. What may be noted, however, is that the Flags of Convenience on the white list are what may be called "traditional" Flags of Convenience, i.e., they have been used for this purpose for many years.

The black list largely consists of small nations, with limited shipping traditions. It may also be noted that the black list includes countries without a coastline, Mongolia and Moldova.

The flag lists are snapshots of the state of the flag in a given year and there will be changes, although not necessarily large. If the lists in Table 2.3 are compared with the lists from the annual report in 2010 (Paris MoU, 2011), we find that eight of the flags on the black list in 2020 also were on the list in 2010. Of the remaining five, four were on the gray list and the last (Palau) was not on the list at all in 2010.

2.1.5 Age of ship

The age of the ship can be expected to be a factor that influences the risk, but studies are not entirely conclusive on this issue. Eliopoulou et al. (2016) looked at the relationship between accident frequency and age and found varying results. The results were based on accidents in the period of 2000–2012:

- Fishing vessels and passenger vessels do not show significant differences between different age groups.
- Bulk carriers show a linearly increasing frequency with age. If 20-year-old ships are compared with newbuilds, the frequency is more than three times higher for old ships.
- LNG (liquefied natural gas) ships have fairly constant accident rate up to 20 years but it is significantly higher for the age group above 20 years.
- RORO (roll-on/roll-off) cargo ships show perhaps the most surprising trend, with ships less than five years of age having the highest accident rates. This could indicate that there have been quality problems with newbuilds and operational factors in the relevant period.

De Maya et al. (2019) looked at MAIB (Maritime Accident Investigation Branch) data for the period 1990–2016, a total of nearly 25,000 accidents. However, this study only analyzed the distribution of accidents, not taking into account the number of vessels in each age category. Accident rates per ship could thus not be established. They found that almost 40% of accidents had occurred with ships in the age group over 20 years. For the age groups 0–5, 6–10, 11–15, and 16–20, the proportion was more or less the same, at around 15% in each group.

In an earlier section (Section 2.2), it was mentioned that the average age of ships in 2010 was 22.7 years. This is the average value, not the median, but it would still indicate that a large proportion (highly likely more than 50%) is more than 20 years. If 40% of accidents are associated with this group, it would in fact indicate that the accident rate for this group is lower than average. However, a more detailed analysis would be necessary to confirm this.

Li et al. (2014) looked at a number of hypotheses regarding factors influencing the risk of ships. They looked at just over 800 accidents in the period 1993–2008 and among others found that accident rates did not increase with age. On the contrary, they found a small decrease with increasing age.

2.1.6 Discussion

In the previous sections, we have looked at a few selected factors that one could hypothesize would influence the risk level of ships. As has been shown, the evidence presented in the literature varies significantly. This variation in the results may be explained by differences in the data selected, the method used for analyzing the data, and the limitations of the studies. However, it may also be an indication that we need to look for other factors than those that have been considered in this chapter, like flag and age.

For flags, the distinction between Flags of Convenience (FoC) and traditional flags has long been regarded as a way of distinguishing between high- and low-risk vessels. However, the evidence for this is not clear. We should then try to go "behind" this assumption and ask ourselves why this distinction has been assumed to be important. The underlying assumption is that regulation is not as strict, that the follow-up of the regulations is more relaxed and that lack of maritime traditions has led to increased risk. However, if we look at some of the FoCs that are on the white list in Paris MoU (Memorandum of Understanding), they have been labeled as FoCs for many years. An explanation for their good safety records may thus be that they have actually improved the state for many of those factors that have been considered to be weak points in FoCs. Instead, it is "new" maritime nations that have taken their place as the work performing with respect to safety.

This has not been investigated in detail so it is still just an assumption. However, it illustrates that it usually is necessary to study the facts in more detail before we can conclude about correlation between various factors and risk levels.

2.2 ACCIDENT STATISTICS

A general problem in the maritime industry is varying and inadequate reporting of accidents, incidents, fatalities, and injuries. This can be observed when different sources of data are compared, often showing significant discrepancies. Efforts are being put into improving this, but this is a process that takes a long time. It should also be noted that this is not unique for the maritime industry but a general problem in most industries.

This will of course mean that some of the statistics that are presented in the following sections are uncertain and that different numbers can be

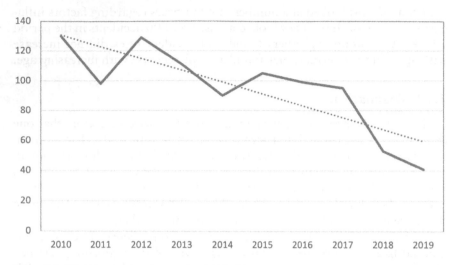

Figure 2.2 Worldwide ship losses for 2010–2019 (number of losses per year) (AGCS, 2020).

found if other sources are consulted. This just underlines the need for carefully considering the quality of the data sources that are being used, who are collecting the data and the purpose of the data collection.

In Figure 2.2, the development in total losses per year in the from period 2010 to 2019 is shown. The figure also includes a linear trend line, clearly showing a strong downward trend during this decade. Considering that the number of ships has increased in the same period, the probability of loss per ships has decreased even more. In 2019, the average probability of a ship being lost was $4.2 \cdot 10^{-4}$, down from $1.2 \cdot 10^{-3}$ in 2011.

It is also interesting to consider the types of ships that are lost. A breakdown of the total number of ships lost in the period of 2010–2019 is shown in Figure 2.3.

The figure shows that 41% of the lost vessels are general cargo vessels, followed by fishery (14%), bulk carriers (9%), and passenger vessels (7%). Cargo vessels is also the largest group of vessels, but they are still over-represented in the accident statistics. If the whole period is considered, the differences between the accident rates for the different ship types are quite large. On average for this period, the accident rate for cargo ships is more than ten times higher than for tankers. It is also two to three times higher than the accident rates for bulk carriers and container ships.

However, concluding that this is because cargo vessels have inherent characteristics that make them less safe is too easy. The average size of the vessels will be different, the same with the age and the flags that they are operating under, etc. These are all factors that are known to influence the risk level. This illustrates that one has to be careful when drawing

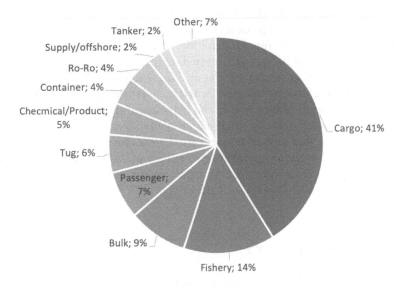

Figure 2.3 Worldwide ship losses per ship type – 2010–2019 (AGCS, 2020).

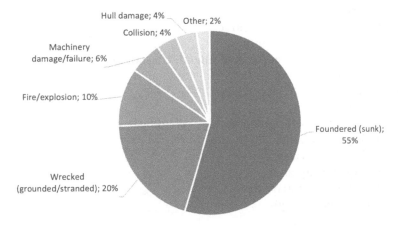

Figure 2.4 Worldwide ship losses per accident type – 2010–2019 (AGCS, 2020).

conclusions from statistics. The causes of differences may be more complex to identify than the first impression may indicate.

Another interesting breakdown is to look at the type of accidents that have led to losses. The breakdown is shown in Figure 2.4.

More than half the losses are due to foundering, followed by grounding/ stranding (20%) and fire/explosion (10%). Other accident types that are represented are machinery damage/failure, collision, and hull damage.

It is important to note that this is a breakdown of accidents involving losses of the ship, i.e., the most serious accidents. In 2019, 2,815 incidents (including 41 losses) were reported to Lloyd's Intelligence Casualty Statistics (AGCS, 2020), and of these, machinery damage/failure represent 37%, collision 10%, grounding/stranding 9%, and fire/explosion 7%. Looking only at losses will therefore not necessarily tell us everything about where focus should be directed to reduce risk in the maritime industry.

Another comment relating to number of incidents is that the average probability that a ship had an incident in 2019 is close to 0.03 or 3%. This can be compared to the value mentioned earlier of 0.042% ($4.2 \cdot 10^{-4}$) average probability of losses per year. In other words, there is on average one loss per approximately 70 reported incidents.

2.3 MARITIME ACTIVITY

Maritime activities have had, and still have, an important role in the business, trade, and economy of many countries. The key areas of maritime activities include:

- Maritime transport:
  - Coastal shipping of cargo
  - International shipping
  - Transport of people both inland, coastal, and overseas
  - Cruise shipping
- Fishing
- Marine farming
- Continental shelf operations (i.e., oil and gas):
  - Rig operations
  - Supply services
  - Pipeline laying
  - Underwater activities
- Science and survey

These activities have several positive attributes, such as employment, production, creation of values and fortune, spreading economic consequences, and positive influences on currency and exchange transactions. There is, however, a price for these benefits in terms of negative effects. Some of the typical accident types found in maritime activities are outlined in Section 2.4.

In this book, focus is on shipping in general and in particular maritime transportation. In Figure 2.5, an overview of the number of ships in the world in the period of 2011–2020 is shown (UNCTAD, 2020). In total, there were nearly 100,000 merchant ships in the world in 2020. There has been a steady increase since 2011, from around 83,000 ships. In total, the

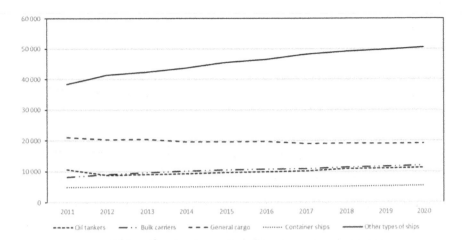

Figure 2.5 World merchant fleet (number of ships) 2011–2020 (UNCTAD, 2020).

fleet has grown by approximately 18% or roughly 2% per year. The largest percentage increase in the period is for bulk carriers, where the number of ships has increased by 45% while for general cargo there is a decrease of 9%.

The largest proportion of ships is "Other", representing about 50% of the total number of ships in 2020, followed by general cargo with close to 20% of the fleet. Bulk carriers and oil tankers represent 11–12% each of the total fleet.

Another way of illustrating the world fleet is to look at tonnage instead of number of ships. This is also a better reflection of the capacity for transporting goods than the number of ships. This is shown in Figure 2.6 for the same period.

The first observation to be made is that the total tonnage has increased far more than the number of ships. The tonnage increased by 46% from 2011 to 2020 (compared to 18% for number of ships). The average size of ships is therefore increasing, implying that the average consequences of an accident also can be expected to increase. The largest increase is seen for bulk carriers (about 60%), followed by container ships (approximately 50%). The increase for container ships is particularly noticeable, since the number of ships increased by only 8% in the period. The average size has therefore increased from 37.000 DWT to 51.000 DWT in the period.

From this figure, we further find that bulk carriers and oil tankers represent the largest proportion of the total fleet measured in cargo capacity. This is not surprising considering the cargo these are built to transport. The average size of these vessels is therefore also the largest, although container ships are close behind oil tankers in terms of size.

Statistics also show that the flag where a ship is registered may be important for the risk. In the following two figures, the world fleet is therefore

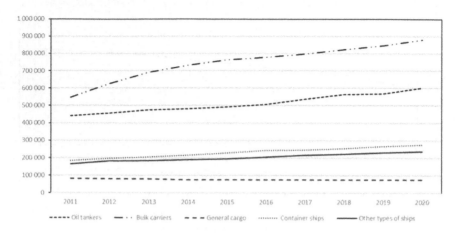

Figure 2.6 World merchant fleet (DWT) 2011–2020 (UNCTAD, 2020).

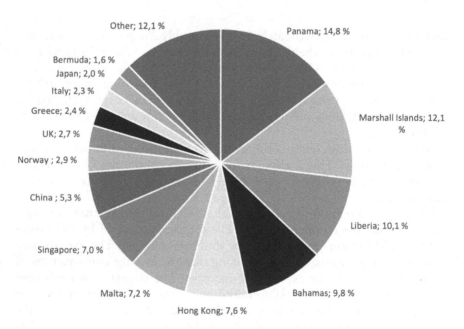

Figure 2.7 Breakdown of world fleet on flag according to value of ships (UNCTAD, 2020).

divided on the flag where the ships are registered. The first figure (Figure 2.7) shows the fleet split on flags according to the value of the ships while the second figure (Figure 2.8) illustrates the split according to number of vessels.

This also illustrates large differences. From the first figure, we see that 14 nations have registered approximately 88% of the total fleet value. In

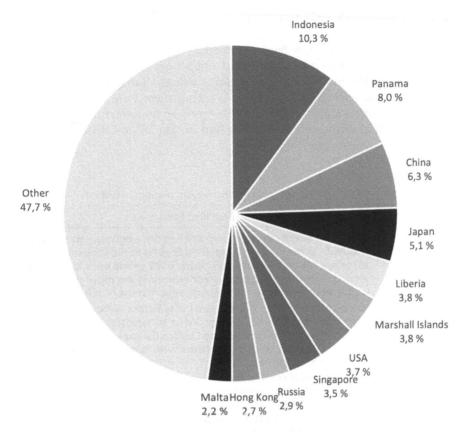

Figure 2.8 Breakdown of world fleet on flag according to number of ships (UNCTAD, 2020).

Figure 2.8, if we look at the number of ships, the 11 largest flags only have about 54% of the total number of ships. The flags at the top of the list also change. In terms of value, Panama, Marshall Islands, Liberia, and Bahamas are the four largest, having 47% of the fleet registered. On the other hand, if the number of ships is considered, Indonesia, Panama, China, and Japan are the largest (comprising just under 30% of the total).

The age of the vessel may also be important for risk, and according to "Review of Maritime Transport 2020" (UNCTAD, 2020), the average age of all ships in 2020 was just under 21 years. For comparison, in 2010 (UNCTAD, 2011), the average age was 22.7 years, i.e., average age has slightly decreased. However, if age is weighted with size (DWT), the average age is 10.4 years. This confirms that the average size of the vessels has increased in recent years. If different ship types are considered, general cargo had an average age of 26.3 years while bulk carriers on average were 9.7 years.

2.4 IMPORTANT ACCIDENTS

The maritime industry has been troubled by many serious accidents and many of these have also had significant impact on legislation and requirements within the field. Some influential accidents (internationally and in Norway) are described in Table 2.4. Several of these have triggered major changes to international legislation. This has been an unfortunate tendency in shipping that serious accidents have had to take place before significant safety improvements are made.

Table 2.4 Some influential international and Norwegian maritime accidents

Vessel/date	Brief description
Titanic 11.4.1912	*Titanic* was a British passenger ship and was the world's largest ship when launched. On her maiden voyage from England to the USA, she struck an iceberg and sank within a few hours. 1,518 people were lost and 706 survived. The sinking triggered major reforms in the shipping legislation, leading among others to the eventual establishment of SOLAS.
Mont Blanc 6.12.1917	*Mont Blanc* collided with the ship *Imo* inside Halifax harbor. This caused a fire on the Mont Blanc which later led to an explosion in the munition cargo of the ship. The crew escaped before the explosion occurred, but about 2,000 people on shore in Halifax were killed in the explosion and more than 9,000 injured.
Torrey Canyon 18.3.1967	*Torrey Canyon* was an oil tanker carrying 119,000 tons of crude oil. She was shipwrecked off the western coast of Cornwall, England, and subsequently released the oil cargo and sank. When the accident occurred, she was the largest vessel ever to be wrecked. It was later concluded that the accident was caused by the captain choosing to take a shortcut through the Isles of Scilly to save time. The accident was related to lack of fairway knowledge, unprecise position fixing, and affected by fishing vessels. The accident is considered to be the trigger for IMO to introduce MARPOL.
Amoco Cadiz 16.3.1978	*Amoco Cadiz* was an oil tanker of 233,690 DWT. The problems started when the steering engine failed and the rudder was disconnected and out of control, and she had to shut down the engines for repairs. This was not successful, and she ran aground 12 hours later on Portsall Rocks, 2 km from the coast of France, after attempts to tow were unsuccessful. Ultimately, she split in three and sank, resulting in the largest oil spill of its kind in history to that date. The entire cargo and fuel oil was released, causing a total spill of 240.000 t.
Berge Istra/ Berge Vanga 30.12.1975 / 29.10.1979	*Berge Istra* and *Berge Vanga* were combination carriers carrying ore on one voyage and oil on the next. Release of gas from oil residues when carrying ore was solved by using inert gas in the tanks. *Berge Istra* sank after severe explosions and it is likely that the same happened with *Berge Vanga* although it was not possible to confirm. Problems with the inert gas operation have been indicated as a cause. Two people survived the *Berge Istra* explosion while none survived Berge Vanga. In total, 70 were killed in these two accidents.

(continued)

Table 2.4 (Continued)   Some influential international and Norwegian maritime accidents

Vessel/date	Brief description
Herald of Free Enterprise 6.3.1987	*Herald of Free Enterprise* was a RORO ferry used for crossing the English channel. It was designed for rapid loading and unloading and had no watertight compartments on the car deck. The ship left Zeebrugge in Belgium with the bow door open and water immediately started to flood the car deck. Minutes later, she capsized but came to rest on the side due to the shallow water. 193 passengers and crew perished. The immediate cause of the sinking was found to be crew negligence in not closing the bow door. However, the official inquiry placed considerable critique on supervisors and a general culture of poor communication in the company. The accident was among those factors that contributed to the introduction of the ISM code.
Exxon Valdez 24.3.1989	*Exxon Valdez* was an oil tanker that was carrying around 180,000 t of oil when she grounded in Prince William Sound, 2.4 km west of Tatitlek, Alaska. The total spill was about 37,000 t over the next days. It is considered the worst oil spill worldwide in terms of damage to the environment. The spill is the largest in US waters from a ship. The spill affected 2,100 km of coastline. Contributing factors to the accident were failure of Exxon Shipping to ensure the crew was fit and competent, failure to maneuver the vessel properly, and a non-functional VTS (Vessel Traffic System).
Scandinavian Star 7.4.1990	*Scandinavian Star* was a RORO passenger ship, recently rebuilt, sailing on her first voyage between Oslo in Norway and Fredrikshavn in Denmark. During the crossing, several fires were started in the accommodation area (likely by an arsonist). The ensuing firefighting, evacuation, and rescue were poorly organized, and out of a total of 482 people onboard, 159 were killed. Contributing factors to the catastrophe were unfinished reconstruction work and lack of time to train the crew. Certain aspects of the events and causes leading to the fire are still not fully understood.
Estonia 28.9.1994	*Estonia* was a RORO passenger ship sailing between Stockholm in Sweden and Tallinn in Estonia. The vessel capsized and sank in hard weather in the Baltic Sea, resulting in the loss of 852 people and only 137 surviving. The investigation concluded that the cause of the accident was failure of the locking mechanism of the bow door, causing water to flood the ship. Recently, investigations have been initiated related to later detected hull damage.
Sleipner 26.11.1999	*Sleipner* was a high-speed catamaran ferry traveling between the Norwegian cities of Stavanger and Bergen. It ran aground in heavy seas and the vessel was a complete loss. Sixteen people were lost, and 69 were rescued. Causes identified were violation of operation restrictions and inadequate navigational control. It was also found that the emergency equipment and organization were inadequate.
Costa Concordia 13.1.2012	The cruise ship *Costa Concordia* was on a voyage along the west coast of Italy when it ran aground off the island of Giglio and reached 80 degrees list before coming to rest on the seabed. In total, there were 4,229 people onboard and 32 were killed in the accident. The Master was sentenced to 16 years in prison as the ship had sailed too close to shore when passing the island and having left the ship before all passengers had been rescued.

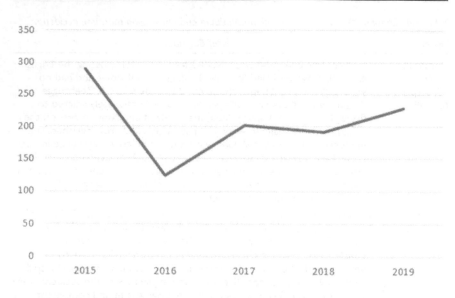

Figure 2.9 Number of lost and missing seafarers per person-year – 2015–2019 (IHS, 2020).

2.5 FATALITIES AMONG SEAFARERS

Figure 2.9 shows the total number of lost and missing seafarers in the period from 2015 to 2019 (IHS Markit, 2020). The number lost and missing is roughly the same. Of those missing, more than 90% are missing at sea and there is thus every reason to believe that they also are dead.

If only those lost are considered, the average individual risk (probability of being lost per year) for seafarers in this period is $1.2 \cdot 10^{-4}$ per person-year. If missing seafarers also are included, the risk increases to $2.3 \cdot 10^{-4}$ per person-year. It should be noted that missing seafarers likely also include suicides, although the proportion is not possible to verify.

Comparatively, the average occupational accident fatality rate for workers in the EU was $1.8 \cdot 10^{-5}$ per person-year (data from Eurostat), i.e., more than a factor 10 lower. Further, the highest fatality rate in EU in 2018 was observed in the mining and quarrying industry, with close to $1 \cdot 10^{-4}$ per person-year. This shows that the maritime transport is a high-risk industry in terms of occupational risk.

2.6 OIL SPILLS AND POLLUTION

One of the major concerns related to environmental damage from maritime accidents is oil spill, primarily of cargo. Spills of bunkers, fuel oil, hydraulic

Table 2.5 Ten largest oil spills from ships

Year	Ship	Location	Cause	Spill size (tons)
1979	Atlantic Empress	Off Tobago, West Indies	Collision	287,000
1991	ABT Summer	Angola	Explosion	260,000
1983	Castillo de Bellver	South Africa	Fire	252,000
1978	Amoco Cadiz	Brittany, France	Grounding	223,000
1991	Haven	Genoa, Italy	Explosion	144,000
1988	Odyssey	Nova Scotia, Canada	Explosion	132,000
1967	Torrey Canyon	Scilly Isles, UK	Grounding	119,000
1972	Sea Star	Gulf of Oman	Collision	115,000
2018	Sanchi	Off Shanghai, China	Collision	113,000
1980	Irenes Serenade	Navarino Bay, Greece	Explosion	100,000

ITOPF (2021).

oil, etc. are also of concern, but the volumes will in most cases be much smaller compared to if there is a spill of cargo, in particular from crude oil tankers. The ten largest oil spills that have been recorded from ships are listed in Table 2.5.

It may be noted that the two largest accidental oil spills to sea that have occurred since 1970 not are related to maritime transport but to offshore oil production. These are Deepwater Horizon in 2010 (estimated spill size 560–585.000 tons) and Ixtoc 1 in 1979–1980 (estimated spill size 454–480.000 tons). Both took place in the Gulf of Mexico. Another observation from this table is also that only one of these spills occurred after 1991, i.e., nine were observed in the 24-year period (1967–1991) and only one in the following 30 years (until 2021).

Every year, more than 1 billion tons of crude oil are transported by sea. In Figure 2.10, the number of spills per year greater than 7 tons is shown. In total, more than 1,800 spills have been recorded since 1970. However, the trend has been sharply downward during the period, from close to 80 spills per year in the 1970s to just 6.2 per year in 2010s. This decline becomes even more impressive when considering that the volume of transported crude oil and other tanker trade has roughly doubled in the same period.

The lowest number of spills was recorded in 2019 and 2020, the two last years of the time series. From the figure, it is hard to discern a clear downward trend in the past decade, but if the period from 2010 to 2020 is examined separately, a linear trend line will show a decrease from around eight spills per year in 2010 to around four in 2020.

The volume of spills has also decreased accordingly. In the 1970s, on average nearly 320,000 tons were spilled each year while this had dropped to just over 16,000 tons per year in the 2010s. Largely, this is due to the reduction in number of spills, since the average spill size was reduced from approximately 4,100 tons in the 1970s to 2,600 tons in the 2010s.

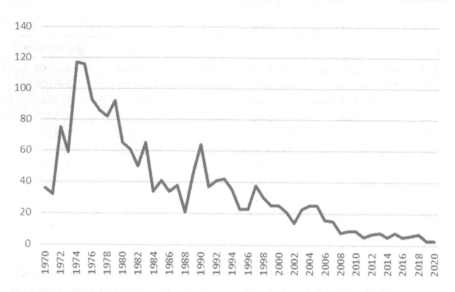

Figure 2.10 Number of spills (>7 tons) in the period of 1970–2020 (ITOPF, 2021).

Interestingly, the average spill size was smaller in the 2000s, at only 1,100 tons. This is a good illustration of the effect that large accidents can have on our statistics. In 2018, the ninth largest oil spill from a ship occurred, Sanchi, at 113,000 tons. Comparatively, there were no large accidents similar to this in the 2000s. If Sanchi is excluded, the average spill size in the 2010s drops from 2,600 t to only about 800 t.

This is an effect that can be seen in many situations. Very large accidents will tend to distort the statistics for long periods of time. The statistics for fatalities in the Norwegian oil and gas industry is a good example of this. In the period 1966–2018, a total of 283 fatalities were recorded. Of these, 123 people were killed in one single accident, the Alexander Kielland accident in 1980. This means that more than 40% of the total number of fatalities in this period is due to one accident. When calculating the average fatality risk per year, it may be questioned whether it is reasonable to include this accident. Clearly, an accident like this is not typical for an "average" year. On the other hand, the fact that it has occurred shows that this is also part of the risk that the offshore workers are exposed to.

2.7 EFFECT OF SOME FACTORS ON THE RISK LEVEL

There are many factors that influence the risk level for ships. In Section 2.5, the effect of ship type was illustrated, showing how some ship types had a significantly higher accident rate than the average. In this section, some other factors will be considered.

Port state control is based on a ranking of ships according to risk. Under the Paris MoU, a risk calculator is used that considers a set of factors which are considered to influence risk. Based on this, the ships are classified as High Risk, Low Risk, or Standard Risk. The factors taken into account in the risk calculator are as follows:

- Ship type (chemical tankers, gas carriers, oil tankers, bulk carriers, and passenger ships are considered to potentially be High-Risk ships)
- Ship age (two groups: less than and more than 12 years)
- Flag – a list of white, gray, and black flags are established, where black represents highest risk
- IMO audit of flag state performed
- The recognized organization (in practice the classification society)
- The ship owner
- The number of detentions and deficiencies for the particular ship

More details about the risk calculator can be found in Chapter 4.

REFERENCES

AGCS. (2020). *Safety and shipping review 2020*. https://www.agcs.allianz.com/news-and-insights/reports/shipping-safety.html

Alderton, T., & Winchester, N. (2002). Flag states and safety: 1997–1999. *Maritime Policy & Management, 29*(2), 151–162.

de Maya, B. N., Ahn, S. I., & Kurt, R. E. (2019). Statistical analysis of MAIB database for the period 1990–2016. In P. Georgiev & C. G. Soares (Eds.), *Sustainable development and innovations in marine technologies: Proceedings of the 18th international congress of the Maritme Association of the Mediterranean (IMAM 2019)*. CRC Press.

Eliopoulou, E., Papanikolaou, A., & Voulgarellis, M. (2016). Statistical analysis of ship accidents and review of safety level. *Safety science, 85*, 282–292.

EMSA. (2019). *Annual overview of marine casualties and incidents 2019*. http://www.emsa.europa.eu/publications/download/5854/3734/23.html

Heij, C., & Knapp, S. (2015). Effects of wind strength and wave height on ship incident risk: Regional trends and seasonality. *Transportation Research Part D: Transport and Environment, 37*, 29–39.

IHS Markit. (2020). *The state of maritime safety 2020*. https://ihsmarkit.com/info/1119/state-of-maritime-safety-digital-report-access.html

ISO 12100. (2010). *Safety of machinery – general principles for design: risk assessment and risk reduction*. https://www.iso.org/standard/51528.html

ITOPF. (2021). *Oil tanker spill statistics 2020*. https://www.itopf.org/fileadmin/uploads/itopf/data/Documents/Company_Lit/Oil_Spill_Stats_publication_2020.pdf

Lewison, G. R. G. (1980). The estimation of collision risk for marine traffic in UK waters. *The Journal of Navigation, 33*(3), 317–328.

Li, K. X., & Wonham, J. (1999). Who is safe and who is at risk: A study of 20-year-record on accident total loss in different flags. *Maritime Policy & Management*, 26(2), 137–144.

Li, K. X., Yin, J., & Fan, L. (2014). Ship safety index. *Transportation Research Part A: Policy and Practice*, 66, 75–87.

Paris MoU. (2011). *Annual report 2010*. https://www.parismou.org/2010-annual-report-voyage-completed-new-horizon-ahead

Paris MoU. (2020). *Annual report 2019*. https://www.parismou.org/2019-paris-mou-annual-report-port-state-progression-detention-rate-down

Reason, J. (1997). *Managing the risks of organizational accidents* (1st ed.). Ashgate.

UNCTAD. (2011). *Review of maritime transport 2011*. https://unctad.org/webflyer/review-maritime-transport–2011

UNCTAD. (2020). *Review of maritime transport 2020*. https://unctad.org/webflyer/review-maritime-transport–2020

Chapter 3

Terminology

3.1 INTRODUCTION

The terminology within the field of safety management and risk analysis is a minefield and even apparently simple terms like *risk* and *safety* can spark heated discussions, in particular among academics. However, of more practical concern is the fact that the diverging definitions also create confusion among practitioners. When regulations, guidance, textbooks, and other material use the terms differently, it is not easy to navigate through this and understand what the different authors have meant. This confusion is increased by the fact that many authors do not define the terms that they are using. In a study by Goerlandt and Montewka (2015), it was found that out of a total of 58 publications that they investigated, about half did not explicitly define the term risk.

All details of the various discussions about different terms will not be included here, but it is necessary to dedicate a chapter to defining some of the key terms that are used in this book. The main purpose is to help clarify the terms and guide users and readers through the many terms that are commonly used.

The starting point has been that we should try to define terms in the way that they are defined within the maritime community, and in particular the way that key regulations, guidelines, and standards use the terms. However, there are many examples of different uses here as well and it has been necessary to deviate in some cases.

In this chapter, terms that are used throughout the book or at least in several chapters are defined, such as risk, safety, risk analysis, and safety management. Terms that primarily have been used in just one chapter will normally be defined in that chapter. All terms that have been defined can however be found in the index. If the index refers to several places in the book where a term has been used, the definition will normally be found on the first page mentioned.

DOI: 10.4324/9781003055464-3

3.2 RISK AND SAFETY

Risk and safety are two words that often are interchanged, even if they clearly do not mean the same thing. Even in the title of the book, we have used both these terms, in *safety* management and *risk* analysis. Since these terms are so central to the whole field, it is natural to start by defining these.

3.2.1 Risk

The term risk is being used in many different ways by different organizations. The definition of risk suggested by the International Maritime Organization (IMO) in the Formal Safety Assessment guidelines (FSA) is as follows:

Definition 3.1: Risk: The combination of the frequency and the severity of the consequence (IMO, 2018a).

According to this definition, risk is comprised of two elements, frequency and consequence. It is defined as a combination of these two elements, without specifying how this combination should be done. A common way of expressing risk is to use the product of these two elements:

$$R = f \cdot C \tag{3.1}$$

Further, frequency is often also replaced by probability in this expression:

$$R = p \cdot C \tag{3.2}$$

When we use this definition, we can also interpret risk as a statistically expected loss.

Calculating risk of oil spills

Let us assume that we have found that the probability of an accident occurring is 0.002 per year. Further, we have found that if this accident occurs, we can expect that the consequence will be that an oil spill of 100.000 tons will occur and that one person will be killed. From Eq. 3.1, we can now calculate the risk:

$$R_{\text{Spill}} = \frac{0.002}{\text{year}} \cdot 100.000 \text{ tons} = 200 \text{ tons/year} \tag{3.3}$$

$$R_{\text{persons}} = \frac{0.002}{\text{year}} \cdot 1 \text{ fatality} = 0.002 \text{ fatalities/year} \tag{3.4}$$

First, we may note that we need to calculate two risk values in this case, to describe both spill risk and risk to persons. This shows that different types of consequences require separate calculations. Second, in the first case, we have calculated that the average (statistically expected) consequence is 200 tons per year. This does not mean that this accident causes 200 tons of spills each year. The reality is that either the accident occurs – and the spill is 100.000 tons – or the accident does not occur – and there is no spill. Similarly with the risk to persons. We can of course never experience 0.002 fatalities – either someone is killed or not. This illustrates that risk defined in this way is a statistically expected loss rather than an experienced loss.

Another illustration of risk is to assume that we have a different situation, with an accident that has a probability of 0.5 per year and a consequence of 400 tons spill. If we calculate the risk for this accident, we find:

$$R_{\text{Spill}} = \frac{0.5}{\text{year}} \cdot 400 \text{ tons} = 200 \text{ tons/year} \qquad (3.5)$$

We can see that this gives exactly the same risk as in the first case, with 200 tons/year. However, our judgment of the two situations may not be the same. A spill of 100.000 tons will be considered as a disaster that may be very hard to manage properly, while a spill of 400 tons normally is regarded as far less dramatic. This shows that calculating risk according to the definitions given above not necessarily tells us everything about how we want to prioritize management of risk. More about that in Chapter 6.

Another definition of risk that is very useful is the one proposed by Kaplan and Garrick (1981):

Definition 3.2: Risk: The combined answer to the three questions: (1) What can go wrong? (2) What is the likelihood of that happening? and (3) What are the consequences?

Questions 2 and 3 in this definition cover the same factors as included in Definition 3.1, but in addition, the first question explicitly specifies that risk is associated with "something that can go wrong". Implicitly, this is also part of the IMO definition, since the focus is on potential loss. An important aspect of this definition is that it also forms the basis for the way that we do risk analysis. This will be elaborated in Chapter 9. This definition will therefore be used in the rest of the book. For practical purposes, this is however not very different from the IMO definition.

It only makes sense to talk about risk in the future, not in the past. This is because probability is part of the definition and it has no meaning to talk about the probability that something can go wrong in the past. Either an accident occurred (and we experienced a negative consequence) or an accident did not occur. In spite of this, the term risk is commonly used both to describe the past and the future. An example is that we can say "Historically, the risk has been much higher than it is today". In reality, what we are then saying is that the losses that we have experienced in the past (the consequences) are higher than what we expect to experience in the future.

In this context it may also be noted that we very often base our risk analysis on past experience, i.e., it is quite natural to use the same term. For this reason, we will in this book also sometimes use the term risk to describe the past, e.g., in Chapter 2 Risk picture. We try to either ensure that the context makes it clear that we are using risk in this way, or we explicitly describe it in the text.

3.2.2 Positive risk

In financial applications and also some other areas, risk can be a positive thing. When we are investing money in the stock market, there is a risk involved in this because we risk losing our money if the share prices drop. However, the reason why we are willing to take this risk is that we believe or hope that the share prices will increase, meaning that we make a profit. In this context, risk therefore also comes with a (potential) reward. When we take a risk, it is done in the belief that the probability of making money is greater than the probability of losing money.

This is of course not directly relevant when we talk about accidents and potential losses. The best we can hope for then is that we avoid accidents and thus incur no losses. However, in practice, the situation is similar because we seldom accept accident risk unless there is a potential reward associated with it. The reward is however of a different nature than the consequence. An example is that we know that accidents may occur when we use ships to transport goods. Ships may sink, the crew may be killed, and the environment may be affected if oil or toxic chemicals are spilled. The reason why we accept this risk is that ships provide an efficient and cheap means of transport. This is thus the "reward" that we get.

There is also a difference from financial decisions in that with transport, the reward is certain, but the losses due to accidents are uncertain. If risk is well managed, cheap and efficient transport may be enjoyed without any losses. In financial investment, this is different since risk and reward are of the same type and uncertainty about the loss also means that there is uncertainty in the reward.

3.2.3 Other definitions of risk

It is worth mentioning that there are many other definitions of risk and also that the term often is being used differently in everyday language compared to the way that it is defined here.

Some examples of the use of risk from news articles on the internet (in February 2020):

- "AI systems claiming to 'read' emotions pose discrimination risks"

- "How your height can predict your risk of cancer and heart disease"
- "More than 1,200 British coastal landfill sites are at risk from erosion"

In neither of these cases is risk being used in the way that it is defined in this book. In all examples, it is more correct to interpret risk as probability or possibility.

In the literature, many other definitions of risk can be found, e.g.:

- "The possibility that human actions or events lead to consequences that harm aspects of things that human beings value" (Klinke & Renn, 2002).
- "Uncertainty about and severity of the consequences (or outcomes) of an activity with respect to something that humans value" (Aven & Renn, 2009).
- "Effect of uncertainty on objectives" (ISO 31000, 2018).

In the first example, risk is a possibility (corresponding to probability), while in the other two cases, we see that uncertainty rather than probability is part of the definition. Uncertainty clearly is an element of risk since we are trying to express that we are uncertain about what may happen in the future.

Goerlandt and Montewka (2015) have looked at how risk is defined in scientific literature concerned with maritime applications. Their starting point was a set of nine alternative definitions of risk, shown in Table 3.1.

As already mentioned, risk was defined explicitly in only about half the cases and where it was defined, 1 and 6 were the most common. It may also be noted that the IMO definition and that of Kaplan and Garrick (1981) both belong in category 6.

We do not go into more details on alternative definitions here, but only conclude that risk can be and is being interpreted and defined in very many different ways.

Table 3.1 Alternative definitions of risk

No.	Risk definition
1	Expected value
2	Probability of an undesirable event
3	Objective uncertainty
4	Uncertainty
5	Potential/possibility of loss
6	Probability and scenarios/severity of consequences
7	Event or consequence
8	Consequence/damage/severity & uncertainty
9	Effect of uncertainty on objectives

Goerlandt and Montewka (2015).

3.2.4 Safety

Safety is normally what we want to achieve when we are designing, building, and operating a ship or any other maritime system, but it is not a concept that is necessarily easy to define. Relatively few definitions can therefore also be found (as opposed to risk). MIL-STD-882E (2012) defines safety as "freedom from those conditions that can cause death, injury, occupational illness, damage to or loss of equipment or property, or damage to the environment". In practice, this definition means that there is zero risk, a target which in virtually all situations is impossible to achieve. This definition is therefore not very useful.

A more practical definition is the following:

Definition 3.3: Safety: Absence of unacceptable levels of risk to assets that we value and want to preserve, such as life and health, the environment, and economical assets.

This is more in line with how safety is used in everyday language. If we say that "I am safe", we will typically mean that we are in a situation where we consider the risk to be negligible or acceptable, not necessarily that the risk is eliminated.

3.2.5 Perceived risk vs calculated risk

When we perform a risk analysis, the results that we arrive at can be termed "calculated risk" and sometimes the term "objective risk" is also being used. We also use the terms "perceived risk" or "subjective risk".

Definition 3.4: Perceived risk: An individual's perception of risk, based on values, preferences, knowledge, experience, and other individual factors.

Since this depends on each individual, there can be large differences between the perceived risk of different individuals and there can also be large differences from the calculated risk. Since knowledge is one of the factors influencing our perception, information about the results from risk analysis may change our perception.

Perceived risk is not considered in risk analysis, but it is clearly important in decision-making. We will revert to factors affecting our perception of risk in Section 6.2.

3.2.6 Use of risk vs safety

In view of the fact that risk and safety are "opposite" terms it is perhaps not surprising that they are used interchangeably in some cases. A good example is the two terms risk management and safety management that are often used to signify the same thing. Likewise, we may say that we are "concerned about the safety of the crew on the ship" or that "we are concerned about the risk to the crew on the ship". In both cases, we are expressing more or less the same.

Within the maritime domain, the term *safety management* has become the most commonly used term, notably through the introduction of the International Safety Management Code (ISM code) (IMO, 2018) that is implemented as part of SOLAS Convention. In many other industries and applications, *risk management* is however the more commonly used term, as illustrated by the fact that the most well-known international standard in this field is called "Risk Management – Guidelines" (ISO 31000, 2018).

We are not suggesting that one term should be preferred over the other, but it is noted that what we normally try to manage is *risk*, with the purpose of reaching a low level of risk (or high level of safety). Similarly, it is also risk that we analyze, not safety. In the following, we will however use safety management as the main term, in accordance with the ISM code. On the other hand, we will use risk analysis and risk assessment rather than safety analysis, since risk analysis is by far the most common term used.

3.3 HAZARD AND ACCIDENT

If we revert to Kaplan and Garrick's (1981) definition of risk, the first question was "What can go wrong?" In this section, some of the key terms used when answering this question are defined. Before defining the terms, it is however useful to look at one of the most common ways of describing how accidents occur, the so-called bow-tie model.

3.3.1 An introduction to the bow-tie model

The bow-tie model is illustrated in Figure 3.1. This is a way of illustrating how accidents occur and forms a basis for risk analysis and subsequently safety management. The name of the model comes from the shape which is comparable to the bow-tie that men wear at formal dinner parties.

The main principles are that hazardous events (in the middle of the bow-tie) can have a number of different causes that lead up to this event. Similarly, there may also be different consequences resulting from the event. If we use collision between ships as an example, there may be many causes of this, e.g., poor look-out, misunderstanding of the other ship's intentions, or technical failure. Similarly, a collision can lead to only minor damage, severe damage to one of the ships or both ships and even sinking of one or both ships in extreme cases. This is illustrated with the "widening" of the bow-tie both on the causal side (left) and the consequence side (right) of the hazardous event.

For now, no further comments are given to the individual terms illustrated in the bow-tie, but all are defined in the following sections.

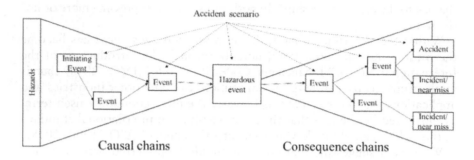

Figure 3.1 The bow-tie model with an illustration of important terms.

3.3.2 Hazard

Hazard is a seemingly simple term that is being defined in different ways and that also is being used to cover many different things, under the heading "things that can go wrong".

IMO have defined it as follows:

Definition 3.5: Hazard: A potential to threaten human life, health, property, or the environment (IMO, 2018).

This definition does not specify what a hazard <u>is</u> but defines the term by stating what a hazard can <u>do</u>, namely threaten something that we value and want to preserve. This is also reflected in the example list of hazards, that is, included in the FSA guidelines (IMO, 2018). The list is shown in Table 3.2.

From this list, hazards may include events (asbestos inhalation, falling overboard), activities (pilot ladder operation), materials (combustible materials, cleaning materials, fuel oil), equipment (cabling, electronic navigation equipment), external influences (storm, lightning), and sources of energy (friction, hot surface). There is even one example of a consequence being listed as a hazard (burns).

It may be argued that this is fine, the main thing after all is to identify anything that can lead to consequences that we do not want and what we call it is of less importance. This may well be, but on the other hand a very wide definition like the one applied can create problems and confusion when we are going to structure the information into a systematic analysis of risks.

Hazards

Based on the definition of hazard, it can be argued that flammable gas, release of flammable gas, ignition of flammable gas, fire, and explosion all are hazards. When we are analyzing this, we see that this is in fact different "aspects" of the same risk, represented by the source of risk (flammable gas) and several events that can develop when we

Table 3.2 Examples of hazards

Examples of hazards	
– asbestos inhalation	– friction
– burns from caustic liquids and acids	– hot surface
– electric shock and electrocution	– incendiary spark
– falling overboard	– naked flame
– pilot ladder/pilot hoist operation	– radio waves
– combustible furnishings	– electronic navigation equipment
– cleaning materials in stores	– laundry facilities
– oil/fat in galley equipment	– deck lighting
– cargo	– funnel exhaust emissions
– paint, oils, greases, etc. in deck stores	– hot work sparking
– cabling	– air compressor units
– fuel and diesel oil	– generator engine exhaust manifold
– fuel, lubricating and hydraulic	– storms
– refrigerants	– lightning
– thermal heating fluid systems	– uncharted submerged objects
– electrical arc	– other ships

IMO (2018).

have flammable gas (release, ignition, fire, explosion). If all of these are analyzed separately, the same risk will be counted several times, exaggerating the risk level.

From this example, we see that if we avoid use of flammable gas, we also avoid all the other hazards that we have identified. To understand these sequences of events (see definition of accident scenario in Section 3.3.4) is therefore important in managing risk. For this purpose, it may be useful to define not just what a hazard can lead to, but more precisely what it is.

Rausand and Haugen (2020) have proposed the following definition:

Definition 3.6: Hazard: A source or condition that alone or in combination with other factors can cause harm.

The definition is similar to the IMO definition in the sense that a hazard is something that can cause harm, but in addition it specifies that this has to be a "source" or a "condition". These terms may also be quite vague, but under this definition, it is clear that events are not hazards. It is also common in some applications to say that hazards are sources of energy, e.g., thermal energy, kinetic energy, potential energy, and electrical energy. This narrows the scope of hazards even further.

An additional comment is that hazards also can be regarded as "normal" in the sense that we are surrounded by hazards all the time. When we are driving a car, the speed represents a hazard that can kill us if we crash.

Table 3.3 Some examples of hazards and hazardous events

Hazard	Hazardous events
Flammable gas stored under pressure	Release of flammable gas
Storage of fuel oil	Leak of fuel, ignition of fuel
Electricity	Overheating of cabling, short circuit
Other ship heading toward own ship	Not changing course, collision
Own ship heading toward shore	Not changing course, grounding
Asbestos	Asbestos inhalation
Hot surface	People or flammable gas exposed to hot surface

When we are working on a computer, we are in close contact with electricity which also is a hazard that in the worst case can kill us if there is something wrong with the power supply. As long as the hazard is controlled, no consequence will however be experienced.

A couple of other terms that also are being used in many contexts are *hazardous situation* and *hazardous event*. International Association of Classification Societies (IACS) (not dated) have defined a hazardous situation as "a situation with a potential to threaten human life, health, property or the environment". This is in line with the IMO definition of hazard, with the addition that it is a situation. Hazardous event is not defined by IMO nor IACS, but can be interpreted as a specific event that potentially can cause harm. However, the use of these terms vary between different sources and other authors may use other definitions.

Hazardous event is sometimes also defined as the first event where we move from a "normal" situation (being surrounded by hazards that are under control) to an "abnormal" situation (where one of the hazards is out of control). An example can be that flammable gas under pressure is a hazard, while a leak of flammable gas is a hazardous event. Some more examples are given in Table 3.3.

3.3.3 Initiating event

Initiating event can be defined as follows:

Definition 3.7: Initiating event: The first event in a sequence of events leading to an incident or accident (adapted from IMO (2018)).

There are a couple of comments worth adding to this definition. Firstly, it is noted that it is the first event in a sequence. This sequence of events will typically be called an accident scenario (see Section 3.3.4). It is noted that it may be hard to define precisely what is the "first" event and the choice is often governed by the purpose and degree of detail of the analysis. It is therefore more commonly used as a term to describe the first event that we choose to model in our risk analysis. Secondly, the outcome may be either an accident or an incident. An accident is a situation where unwanted consequences have occurred (Section 3.3.5), while an incident has a potential to cause unwanted consequences (Section 3.3.6). Initiating events can lead to unwanted consequences but will not necessarily do so.

3.3.4 Accident scenario

Accident scenario can be defined as follows:

Definition 3.8: Accident scenario: A specific sequence of events from an initiating event to a hazardous situation or an undesired consequence (adapted from IMO (2018)).

This is more specific than the IMO definition and ties in with the previous definition of initiating event. It is noted that an accident scenario is not a specific event or a state/condition, but a sequence of events. This is sometimes also called an *event chain*, and this may be split into *causal chains* and *consequence chains* (see bow-tie in Figure 3.1).

3.3.5 Causal factor

A causal factor may be defined as follows:

Definition 3.9: Causal factor: A condition or a state related to a system, physical environment, or human operator that explains a technical failure or operator error.

An accident scenario may alternatively be defined as a sequence of interacting causal factors and events. Causal factors in a narrow definition "explains" why the succeeding event takes place. Causal factors related to technical systems may be inadequate maintenance, overloading of capacity or wrong operation, and subsequently result in reduced function, abnormal response, or breakdown. Extreme environmental factors like strong wind, high waves, strong current, and restricted visibility are all causal factors that may influence both human operation and the performance of technical systems. The non-performance of the operator "on the sharp" may be explained by exhaustion (lack of rest), lack of motivation, or incompetence. These causal factors may result in omissions or commissions (wrong actions).

It should finally be pointed out that the terms event and causal factor are used in different ways in risk analysis and accident analysis.

3.3.6 Accident

Accident is defined as follows:

Definition 3.10: Accident: A sudden, unwanted, and unplanned event or event sequence that has led to harm to people, the environment, or other tangible assets (Rausand & Haugen, 2020).

Accidents are events, but they are events that we do not want to happen, that we plan to avoid and that occur suddenly. There may also be slowly developing events, e.g., corrosion, that can lead to harm, but we would not normally call it an accident. Corrosion may lead to failure of a structure or a component and the failure may be called an accident. Further, accidents lead to harm to one or several assets that we want to protect. This is an important distinction from e.g. hazard, which can cause harm. Accidents are events where harm actually has occurred.

Accidents can be categorized in many different ways, depending on type of accident, source of accident, location of accident, etc. Most common is perhaps to categorize according to the type of accidental event, e.g., collision, grounding, stranding, fire, and capsizing. This is useful because the measures that we need to put in place to prevent these accidents or control the consequences in many cases will be quite different. A potential problem with this is that different sources may use different definitions of the accident types. Is "Collision" only collisions between two moving ships or does it also include impact between a stationary ship and a moving ship? Impact between a ship and any stationary object? Impact between ship and a quay? Allision is sometimes used to describe impact between a ship and a stationary object. It is important to understand what the definitions are since it may influence both the modeling of risk and in particular the available data sources.

Another common categorization is to distinguish between organizational and occupational accidents (see Section 2.4). The terms process and personal accidents are also used. The main difference between the two lies in the potential for harm. Occupational accidents are spatially limited and usually involve just one or two persons, normally those that perform an activity or operate equipment. For these, the consequence may in the worst case be fatalities, but the accident will not have a potential to involve more people or large areas. Typical examples are slips, trips, and falls.

Organizational accidents will on the other hand have a large potential for harm, potentially involving large groups of people and possibly also spreading over large areas. Examples of this can be large fires, collisions between ships, spills of total oil cargo, etc. Another characteristic of organizational accidents is that their causes can be traced back to multiple decisions and events involving large parts of an organization and that they often lead to multiple and extensive changes to the way we think about accidents and protect ourselves against accidents. An example is the sinking of Titanic which led to the introduction of SOLAS.

3.3.7 Incident and near miss

Like many of the other terms that are described, incident is also being used in different ways. On definition is as follows:

Definition 3.11: Incident: A sudden, unwanted, and unplanned event or event sequence that could reasonably have been expected to result in harm to one or more assets, but actually did not (Rausand & Haugen, 2020). An equivalent term is a non-conformity.

This is identical to the way that accident was defined, with the major difference that accidents cause harm while incidents do not. Sometimes, an incident is described as an event that, under slightly different circumstances, would have been an accident. In the bow-tie model in Figure 3.1, this has been illustrated by accident scenarios (event sequences) that can end either in an accident or an incident, depending on the precise events and conditions being part of the scenario. Another term that often is used to mean the same thing as incident is *near miss*.

Within the maritime industry, incident is partly used more or less synonymously with accident. This just serves to illustrate that we need to make sure that we understand precisely what is meant when various terms are being used; otherwise, we easily misunderstand and draw wrong conclusions.

3.4 FREQUENCY AND PROBABILITY

We now revert to Kaplan and Garrick's (1981) definition of risk and look at the second question: What is the likelihood of something going wrong? In this section, we will look at how we measure this, namely through probabilities or frequencies.

3.4.1 Probability

Probability is a complex topic that has been discussed extensively and where there exists no common agreement on how to interpret it. For this purpose, we will limit the discussion to only consider two interpretations that are of main relevance to the topic of this book.

A frequentist (or objective) interpretation of probability: This is based on the assumption that the probability of an event occurring is a property of the situation, system, or event that we are considering. By repeating the situation or the event ("the experiment") a sufficiently large number of times and calculating the ratio between the number of times the event occur to the total number of outcomes, the probability of the event can be found. It is generally acknowledged that in risk analysis, it is not realistic to fulfill the assumption about repeatable experiments. The frequentist interpretation is therefore not applicable.

A Bayesian (subjective) interpretation of probability: The probability is seen as our degree of confidence in the occurrence of an event. This confidence is measured on a scale from 0 to 1. An event with probability 0 is believed to be impossible; an event with probability 1 is believed to occur with certainty (IMO, 2018). One implication of this is that different people may have different views on the probability, and it can be argued that all views are equally correct!

For the purpose of performing risk analysis, the different interpretations do not have any practical implications. The rules for calculating risk are not affected by the interpretation and the way of performing a risk analysis is therefore the same. However, it has important implications for the *interpretation* of the results from a risk analysis and therefore also for how we use the results when making decisions about risk. If we accept that probability is subjective, that also means that risk is subjective.

3.4.2 Frequency

In many cases, it is more relevant to use frequency instead of probability to measure the likelihood of an event occurring. The frequency is usually an expression of the number of occurrences (of an event) per unit time (e.g., per year). Frequency can also be understood as how often an event occurs.

3.5 CONSEQUENCE

Moving to the third question in the definition of risk, "What are the consequences if something goes wrong", we also move into somewhat less troubled waters in terms of how to define it and how it is being used.

Definition 3.12: Consequence: The outcome of an accident (IMO, 2018).

This is a very simple definition and can in principle cover all types of outcomes, being negative or positive. In practice it will be negative consequences that are relevant to consider. The consequences are associated with one or several *assets* that we want to protect. Assets may be people, but it may also be the environment, economic values, reputation, etc. When the consequences of an accident are considered, consequences for different assets are normally described separately. For a given accident, the outcome may be, e.g., that "the consequence is 1 fatality, spill of 1,000 tons of oil and property damage valued at 1 mill USD".

A number of other terms are also being used, e.g., *damage*, *severity*, and *loss*, meaning more or less the same.

3.6 SAFETY MANAGEMENT

As already pointed out earlier in this chapter, safety management and risk management seem to be used more or less interchangeably. Within the maritime community, safety management is the most commonly used term, among others because of the ISM code.

Definition 3.13: Safety management: Coordinated activities aimed at fulfilling a set of safety policies and objectives.

Definition 3.14: Safety management system: A structured and documented system of activities aimed at fulfilling a set of safety policies and objectives.

Safety management is described in Chapter 5.

3.7 STAKEHOLDER

IMO does not define this term and other sources must be consulted. ISO 31000 provides the following definition:

Definition 3.15: Stakeholder: Person or organization that can affect, be affected by, or perceive themselves to be affected by a decision or activity.

Other terms are also being used, e.g., interested party (ISO 45001, 2018).

Stakeholder is here defined generally and is not specifically related to risk, but covers any effect of a decision or an activity. This means that, e.g., persons or organizations that are affected by a decision by a government to raise taxes could be regarded as stakeholders. An example of a decision could be to build and operate a fleet of autonomous ship. Some of the stakeholders in the risk associated with this activity could be the maritime authority in the country where the ships will be registered, the ship owner, crew, seamen's unions, those who want to transport goods on these ships, ports they will be visiting, other ships, environmental groups, and so on. The stakeholder concept is discussed in Chapter 4.

3.8 RISK ANALYSIS AND RISK ASSESSMENT

Risk analysis and risk assessment are two additional terms that also tend to be mixed up in many applications. However, common today is to view risk assessment as a wider process than risk analysis, including more steps. The following definitions illustrate this.

Definition 3.16: Risk analysis: A systematic study to identify and describe what can go wrong and what the causes are, the likelihoods, and the consequences might be (Rausand & Haugen, 2020).

Definition 3.17: Risk assessment: The process of planning, preparing, performing, and reporting a risk analysis, and evaluating the results against risk acceptance criteria (Rausand & Haugen, 2020).

Risk analysis is in simple terms the process that we apply to describe or calculate risk, by answering the three questions that form our definition of risk. *Risk assessment*, on the other hand, also includes preparing this process and evaluating the results of the process. The definitions found in ISO 31000 (2018) are similar.

IMO (2018) use Formal Safety Assessment rather than risk assessment. However, risk analysis is one of the steps in the Formal Safety Assessment process. In practice, FSA is therefore a risk assessment the way that it is defined here.

In this book, the following terminology will be used:

- When discussing the process described by IMO, the term Formal Safety Assessment will be used. This will always mean the process that they have described in their guidelines.
- When discussing the process of assessing risk more in general terms (not necessarily as part of the FSA process), the term risk assessment will be used.
- Risk analysis will always be used (not safety analysis), to describe risk analysis in general but also to describe the step in the FSA procedure that this refers to (in accordance with IMO).

This may be a bit confusing at times, but we believe this is the best way of solving this terminology problem.

3.9 RISK CONTROL AND REDUCTION

There are several terms that are important here and that should be defined. Within safety management, the term "barrier" is particularly important.

Definition 3.18: Barrier: Physical or engineered system or human action (based on specific procedures or administrative controls) that is implemented to prevent, control, or impede energy released from reaching the assets and causing harm (Rausand & Haugen, 2020).

In practice, this describes all measures put in place to control risk, be it technical systems or human actions that contribute to prevent accidents from occurring or reducing the consequences. The terms risk reduction measures, safety measures, and layers of protection together with many others are also used to signify more or less the same.

IMO has defined two particular terms that are used to describe barriers:

Definition 3.19: Risk control measure (RCM): A means of controlling a single element of risk.

Definition 3.20: Risk control option (RCO): A combination of risk control measures.

This is discussed in more detail in Chapter 15.

3.10 RISK ACCEPTANCE CRITERIA

Since we are very seldom able to remove all risks, we usually end up in a situation where we have to make a decision about whether the risk is sufficiently low or not. For this purpose, we use decision criteria. The most common term used to describe these decision criteria are risk acceptance criteria.

Definition 3.21: Risk acceptance criteria: Criteria used as a basis for decisions about acceptable risk.

The word "acceptance" is sometimes contended, because it may be interpreted as meaning that we accept that accidents can happen and that serious injury and fatalities may occur. IMO has chosen to use risk evaluation criteria instead of risk acceptance criteria. In practice, they have the same meaning.

Another term that is also being used quite commonly is *risk tolerability criteria*. By using this, we signal that risk is not something that we accept unconditionally, but we may be willing to tolerate risk provided the returns are proportional to the risk that we take.

REFERENCES

Aven, T., & Renn, O. (2009). On risk defined as an event where the outcome is uncertain. *Journal of Risk Research, 12,* 1–11.

Goerlandt, F., & Montewka, J. (2015). Maritime transportation risk analysis: Review and analysis in light of some fundamental issues. *Reliability Engineering and System Safety, 138,* 115–134.

IACS. (not dated). *FSA Glossary.* IACS.

IMO. (2018a). *ISM Code - International Safety Management Code with Guidelines for Its Implementation.* London: International Maritime Organization.

IMO. (2018b). *Revised Guidelines for Formal Safety Assessment (FSA) for use in the IMO Rule-Making Process.* London: International Maritime Organization.

ISO 31000. (2018). *Risk Management - Guidelines.* https://www.iso.org/standard/65694.html

ISO 45001. (2018). *Occupational Health and Safety Management Systems.* Geneva: ISO.

Kaplan, S., & Garrick, B. (1981). On the quantitative definition of risk. *Risk Analysis, 1,* 11–27.

Klinke, A., & Renn, O. (2002). A new approach to risk evaluation and management: Risk-based, precaution-based, and discourse-based strategies. *Risk Analysis, 6,* 1071–1094.

MIL-STD-882E. (2012). *Standard Practice for System Safety.* Washington, DC: US Department of Defense.

Rausand, M., & Haugen, S. (2020). *Risk Assessment: Theory, Methods and Applications, 2nd Ed.* New Jersey: Wiley.

Chapter 4

Stakeholders, rules, and regulations

4.1 INTRODUCTION

Waterborne transport of basic materials and products has for centuries been the main prerequisite for trade between nations and regions and has without doubt played an important role in creating economic development and prosperity. The cost of maritime transport is very competitive compared with land and airborne transport, and the increase to the total product cost incurred by shipping represents only a few percent. Negative aspects of waterborne transport include longer transport time because of relatively low ship speed, congestion in harbors resulting in time delays, as well as less efficient integration with other forms of transport and distribution.

Shipping has from time to time been criticized for unacceptable safety and environmental performance, and this will be discussed in the next chapter. At this point, we only make the following remark: in view of the relatively low cost of transport, it is a paradox that some areas of shipping have a relatively low standard of safety. Efficient transport should be able to pay for an acceptable safety standard.

It has been discussed for some time whether basic economic mechanisms could ensure safe shipping. In this context, the following questions are relevant:

- Is there any economic motivation for high levels of safety?
- Who should pay for increased safety?
- Are there any trade-offs between safety and efficiency?

These questions will be addressed briefly in this chapter, as well as throughout this book.

DOI: 10.4324/9781003055464-4

4.2 INTERNATIONAL TRADE AND SHIPPING

4.2.1 Seaborne transport

Maritime transport is crucial to the development of the global economy. Today raw materials, components, semi-finished products, and finished products are transported by sea. Due to the increased globalization of the production processes international maritime trade has grown steadily during the last decades. Measured in terms of volume seaborne transport represent 80% and 67% by economic value (2017). Since 1970 up to 2018, the total volume of world trade has increased more than four times. This means that the average annual growth has been 2.5%. It shall however be kept in mind that it has been a considerable variation in growth of trade seen year by year as indicated in Figure 4.1. This fact has a strong impact on seaborne transport in terms of demand for shipping services. It is interesting to note that the relative importance of tanker shipping has decreased from 55% in 1970 to 29% in 2018. Other dry cargo represents the largest category today with 42% (see Figure 4.2).

One of the challenges of improving shipping safety is that shipping business is subject to strong market fluctuations. This is because the demand for seaborne transport fluctuates significantly from one year to the next, while the supply of merchant vessels is more or less constant, with a stable growth. This means that we will see periods with both surplus and lack of supply of shipping capacity. The financial situation of the shipping companies varies accordingly and this also influences their willingness and ability to spend on improving safety.

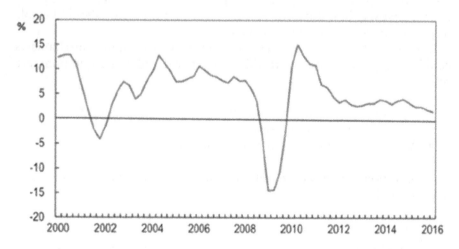

Figure 4.1 World trade growth. Percentage change annually (OECD Outlook 99 Database, 2016).

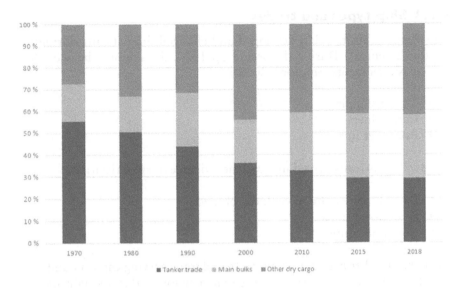

Figure 4.2 Relative size of the main shipping trades 1970–2018 (UNCTAD Review of Maritime Transport, 2019).

4.2.2 Shipping markets

The supply of shipping services will to a large degree be determined by the demand for seaborne transport services. The shipping industry can be divided into five markets:

1. Newbuilding market – where ships are being ordered
2. Freight market – where they are being chartered (used for transportation)
3. Sale and purchase market – where they are being sold to other shipowners
4. Leasing market – where a vessel is rented for a given period
5. Demolition market – where they are being sold to scrap yards

New ships will be ordered and built in order to meet an increased demand for shipping services. Similarly, new ships are built when they are becoming competitive compared to old vessels that are gradually scrapped. The newbuilding activity is also determined by innovation in ship design.

The shipper of goods arranges for transport with a shipowner in the freight market by fixing a contract. The freight contract may take several different forms determined by the type of goods, ship, and service. The ship owning or management business is dynamic in itself: Companies are started, are changing their market focus, and are terminated over time. In this sense, ships are sold and purchased between these companies.

4.2.3 Ship types and trades

The shipping business has developed and changed dramatically during the twentieth century. During the period up 1950, the shipping business had three basic segments (Stopford, 2009):

- Tramp shipping
- Cargo liners
- Passenger liners

Gradually this transformed into a greater degree of specialization:

- Bulk shipping: Grain, coal, oil, ore, etc.
- Specialized ships
- Container ships

There has further been a great increase in cargo carrying capacity and even speed for certain ship types. And the tendency for further specialization of trades and ship types has continued. Figure 4.3 gives an overview of how the world fleet has developed during the period of 2009–2018. The annual growth rate in terms of number of vessels has been 1.6% and in terms of deadweight 5.3%. This means that the mean vessel size has increased. The most important ship types in terms of number of vessels are conventional cargo ships, bulk carriers, crude oil tankers, chemical tankers, and container ships. However, there are several other specialized vessels too. When focusing on deadweight there are three dominating types, namely bulk carriers, crude oil tankers, and container ships. The strongest annual growth has been seen for bulk carriers (5.4%), gas tankers (3.5%), chemical tankers (3.4%), and special cargo (3.1%).

Conventional cargo ships represent a challenge to the improvement of maritime safety. This ship category is the largest in terms of number and has the highest average age compared to other ship types. These ships show up relatively often in inspection reports as for instance in Port State Control (see Section 4.9).

4.2.4 Economics of shipping

It is reasonable to assume that one of the prerequisites for attaining an acceptable safety standard is that a company has a sound economy and thereby being able to work systematically with safety-related matters such as training of personnel, raising the technical standard, and improving management routines. At first look that might seem difficult in the shipping business where the income and revenues are fluctuating dramatically.

To give a better view of the volatility of the shipping business, the operation of VLCC tankers will be taken as a case and focused on the development in

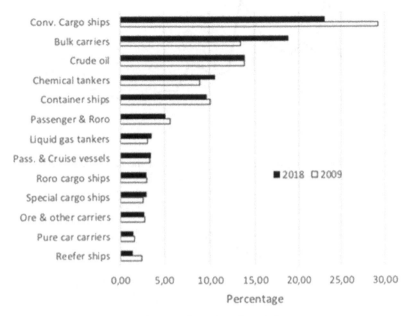

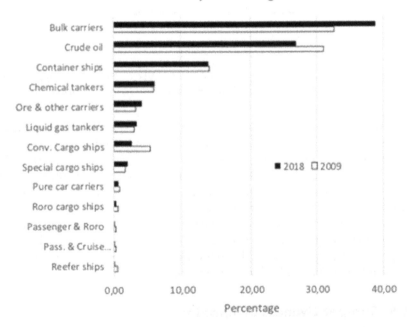

Figure 4.3 Distribution of world merchant fleet by main ship types. Ships greater than 300 gross tons (ISL – Shipping Statistical Yearbook, 2018).

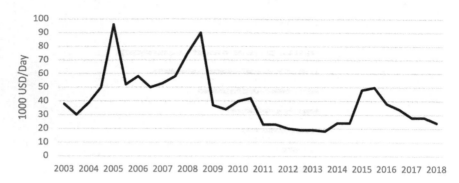

Figure 4.4 Fearnleys – 12 months time charter rates for modern VLCCs (ISL – Shipping Statistics Yearbook, 2010, 2014 & 2018).

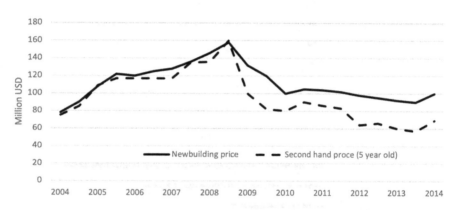

Figure 4.5 Newbuilding and second-hand prices for VLCC (Ship Operating Costs – AnnualReviewandForecast.AnnualReport2014/15.DrewryMaritimeResearch).

first years of this century. It has been a dramatic fluctuation in the time charter rates as illustrated in Figure 4.4. Contracts have been fixed at rates as low as 19,000 USD/day and as high as 96,000 USD/day on the top, i.e., a variation with a factor of five. This is also reflected in the price for newbuildings and second-hand vessels as shown in Figure 4.5. The fluctuations are however less dramatic compared with the freight rate. This can be explained by the fact that the demand side adjust slower: The time to order and build the vessels is relatively long compared to the timescale for market fluctuations. The scrapping price is also influenced by the market situation. Figure 4.6 shows how a depressed freight market lead to more scrapping and higher prices.

4.2.5 Competitiveness of shipping

It is also possible to argue for the opposite view, namely that the low cost of sea transport should put shipping in a favorable position compared with

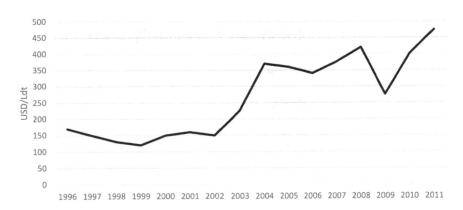

Figure 4.6 Average ship scrapping price in USD per lightweight tonnes (Ldt) (World Shipbuilding – Market Review & Forecast. Annual Report 2012/13. Drewry Maritime Research).

Table 4.1 Relative price of seaborne transport, Far East – Europe/US in 2002 (distance 9,000 nm)

Product / vessel / capacity utilization (%)	Sales price/ unit (USD)	Freight cost/ unit (USD)	Relative freight cost
I barrel crude oil /VLCC / 50%	30	1.5	5.0%
I tonne wheat / Bulk 52,000 DW / 100%	220	14	6.6%
I car / Multipurpose Ro-Ro / 50%	21,000	558	2.7%
I refrigerator/ Container 6600 TEU / 80%	550	9	1.6%

more expensive transport modes. As shown in Table 4.1, the relative sea transport cost lies in the range of 2–5% of the product cost. Let us make a thought experiment: If we assume that the safety of shipping could be improved significantly at a cost of 50% on the freight rate, this would only have resulted in 1–2.5% higher product price, which is marginal.

The efficiency of seaborne transport becomes even more clear when considering the case of crude oil transport. As shown in Figure 4.7, the long-haul transport only stands for 1% of the cost of gasoline on the market.

The relative competitiveness of seaborne transport to the other main forms (rail, air, and road) is illustrated in Figure 4.8 based on the transportation of a 40 feet container from Shanghai to Western Europe. The greatest contrast is sea versus air transport. Air transport takes 5 days versus 28 days for the ship. On the other hand, air transport costs eight times as much.

An argument against sea transport in these kinds of discussions is the relatively longer transport time due to low steaming speed and delays in ports. Despite this, the competitiveness of shipping is increasing even on shorter routes. A comparison of sea and air transport is shown in Table 4.2.

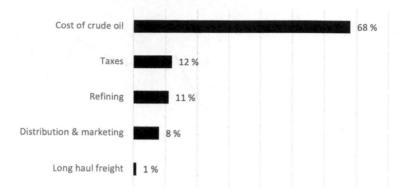

Figure 4.7 Break down of the main cost elements for gasoline.
Source: API.

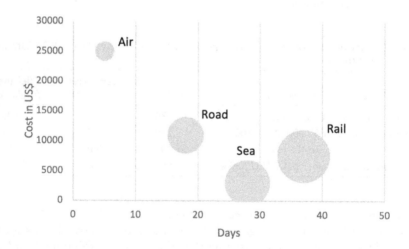

Figure 4.8 Comparison of cost and time for transportation of 40 ft container from Shanghai to Western Europe, 2020.

Seaborne transport is cheaper, more environmentally friendly but mainly competitive for bulk materials. Air transport is better suited for high cost products where short transport time is vital.

Transport is increasingly subject to concern about the environmental consequences. Road transport is a significant environmental problem in Central Europe and the road system is more or less exhausted. In Table 4.3 is given a crude comparison of the environmental effects to air, water, and land resources, solid waste, noise, and in terms of accident risk. Marine transport is indirectly influencing water and land resources through construction of port and canals. Secondly, the risk of spill of petroleum and hazardous chemicals because of accidents is significant. It is a growing concern for the emission of

Table 4.2 Typical cargo types for seaborne and airborne transport

Reasons for sea transport	Reasons for shipping by air
Cost is definitely lower both with respect to weight and volume	Speed: Air transport is 30 times faster. High value: Lower "warehouse" cost
CO_2 emissions are lower: 3% compared to air for an item of 2 tonnes over 5,000 km	Reliability: Better tracking and less delays Protection: Less damage to cargo
Typical cargo	**Typical cargo**
Heavy objects	Electronics
Bulk material: Oil, ore, grain, cement, etc.	Apparel: Primarily seasonal items
Semi-manufactured items	Pharmaceuticals
Finished products in large volumes	Documents and samples
	Other seasonal products

greenhouse gases (GHG) related to burning of fossil fuels and this also affects shipping although in a less critical way than other economic branches. As shown in Figure 4.9 the maritime sector contributed only 3% and the same figure for air transport. However, these two sectors have the highest growth rate. But despite the relatively low emission from shipping, the sector now takes this problem seriously and is working toward alternative energy systems.

Within seaborne transport, the largest contributions to GHG emissions are container ships, bulk carriers, and tankers as shown in Figure 4.10.

4.3 THE SHIPPING INDUSTRY SYSTEM

4.3.1 Stakeholders

In shipping there are several actors and stakeholders that have an influence on safety. By actors we understand parties that are directly in seaborne transport. Today, there is however also increasing focus on the broader concept, stakeholders, that both involve actors and other parties affected by the shipping business. The most important stakeholders are presented in Figure 4.11. The role of the actors is outlined briefly in Table 4.4. It should be evident that the different actors within the shipping domain to some extent have competing interests that may complicate the issue of promoting safety. The competing interests are a result of various factors such as the following:

- Who is controlling who?
- Who sets the safety standards?
- What is the motivation for safe operation?
- Who is picking up the bill for safety improvements and accidents?

We will return to the questions of safety management and the regulation of shipping in later chapters.

Table 4.3 Main environmental effects of transport modes

Effect of transport	Mode of transport			
	Rail	Road	Air	Marine and inland water
Air		Local greenhouse gases, fuel additives, and particulates. Global greenhouse gases.	Emissions from engine	Emissions from engine
Water resources		Pollution of surface water and groundwater. Modification of water system by road building	Modification of water tables, river courses, and field drainage in airport construction	Modification of water system during port construction and canal cutting and dredging
Land resources	Land taken for rights of way, terminals, and obsolete facilities	Land taken for infrastructure; extraction of road building materials	Land taken for infrastructures; abandoned sites	Land taken for infrastructure and obsolete port facilities and canals. Beaching and scrapping of ships
Solid waste	Abandoned lines, equipment, and rolling stock	Abandoned spoil tips and rubble from road works. Vehicles withdrawn from service and waste oil	Scrapped aircraft	Vessel and craft withdrawn from service
Noise	Noise and vibration at terminals and along lines	Noise and vibration in cities and along main roads	Noise around airports from aircraft and road traffic	
Accident risk	Derailment or collision of trains carrying hazardous substances	Death, injuries, and property damage. Risk of hazardous substances and structural failure of worn facilities		Bulk transport of petroleum and hazardous substances
Other impacts	Partition or destruction of neighborhoods, farmland and wildlife habitats	Partition or destruction of neighborhoods, farmland and wildlife habitats	Congestion on access routes to airports	

Source: www.geographynotes.com/

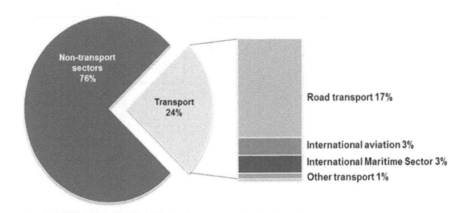

Figure 4.9 Relative contribution of transport to GHG emission in 2013 (Maragkogianni et al., 2013).

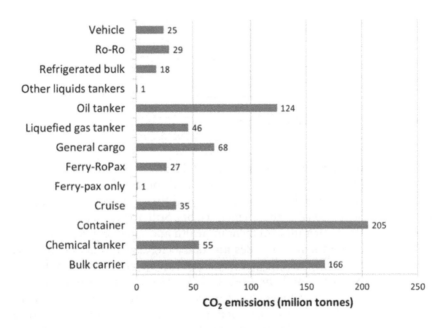

Figure 4.10 CO_2 emission by ship type in 2012 (IMO, 2014).

4.3.2 Corporate social responsibility

The loss rate expressed as the annual number of ships lost relative to the number of ships at risk has steadily decreased for more than a century. There is therefore no doubt that the safety levels have improved. The driving forces have been stricter regulation of ship design, manning of vessels, and systematic

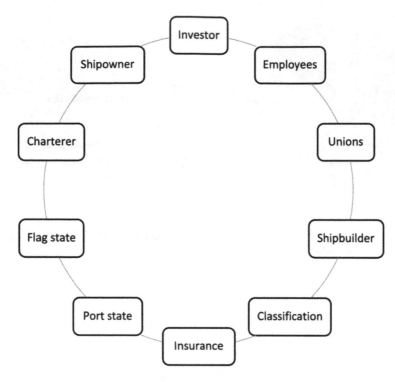

Figure 4.11 Primary stakeholders in shipping.

management by shipping companies. The Classification Societies have also played a key role in developing knowledge-based construction rules. The conditions for seafaring people have been improved through better working conditions and more reasonable employment rights. Still, there is a consensus that there is room for further improvement in health and safety conditions.

Secondly, during the last decades an increasing concern about the environmental aspects of shipping has been seen. Ship accidents may lead to pollution of the ocean and the operation of vessels lead to both air and sea pollution. These challenges may of course be met with strengthened regulations but there is an increased realization that Corporate Social Responsibility (CSR) may be the answer. The philosophy behind CSR is the belief that commercial success in a longer perspective must work with the affected community. That means that the shipping business must understand the stakeholders they are working with and the ones that are affected by their operation. *World Business Council for Sustainable Development* has defined CSR as follows (WBCSD, 1999):

> Corporate social responsibility is the continuing commitment by business to behave ethically and contribute to economic development while

Table 4.4 Actors in shipping that influence safety

Actor	Influence on safety
Shipbuilder	Technical standard of vessel
Shipowner	Decides whether technical standards will be above minimum requirements
	Selects crew or management company for crew and operation
	Makes decisions regarding operational and organizational safety policies
Cargo owner	Pays for the transport service and thereby also the quality and safety of the vessel operation
	May undertake independent assessments of the quality of the shipper
Insurer	Takes the main part of the risk on behalf of the shipper and cargo owner (i.e., vessel, cargo, third party – P&I)
	May undertake independent assessment of the quality of the shipper
Management company	Responsible for crewing, operation, and upkeep (i.e., maintenance of the vessel on behalf of the shipowner)
Flag State	Control of vessels, crew standards, and management standards
Port state	Control of vessels, crew standards, and management standards
Classification Society	Control of technical standards on behalf of insurer
	Undertakes some control functions on behalf of the Flag State
Port administration	Responsible for safety in port and harbor approaches
	May control safety standard of vessels, and in extreme cases deny access for substandard vessels

improving the quality of life of the workforce and their families as well as of the local community and society at large.

A theoretical basis and motivation for CSR has been given by Carroll (1991, 2016). He introduced the so-called pyramid of corporate social responsibility as illustrated in Figure 4.12. The basis for any business activity is *economic responsibility* in terms of providing goods and services which benefit both the shareholders and society. The second level, called *legal responsibility*, simply implies that the business shall operate in accordance with laws and regulations or what may be termed the *codified ethics*. Already today and even more in the future, it will be expected that businesses operate in accordance with norms and values beyond legal standards and thereby demonstrate *ethical responsibility*. The ultimate form is *philanthropic responsibility*. It will be expected that the businesses will contribute more directly with their resources to the society at large. The *philanthropic responsibility* may take the form of donations and participation in community development.

CSR developed initially from external pressure from society being subject to environmental and medical harm by businesses. In the future it is

Figure 4.12 Carroll's pyramid of corporate social responsibility.

expected that CSR will be an integrated part of the business philosophy. On the other hand, it is realistic to acknowledge that the full potential of CSR is yet to be seen. Attitudes must change and demonstrate a will to pay the price for use of the environment.

4.3.3 The shipowner

In the case of severe shipping accidents and losses, the shipowner will be subjected to particular attention. This is natural given the fact that the shipowner has responsibility for the vessel, as well as manning, maintaining, and operation of it; directly or indirectly through external management.

Questions that are raised in connection with maritime accidents are whether the shipowner has demonstrated a genuine concern for safety, and whether the standards of the vessel and its crew have been sacrificed for profit. The shipowner may counter such questions by claiming that the standards will not be better than what the market is willing to pay for.

Shipowners take some key decisions that have profound consequence for safety. The choice of Flag State for registration of vessels, choice of Classification Society, and arrangements for insurance are some key decisions. There exist international markets for these services in which different standards and corresponding fees can be found. The safety standard will therefore to a large degree be a result of what the owner is willing to pay for these services. A much discussed and fairly controversial topic is the practice of "flagging out", in which the shipowner registers a vessel in a country other than where it operates. Flagging out is mainly done for economic reasons, as shown in Figure 4.13. Availability of cheap labor, and the costs and strictness of safety control, seem to be key concerns for the

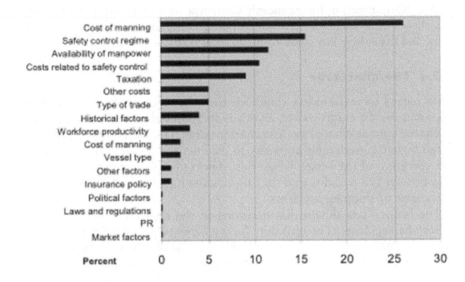

Figure 4.13 Reasons for flagging out. Distribution of answers from questionnaire study. (Adapted from Bergantino and Marlow, 1998.)

owner. Based on this, it must be asked whether shipowners, through their choice of flag, sacrifice safety.

The safety aspect of ship operation should also be seen in a wider context as shipowners have several different objectives that need to be balanced. These objectives include the following:

- Stay in business: return on investment.
- Marketing: win well-paying freight contracts.
- Service: minimize damage on cargo, keep on schedule.
- Efficiency: operate and maintain the vessel.
- Employer: attract competent personnel.
- Subcontracting: select efficient service providers.
- Availability: minimize unplanned off-hire.

It is not necessarily obvious that these objectives and priorities for a given company are always consistent with a high safety standard. In this view it is of great importance that shipowners have clearly defined policies that never compromise on safety. An alternative view is that there is no conflict between cost, efficiency, and safety. The main argument for this position is that in order to stay in business and thrive in the long term, shipowners have to operate safely and keep their fleets well maintained and up to standard. This view may, however, be a little naive, as substandard shipping companies may not necessarily have a long-term perspective of their business. An

OECD (Organization for Economic Cooperation and Development) study has shown that such substandard shipping companies are competitive on price and take their fair share of the market (OECD, 2002).

4.3.4 The charterer

With regard to vessel safety standards there has recently been an increasing focus on the cargo owner, as this is the party that decides which ship to charter and at what price. The charter party (i.e., the contract) gives the cargo owner considerable authority to instruct the Master with respect to the operation of the vessel. Given this important role in terms of safety, it may be seen as a paradox that the cargo owner has minimal, if any, liability in the case of shipping accidents.

The history has shown that decisions of the charterer may contribute to serious accidents. The tankship *Torrey Canyon* grounded on the Seven Stones in 1967 partly because of a charterer decision. The vessel was headed for the port of Milford Haven but due to inadequate navigation the vessel had been drifting easterly from the intended route. The charterer instructed the vessel to reach the port on high tide as the alternative was to be delayed for 12 hours. The Master was then forced to take an unplanned course between Isles of Scilly and Seven Stones rather than taking the longer route by heading out to open sea. Due to lack of passage planning and local fishing vessels in the fairway, the tanker grounded.

Another much discussed issue is the chartering of substandard vessels. Despite stricter regulations for vessels and crew, this is still a problem. As a result, charterers hiring substandard vessel have been subject to increased criticism.

There are different forms of charter contracts: bareboat, time, and voyage charter. In the case of bareboat charter and some other arrangements, the charterer is in reality the legal carrier with the same responsibilities as a shipowner. As a result of legislation such as, e.g., the United States Oil Pollution Act 1990 (OPA, 1990), the charterer holds a greater responsibility. The charterer will seek cover for third-party liability with standard P&I insurance (Protection & Indemnity). Relevant risks are dangerous cargo, pollution, unsafe ports, stowage, and crew accidents. Due to different laws, regulations, and legal practice in different countries, the charterer is not necessarily fully protected.

Oil Companies International Marine Forum (OCIMF – www.ocimf. org) introduced their *Ship Inspection Report Programme* (SIRE) in 1993 to meet the negative impact of major tankship accidents and substandard shipping. Based on inspection reports, a database of more than 180,000 entries has been accumulated. At the outset, the SIRE system was intended as a tool for oil companies when chartering ships but today has a broader use by the petroleum industry and even government bodies.

Seaborne transport may be seen as a part of production and logistic processes. It is a fact that products to an increasing degree are marked or classified according to different environmental and CSR standards. This will also have an impact on the sea transport element and thereby a push toward higher standards. It is possible to foresee that the transport phase increasingly must document environmental friendliness and acceptable workplace conditions. It can therefore be concluded that the charterer ought to have greater focus on both safety and CSR standards.

4.4 THE MARITIME SAFETY REGIME

Given the fact that shipping is exposed to the fluctuations in international trade, focus on safety may be demanding in terms of manpower and economy. This is true both during periods with increasing demand for shipping services and in periods with decreasing demand. With increased demand the shipowner may focus on marketing, investment in new vessels, and operation at the expense of safety. During market lows, the economic basis for maintaining the same priority on safety may be threatened. It is not immediately clear that safety still is an important issue to maritime transport. One may argue that ship design has reached an advanced level, crew training has a high standard, and shipping companies are as advanced as other businesses in terms of safety management. Secondly, shipping is subject to rigorous control and continuously has the attention of governments and the public. Table 4.5 shows that seaborne transport today is strictly regulated because of a series of internationally adopted safety conventions.

The average loss rate for the world fleet has been reduced significantly during the same period measured in percent annually relative to the fleet at risk: in 1900, the average loss rate was 3%. This was reduced to 0.5% in 1960, and by 2019, it had been further reduced to 0.04% (refer Section 2.5).

It is too early at this stage of the book to discuss whether the safety level in maritime transport is acceptable or not. However, it can on the other hand be argued that there is a case for increased safety effort unless it can be shown that the result cannot be defended with regard to the resources spent. Another way of thinking is that safety should be on the agenda as long as accidents are rooted in trivial errors or failures (very often human error). Thirdly, ship accidents should have our attention as long as they lead to fatal outcomes and the consequences for the environmental are unknown.

4.4.1 Why safety improvement is difficult

Despite the fact that safety is at the top of the agenda both in the shipping business itself and by regulators, it may appear that the pace of safety improvements is rather low. The degree to which this general observation is true will not be discussed in any depth here. However, some explanations for such a view are presented below:

Table 4.5 Milestones in maritime safety (date for adoption)[a]

Year	Initiative or regulation
1914	Safety of Life at Sea (SOLAS) : Ship design and lifesaving equipment
1929	First international conference to consider hull subdivision regulations
1948	The International Maritime (Consultative) Organization (IMO) is set up as a United Nations agency
1966	Load Line Convention: Maximum loading and hull strength
	Rules of the road
	The International Association of Classification Societies (IACS): Harmonization of classification rules and regulations
1969	Tonnage Convention
1972	Int. Conv. on the International Regulations for Preventing Collisions at Sea (COLREG)
1973	Marine Pollution Convention (MARPOL 73)
1974	IMO resolution on probabilistic analysis of hull subdivision
1978	International Convention on Standards for Training, Certification and Watchkeeping for Seafarers (STCW)
1979	International Convention on Maritime Search and Rescue (SAR)
1988	The Global Maritime Distress and Safety System (GMDSS)
1993	International Safety Management (ISM) Code established by IMO
2004	International Ship and Port Facility Security (ISPS) Code established

[a] An excellent summary is given by Vassalos (1999).

- *Short memory*: When safety work is successful, few accidents tend to happen. This lack of feedback can make people believe, both on a conscious and subconscious level, that they can relax on the strict requirements they normally adhere to.
- An even simpler explanation is complacency, i.e., that people tend to forget about the challenges related to safety if no accidents or incidents give them a "wake-up call". This weakness seems to degrade the safety work effectiveness of both companies and governments.
- *Focus on consequences*: People tends to focus on the consequences of an accident rather than its root causes. There is, for instance, great uncertainty attached to whether oil pollution is reduced in the best way by double-hull tankers or heavy investment in containment and clean-up equipment. Doing something about consequences is generally much more expensive compared to averting or reducing the probability and the initiating causes of an accident.
- *Complexity*: Safety involves technological, human, and organizational factors, and it can be very difficult to identify the most cost-effective set of safety-enhancing measures across all potential alternatives. There is also a tendency among companies, organizations, and governments to go for technical fixes, whereas the root

causes in a majority of cases are related to human and organizational factors. It seems to be easier to upgrade vessels than to change people's behavior.

- *Unwilling to change*: Humans have a tendency to avoid changing their behavior, also when it comes to safety critical tasks. People sometimes express their understanding of the need for change, but in practice use all means to sabotage new procedures. In some companies, "cutting corners" is unfortunately part of the culture of the company.
- *Selective focus*: Formal safety assessment (i.e., a risk analysis and assessment methodology described later) is in general seen as a promising tool for more efficient control of risk. However, such methods may be criticized in several ways: they oversimplify the systems studied, a number of failure combinations are overlooked due to the sheer magnitude of the problem, and operator omissions (e.g., forgetting or overlooking something) are not addressed in such models. In spite of these weaknesses, use of FSA is still a better option than doing nothing.

4.4.2 Rules and regulations

This chapter will give an outline of the regulation of safety in seaborne transport. The control of safety is primarily based on the rules (conventions and resolutions) given by the United Nations agency the International Maritime Organization (IMO). These rules have international application, but some reference will also be made to national regulations by taking the Norwegian legal regime as an example.

When we use the term *safety*, it will encompass:

- Safety and health of persons
- Safety of the vessel
- Environmental aspects

Safety is regulated based on different *legal sources*, the key ones of which are the following:

- International and regional laws and regulations
  - UN Law of the Seas (UNCLOS)
  - European Union (EU) Directives
- International conventions and resolutions
- National laws and regulations
- National territorial zones
- Case law (court rulings)
- Port State control guidelines
- Construction rules (ship classification)

It should also be kept in mind that there are a number of *actors* that have an impact on safety. The primary ones are as follows:

- International Maritime Organization (IMO)
- International Labour Organization (ILO)
- Flag States (Government authorities)
- Port State control (Regional authorities)
- Classification Societies
- Voluntary safety control (Vetting)

4.4.3 The structure of control

Seen from a national point of view, the regulation of safety in maritime transport is based on a set of international rules that are adopted by the legislative assembly (Parliament). The concrete rules and regulations are written or translated by the responsible government branch (Foreign and International Trade Department). The role of the Maritime Administration is to ensure that regulations are followed by the shipowner through proper control and certification. This is what is termed Flag State Control (FSC). The Classification Society has a role in the certification process, although this is primarily motivated by the insurance business, cargo owner, and third-party interests.

The control of safety in shipping is complex for several reasons:

- International, regional, and national laws and regulations.
- Control is exercised by several agencies.
- Control affects the various life cycles of the vessel.

A simple outline of the number of actors and interactions is shown in Figure 4.14. It should also be kept in mind that shipping as an internationally oriented business, is highly competitive and is also influenced by dramatic economic cycles. Seen from the shipowner's perspective, the safety standard is a result of the cross-pressure between control and commercial competition.

4.4.4 International Maritime Organization (IMO)

International cooperation to improve the safety at sea was strengthened after the *Titanic* loss in 1912. The Inter-Governmental Maritime Consultative Organization (IMCO) was created in 1948 and went into force in 1958. This is an organization under the United Nations system. Its main objective is maritime safety, and it has later changed its name to IMO. Its prime function is to establish rules based on participation by the member states. The main principle in the regulation of shipping is harmonized national rules based on conventions and resolutions given by the IMO. IMO has a

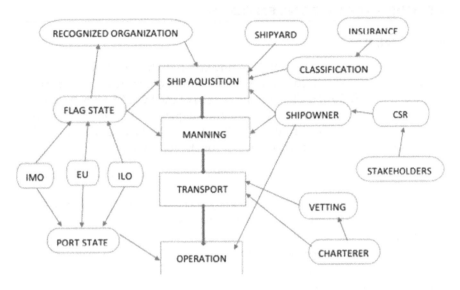

Figure 4.14 Actors and interactions in safety control.

complex set of committees that draft and revise regulations which finally are adopted by the General Assembly consisting of representatives from all member states. IMO has four main committees:

1. Maritime Safety Committee (MSC): Deals with all aspects of maritime safety, both technical and operational.
2. Marine Environment Protection Committee (MEPC): Main task is to prevent pollution of the marine environment and inspection of polluting spills from ships.
3. Legal Committee (LEG): Deals with legal matters within IMO.
4. Technical Co-operation Committee (TC): Acts on behalf of IMO in cases of technical cooperation with other agencies.

Regulations adopted by IMO are normally given in the form of Conventions. A new regulation has to be ratified by minimum two-thirds of states before it enters into force. This form of ratification is termed "explicit acceptance procedure". Its weakness is that the time before the convention enters into force may be lengthy. A more effective procedure for technical regulations is "tacit acceptance procedure" which means that the regulation is effective from a specified date. Apart from conventions, regulation may also be given as recommendations and guidelines. IMO has no power to enforce the international safety regulations. This is the task of the member states in their role as the so-called Flag States.

4.5 SHIP SAFETY CONVENTIONS

4.5.1 SOLAS

The main objective of the SOLAS Convention (Safety of Life at Sea) is to specify minimum standards for the construction, equipment, and operation of ships (SOLAS, 2001). The present version of the SOLAS Convention was adopted in 1974 and was later revised and supplemented with the so-called Protocols. It entered into force in 1980. SOLAS-74 has 12 articles and 14 chapters with the specific requirements for the following areas:

I	General provisions
II-1	Construction – Subdivision and stability, machinery and electrical installations
II-2	Fire protection, fire detection, and fire extinction
III	Lifesaving appliances and arrangements
IV	Radiotelegraphy and radiotelephony
V	Safety of navigation
VI	Carriage of cargoes
VII	Carriage of dangerous goods
VIII	Nuclear ships
IX	Management for the safe operation of ships (ISM Code)
X	Safety measures for high-speed craft (HSC Code)
XI-1	Special measures to enhance maritime safety
XI-2	Special measures to enhance maritime security
XII	Additional safety measures for bulk carriers
XIII	Verification of compliance
XIV	Safety measures for ships operating in polar waters (Polar Code)

The convention has been amended a number of times since its adoption in order to be in accordance with the development of new technology and new safety knowledge. The regulation is to a large degree prescriptive by specifying solutions in minute technical detail. Performance criteria are only applied to a limited degree. This has two main drawbacks: technical solutions specified in SOLAS may become obsolete even before it enters into force, and the lack of focus on performance criteria does not stimulate the designer to find or invent better solutions.

The SOLAS-74 Convention has been ratified by most nations. To become effective, the convention has to be translated into the official national language and be formally adopted by the government branch.

4.5.2 UNCLOS

The basis for international shipping is the principle of freedom of the seas. The international legal basis is defined in the *United Nations Convention on the Law of the Sea* or *UNCLOS* (AMLG, 2004). The principle has the following key elements:

- Ships may sail without restriction in all waters on "innocent passage" (Article 17).
- The country of registration (Flag State) has the sole jurisdiction over the ship (Article 91).
- Countries have limited jurisdiction even in their own territorial sea.
- "The freedom of the high seas" (Article 87) states the right to sail on open sea.

The coastal state has at the outset the following rights:

- The outer limit of the territorial sea is 12 nm from the coast (baseline) within which it has full jurisdiction.
- The exclusive economic zone stretches out to 200 nm:
  - Very limited control jurisdiction.
  - Certain rights to take measures to preserve the marine environment.
  - However, the control should be exercised in accordance with international practice or non-discrimination against foreign vessels (Article 227).

The above means that the coastal states have to exercise their rights with respect to pollution hazards with delicacy. This becomes even more complicated when a state has both a substantial international trading fleet and a threatened coast. A good example is one of the initiatives of Spain and France in the aftermath of the *Prestige* accident. In an EU communication, the following is stated:

> ... INVITES Member States to adopt measures, in compliance with international law of the sea, which would permit coastal States to control and possibly to limit, in a non- discriminatory way, the traffic of vessels carrying dangerous and polluting goods, within 200 miles of their coastline...

This position has been strongly opposed by INTERTANKO, which stresses that any measure in this area must adhere to international law and more specifically UNCLOS. When UNCLOS and other conventions are ratified by a country, the government has to implement them in national laws and regulations.

4.5.3 International convention on load lines, 1966

It has long been recognized that limitations on the draught to which a ship may be loaded make a significant contribution to her safety. These limits are given in the form of a freeboard, which, besides external weather-tightness and watertight integrity, constitute the main requirement of the Convention.

The first International Convention on Load Lines (IMO, 2002), adopted in 1930, was based on the principle of reserve buoyancy. It was also recognized then that the freeboard should ensure adequate stability and avoid excessive stress on the ship's hull because of overloading.

The regulation takes into account the potential hazards present in different geographical zones and different seasons of the year. The technical annex contains several additional safety requirements concerning doors, freeing ports, hatchways, and other items. The Convention includes Annex I with the following four chapters:

I. General
II. Conditions of assignment of freeboard
III. Freeboards
IV. Special requirements for ships assigned timer freeboards

Annex II covers zones, areas, and seasonal periods, and Annex III certificates, including the International Load Line Certificate.

4.5.4 STCW convention

The International Convention on Standards of Training, Certification and Watchkeeping for Seafarers (STCW) was the first step to establish basic requirements on training, certification, and watchkeeping for seafarers at an international level. The technical provisions of the Convention are given in an annex containing six chapters:

1. General provisions.
2. Master-deck department: This chapter outlines basic principles to be observed in keeping a navigational watch. It also lays down mandatory minimum requirements for the certification of masters, chief mates, and officers in charge of navigational watches on ships of 200 grt or more.
3. Engine Department: Outlines basic principles to be observed in keeping an engineering watch. It includes mandatory minimum requirements for certification of officers of ships with main propulsion machinery of 3,000 kW.
4. Radio Department.
5. Special requirements for tankers.
6. Proficiency in survival craft.

The 1995 amendments represented a major revision of the 1978 Convention (STCW, 1996). The original Convention has been criticized on many counts. It referred to vague phrases such as "to the satisfaction of the Administration", which admitted quite different interpretations of minimum manning standards.

Others criticized that the Convention was never uniformly applied and did not impose strict obligations on the Flag States regarding its implementation.

4.5.5 MARPOL

Both SOLAS and ICCL have an indirect effect on preventing pollution from ships. However, there was a dramatic development of specialized tankers after the Second World War in terms of ship size and complexity of operation. The International Convention for the Prevention of Pollution from Ships (MARPOL) seeks to address the environmental aspects related to design and operation of these ships more directly (IMO, 2002).

The Convention prohibits the deliberate discharge of oil or oily mixtures for all seagoing vessels, except tankers less than 150 gross tons and other ships less than 500 gross tons, in areas denoted "prohibited zones". In general, these zones extend at least 50 nautical miles from the coastal areas, although zones of 100 nm and more were established in areas which included the Mediterranean and Adriatic Seas, the Gulf and Red Sea, the coasts of Australia and Madagascar, and some others.

MARPOL introduces several measures:

- Segregated ballast tanks (SBT): ballast tanks only used for ballast as cargo oil is prohibited. Reduces the cleaning problem.
- Protective location of SBT: Are arranged in bottom or sides to protect cargo tanks against impact or penetration.
- Draft and trim requirements: to ensure safe operation in ballast condition.
- Tank size limitation to limit potential oil outflow.
- Subdivision and stability in damaged condition.
- Crude oil washing (COW).
- Inert gas system (IGS) for empty cargo tanks.
- Slop tanks for containing slop, sludge, and washings.

The implementation of MARPOL is based on a complex scheme where ship size and whether it is an existing or a new building determine which requirements apply.

4.5.6 The ISM Code

The introduction of the International Management Code for Safe Operation and Pollution Prevention (IMO, 1994) represented a dramatic departure in regulatory thinking by the IMO. It acknowledges that detailed prescriptive rules for design and manning have serious limitations. Inspired by principles from quality management and internal control, the ISM Code shall stimulate safety consciousness and a systematic approach in every part of the organization both ashore and onboard.

The main intention with ISM is to induce the shipping companies to create a safety management system that works. The Code does not prescribe in detail how the company should undertake this, but just states some basic principles and controls that should be applied. The philosophy behind ISM is commitment from the top management, verification of positive attitudes and competence, clear placement of responsibility, and quality control of work processes.

The ISM Code is described in more detail in Chapter 5 Safety Management System.

4.6 INTERNATIONAL LABOUR ORGANIZATION

The mission of the International Labour Organization (ILO) is to promote social justice and internationally recognized human and labor rights. As a UN agency it brings together governments, employers, and workers representatives. ILO also has the competence to regulate commercial shipping. The four strategic objectives are:

- Set and promote standards and fundamental principles and rights at work.
- Create greater opportunities for women and men to decent employment and income.
- Enhance the coverage and effectiveness of social protection for all.
- Strengthen tripartism and social dialogue.

It is a fact that historically the working and social conditions of seamen have not been up to a reasonable standard. The situation was characterized by uncertain employment, unsafe working conditions, high accident rates, and high mortality. ILO adopted the Maritime Labour Convention (MLC) in 2006. It has five sections:

1. Minimum requirements for seafarers to work on a ship
   - Minimum age
   - Medical certificate
   - Training and qualifications
   - Recruitment and placement
2. Conditions of employment
   - Seafarers' employment agreement
   - Wages, minimum wages
   - Hours of work and rest
   - Entitlement to leave, repatriation
   - Compensation for ship loss
   - Manning levels
   - Career and skill development

3. Accommodation, recreational facilities, food, and catering
   • Design, ventilation, heating, prevention of noise and vibration
   • Sanitary and hospital accommodation
   • Food and catering
4. Health protection, medical care, welfare, and social security protection
5. Compliance and enforcement

Similar to the IMO Conventions, the Flag States have to ratify the MLC Convention. The convention is implemented by requiring that the commercial ship shall have a *Maritime Labour Certificate*.

4.7 EUROPEAN UNION

The European Union has been increasingly involved in the control of international shipping realizing to be both a region with both a Flag State and Port State perspective. Shipping is important for EU by the fact that it represents 74% of import/export and 37% of goods within the union. An action plan – *A Common Policy on Safe Seas* – was adopted in 1993. The objective was to have a unified implementation of international regulation, more efficient Port State Control, and strengthening of vessel traffic monitoring. To improve the standard of vessels, EU initiated a plan for approval of Classification Societies or what called "recognized organizations". The Commission has through the years adopted several Directives that have had strong impact on the implementation of safety regulations. The presence of EU has become more and more noticeable within IMO. In contrast to IMO conventions, directives need not to go through a ratification process but become effective immediately. Later EU has taken several measures:

• *ERIKA I package (2000)*: Speeding up the out-phasing of single-hull tankers. Stricter rules for Classification Societies.
• *ERIKA II package (2000)*: Increased focus on single-hull tankers and establishment of EMSA (*European Maritime Safety Agency*).
• *Erika III package (2004)*: New Inspection Regime, increased application of Paris MOU, establishing maritime information system *THETIS*, Insurance directive, Liability of carriers of passengers, and amendment of directive on Vessel Traffic Monitoring and Information System (VTM).
• 2006: "Green Paper for a Future Union Maritime Policy. A European Vision for the Oceans and Seas".
• Limassol Declaration (2012): Sets 2020 targets on green maritime transport, renewable marine energy, aquaculture farming, coastal and maritime tourism, and technologies for the safe and sustainable exploitation of marine mining resources.

- Parliament resolution (2018): International Ocean Governance: an agenda for the future of our oceans in the context of the 2030 Sustainable Development Goals.

In the late 1990s, the idea of a joint European regulatory agency for the maritime domain was put forwards, resulting in the establishment of the European Maritime Safety Agency (EMSA) in 2002. According to their website (emsa.europa.eu), "EMSA's mission is to serve EU maritime interests for a safe, secure, green and competitive maritime sector and act as a reliable and respected point of reference in the maritime sector in Europe and worldwide." EMSA thus has a wider scope than just safety, although originally focusing on oil spills and maritime accidents.

4.8 ENFORCEMENT OF SAFETY REGULATION

4.8.1 Flag State Control

As already pointed out, the set of internationally accepted safety rules and regulations are not enforced by the IMO itself but by the so-called Flag States. The national maritime administration is acting as Flag State on behalf of the country in question. Based on plans, technical documentation, and inspections, a ship is subject to registration and awarded the necessary safety-related certificates.

Since IMO has no power to enforce its conventions, all ratifying countries must implement the conventions in their national laws and regulations. The Norwegian government has for instance given the *Ship Safety and Security Act* for Norwegian vessels. The law gives the basis for The Norwegian Maritime Authority to act as a Flag State. The law is organized in 11 chapters:

1. Introductory provisions
2. Owners' duties. Safety management
3. Technical and operative safety
4. Work conditions and personal safety
5. Environmental safety
6. Safety and environmental preparedness
7. Control
8. Administrative measures
9. Administrative sanctions
10. Criminal liability
11. Other regulation

The purpose of the act is stated as follows:

> This Act shall safeguard life, health, property and the environment by facilitating a high level of ship safety and safety management, including

preventing pollution from ships, ensuring a fully satisfactory working environment and safe working conditions on board ships as well as appropriate public supervision of ships.

The law basically applies to vessels with length greater than 24 meters, but the Administration (Flag State) may decide that other vessels also must be built in accordance with the rules under the law. The jurisdiction of this law is in principle limited to Norwegian vessels. The maritime administration primarily acts as *Flag State*. However, international law has developed during the last decades and today accepts that a nation may exercise some control and, if necessary, detain a foreign vessel viewed as a risk to human life (passenger transport) and coastal environment (oil pollution). The maritime administration in that sense also acts as a *Port State*. We will return to this role later.

Shipping activity in Norway or more precisely the Norwegian national register vessels (NOR) are subject to both private and public law. The international register in Norway (NIS) is regulated through a separate act.

All member states of IMO have implemented the Conventions and other regulations through national laws and regulations. The laws take quite different forms and differ in degree of prescriptiveness.

4.8.2 Delegation of flag State Control

Some Flag States accept foreign vessels and have become what is commonly termed international or offshore registers. The standard of some of these registers has been questioned and they have been branded as *Flags of Convenience* (FOC). They are suspected to offer registration to foreign owners mainly for economic reasons and are viewed as having a lenient enforcement of safety regulations. Another characteristic is the lack of own maritime competence or minimal maritime administration. A common practice is to delegate the control to an independent certifying authority, primarily Classification Societies and even consultants.

Flag State control (FSC) has for years been the main approach to the safety control of shipping. It has, however, become evident that different Flag States have varying competence and motivation to undertake their role. This was clearly demonstrated in a small survey of the *SAFECO I* project (Kristiansen and Olofsson, 1997). Table 4.6 shows the loss rate for selected Flag States. The annual loss rate may vary by a factor of more than 10. This great variation can even be observed among European Flag States, as shown in Figure 4.15.

4.8.3 The Flag State audit project

The Seafarers International Research Centre (SIRC) at Cardiff University has undertaken an assessment of the performance of the main Flag States

Table 4.6 Total loss rate by flag for vessels greater than 100 grt

Flag	Fleet size 1993	Loss rate per 1,000 shipyears 1994–95
Denmark	599	3.1
France	769	1.2
Germany	1,234	1.3
Netherlands	1,006	0.84
Norway	1,691	1.3
United Kingdom	1,532	2.7
North Europe selected	6,831	1.8[a]
Cyprus	1,591	5.3
Greece	1,929	1.8
Italy	1,548	1.7
Malta	1,037	5.5
Portugal	307	—
Spain	2,111	3.0
Mediterranean selected	8,523	3.2[a]
Japan	9,950	1.3
Korea (South)	2,085	4.2
Philippines	1,469	2.3
Singapore	1,129	0.4
USA	5,646	2.4
Bahamas	1,121	2.6
Liberia	1,611	1.9
Panama	5,564	3.9
Worldwide	80,655	2.4

[a] Weighted estimation on the basis of the selected countries.
Source: *World Fleet Statistics* and *Casualty Return*, Lloyd's Register, London.

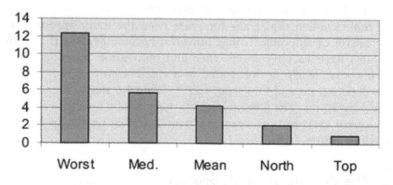

Figure 4.15 Loss rate per 1,000 shipyears of European fleet segments, 1994–95 (Med.: Mediterranean countries, North: Northern and Central Europe).

(Alderton and Winchester, 2001). Some shipowners prefer to register their fleet under a flag other than the national one. This has been a practice for years but has gained renewed importance during the present trend toward globalization and deregulation of industry and trade. Some of these flags lack both motivation and competence to enforce the international safety standards set by IMO. These flags have been termed Flags of Convenience (FOC). However, today it seems too simple to distinguish between national flags and FOCs. The International Transport Workers' Federation therefore commissioned a study of the performance of the various flags operating today.

The first step in the study was to define a set of criteria for ranking flags. It was decided to create an index (FLASCI) based on a weighted ranking of the following factors:

1. The nature of the maritime administration
2. Administrative capacity
3. Maritime law
4. Seafarers' safety and welfare
5. Trade union law
6. Corruption
7. Corporate practice

The relative weighting and detailed factors assessed are summarized in Table 4.7. Data were retrieved from a literature search, and review of Internet sources and other available information on the Flag States such as Port State Control statistics. The Flag States got scores of between 19 and 84 and inspection of the findings suggested that the Flag States might be grouped into five categories, as shown in Table 4.8. The study clearly shows that flags show greater variation in performance than has generally been accepted. Some of the main findings were as follows:

- Some of the so-called second registers perform as well as the best national registers: Norway (NIS), Denmark (DIS), Germany (GIS), and France (Kerguelen Islands).
- A few of the established FOCs are performing relatively well: Bermuda (63). Other FOCs such as Bahamas (43) and Liberia (43) are ranked lower but are still better than the worst performing.
- There seems to be a clear correlation between low performance and short operation as flag (new entrants). Port State control of these flags shows a quite high detention rate as shown in Table 4.9.

The last point can be explained by the apparent dynamics in the "market" of Flag States. FOCs will, after some time when they are more established, be under pressure to improve their performances. As they gradually do this, it will open a market for new flags that will offer a more lenient safety

Table 4.7 Flag State Conformance Index (FLASCI)

	Port State Control rates Casualty rates Pollution incidence Own-citizen labor force participation	Own-citizen beneficial ownership Abandonment of crews Appearance in crew complaints DB
FS administrative capacity 30%	Death records Crew records of service Health screening procedures & records Accessibility of consular services Enforcement of IMO & ILO Conventions	Casualty investigation capacity Statistics of ships, owners, & labor force Certification of seafarers Involvement in training & education
FS maritime law 20%	Ratification of IMO & ILO Conventions Provisions of maritime legal code Publication of relevant law reports	Specialist law practitioners Location of registry "Ownership" of registry Vessel registration requirements
Miscellaneous Maritime 5%	Maritime welfare support & maritime charities Maritime interest groups	Government ministries with maritime remit Stock exchange maritime listings State-owned shipping
Trade union law 10%	Legal rights for migrant labor Independent trade unions Mediation/arbitration procedures	Provision for trade union recognition Enforcement of trade union recognition procedures. Industrial tribunal / labor courts
Corruption 10%	Probity of public officials Misapplication of public funds	Integrity of political institutions & legal process Corporate integrity
Corporate practice 10%	Regulation of financial institutions Regulation of non-resident companies	Regulation of accounting standards Legal definition of corporate public responsibility

Source: Alderton and Winchester (2001).

regime. The SIRC study also showed that the fleets of the new entrants have a much higher growth rate than the average rate for the world fleet.

The SIRC study also analyzed the working conditions on board, and it was confirmed to be a less attractive climate on new entrant flag vessels.

4.9 PORT STATE CONTROL

4.9.1 UNCLOS

The basis for international shipping is the principle of freedom of the seas. The international legal basis is defined in the *United Nations Convention*

Table 4.8 Ranking of selected Flag States

Category	Selected Flags (score)	Score range
Traditional maritime nations Centrally operated second registers	NOR (84) UK (80) DIS (77) NIS (77) Netherlands (76) GIS (75) Kerguelen Islands (72)	84-72
Semi-autonomous second registers	Hong Kong (64) Bermuda (63) Latvia (60) Cayman Islands (62) Estonia (58)	64-58
Established open registers (seeking EU membership) National registers	Cyprus (50) Malta (49) Russia (48) Bahamas (43) Liberia (43) Panama (41)	50-41
New open registers	Marshall Islands (36) Ukraine (36) Honduras (35) Lebanon (35)	36-35
New entrants to the open register markets	Saint Vicent & Grenadines (30) Bolivia (30) Belize (27) Equatorial Guinea (24) Cambodia (19)	30-19

Alderton and Winchester (2001).

Table 4.9 Detention rate for 'new entrant' flags

	Belize	Bolivia	Cambodia	Equatorial Guinea
Asia–Pacific MOU (average 7%)	24.7%	No data	30%	11.1%
Paris MOU (average 9%)	31.4% Black listed	70%	24.8% Black listed	14.3%
USCG (average 5%)	50.6% Targeted	No data	Too few inspections	28.6% Targeted

Alderton and Winchester (2001).

on the Law of the Sea or *UNCLOS* (AMLG, 2004). The principle has the following key elements:

- Ships may sail without restriction in all waters on innocent passage (Article 17).
- The country of registration (Flag State) has the sole jurisdiction over the ship (Article 91).
- Other countries have limited jurisdiction even in own territorial sea.

The coastal state has at the outset the following rights:

- The outer limit of the territorial sea is 12 nm from the coast (baseline) within which it has full jurisdiction.
- The exclusive economic zone stretches out to 200 nm:
  - Very limited control jurisdiction.

- Certain rights to take measures to preserve the marine environment.
- However, the control should be exercised in accordance with international practice or non-discrimination against foreign vessels (Article 227).

The above means that the coastal states have to exercise their rights with respect to pollution hazards with delicacy. This becomes even more complicated when a state has both a substantial international trading fleet and a threatened coast. A good example is one of the initiatives of Spain and France in the aftermath of the *Prestige* accident. In an EU communication, the following is stated:

... INVITES Member States to adopt measures, in compliance with international law of the sea, which would permit coastal States to control and possibly to limit, in a non- discriminatory way, the traffic of vessels carrying dangerous and polluting goods, within 200 miles of their coastline...

This position was strongly opposed by INTERTANKO, stressing that any measure in this area must adhere to international law and more specifically UNCLOS.

4.9.2 MOU Port State Control (PSC)

An overriding principle is that under the international safety conventions a certificate issued by Flag State A is equivalent to a certificate issued by Flag State B. However, a Port State may challenge a certificate if there are indications that the condition of the foreign vessel is not in accordance with the particulars of the certificate. The legal basis for Port State Control (PSC) in Europe is found in the so-called Paris MOU (Paris MoU, 2004), the "Memorandum of Understanding on Port State Control" signed in 1982 by 19 European states and Canada.

The introduction of PSC was initially heavily opposed by shipping interests who feared that it would have a negative impact on the principle of equal market access and free competition. But in the end, all involved parties acknowledged the shortcomings of Flag State Control and the necessity of giving Port States authority to control shipping in their own waters.

The MOU has been given legal basis in national and international law, for instance in Norway by Regulation regarding control of foreign vessels, and similarly in Europe by Council Directive 95/21/EC of 19 June 1995 (Directives of the European Commission have status as law). The objective for each Port State is to control 25% of the foreign flagships calling at their ports on an annual basis. An inspection may result in:

- *Deficiency:* a non-conformity, technical failure or lack of function. A deadline for correction will be given.

- *Detention:* a serious deficiency or multitude of deficiencies that must be corrected before the vessel is allowed to leave the port.
- *Refusal:* ships having a multitude of detentions or lacking an ISM certificate may be banned from European waters.

Since the Paris MOU was established, a number of similar MOUs have been set up in other parts of the world. More information is available on the EQUASIS homepage (EQUASIS, 2004). The findings and actions of Paris MOU are published in yearbooks (Paris MoU, 2019). A summary of the number of inspections, deficiencies, and detentions is shown in Figure 4.16.

The average number of inspections with deficiency has been 52% (every second inspection) during the last three years. The relative number of detentions has been in the range of 3–4% in the recent period. The following areas have the highest deficiency rate:

- Fire safety
- Safety of navigation
- Lifesaving appliances
- Labor conditions, health protection, medical care, social security
- Certificates and documentation

The vessel type with the highest number of detentions is general cargo vessels. The detention rate is twice the average rate (6.6%). This situation is especially demanding as general cargo vessels are the largest group in the world fleet. An overview of the performance of the ship types is given in Table 4.10. The Flag States show quite different performance in terms of deficiencies and detentions. A selected number of Flag States are shown in Table 4.11 with respect to inspections and detentions. The worst performing states are so-called new entrants with limited experience as Flag State. Panama is a traditional Flag of Convenience (FOC) but is still performing below average. The states with average performance are not easy to characterize but consist of minor European States, Russia, and the United States. The countries with lowest detention rate are traditional European shipping nations, traditional FOC and Hong Kong/China.

Based on the inspection findings for the last three years, Paris MOU categorizes the performance of the Flag States based on the detention rate:

- White list (0.8–1.3%)
- Gray list (1.6–10%)
- Black list (11–26%)

The numbers indicate typical ranges for the detention rate. It is assumed that the number of detentions relative to the number of inspections is following a binominal distribution. Paris MOU therefore applies statistical methods to decide whether a Flag State belongs to any of the three categories. The maximum

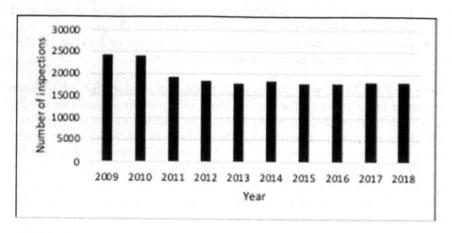

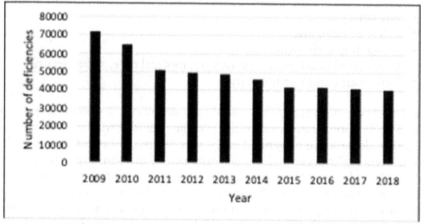

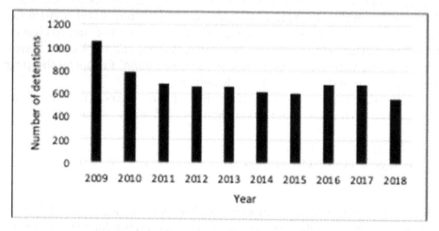

Figure 4.16 Inspections, deficiencies, and detentions in 2018 (Annual Report 2018. Paris MOU on Port State Control).

Table 4.10 MOU Port State findings for ship types in 2018

Ship type	Nr. of inspections	Nr. of detentions	Detentions in % of inspections
Bulk carrier	887	29	3.3
Chemical tanker	426	4	0.9
Commercial yacht	32	1	3.1
Container	402	3	0.7
Gas carrier	100	1	1.0
General cargo/multipurpose	1,172	77	6.6
Heavy load	11	0	0.0
High-speed passenger craft	3	1	33.3
NLS tanker	5	0	0.0
Offshore supply	100	0	0.0
Oil tanker	355	7	2.0
Other	55	1	1.8
Other special activities	112	2	1.8
Passenger ship	38	0	0.0
Refrigerated cargo	54	2	3.7
Ro-Ro cargo	163	1	0.6
Ro-Ro passenger ship	21	0	0.0
Special purpose ship	22	0	0.0
Tug	63	2	3.2
Total	4,021	131	3.3

Annual Report 2018. Paris MOU on Port State Control.

Table 4.11 Detention rate for a selection of Flag States

Flag State	Number of inspections	Detention rate in %
Antigua and Barbuda	784	5.2
Panama	2,101	4.7
Sierra Leone	122	15.6
Togo	175	14.9
Cyprus	707	2.8
Malta	1,531	2.5
Portugal	378	2.9
United States	71	2.8
Russian Federation	448	3.1
Bahamas	731	1.2
Greece	273	1.1
Hong Kong, China	635	1.3
Liberia	1,436	2.0
Marshall Islands	1,501	0.9
Netherlands	988	1.9
Norway	536	1.3
United Kingdom	386	0.5

Annual Report 2018. Paris MOU on Port State Control.

accepted value of detentions has been set to 7%. The significance level is set to 95%. The corresponding critical values for the number detentions are given by:

$$\text{Black to grey limit: } u_{b-g} = N \cdot p + 0.5 + z\sqrt{N \cdot p \cdot (1-p)} \qquad (4.1)$$

$$\text{White to grey limit: } u_{w-g} = N \cdot p - 0.5 - z\sqrt{N \cdot p \cdot (1-p)} \qquad (4.2)$$

where
 N=number of inspections
 p=relative number of detentions
 z=corresponding value of p for the confidence interval=1.645

Let us demonstrate the approach. A Flag State has 108 inspections that resulted in 25 detentions. The relative number of detentions is p^x=25/108=23%. This indicates that the flag most likely will be on the black list. However, to take the statistical uncertainty into consideration the black–gray limit should be checked. The critical value is given by:

$$u_{b-g} = 108 \cdot 0.07 + 0.5 + 1.645\sqrt{108 \cdot 0.07 \cdot (1-0.07)} = 12 \qquad (4.3)$$

The number of detentions (25) is greater than critical value 12. It can be concluded that the state should be put on the black list. The Port State Control has also introduced a measure for the degree of deviation from the acceptable number of detentions. The excess factor may be found by means of a calculator found on the webpage of Paris MOU. In the current case the excess factor was 4.26 which indicates a very high risk.

4.10 CLASSIFICATION SOCIETIES

Classification Societies are independent bodies which set standards for design, maintenance, and repair of ships. They were on the outset formed to serve marine insurance that lacked the necessary technical competence to assess the safety standard of vessels. The classification building rules cover:

- Hull strength and integrity
- Materials
- Main and auxiliary machinery
- Electrical installations
- Control systems
- Safety equipment

A vessel is given a *class certificate* based on drawings, engineering documentation, and inspections during construction and tests. A classed vessel will further be surveyed on a regular basis and given recommendations for necessary maintenance and repair to keep its class.

The class is the basis for negotiating insurance of the vessel. The class in this sense is a kind of quality check for the insurance company. The Classification Society has historically had no official role relative to international and national safety regulation. This has however changed during later years. Today they also offer statutory services and assistance to the maritime industry and regulatory bodies with respect maritime safety and pollution prevention. Both IMO and ILO have accepted that Class Societies act on behalf of the Flag State and in this role, they are called *Recognized Organizations* (RO). The following are the approved ROs at the time of writing (2021):

- American Bureau of Shipping
- Bureau Veritas
- China Classification Society
- DNV
- Korean Register of Shipping
- Lloyd's Register
- Nippon Kaiji Kyokai
- Registro Italiano Navale
- Russian Maritime Register of Shipping

There are about 40 class institutions in all, and the owner is, in principle, free to select a class among those institutions. As the owner must pay for the class and associated services, it may become a matter of trade-off between safety and cost:

- Class institutions offer different standards, control regimes, and tariffs.
- They compete on price.
- Some are not serious in enforcing control and follow-up maintenance.
- Owners may "jump between institutions" to avoid costly maintenance.
- Change of class means that an outstanding survey is delayed for three months.

Port State control clearly documents that the performance of the Classification Societies differs quite substantially. The best classes have a detention rate in the order 0.3% compared to the average, that is, 3.3%. The worst performing Class Societies have a performance as bad as up to 12%. However, it should be kept in mind that the worst performers have a relatively small part of the classification market.

Over the recent year it has been a continuous process of harmonization and upgrading of class rules. The serious class institutions are organized in IACS (International Association of Classification Societies). The members cooperate to attain a harmonized standard for the serious institutions.

Finally, the serious institutions maintain a high professional standard and contribute in many ways to the advancement of the safety standards.

4.11 CIVIL MARITIME LAW

Shipping as a commercial activity is also regulated by national and international laws and conventions. Several aspects of shipping are covered:

- Registration of vessel
- Unseaworthiness
- Seaman status, claims
- The Master of the vessel
- Liability, and limitation of liability
- Contracts of affreightment, bill of lading
- Carriage of passengers
- Property damage
- Cargo
- Longshoremen claims
- Oil pollution and environmental liability
- Liability for collisions
- Salvage
- Marine insurance
- Maritime inquiries
- Ship mortgage
- Admiralty jurisdiction

The Master of a vessel has a unique role in comparison to the other crew members.

The law states that the Master:

- Has the highest authority on board
- Is responsible for seaworthiness
- Is responsible for seaworthiness in relation to the cargo
- Has the power to enter a contract with a salvage tug

This philosophy has a historical background because in the age of the sailing ships, the shipowner had no daily control over his vessel and had to rely on the trust and competence of his representative on board, namely the Master. In today's world of modern communication, the Master can report and confer daily with the manager. Likewise, the manager has the complete freedom to instruct and control his vessel in detail. This means that the role of the Master has changed significantly. This fact makes his unique authority and responsibility somewhat outdated. As a reflection of the changed situation revised legislation has put more weight on the role and responsibility of the *Company*.

Liability in case of sea transport is a large and complex topic. Here we shall only comment on liability in relation to passengers and environment.

The Norwegian Maritime Code covers the matter of liability in case of personal injury and death to passengers. Liability will be imposed under circumstances where the injury is caused by fault or neglect of the carrier. An important requirement, however, is that the claimant must prove that:

- The harmful event took place during the voyage, and
- Was the result of fault or neglect by the carrier.

In other words, liability stemming from personal injury is *objective*. Under EU law (EU, 2009), the liability for personal injury or fatality is limited upwards to 400.000 SDR (corresponding to about 550.000 USD). Presently 422 limits the compensation to NOK 1,622,500 per passenger. During the last 20–30 years, the world has witnessed a number of serious ship accidents with massive spills like *Torrey Canyon*, *Amoco Cadiz*, and *Exxon Valdez*. This soon raised the matter of liability related to environmental harm. The so-called CLC Convention (the Intervention and Liability Convention) was approved in 1969. It represented a radical change by stating that:

- Owners of ships transporting oil as bulk cargo are made strictly liable for oil pollution, with virtually no exceptions, and
- The amounts could only be limited to sums much larger than the general rules.

This means that, contrary to personal injury, in the case of an environmental accident a claim can be put forward without proving negligence.

The United States has introduced its own rules through the Oil Pollution Act (OPA, 1990), which gives the plaintiff almost unlimited right to make the shipowner liable.

Activities on board a vessel may also be subject to prosecution under the Norwegian Criminal Code of 1902. §12 contains rules with respect to personal acts and §48 covers the provisions for companies. The main principle is that the same laws that apply ashore also apply to Norwegian vessels.

REFERENCES

Alderton, T. and Winchester, N. (2001). The flag state audit. *Proceedings of SIRC's Second Symposium*. Seafarers International Research Centre.

AMLG. (2004). *Admirality and maritime law guide – International conventions*. http://www. admiraltylawguide.com/conven/unclospart2.html

API. American Petroleum Institute.

Bergantino, A. and Marlow, P. (1998). Factors influencing the choice of flag: Empirical evidence. *Maritime Policy and Management, 25*(2), 157–174.

Carroll, A. B. (1991). The pyramid of corporate social responsibility: Toward the moral management of organizational stakeholders. *Business Horizons, 34*(4), 39–48.

Carroll, A. B. (2016). Carroll's pyramid of CSR: Taking another look. *International Journal of Corporate Social Responsibility, 1*(1), 1–8.

Drewry. (2015). *Ship operating coasts annual review and forecast 2014/15.* https://commons.wmu.se/lib_reports/5/

EQUASIS. (2004). *A public site promoting maritime safety and quality.* http://www.equasis.org/

EU. (2009). *Liability of carriers of passengers by sea in the event of accidents.* No 392/2009. https://puc.overheid.nl/nsi/doc/PUC_2325_14/2/

IMO. (1994). *International Safety Management Code (ISM Code).* https://www.imo.org/en/OurWork/HumanElement/Pages/ISMCode.aspx

IMO. (2002a). *International convention on load lines.* https://www.imo.org/en/About/Conventions/Pages/International-Convention-on-Load-Lines.aspx

IMO. (2002b). *MARPOL 73/78 Consolidated Edition 2002.* https://wedocs.unep.org/handle/20.500.11822/2381?show=full

IMO. (2014). *Third IMO GHG study 2014.* https://www.imo.org/en/OurWork/Environment/Pages/Greenhouse-Gas-Studies-2014.aspx

ISL. (2010). *Shipping statistics yearbook 2010.* https://shop.isl.org/ssyb–2010/

ISL. (2014). *Shipping statistics yearbook 2014.* https://shop.isl.org/ssyb–2014/

ISL. (2018). *Shipping statistics yearbook 2018.* https://shop.isl.org/isl-ssyb/ssyb–2018/

Kristiansen, S. and Olofsson, M. (1997). *SAFECO – Safety of Shipping in Coastal Waters. Operational Safety and Ship Management – WPII.5.1: Criteria for Management Assessment* (Report No. MT23 F97–0175). MARINTEK AS.

Maragkogianni, A., Papaefthimiou, S. and Zopounidis, C. (2013). Emissions trading schemes in the transportation sector. In (Anonymous): *Sustainable technologies, policies, and constraints in the green economy* (pp. 269–289). USA: Information Resources Management Association, Hershey, PA, USA.

OECD. (2002). Shipping and environment: Dealing with substandard ships. Paper submitted by Transport Canada. *OECD Workshop on Maritime Transport.* Organisation for Economic Co-operation and Development, DSTI/DOT/MTC(2003)18. Paris.

OECD. (2016). *Economic Outlook 99 database inventory.* https://stats.oecd.org/Index.aspx?DataSetCode=EO99_INTERNET

OPA. (1990). *Oil Pollution Act of 1990* (33 U.S.C. 2701–2761). United States Coast Guard. https://www.uscg.mil/Mariners/National-Pollution-Funds-Center/About_NPFC/OPA/

Paris MoU. (2004). *Annual report 2003.* https://stats.oecd.org/Index.aspx?DataSetCode=EO99_INTERNET

Paris MoU. (2019). *Annual report 2018.* https://www.parismou.org/system/files/2018%20Annual%20Paris%20MoU.pdf

SOLAS. (2001). *SOLAS consolidated edition 2001.* https://books.google.no/books/about/SOLAS.html?id=Nhnh_h9nBQsC&redir_esc=y

STCW. (1996). International *Convention on standards of training, certification and watchkeeping for seafarers* (1996 Edition). https://treaties.un.org/pages/showDetails.aspx?objid=08000002800d6d42

Stopford, M. (2009). *Maritime economics* (3rd ed.). Routledge.

UNCTAD. (2019). *Review of maritime transport 2019*. https://unece.org/fileadmin/DAM/cefact/cf_forums/2019_UK/PPT_L_L-UNCTAD-RMT.pdf

Vassalos, D. (1999). Shaping ship safety: The face of the future. *Marine Technology and SNAME News, 36*(02), 61–76.

WBCSD. (1999). *Corporate social responsibility: Meeting changing expectations.* World Business Council for Sustainable Development, Geneva.

Chapter 5

Safety management system

5.1 INTRODUCTION

In Chapter 3, safety management is defined as "Coordinated activities aimed at fulfilling a set of safety policies and objectives". In practical terms, safety management can be described as all the activities that are done to ensure that the risk is at an acceptably low level throughout the lifetime of the system being considered.

The traditional approach to safety management in the maritime industry, and in virtually all other industries, has been what may be called a *reactive approach*. This means that we respond whenever accidents, mishaps, or negative consequences occur and try to fix the problems that caused the accident to avoid the same event in the future. This is a sensible approach since we try to learn from what has gone wrong and improve on that. The major disadvantage is of course that accidents and major losses must occur before action is taken. In the maritime industry, virtually all major regulations have been introduced as a result of serious accidents.

The *proactive approach* is based on trying to predict what may go wrong in the future and prepare for this before any accidents have occurred. Provided this is done in a systematic manner, we should be able to avoid many of the accidents that otherwise would have occurred.

Modern safety management is based on a combination of proactive and reactive approaches. Risk analysis is used to support the proactive approach, but at the same time, accidents and incidents are investigated to ensure that we can learn how to avoid similar occurrences in the future.

In this chapter, a general description of what a modern safety management process consists of is given. The description is based on a variety of sources and covers the main elements on a high level. Two international standards that describe risk management in general and occupational health and safety management specifically are ISO 31000 (2018) and ISO 45001 (2018), respectively. Most of the elements in the general process are described in more detail in subsequent chapters in the book, but in this chapter an overview of the whole process is provided.

DOI: 10.4324/9781003055464-5

This is followed by a more specific introduction to the International Safety Management (ISM) Code (IMO, 1994b). The ISM Code has been implemented as part of SOLAS and is applicable to the maritime industry.

5.2 PRESCRIPTIVE VS FUNCTIONAL RULES AND REGULATIONS

It is common to distinguish between prescriptive and functional rules and regulations in safety management. Prescriptive rules are characterized by being very specific in their requirements, e.g., stating that there should be at least as many lifeboat seats as the maximum number of persons onboard a vessel. The purpose of this rule is of course to ensure that safe evacuation can be performed. A functional rule would instead focus on what we want to achieve or the function. An example could be that it is stated that there should be evacuation means available to ensure safe evacuation of all personnel onboard within a specified time limit. This does not specify how this goal should be reached, only what the goal is. Functional rules are also often goal-based rules.

Prescriptive rules are simple to use and to follow, since they specify what the ship should look like, what it can do, and what equipment it has to contain. This is a clear advantage of prescriptive rules, but at the same time it is also a weakness. If the rules specify in detail what solution should be used, it restricts innovation and makes it difficult to get alternative solutions, that may be better than the rules specified and approved.

Functional rules will on the other hand give designers, shipyards, and owners greater freedom with respect to how they meet the requirement, allowing for innovative solutions to be implemented. However, the disadvantage is of course that the rules often have to be interpreted and that this interpretation may vary significantly from one application to another. In the example above, the term "safe evacuation" was used, but what does this mean? It is not possible to achieve a solution that guarantees that all personnel can be safe in all situations, therefore we must decide what is meant by "safe".

Hopkins (2011) argues that functional rules need to be supplemented with prescriptive guidance. The Norwegian offshore safety regime is a good example of this. The regulations are functional, but the guidance to the regulations refers to recognized standards that may be used to meet the functional requirements. This helps the designers to find a way to meet the regulations by applying recognized solutions, but at the same time opens for alternatives.

5.3 SAFETY MANAGEMENT IN A WIDER PERSPECTIVE

Safety management is an important process, but it is underlined that for this process to be successful, it has to be an integrated part of the overall management processes of the company. From an overall perspective, safety is just one of the objectives of a company, along with making a profit and other objectives. It is necessary to see all of these objectives together and not separate out safety as a "special" objective.

This is important in view of the continuous balancing of among other profits vs safety. Even if a common statement is that "safety is always our number one priority", this will never be the case in practice. Risk acceptance criteria that help us make decisions are established, but decisions can never be made in a vacuum, without considering costs. If the benefits of taking a risk are sufficiently high, even risks that we normally would consider to be too high may be accepted. Safety may be high on the agenda, but we cannot claim that it is always prioritized over other objectives. By integrating safety management with management of other objectives, safety can be given a weight in decisions in accordance with company policies and expectations from society.

If safety is seen as a separate "issue", the balancing of safety vs profits and costs becomes more difficult to do, and there may be a tendency that top management does not fully take the responsibility for making decisions about safety. Instead, it is left to the "safety management people" to make sure that safety is taken care of while top management can continue with their business as usual focusing on income and profit. The integration of safety management with general management processes is therefore considered to be very important.

According to ISO 31000 (2018), there are several other principles that are important for success with safety management:

- It should be systematic and cover all relevant risks.
- It should be customized to fit the organization and its context. The ISM Code can be regarded as a customization to fit shipping companies and ships.
- It should ensure that relevant stakeholders are informed and included.
- It should be able to adapt to changes in context and changes in risk, to always be fit for purpose.
- It should be based on the best available information.
- It should consider human, organizational, and cultural factors.
- It should be improving continuously.

ISO 45001 (2018) also lists many of the same success factors.

When later looking at the steps in a typical safety management process and also the ISM Code, it may be worth noting that there are many similarities with ISO 9001 (2015) and processes for quality management. The overall principles are the same, but the terminology used is more general in ISO 9001. The fact that these processes are similar should ease the process of integrating safety management in the overall management of the company, although it is noted that the differences in terminology may make it difficult to see the differences and similarities.

Safety management is an activity that is difficult to keep up and stay focused on. When we are doing something, we are used to seeing the result of our efforts, in terms of a new ship being built, profits raising, more efficient systems being implemented, and so on. Good safety management on the other hand implies that nothing happens, in the sense that no accidents occur. If no accidents occur, it is also easy to fall into the trap of assuming that it is not possible to improve any further. In a situation like this, it can be hard to keep up the motivation to continue the effort required to maintain this state. However, zero accidents is not the same as zero risk and there is always scope for improvement. A joke about safety management is that it can be compared to cleaning your house – it is not until you stop doing it that you really see the effect of it!

5.4 SAFETY MANAGEMENT PROCESS

There are many descriptions of what is included in safety management, but internationally, the standard ISO 31000 Risk Management Guidelines (ISO 31000, 2018) is widely recognized as a source that is generally applicable. A similar process is also described by IALA (2013) in their Risk Management Guidelines. In Figure 5.1, a general safety management process, adapted from the IALA guidelines, is shown. In the following subsections, each of the steps in the process is briefly described.

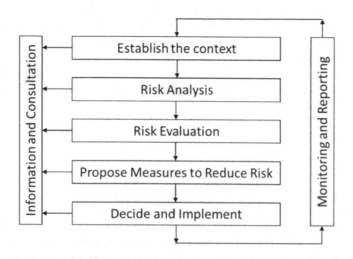

Figure 5.1 Safety management process (adapted from IALA (2013)).

5.4.1 Establish the context

The context is all the factors that represent the framework for the safety management process, aspects that impose limitations on the process, and other factors that may influence the process. It is common to divide this into external and internal context.

The foremost factors in the external context are all laws, rules, and regulations that impose restrictions or requirements on the safety management of a company. In shipping, IMO is the primary organization that develops regulations, like SOLAS, STCW, MARPOL, etc. The national maritime authorities and the port states enforce these regulations, together with classification companies. There may also be national and local regulations that have to be complied with.

Other external factors may be market conditions and requirements from clients (e.g., from oil companies related to transport of crude oil). Labor organizations, NGOs (Non-Governmental Organizations), and the public are also stakeholders that are part of the external context and that may have to be taken into account to a larger or smaller degree.

The external context can also be divided into a mandatory and a voluntary context. Mandatory context is the part of the external context that all actors have to relate to and comply with. Laws, rules, and regulations are the most obvious element of this. The voluntary context is the elements that the company may choose to take into account. An example may be introduction of (non-mandatory) measures to reduce emissions, spills, and other negative impact on the environment. Some companies may choose to apply stricter standards than required, often to improve their reputation with the public and clients.

The internal context is related to the company itself. Stakeholders that are relevant to consider are owners, management, and other employees. Systems and structures that form part of the internal context are policies and objectives, company culture, organizational structure, procedures and practices, and responsibilities.

As can be seen, the context that safety management is performed within can be quite complex and contains many elements that need to be understood and considered both when establishing the safety management process and when this is put into practical use.

5.4.2 Risk analysis

In the process shown in Figure 5.1, Risk Analysis is shown as a single step. Some sources also show this as two steps, called hazard identification and risk analysis, respectively. In this book, hazard identification is defined as part of the risk analysis. It is therefore natural to see these together. Normally, they are also performed as an integrated activity.

In Chapter 3, we give one possible definition of risk as the answer to the following three questions: (1) What can go wrong?, (2) how likely is it?, and (3) what are the consequences? The risk analysis is aimed at answering these three questions and thus describing risk, qualitatively or quantitatively.

This is an essential part of the safety management process because it provides an understanding of what the risk is. Without this understanding, we will not be able to identify relevant and targeted measures to control risk and safety management will be more or less a "shot in the dark". This is thus also the step that provides the basis for proactive management of risk.

More details about how a risk analysis is done and specific methods can be found in Chapters 8–12.

5.4.3 Risk evaluation

The risk evaluation uses the results from the risk analysis and compares these results with risk acceptance criteria. This is the basis for determining what should be done with the risk, including whether measures to reduce risk need to be identified and implemented.

Risk acceptance criteria are discussed further in Chapter 6.

5.4.4 Propose measures to reduce risk

Based on the risk analysis and risk evaluation, measures that may be implemented to reduce risk have to be identified. This can be measures aimed at removing or reducing hazards, preventing or reducing the probability of hazardous events, or mitigating or otherwise reducing the consequences. A good understanding of the accident scenarios is vital for identifying risk-reducing measures.

A second element of this step is to evaluate the effect of proposed measures. The effect on risk is of course important, how much reduction can be achieved if this measure is implemented? In addition, also the cost and other implications of the measure need to be considered before a decision can be made.

Chapter 15 discusses risk reduction measures, how to identify and evaluate them and specifically describe cost-benefit analysis as a tool for evaluating risk reduction measures.

5.4.5 Decide and implement

The preceding steps form the basis for making decisions about what to do with risk. The decisions will normally not be made purely based on these results. Risk levels and cost-benefit analysis will be important input, but a number of other aspects may also be considered, like industry practice, expectations from stakeholders, practicality of implementing the risk-reducing measures, considerations of reputation (very important for, e.g., the cruise industry), etc. This underlines the importance of integrating safety management with general management, to enable a broad consideration of all aspects that may influence the decision.

In simple terms, there are two main options that are relevant:

- Accept risk as is.
- Introduce measures to reduce risk. And given this option, it will also have to be decided which measures to implement.

The implementation of the decision(s) is the second part of this step. This covers all the practicalities related to making sure that the decision is put into operation. Many types of activities can be involved in this, e.g., modifying work procedures, modifying designs, buying and installing new equipment, and preparing and implementing training programs.

5.4.6 Monitoring and reporting

In Figure 5.1, this activity closes the loop to start the process from the top again. Safety management is a circular process, where continuous verification that it is working and continuous improvement are necessary. In practice, the process will not run in a continuous loop like this since there will be iterations and jumps back and forth to achieve the objectives.

This step covers a range of different activities, with the overall purpose of verifying that the safety management system is working as intended. Activities can involve:

- Monitoring the outcome of the safety management process, by measuring trends in various indicators of safety performance. Examples of such indicators are the number of lost time incidents, reported deviations from safe practices and procedures, and reported failures of safety systems. Monitoring major accident risk is a particularly challenging topic, since the number of accidents is low. Finding good indicators that measure the risk is therefore important.
- Testing, inspection, and maintenance of equipment that is important for maintaining safety. This can be all types of life-saving appliances such as lifeboats, life rafts, and survival suits but also all other safety equipment that is in place to detect, prevent, control, and mitigate hazardous events. Some examples are fire and smoke detectors, fire alarms, fire suppression systems, radar systems, communication systems, emergency power systems, etc. Identification and follow-up of these systems are briefly discussed in Chapter 15.
- Observations and safety tours are used to verify that all safety critical work operations are performed in a safe manner and in accordance

with procedures. Questionnaire surveys may also be used for this purpose.

- Accident and incident investigation – An important part of monitoring is also to do thorough investigations of accidents and incidents and other non-conformities. The objective is to understand why they have happened and use the findings as a basis for learning and improving. This is part of the reactive approach to safety management. Accident investigation is covered in Chapter 19.

5.4.7 Consultation and reporting

An activity that runs in parallel with the main process and that may be activated through various steps in the process is consultation and reporting. This step is sometimes also described as communication and consulting. The purpose of this activity is twofold: to inform and to engage relevant stakeholders in the safety management process. Information about the process and the results must be communicated and stakeholders are given an opportunity to discuss and give feedback on the information that they get. This is used to adjust and improve the process as necessary. An example of this is when risk analysis is performed of new major hazard facilities. It is common in many cases to allow the public to review and comment on this before final decisions are made.

Consultation and reporting can be relevant to all the other activities in the process, although different stakeholders may be relevant to include in different steps. The following are some examples of stakeholder involvement:

- Context – The external context is to a large extent given for a company (although they may choose to operate in different markets, using different flag, etc.) and consultation on this part of the context is not very relevant. However, the internal context, e.g., policies and objectives are defined by the company itself and it will be relevant to include stakeholders like management, employees, and perhaps also external parties verifying compliance with the ISM Code.
- Hazard identification and risk analysis – This step may require input from various stakeholders, both internal and external, and if the results show that the risk may affect parties outside of the company, it is often also necessary to communicate results from risk analysis to these external stakeholders. Authorities will typically also be interested in results from risk analysis.
- Propose measures to reduce risk – this is a step where it is important to have input from different stakeholders who can contribute to identify good measures for different types of risk. Personnel close to the "sharp end" of the business may, e.g., have very good proposals for measures that can contribute to reduce risk since they are closely familiar with the equipment and operations involved.

- Monitoring and review – This step also requires discussion with key stakeholders, mainly internally, but there may also be external stakeholders that are interested in this.

Communication of information about risk can be quite demanding and, as was discussed in Chapter 3, the terminology or language that we use may hinder good communication. It is particularly important to adapt the information to suit the level of expertise and the background of each stakeholder. Information that is well suited for, e.g., experts in the field of risk analysis may be completely meaningless for the general public because of the technical details and the terminology used.

5.5 ISM CODE

5.5.1 Background

In the late 1980s and early 1990s, there was an increasing recognition of how organizations, their decisions, and their culture influenced safety. Human factors had long been regarded as important for safety, but focus was now also turned toward how organizational factors influenced the humans working in the organization. Companies should ensure that the staff are properly informed and equipped to fulfill their operational responsibilities safely. Decisions taken ashore can be as important as those taken at sea, and there is a need to ensure that every action affecting safety, taken at any level within the organization, is based on a sound understanding of its consequences.

This is the motivation behind the adoption by the International Maritime Organization (IMO) of the "International Code for the Safe Operation of Ships and for Pollution Prevention", normally referred to as the International Safety Management (ISM) Code. The code is the reflection of this objective on the part of the member states (IMO, 1994a, 1994b). The ISM Code establishes an international standard for the safe management and operation of ships by setting requirements for the organization of company management in relation to safety and pollution prevention, and for the implementation of a safety management system (SMS). The ISM Code addresses the very important issues relating to human factors, and it has been argued that this is one of the most significant documents produced by IMO.

The Assembly of IMO has adopted a series of resolutions dealing with guidelines on management procedures to ensure the safest possible operation of ships and the maximum attainable prevention of marine pollution. These resolutions culminated in the ISM Code, which was adopted in November 1993 by resolution A 741 (18). In May 1994, a SOLAS (i.e., the International Convention for Safety of Life at Sea) Conference decided on a new Chapter IX of SOLAS which makes the ISM Code mandatory for ships, regardless of the date of construction. For most ship types the code was implemented on the 1st of July 1998, but from the 1st of July 2002 the code is mandatory for all ships, including mobile offshore drilling units. For passenger ships and high-speed passenger crafts there is no lower limit in terms of vessel size, but for all other ship types there is a lower limit of 500 gross register tons (grt).

After the introduction of the code, it has been amended several times and the most recent version at the time of writing this book was implemented in 2018 (IMO, 2018).

The introduction of the ISM Code represented a dramatic departure in regulatory thinking by the IMO. It acknowledges that detailed prescriptive rules have serious limitations. Inspired by principles of quality management and internal control, the ISM Code was aimed at stimulating safety consciousness and a systematic approach in every part of the organization both ashore and onboard. Developing a good safety culture is a key element in safety management. The code was also triggered by an increasing realization that management could and should play a major role in preventing accidents.

The ISM Code is a creation of the so-called "culture of self-regulation" or a functional regulatory regime in which regulations go beyond the setting of externally imposed compliance criteria, often called a prescriptive regime. The ISM Code concentrates on internal management and organization for safety, which means that safety is organized by those who are directly affected by the implications of failure.

5.5.2 The content of the ISM Code

The ISM Code is based on a set of general principles and objectives to be achieved. It comprises of two parts divided into 16 sections over 13 pages and is hence a fairly short document. In addition comes appendixes and also separate guidelines for the implementation of the code.

The main purpose of the ISM Code is to demand that individual ship operators create a safety management system that works. The code does not describe in detail how this should be done but lists several requirements to the SMS. The underlying philosophy behind the code is commitment from the top, verification of competence, clear placement of responsibility, and quality control of work.

IMO has stated that the high-level objectives "are to ensure safety at sea, prevention of human injury or loss of life, and avoidance of damage to the environment, in particular to the marine environment and to property". More specifically, the objectives of safety management in a company should be to (IMO, 2018):

- provide for safe practices in ship operation and a safe working environment;

- assess all identified risks to its ships, personnel, and the environment and establish appropriate safeguards; and
- continuously improve safety management skills of personnel ashore and aboard ships, including preparing for emergencies related both to safety and environmental protection.

These are also elements of a strong safety culture.

These objectives show that the ISM Code has relations to existing or traditional safety management approaches such as technical solutions to improve safety, training, emergency preparedness, and risk analysis. The 16 chapters of the code are listed in Table 5.1.

Table 5.1 The ISM Code – Main content of each chapter

		Chapter	Content of chapter/Requirements
Part A	1	General	Definitions of key terminology, objectives of the code, and where it is applicable. Functional requirements for a safety management system.
	2	Safety and environmental protection policy	Requirements to establish and implement a policy that describes how the objectives of the code will be achieved.
	3	Company responsibilities and authority	Identification of the company responsible for the operation. Specify responsibility, authority, and interrelation of key personnel. Ensure adequate resources and shore-based support.
	4	Designated person(s)	Identity/assign person(s) serving as a link between vessel and company. Monitor safety and secure resources and support.
	5	Master's responsibility and authority	Statement of Master's responsibility for the safety onboard. The Master's responsibilities include implementation of the policy, motivating the crew, issuing orders, verifying adherence, review the SMS, and reporting deficiencies in the system.
	6	Resources and personnel	Master must be properly qualified, know the SMS and have the support to perform his duties. The ship should be manned with qualified, certified, and medically fit seafarers. Manning must be appropriate to maintain safety. Company should have procedures for familiarization of personnel with new duties, all personnel should have adequate understanding of relevant rules, regulations, codes and guidelines and necessary training must be provided. Requirements for providing information to the crew in a language that they understand. Company must ensure effective communication between personnel.

(continued)

Table 5.1 (Continued) The ISM Code – Main content of each chapter

		Chapter	Content of chapter/Requirements
	7	Shipboard operations	Procedures, plans, instructions, and checklists must be prepared for key safety critical shipboard operations. Definition of tasks and assignment of personnel.
	8	Emergency preparedness	Emergencies to be identified, emergency procedures, and program for drills and exercises in emergency response must be prepared.
	9	Reports and analysis of non-conformities, accidents, and hazardous occurrences	Develop procedures for reporting, investigating, and analyzing non-conformities, accidents, and hazardous situations. Establish procedures for implementation of corrective actions.
	10	Maintenance of the ship and equipment	Maintenance in accordance with rules and company-based requirements. Ensure inspections, non-conformity reporting, corrective action, and record keeping. Identify safety critical systems. Ensure reliable operation and testing.
	11	Documentation	Develop procedures controlling all documents and data relevant to the SMS. Ensure that valid documents are available where relevant, changes to documents are reviewed and approved and that obsolete documents are removed.
	12	Company verification, review, and evaluation	Carry out internal safety audits to verify compliance with SMS. Periodically evaluate efficiency of SMS. Auditing personnel should have an independent role. Feedback of findings to involved personnel and responsible management. Timely corrective actions should be taken.
Part B	13	Certification and periodical verification	The company should have a Document of Compliance (DOC), and the vessels a Safety Management Certificate (SMC).
	14	Interim certification	Interim DOC can be issued to newly established companies and to companies operating new ship types. Interim SMC can be issued to new ships, and to ships that change owner or flag.
	15	Verification	Procedures acceptable to the Administration should be established for all verifications required according to the code.
	16	Forms of certificates	DOC and SMC should follow the format of the documents in the appendix to the Code.

Compared to the general SMS that was presented in Figure 5.1, only a few of the elements are covered by the ISM Code. This is shown in Figure 5.2, where the shaded elements are those that are covered in the code.

As can be seen, the code addresses only three of the elements in the overall process, establishing the context, decide and implement, and monitoring

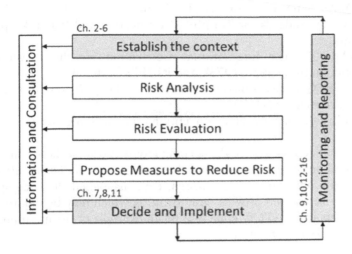

Figure 5.2 Chapters in the ISM Code seen in relation to a general SMS.

and reporting. In particular, it is noted that the primary element of proactive safety management, risk analysis, is not explicitly covered by the code. This implies that it is not sufficient to comply with the requirements in the code to establish a "complete" safety management system in accordance with what was described in Section 5.3. On the other hand, there are no requirements in the ISM Code that contravenes "good practice" for safety management systems in general.

Since the ISM Code primarily gives functional requirements, it is not obvious what an acceptable SMS should look like. Some examples of the functional requirements are in Section 6.5 where the company shall identify "any training which may be required" and that personnel should be able to "communicate effectively". No specification is provided for what training is required or as to how effective communication should work. Also, it is not specified what "effective communication" actually is. Two comments are relevant to make at this stage. First, individual companies may have very different opinions about what constitutes "effective communication" (or any other functional requirements in the code). This opens for significant variations in what is "good enough" between companies. Implicitly, this can also lead to large variations in risk level. Secondly, there may be many ways of reaching this goal, and this allows companies to find the way that is most efficient for them and provides the best solution. This flexibility is obviously positive.

It is important to recognize that the ISM Code and its requirements to company safety management systems must be seen in the context of already existing international safety regulations. The main safety conventions in this respect are as follows:

- SOLAS: Safety of Life at Sea (1974) and SOLAS Protocol (1978, 1988)
- STCW: Standards for Training, Certification and Watch-keeping for Seafarers
- MARPOL: The International Convention for the Prevention of Pollution from Ships (1973), and its 1978 Protocol
- COLREG: Convention on the International Regulations for Preventing Collisions at Sea (1972)
- ICLL: The International Convention on Load Lines (1966)

ISM does not address any of the specific requirements in these conventions but is based on the assumption that the management system should ensure that they are adhered to. This means, e.g., that when the code specifies that companies should provide "any training which may be required", the requirements in the STCW code will be an example of the minimum standards that they have to meet.

5.6 EFFECT OF THE ISM CODE

The ISM Code became mandatory for most ships in 1998, i.e., more than 20 years ago at the time of writing this book. If the code has had the effects that were intended when it was introduced, we would expect to see signs of that now, after so many years.

On the face of it, the easiest way to measure the effectiveness of the code should be to look at accident statistics and see if the accident rates have been reduced over this period. This can be an indication, but it cannot really give us the answer since there are several other factors that also have changed over this period. One aspect is that there are other changes to rules and regulations that also can be expected to influence the risk level. One important example is the Maritime Labour Convention (ILO, 2006) that came into force in 2013. A more general trend is the increased awareness of safety in society. This is also likely to have influenced the maritime industry.

Because of this, it is hard to tell what the effect of the ISM Code is and what is the result of other changes in the same period.

A number of studies have been performed, aimed at determining the effect of the ISM Code. Partly, these have looked at accident rates, but also at the degree of implementation of the code. The results are ambiguous, and based on published results, it is hard to determine what the effect has been.

Tzannatos and Kokotos (2009) found that the accident rates in Greek shipping were significantly different before and after the introduction of the ISM Code. The effects were particularly clear for accidents involving human elements in tankers and Ro-Pax vessels.

Bhattacharya (2012) looks at the effectiveness of the ISM Code from a different perspective. The perception of the effectiveness of the code among managers and seafarers is investigated. The main conclusion is that these two groups have very different views on the effectiveness. Managers viewed it as a tool that would be beneficial for the health and safety of seafarers, the seafarers themselves looked upon this as a "paper exercise" that added little or no value to the management of risk. According to the author, this could be traced back to a lack of trust between seafarers and managers. Without this trust, developing a good safety culture, which is at the bottom of the ISM Code, is also very hard if not impossible.

Another source of the degree of implementation of the ISM Code on ships is the inspection results from port state controls. As an example, the results from Paris MoU (Memorandum of Understanding) for the period from 2017 to 2019 are considered. In this period, nearly 51,000 inspections were performed, and a total 5,140 ISM-related deficiencies were found. It is noted that there may be more than one ISM-related deficiency identified in each inspection. In 2018, ISM deficiencies accounted for less than 5% of all deficiencies identified in the inspections. The Paris MoU data that are publicly available do not specify more in detail which chapters in the Code the deficiencies are related to.

A study by Batalden and Sydnes (2014) gives some information about where the weaknesses are. They looked at 85 accidents investigated by MAIB in the period of 2002–2010. A total of 478 causal factors were identified. Of the ISM-related causes, the most important are related to Master's authority and responsibility (Section 5), Resources and personnel (Section 6), Plans for shipboard operations (Section 7), and Verification (Section 12).

Grays and Sims (1997) report an analysis of non-conformities found in audits on different safety management systems, including ISO- and ISM-based systems. The six dominating areas, which represent 66% of all non-conformities, are as follows:

- Maintenance
- Documentation
- Resources and personnel
- Emergency preparedness
- Management system
- Operational procedures

All these areas are addressed by the ISM Code.

An alternative format is to present the degree of attainment relative to the target values set by the company. This may be more meaningful if one has defined a long-term plan to improve the SMS. Figure 5.3 shows the audit results for a shipping operation within Exxon. The functions with the greatest gap between target and achieved score are listed from the top in the diagram. It is interesting to note that the largest gap was found for "Personnel" and "Reporting, investigation and analysis [of non-conformities]". These

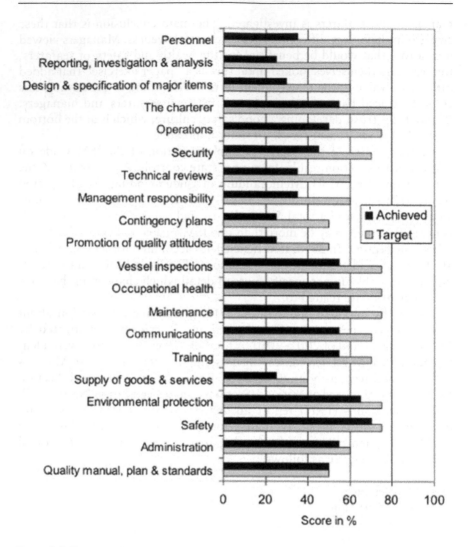

Figure 5.3 Target and achieved score for audited functions.

are typical "soft" aspects of the SMS that are difficult to change. Illustrating this is the finding that the quality plan itself is 100% in accordance with the target. It is a general experience with safety management systems that the easiest part of quality management is to produce plans and documentation.

REFERENCES

Batalden, B.-M., & Sydnes, A. K. (2014). Maritime safety and ISM code: A study of investigated casualties and incidents. *WMU Journal of Maritime Affairs, 13*, 3–25.

Bhattacharya, S. (2012). The effectiveness of the ISM code: A qualitative enquiry. *Marine Policy, 36*(2), 528–535.

Gray, J., & Sims, M. (1997). Management system audits for ship operators - an auditor's experience. *Transactions of Institute of Marine Engineers, 109*, Part 3.

Hopkins, A. (2011). Risk-management and rule-compliance: Decision-making in hazardous industries. *Safety Science, 49*(2), 110–120.

IALA. (2013). *IALA Guideline -1018 Risk Management (3rd ed.).* https://www.iala-aism.org/product/risk-management-1018/

ILO. (2006). *Maritime labour convention.* https://www.ilo.org/global/standards/maritime-labour-convention/lang--en/index.htm

IMO. (1994a). *International code of safety for high-speed craft (HSC Code).* https://puc.overheid.nl/nsi/doc/PUC_2409_14/5/

IMO. (1994b). *International safety management code (ISM Code).* https://digitallibrary.un.org/record/153410

IMO. (1995). *Guidelines on Oil Spill Dispersant Application including environmental considerations.* https://wedocs.unep.org/xmlui/handle/20.500.11822/2403

ISO 31000. (2018). *Risk management - guidelines.* https://www.iso.org/standard/65694.html

ISO 45001. (2018). *Occupational health and safety management systems - Requirements with guidance for use.* https://www.iso.org/standard/63787.html

ISO 9001. (2015). *Quality management systems - requirements.* https://www.iso.org/standard/62085.html

Tzannatos, E., & Kokotos, D. (2009). Analysis of accidents in Greek shipping during the pre- and post-ISM period. *Marine Policy, 33*(4), 679–684.

Chapter 6

Risk acceptance

6.1 INTRODUCTION

In an ideal world, all would agree that the best solution would be if all risk could be removed to achieve a completely accident-free situation. In practice, this is however possible in very few, if any, circumstances. Regardless of what efforts we make, there will be a remaining or residual, risk that we cannot remove completely. There are thus practical limitations on what can be achieved. Equally important in most situations is however the fact that the cost of reducing risk will increase the more effort we put into reducing it. We will therefore arrive at a point where we have to ask ourselves whether it is worth the cost and effort to reduce risk further. Even if statements like "Safety is always our priority number one" flourish in industries and political life, we will eventually come to a point where we have to say that there are other objectives that are more important.

This is the reason why we need risk acceptance criteria or risk decision criteria as it is also called. Risk acceptance criteria can help us:

- Make decisions that are consistent with respect to risk.
- Make decisions that balance risk vs other objectives in an explicit manner.
- Ensure that risk/safety is explicitly considered in decisions. Without criteria it is harder to give safety the priority it needs.
- Make decisions aimed at meeting expectations from regulators, other stakeholders, and society at large.

Different terms are being used to essentially describe the same thing. In this book, the term risk acceptance criteria are used, although other terms also may be relevant:

- Risk decision criteria – this term underlines that these are criteria for making decisions about risk. Implicitly this also means that we use it

DOI: 10.4324/9781003055464-6

to decide what risk is accepted, but the emphasis is on the fact that it helps us to make decisions.
- Risk tolerability criteria – this term focuses on the fact that risk is not something we accept unconditionally, but we can tolerate it in a given context and with a given set of benefits. It may be argued that it is the activity/technology with its associated benefits and risks that we accept, not risk in isolation.

This last comment is important for the understanding of risk acceptance in general. It is tempting to argue that it must be possible to develop a set of "universal" criteria that can be applied in all situations. However, risk acceptance is highly dependent on the benefits that we can get by accepting the risk. If the benefits are high, we are also more willing to accept high risk. Therefore, it is not possible to determine a common level that can be generally applicable.

A second observation that also may be made at this point is that determining and agreeing on what is acceptable in a given situation is still not easy, even if we know the benefits and the risk. A main reason for this is that the risk and the benefits usually are measured in different dimensions or types of consequences. The risks are expressed, e.g., as loss of life, injury, or damage to the environment while the benefits typically will be increased profits, reduced costs, improved efficiency, and so on, all of which can be measured in monetary terms. We are thus faced with a situation where we have to weigh two completely different types of consequences and benefits against each other. In order to make decisions, we have to find a common scale against which these can be measured, and that means that we have to place a monetary value on loss of life, injuries, or environmental damage. This is a controversial issue that can raise a lot of discussion.

In the rest of this chapter, the following topics will be covered:

- First, some discussion of factors that affect risk acceptance is provided. This is a general discussion, not aimed at any particular application.
- Next, some basic principles for deciding how to approach decisions about risk are presented.
- This is followed by a presentation of some commonly used ways of formulating risk acceptance criteria.
- Finally, risk acceptance criteria used in the maritime industry are discussed explicitly.

The focus will be on criteria related to risk to people, although the basic principles and the common ways of formulating criteria can be applied to any type of consequences. Some explicit comments on criteria for environmental risk will be added under the discussion of the maritime industry.

6.2 SOME FACTORS AFFECTING RISK ACCEPTANCE

6.2.1 Risk perception

Our perception of risk is an important factor in risk acceptance and is generally an important factor in decision making about risk.

It is common to distinguish between what we may call calculated risk and perceived risk. Sometimes, these are also called objective and subjective risk, although both in reality are subjective. Calculated risk is the output that we get from a risk analysis. This is based on systematic analysis of the system being considered, careful collection of data and other information, and bringing all this together to give us a description of the risk. Perceived risk is an individual's or a group's beliefs about risk or subjective experience of the risk. This will be based on the knowledge of the individual or group, combined with values, priorities, and preferences, but not placed in a systematic framework like risk analysis is.

Perceived risk and calculated risk can therefore be very different, and different individuals may also have very different risk perceptions. An example of this is fear of flying. Some people are extremely afraid of flying, mainly because they perceive the risk to be too high. However, most people will have no problems with flying at all and consider the risk to be so low that they normally will not even think about it as a problem.

Risk perception is not something we consider in risk analysis, although we have to be aware that the risk perception of the risk analyst will affect the results, even if a systematic method for analyzing risk is applied. This is because there is always a significant element of judgment in any risk analysis, both with respect to assumptions, interpretation, and use of data and input of subjective judgment.

Risk perception however plays a major role in decision making. The decision-maker will of course have access to results from the risk analysis, but his or her risk perception can still influence the decision. In most situations, there will further be other criteria and objectives that are to be met as part of the decision-making process. Weighing of different objectives can also be influenced by risk perception and thus affect the decision.

Differences in risk perception can also be part of the explanation of why there are often conflicts between different stakeholders when decisions about risk are to be made.

Fischhoff et al. (1978) have summarized the various factors that influence perceived and acceptable risk in two dimensions: High/low technology (where high technology can be characterized by being new, involuntary, and poorly known) and certainty of death.

Interestingly, the way that risk is presented (what risk metric we use) may also influence perceived risk (Slovic et al. 2000a). In addition, the frequency of occurrence of events is important as we tend to underestimate risk associated with common events and overestimate risk associated with rare events (Slovic et al. 2000b).

In the following, some of the important factors influencing risk perception and acceptance are briefly discussed.

6.2.2 Benefits

Probably the most important factor influencing our willingness to accept risk is the benefits that we can get from accepting the risk. A baseline assumption can be that if we do not get any benefits from taking risk, we are not willing to accept any risk that is not considered (or perceived) to be negligible. This is also in many cases an important cause of conflict between different stakeholders.

The case of maritime spills is an example of this. For shipowners who make their living out of transporting oil, the willingness to accept risk is primarily tied to the cost that they will incur if a spill occurs. However, fishermen making a living out of the fishing grounds where a spill may occur have no practical benefits from the oil transport through the area and therefore have no benefits. For them, the willingness to accept this risk will therefore typically be much lower. Similar effects are seen in people living nearby hazardous installations. For them, only negative effects and no benefits are relevant. An interesting study of this has been performed on nuclear power plants. Neighbors who worked at the plant were compared with neighbors who worked elsewhere, and it was found that those who worked at the plant were willing to accept higher risk than those who did not. The reason for this is of course that the workers have their income from the plant, a clear benefit to them.

One effect of the importance of this is that risk acceptance criteria with fixed upper limits of what risk we can accept in practice may have limited value. What has been seen in many practical examples is that provided the benefits are large enough, we are usually willing to accept also risks above the defined limits. It is therefore argued that applying upper limits is of no value and that we should apply the ALARP principle (see Section 6.5) regardless of what the risk level is.

6.2.3 Risk aversion

The way that we calculate risk may be as follows. An accident type with frequency 1 per year and consequence of 1 fatality gives a risk of 1 fatality per year. Similarly, an accident type with frequency 0.001 per year and consequence 1,000 fatalities will also give a risk of 1 fatality per year. From a risk analysis point of view, they are thus equal. However, in practice, most people will argue that the latter case is less acceptable than the first case. This is an example of "risk aversion", where our judgment of events with extreme consequences will tend to be overshadowed by the consequence, even if the probability and risk may be low.

One application area where this has been an important issue has been nuclear power. In the worst case, nuclear accidents will have extreme and long-lasting consequences and it is argued that this makes it unacceptable, regardless of how low the probability of such an accident may be. It is also noted that the nuclear accidents that have occurred generally have had far less consequences than described in the worst-case scenarios. At the time of writing, the most recent accident was in Fukushima, an accident that among others caused Germany to abolish plans for further development of nuclear power. However, the nuclear release from the accident has caused and will cause far fewer fatalities than the evacuation did.

Risk aversion could perhaps better have been called "consequence aversion" since it is more related to our acceptance of extreme accidents with large consequences than risk as we define it in this book.

6.2.4 Time since accidents and own experience

The time since accidents have occurred will also influence our perception of risk, and thereby also what we are willing to accept. In particular, own experiences with accidents will be important. Around 2000, there was a serious accident with a high-speed passenger craft, Sleipner, in Norway, in which 17 people were killed. Surveys performed shortly after the accident and then repeated three years later showed that the general perception of risk was much higher shortly after the accident compared to three years later (Bjørnskau 2004).

The fact that an accident recently has occurred will remind us that it is a risk associated with the activity and therefore tend, more or less subconsciously, to adjust upwards our internal "estimates" for how likely this is to happen. We therefore tend to believe that the risk is much higher just after an accident.

6.2.5 Time until effects are experienced

Another factor that also influences our risk perception is the time until the effects are experienced. For accident risk, this may not be very relevant since effects usually are seen immediately. However, for, e.g., lifestyle risks such as smoking and eating unhealthy food, the effects are normally not seen until after a long time. Even if statistics show smoking will give a high probability of getting lung cancer, we may still readily accept that risk. A key factor in explaining this is of course that we get the benefit immediately (by smoking a cigarette), while the consequences will not be experienced until many years into the future. A Norwegian proverb expresses this wisdom: "After the sweet itching comes the sour sting".

6.2.6 Lack of understanding of the risk

In general, if we have limited understanding of what the risk is; our perception of risk will tend to be quite high. Some examples of situations where this factor may influence our perception are:

- When faced with novel technology. This means we have limited experience and usually limited knowledge about what may happen. Our uncertainty is higher, and our perception of the risk is high.
- When faced with technology or situations that we are not personally familiar with. In such situations, uncertainty will also be high and risk perception is high.
- When faced with situations where we have limited control of risk. Partially, this is also related to the two previous points.

6.3 DECISION-MAKING PRINCIPLES

There are three underlying principles that can be applied as a basis for establishing sound acceptance criteria and that are useful to understand (see, e.g., HSE 2001).

- The **utility** principle. This implies that risk reduction should be applied and prioritized based on the utility that we gain from reducing risk. This means that risk reduction effort should be prioritized where the ratio between cost and risk reduction is lowest. Underlying this principle is that we have limited resources to spend and that we should aim to get as much risk reduction as we can with our limited resources. From a societal point of view, this may be an important effect. The principle does not specify how large resources we should spend on reducing risk but gives us a rule for prioritizing where to spend available resources.
- The **equity** (or equality) principle. This principle takes a starting point that all humans should be treated equally and therefore that no-one should be exposed to a too high risk (again without specifying what this is). If individuals are exposed to risk above a specified level, resources should be used to reduce that risk, regardless of cost and effort required to reduce it. From an individual's point of view this is positive but it is also clear that this can lead to different priorities than the utility principle.
- The **technology** principle. This principle is based on a comparison with existing technology where the risk is already generally accepted in society. As long as the risk level is on the same level or lower, it is considered to be acceptable. This is often expressed through "Best Available Technology", "Best practice" or similar terms.

It is noted that the utility and equity principles say nothing about the *level* of risk that is acceptable, but only the principles for establishing the criteria for making decisions. The technology principle may be regarded as implicitly specifying a risk level since it is compared with an existing situation, and the risk associated with that. However, application of a given technology will not necessarily give the same risk level in all situations.

It may be argued that all standards and regulations (e.g., SOLAS) that specify technical solutions to maintain safety support the technology principle. However, history has shown that this does not necessarily ensure that risk is at an acceptable level in all situations.

A final comment is that the utility and equity principle both rely on risk analysis to support the decision-making process whereas the technology principle does not. This is because the technology principle only considers risk implicitly.

6.4 INDIVIDUAL VS SOCIETAL RISK ACCEPTANCE CRITERIA

Risk acceptance criteria can be formulated in different ways, but one important distinction is between individual and societal criteria.

Individual criteria are criteria expressing risk per individual. From an individual's point of view, this is obviously the most important factor to consider when deciding whether risk can be accepted or not. Individual risk can be expressed in different ways, e.g., directly as an annual probability of being killed. This is often called IRPA, or Individual Risk Per Annum. Individual risk can typically be used to ensure that individuals are not exposed to excessive risk in a work situation. It may be noted that even if the risk is expressed per individual, it is common to consider a group of people and calculate an average IRPA for an individual in this group. If the group is large, this average may mask large differences within the group that is being considered. One should therefore try to consider reasonably homogenous groups.

Societal risk is an expression of the risk that a group of people is exposed to. The group can vary greatly in size. A couple of examples can be the crew of a ship (that can be just a handful of people) or the passengers on a cruise ship (that can be some thousands). Societal risk can also be expressed in many different ways. A simple measure is PLL, or Potential Loss of Life (also called Probable Loss of Life), per year in a group. This is an expression of the total number of fatalities that can be statistically expected within the group per year. Since the size of the group can vary, PLL will also vary, and comparing PLL values for different groups is therefore not possible.

PLL is however useful for comparing risk in different situations for the same group. Therefore, it is also commonly used in connection with cost-benefit analysis, where we are comparing risk to the same group with

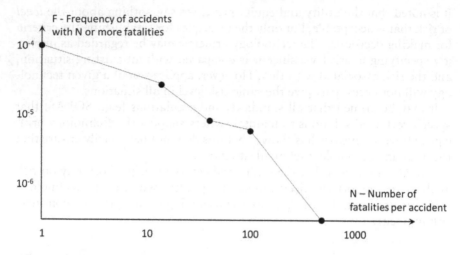

Figure 6.1 Illustration of FN curve.

and without risk reducing measures implemented. The reduction in PLL (reduction in statistically expected number of fatalities) can then be compared with the costs involved in reducing risk. By introducing a valuation of the saved lives, a direct comparison is possible. This is further discussed in Chapter 12 on Formal Safety Assessment.

Another way of expressing societal risk is by use of an FN curve. An FN curve illustrates the relationship between the frequency of accidents and the severity of accidents. Figure 6.1 illustrates an FN curve.

On the horizontal axis, the number of fatalities per accident, N, is given. The number 10 thus represents an accident with 10 fatalities. On the vertical axis, accident frequency is given, but it is specified as the frequency of accidents with N *or more* fatalities. From the figure, we can see that the curve goes through the point $(1, 10^{-4})$ per year. This means that the frequency of accidents with 1 or more fatalities is 10^{-4} per year. This is the same as the frequency of fatal accidents. If we want to know the frequency of accidents with exactly N fatalities, this can be calculated as follows:

$$F_{N \text{ fatalities}} = F_{N+1} - F_N \tag{6.1}$$

Some important points to note about the FN curve:

- The vertical axis represents the frequency and the horizontal axis represents consequences, implying that points on the FN curve can be regarded as representations of risk (the combination of frequency and probability).

- Since N is the number of fatalities per accident, the horizontal axis will only contain integers. It does not make sense to talk about 0.5 people being killed in an accident, even if a risk analysis may calculate that *on average*, 0.5 people are killed.

The FN curve can only be horizontal or fall as N increases. This is obvious from the definition of F. The frequency of accidents with N fatalities or more includes the frequency of accidents with $N+1$ fatalities or more and the latter frequency therefore cannot be greater than the first, only the same or smaller.

6.5 ALARP

ALARP (As Low As Reasonably Practicable) is probably the most commonly applied approach to defining acceptance criteria for risk and for managing risk. ALARP has a long history and has its origin in the UK. It was first introduced in a court case, Edwards vs National Coal Board in 1949 (Court of Appeal 1949), but has later become part of legislation, regulations, and guidance documents in many parts of the world and in many industries and contexts.

The ALARP principle is based on dividing risk into three "zones" or different levels. The principle is illustrated in Figure 6.2.

The zones can be described as follows:

- High risk – This is referred to as "unacceptable risk" or "intolerable risk". In this zone, the risk is so high that we are not willing to accept it regardless of the benefits. The implication of this is that risk reduction measures must be introduced or that the activity causing the risk cannot be permitted.
- Medium risk – This is often called the "ALARP zone". In the ALARP zone, the risk is still too high to accept unconditionally, but at the same time it is not so high that we conclude that this is intolerable. If we find that risk is in this zone, we should therefore try to reduce risk to a level "as low as reasonably practicable". The implication that risk reduction measures should be identified and implemented is reasonably practicable, but it is not an absolute requirement to reduce risk.
- Low risk – The low-risk zone is also termed "broadly acceptable" or "tolerable". This risk level is so low that we normally accept it without any expectations for further reduction of the risk.

The crucial difference between the three zones lies in the strategy for treating risk, from "must reduce", to "reduce to ALARP" to "no reduction" in the three zones respectively.

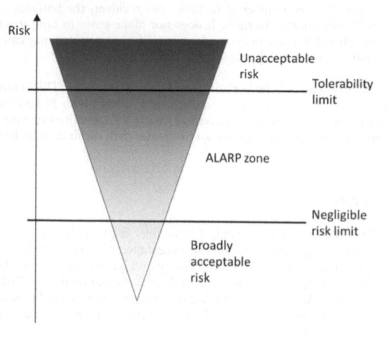

Figure 6.2 ALARP principle.

The ALARP principle is widely applied and we see it implemented, e.g., in the risk matrix, where the red, yellow, and green zones in practice correspond to the high, medium, and low risk in the ALARP principle.

If the three basic principles described in Section 6.3 are considered, it is noted that the ALARP principle combines all three:

– The tolerability limit reflects application of the equity principle, by introducing an upper limit to what risk can be accepted or tolerated.
– The medium (ALARP) zone will in practice take into account both the utility and technology principle. The fact that risk should be "as low as reasonably practicable" implies that a utility thinking can be applied – if the cost is too high, it is not necessary to reduce risk. In addition, it is an implicit requirement in ALARP that best practice should be applied to achieve ALARP. This means that the technology principle also is introduced.

In the discussion of the utility and equity principles, it was noted that neither specifies what an acceptable risk level is. The same applies to the ALARP principle, although the principle means that two limits must be established: the limit between the "unacceptable" and ALARP zones and the limit between ALARP and "broadly acceptable zone". Determining these limits is a complex task, requiring many factors to be considered and typically consulting with a variety of different stakeholders. We will not go into details on this process in this book.

6.5.1 Achieving risk that is ALARP

If risk is in the ALARP zone, further risk reduction should be considered to ensure that risk is as low as reasonably practicable. In practice, this can be ensured by applying a systematic process that covers the following steps:

- Ensuring that relevant regulations and standards are met and that "good practice" is followed. Cost-benefit analysis can thus not be used as an argument for removing risk reduction measures that are part of good practice in industry.
- Systematic identification of measures that can contribute to reduce risk.
- Implementation of all measures where there not are good reasons for not implementing them. This point is important to note. The main rule is that any identified measures should be implemented, unless we have good arguments for not doing it. One such argument can of course be that the cost is too high.
- Application of cost-benefit analysis to determine if remaining risk reduction measures should be implemented or not.

The process is illustrated in Figure 6.3.

When applying cost-benefit analysis, it is normally a requirement that measures should be implemented unless the costs are "grossly disproportionate" to the benefits. This is often expressed through a disproportionality factor d, which may vary from 1 to a considerably larger number. HSE (2001) recommends that d is 10 when the risk is close to the intolerable level (upper boundary of the ALARP zone), reducing to 1 when approaching broadly acceptable risk.

6.6 OTHER PRINCIPLES FOR RISK ACCEPTANCE

The ALARP principle is by far the most common principle, but there are also some other principles applied that are worth mentioning briefly.

Within railway transport, two principles called GAMAB and MEM are being applied (IEC 2002). Both are essentially based on a technology principle, comparing new technology (new transport systems) with existing technology:

- GAMAB (Globalement Au Moins Aussi Bon – globally at least as good) states that new systems should have a risk level equal to or lower than existing systems. This is a direct comparison with existing

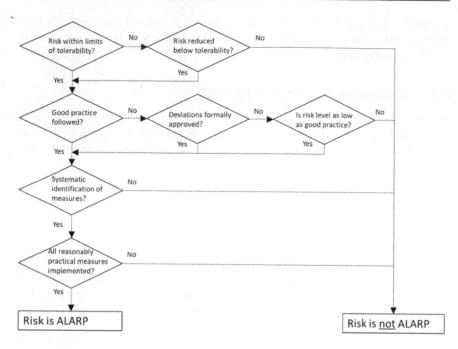

Figure 6.3 Complying with the ALARP principle (based on HSE UK 2001).

technology applied for similar purposes. This is a principle that is applied by IMO, although under a different name, and it is discussed in Section 6.8.

– MEM (Minimum Endogenous Mortality) states that new systems that are introduced should not significantly increase the total "technological risk" that the population is exposed to. Technological risk includes risk related to work, transport, leisure activities, etc., but not diseases and lifestyle risk. "Significant" is defined as an increase in individual risk greater than $1 \cdot 10^{-5}$ per year.

Another important principle is the Precautionary principle. This was originally introduced in relation to environmental risk and states that "where there are threats of serious or irreversible damage, lack of scientific certainty shall not be used as a reason for postponing cost-effective measures to prevent environmental degradation". This principle focuses on the uncertainty about future negative outcomes. In some situations, we only suspect that there may be negative consequences of our actions and decisions but cannot know for certain. The principle states that even if we do not know for certain whether there will be negative effects, we should still try to reduce risk, if cost-effective measures are available.

An example that has been much discussed in recent years is climate change and whether this is the result of natural variation or human activity. It is very hard to prove conclusively that climate change is man-made even if more and more research indicating that this is the case is available. If the Precautionary principle is followed, cost-effective measures to reduce the effect should however be implemented.

6.7 QUALITATIVE RISK ACCEPTANCE CRITERIA

It is probably most common to think about quantitative criteria when risk acceptance criteria are being discussed. However, criteria may also be expressed using various qualitative formulations. These do not explicitly say anything about the risk level, but they will still impose limitations on design and operation that implicitly also affect the risk level. A couple of examples are (based among others on MSC 72/16 (IMO 2000)):

- Ships should not pose risks that could be reasonably avoided.
- No single failure or error should lead to an accident.

It is noted that neither of these criteria requires that a quantitative risk analysis is performed, although it is necessary to identify hazards and failures.

Criteria may also be expressed as comparisons with existing technology or existing operations. In these cases, a quantitative risk analysis will normally be required. Some examples of how this can be formulated are as follows (based on MSC 72/16 (IMO 2000)):

- Ships should be as safe a workplace as manufacturing and process industries on land.
- Passenger ships should offer transport as safe as other means of transport.
- Risks from catastrophic accidents should be a small proportion of total risks.
- New ships should have risk at least as low as existing ships.

It may be noted that the last criterion is identical to the GAMAB criterion described in Section 6.6.

In none of the cases described above is it necessary to explicitly decide what risk level is acceptable. For the comparative criteria, we use the fact that there exist other technologies and activities in society already that are considered acceptable, therefore the risk associated with them is also (implicitly) accepted. However, we still need to calculate the risk to determine if the criteria are met and similarly we also need to know the risk for the activities that we are comparing with.

6.8 ACCEPTANCE CRITERIA IN THE MARITIME INDUSTRY

6.8.1 Criteria for risk to people

Traditionally, risk acceptance criteria have not been widely used in the maritime industry, mainly because risk-based approaches to safety management have been introduced fairly recently. If risk is not described or quantified explicitly, there is no need for criteria either. However, with the introduction of FSA, IMO also recognized that there was a need for risk acceptance criteria that could be used to assess current and proposed rules.

The FSA guidelines (IMO 2018) do not specify limits that shipowners and others need to follow, but provide some guidance on possible formulations and also on levels of risk that may be considered acceptable. It is suggested that criteria can be expressed both in terms of individual risk and societal risk, expressed as IRPA and FN curve, respectively.

6.8.1.1 Individual risk

For individual risk, it is recognized that there should be differences in what can be considered acceptable for different groups of people (IMO 2018). Further, it is also distinguished between existing ships and target levels for new ships, as a means of continuously driving improvements in risk level. Table 6.1 illustrates the proposals presented in the FSA guidelines (IMO 2018). Note that ALARP is the underlying principle.

It is noted that these are not absolute requirements, but if these levels are not met, it is an indication that cost-effective measures to reduce risk may be available. For new ships, this is also the case and the main requirement is that risk should be reduced to ALARP.

In a footnote to this table, it is stated that "for comprehensive FSA studies for new ships a more demanding target is appropriate." This is a clear expectation that further improvements should be aimed for, in particular when developing new ship types and new technology. This can have implications for, e.g., autonomous ships, new energy carriers (hydrogen, batteries), and other novel designs.

Table 6.1 Suggested tolerable risk levels and target levels for new ships

IRPA	Maximum tolerable risk (per year)	Target level for new ships (per year)
Crew members	10^{-3}	10^{-4}
Passengers	10^{-4}	10^{-5}
Public ashore	10^{-4}	10^{-5}
Negligible risk	10^{-6}	10^{-6}

6.8.1.2 Societal risk

For societal risk, IMO proposes to use FN curves to illustrate acceptance criteria, also here applying the ALARP principle. In the FN curve, two (normally straight) lines are drawn, representing the tolerable limit and the negligible limit. The lines can be determined by specifying a point and the slope of the curve:

- The point is called the anchor point and specifies one of the points on the line. This is specified as (N, F), e.g. $(1, 10^{-3})$ or $(100, 10^{-6})$. We need to specify an anchor point for each of the two lines.
- The slope of the line is called the aversion factor (ref the discussion of risk aversion in Section 6.2.3). A slope of −1 means that the acceptance criterion is risk neutral. This means if we multiply the frequency and the number of fatalities at any point on the line, we get the same value. A slope that is smaller than this, e.g. −2, means that we are risk averse in the sense described in Section 6.2.3. We are then less willing to accept accidents with large consequences, even if the product of the frequency and the consequence is the same as for less severe accidents.

As mentioned, the two lines are normally straight and parallel (i.e., both have the same, constant slope). However, the lines may also be divided into, e.g., two sections, where a slope of −1 (risk neutral) is used for the lower part of the curve and a more risk averse slope for the upper part. The two lines may also have different slopes (Figure 6.4).

Vanem (2012) proposes a set of risk acceptance criteria for some ship types, expressed in terms of anchor points:

Ship type	Intolerable risk	Negligible risk
Tankers	$(10, 2 \cdot 10^{-3})$	$(10, 2 \cdot 10^{-5})$
Bulk/ore carriers	$(10, 1 \cdot 10^{-3})$	$(10, 1 \cdot 10^{-5})$
Passenger RO-RO ships	$(10, 1 \cdot 10^{-2})$	$(10, 1 \cdot 10^{-4})$

These are all combined with aversion factors of −1. It is noted that for RO-RO ships, this implies that the limit for intolerable risk also passes through the point $(1, 1 \cdot 10^{-1})$. This means that fatal accident (with 1 or more fatalities) is within tolerable limits as long as the frequency is 0.1 per year. This is clearly quite a high risk level.

6.8.1.3 Converting injuries to fatalities

Risk acceptance criteria for risk to people are commonly related only to fatalities and not injuries. Commonly, quantitative risk analyses also focus

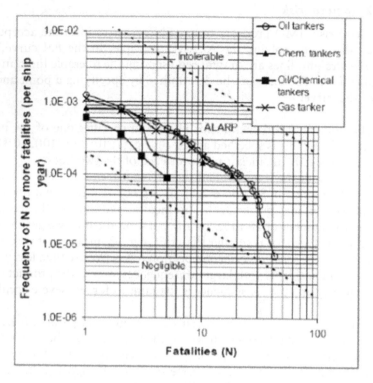

Figure 6.4 Illustration of risk acceptance criteria in FN curve (IMO 2018).

primarily on fatality risk. However, the FSA guidelines (IMO 2018) propose a way of combining injury and fatality risk into a total risk number, called risk equivalence concept. This is based on the following assumptions or rules:

- Ten severe injuries are considered to equate one fatality
- Ten minor injuries are considered to equate one severe injury

By applying these conversion factors, it is possible to calculate one fatality value that includes both fatalities and injuries. Definitions of what constitutes "severe" and "minor" injuries are however not provided.

6.8.1.4 The equivalence principle

IMO, in recognizing that prescriptive regulations can restrain innovation, has also introduced an equivalence principle (IMO 2013). IMO states:

> One approach to the approval of an alternative and/or equivalent design is to compare the innovative design to existing designs to demonstrate that the design has an equivalent level of safety.

This is similar to the GAMAB principle that was briefly mentioned earlier and is based on the technology principle. As long as it can be shown that a new solution has the same level of safety as already existing solutions, the new solution is acceptable. One challenge with this may be to decide what to compare with. This principle also poses some questions related to the criteria for individual risk, where an expectation for reduction in risk for new designs is expressed.

In the same document, IMO also states:

> In order to demonstrate an equivalent level of safety, functional requirements and performance criteria should be established for essential ship functions, which should be met by the alternative and/or equivalent design.

This approach means that a design can be approved even without doing risk assessment. A functional requirement for an evacuation system may, e.g., be that it should be able to evacuate all passengers within 60 minutes under specific weather conditions. As long as it can be shown that a new evacuation system meets these criteria, no risk assessment is required.

6.8.2 Environmental criteria

For the maritime industry, risk to the environment is primarily related to spills to sea, and in particular oil spill from tankers. The consequences associated with spills to the sea can be expressed in various ways:

- In the simplest form one uses the magnitude of the spill, often expressed in categories. This can be, e.g., <5 tons, 5–50 tons, 50–500 tons, and so on.
- The consequences may also be expressed in terms of the length of the coastline affected by a spill. This will be a function both of spill size, distance to shore, environmental conditions, and properties of the spilled oil. Categories are commonly used also here, e.g., <1 km, 1–5 km, 5–20 km, etc.
- A third option is to express the consequences in terms of the time it takes for the environment to recover after a spill. This will take into account even more factors than the previous measure, in particular what resources are affected and how vulnerable they are to spill. Categories that can be used can be, e.g., <1 month, 1 month–1 year, 1 year–10 years, and more than 10 years.

Risk can be expressed by assigning frequencies to each of the categories, based on the results from risk analysis. Similarly, risk acceptance criteria can be established by specifying limits on the frequencies for each category. An example is shown in Table 6.2.

Table 6.2 Illustration of formulation of environmental risk acceptance criteria

Category (spill size)	Upper acceptable limit (frequency per year per ship)
<5 tons	10^{-2} per year
5 t–50 tons	10^{-3} per year
50 t–500 tons	10^{-4} per year
500 t–5,000 tons	10^{-5} per year
>5,000 tons	10^{-6} per year

The level of risk that is considered acceptable is very often tied to the cost of the oil spill. Psarros et al. (2011) proposed to use the Cost of Averting a Tonne of Oil Spilt (CATS) and deciding acceptable risk based on cost-effectiveness considerations. The cost is a combination of the clean-up costs (for removing oil washed ashore) and the compensation costs (to fishermen and others affected by the spill).

This approach means that the valuation of the environment only is tied to compensation costs, i.e., the environment only has value if there are commercial interests in exploiting it. Natural beauty, bio-diversity, etc. will not have any value in itself in this way of determining acceptable risk.

Psarros et al. (2011) propose a value of USD 80.000 and an assurance factor of 1.5, taking into account that society may be willing to spend more resources on preventing spills instead of cleaning up afterwards.

Friis-Hansen and Ditlevsen (2003) have proposed to use a more comprehensive measure, the Nature Preservation Willingness Index (NPWI). This assumes that the environment has a value in itself and that this value increases with the wealth of the nation. This is based on the fact that increasing wealth normally also means more time available for humans to enjoy a clean or unaffected environment.

6.8.3 Criteria for other consequences

It may also be relevant to establish risk acceptance criteria for other types of consequences, e.g., economical loss or damage to reputation. In this case, acceptance is in most cases purely a question of cost-effectiveness, i.e., as long as the cost of reducing risk is greater than the (statistical) benefit of reduced risk, risk is accepted as is.

It may be worth noting that in some cases, the economic consequences may be considered so large that risk-based criteria are not adequate. This may be the case if the consequences are so large that the company cannot tolerate it without going bankrupt. Even if the statistically expected losses (i.e., the risk) are low, the consequences are so large that it may still be concluded that this is an unacceptable risk to take. The decision criterion is then related to maximum losses rather than risk.

REFERENCES

Bjørnskau, T. (2004). *Safety in transport – perceptions of safety when using different means of transport (in Norwegian)*. Trygghet i transport : oppfatninger av trygghet ved bruk av ulike transportmidler. Report 702/2004.Transport økonomisk institutt, Oslo.

Court of Appeal (1949). *Edwards vs National Coal Board*. https://www.xperthr. co.uk/law-reports/edwards-v-national-coal-board/49729/

Fischhoff, B., Slovic, P., Lichtenstein, S., Read, S., & Combs, B. (1978). How safe is safe enough? A psychometric study of attitudes towards technological risks and benefits. *Policy sciences, 9*(2), 127–152.

Friis-Hansen, P., & Ditlevsen, O. (2003). Nature preservation acceptance model applied to tanker oil spill simulations. *Structural Safety, 25*(1), 1–34.

HSE. (2001). *Reducing risks, protecting people – HSE's decision-making process.* https://www.hse.gov.uk/risk/theory/r2p2.pdf

IEC. (2002). *Railway applications: Specification and demonstration of reliability, availability, maintainability and safety (RAMS)*. https://standards.global-spec.com/std/494463/IEC%2062278

IMO. (2000). *Formal safety assessment: Decision parameters including risk acceptance criteria*. http://research.dnv.com/skj/FsaLsaBc/MSC72-16.pdf

IMO. (2013). *Guidelines for the approval of alternatives and equivalents as provided for in various IMO instruments*. https://www.mardep.gov.hk/en/msnote/pdf/msin1339anx1.pdf

IMO. (2018). *Revised guidelines for Formal Safety Assessment (FSA) for use in the IMO rule-making process*. https://wwwcdn.imo.org/localresources/en/Our Work/HumanElement/Documents/MSC-MEPC.2-Circ.12-Rev.2%20-%20 Revised%20Guidelines%20For%20Formal%20Safety%20Assessment%20 (Fsa)For%20In%20The%20Imo%20Rule-Making%20 Proces...%20(Secretariat).pdf

Psarros, G., Skjong, R., & Vanem, E. (2011). Risk acceptance criterion for tanker oil spill risk reduction measures. *Marine Pollution Bulletin, 62*(1), 116–127.

Slovic, P., Fischhoff, B., & Lichtenstein, S. (2000a). Cognitive processes and societal risk taking. In Slovic, P. (Ed.): *The perception of risk* (pp. 32–50). Earthscan, London.

Slovic, P., Fischhoff, B., & Lichtenstein, S (2000b). Rating the risks. In Slovic, P. (Ed.): *The perception of risk* (pp. 104–120). Earthscan, London.

Vanem, E. (2012). Ethics and fundamental principles of risk acceptance criteria. *Safety Science, 50*(4), 958–967.

Chapter 7

Human and organizational factors

7.1 INTRODUCTION

It is often claimed that human errors are the cause of at least 80% of maritime accidents (Coraddu et al., 2020). Although certain functions have been automated and there is an ongoing development toward autonomous shipping, a ship is still largely a human-controlled system. If we want to improve safety in the maritime industry, it is therefore also obvious that we have to understand how and why human errors occur, and also what human errors are. The concept of human error will be discussed in Chapter 11, but in this chapter, we will look more closely at some of the key factors that influence human performance and thus also human error.

Human and organizational factors are important for safety for several reasons:

- Human errors are very often mentioned as the most important cause of accidents
- Human factors (ergonomics) are a source of work-related illnesses/injuries
- Their importance in occupational accidents
- Human and organizational factors are increasingly identified as contributing causes to major accidents.

In this chapter, this topic is mainly explored to provide an understanding of some of the key human and organizational factors relevant in maritime transport and in what ways they can influence risk. Modeling of these factors in risk analysis is addressed in the chapter on Human Reliability Analysis (Chapter 11).

Before moving on to specific factors, it may be useful to look briefly at the terminology. There are two key concepts that sometimes are used more or less synonymously, namely ergonomics and human factors. In addition, we may also discuss the concept of human and organizational factors, as it is used in the title of the chapter. A simple illustration of how these concepts can be related to each other is shown in Figure 7.1.

DOI: 10.4324/9781003055464-7

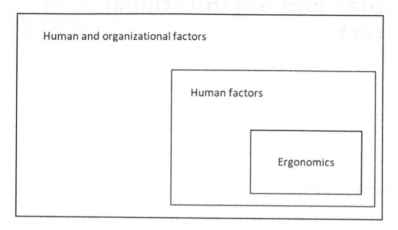

Figure 7.1 Relationship between key terms.

In ergonomics, particular emphasis is put on the design of displays, controls, and the workplace. The human physical dimensions (anthropometry) and our capacity with respect to sensing and control ability are especially taken into consideration. The physical aspects of the environment such as climate, noise, and vibration are also a concern.

Human factors focus primarily on the work situation assessed in the light of psychological factors. A key aspect is the relation between the job or task requirements and the human capacity. Factors like mental capacity to process information, motivation, and interaction with colleagues have to be taken into consideration. The focus is thus somewhat wider than ergonomics, and with more focus on internal processes within the human brain such as sensing, perception, cognition, and response (action and communication).

Human and organizational factors usually have an even wider interpretation, including not only personal but also wider organizational and structural factors that influence our performance. Factors like organizational structure, roles and responsibilities, management and supervision, and safety culture are often brought in as part of this term.

In this chapter, some selected human and organizational factors will be covered, focusing on factors that have been studied and that often are considered to be among the most important. It should be underlined that this is in no way an attempt of covering this field in all its details. What we are concerned about are primarily factors that can influence human performance, either positively or negatively. Human performance will in turn influence the probability that we make errors or whether we are able to perform the tasks that we have been assigned to do in a satisfactory way. Human and organizational factors thus influence the probability of human error.

7.2 HUMAN AND ORGANIZATIONAL FACTORS DATA

As a starting point a brief survey of data that indicate the presence or frequency of human and organizational factors (HOF) is presented.

Schröder-Hinrichs et al. (2013) did an investigation of how IMO-MSC (Maritime Safety Committee) dealt with HOF issues primarily after serious and catastrophic accidents. The main approach was to undertake a text analysis of IMO documents and code accident factors. A similar approach was undertaken for scientific papers published by *Journal of Navigation* and *Maritime Policy & Management*. The findings on a factors group level are shown in Table 7.1. The grouping of factors is environment, organization, personnel, and technical factors. The scatter of data indicates bias both in terms of professional background and the objective of the publications. The lowest figure for HOF is 21.9% (*Journal of Navigation*) and the highest is 54.2%, but still considerably lower than the often cited 80%. The moderate focus on HOF in the professional literature may be seen as a paradox in view the broad acceptance of these factors as a major problem in maritime risk.

Coraddu et al. (2020) give an overview of dominating HOFs based on an analysis of 91 accident cases from MAIB (Maritime Accident Investigation Branch) covering cases related to operation in UK waters in the period of 2011–2016. They ranked the factors both in terms of frequency and in strength of ability to predict an accident category. The latter means a factors ability to understand a particular accident and can be identified with so-called data-driven methods. The authors applied two alternative approaches namely RF (*Random Forests*) and BSVM-BK (*Support Vector*

Table 7.1 Summary of coding results for MSC documents and journal manuscripts

Factor group	Factors	Journal of Navigation (%)	Maritime Policy & Management (%)	IMO-MSC (%)
Environmental context	Weather, ergonomics, finance, market, flag, standards, port state	25.2	35.8	37.7
Organizational infrastructure	Human resources, technical resources, culture, supervision, procedures, operations	5.0	26.1	30.7
Personnel subsystem	Cognitive factors, physiological state, crew interaction, personal readiness	16.9	28.1	11.4
Technical subsystem	Automation, design, lack of equipment, unsuitable equipment, lack of good equipment	52.9	10.0	20.2
Total		100	100	100

Source: Schröder-Hinrichs et al. (2013).

Table 7.2 Ranking of HOF factors

Human & Organizational factors	By frequency	Ranked by RF[1]	Ranked by M-SVM-BK[2]
Inadequate work preparation	4	1	9
Lack of knowledge	1	2	10
Training ignored		3	2
Emergency training program		4	1
Contingency plans not updated		5	4
Inadequate work methods	6	6	8
Improper performance of maintenance and repair		7	5
Anthropometric factors		8	3
Safety awareness, cutting corners	3	9	
Lack of skill	2	10	
Inadequate briefing, instruction	10		6
Inadequate manning			7
LTA physical/physiological capability	5		
Inadequate procedures and checklists	7		
Lack of motivation/morale	8		
Improper supervisory example	9		

Adapted from Coraddu et al. (2020).

Machines with Boolean Kernels). The analysis revealed that ranking based on frequency did not agree with ranking based on ability to predict an accident. The three sets of rankings are shown in Table 7.2. It can be concluded that four of the frequency rankings are not identified by any of the prediction rankings (RF and BSVM-BK). These factors were as follows:

- LTA (Less Than Adequate) physical/physiological capability
- Inadequate procedures and checklists
- Lack of motivation/morale
- Improper supervisory example

These factors are often in focus in the discussion of causal factors. The prediction rankings agree on eight factors but give them different priorities.

7.3 CLASSIFICATION OF HUMAN AND ORGANIZATIONAL FACTORS

Identifying human and organizational factors that can influence human performance and thus the risk is a complex task, requiring understanding of the whole sociotechnical system that the humans are working in. This requires understanding of technology, humans, and organizations and involves many professional disciplines.

A help in identifying factors is various classification schemes that have been proposed. A classification scheme will typically provide categorizations of factors, as well as comprehensive lists of factors that may be relevant. Classification schemes are also useful in accident investigations because they help the investigators to identify and classify causes of accidents. In turn, this will also be useful for doing statistical analysis of accidents, to determine the most commonly occurring causes. Without a classification scheme, the description of causes and the choice of causes will be up to each individual investigator and statistical analysis afterwards is often time-consuming and can be quite hard.

One of the earliest systematic lists of factors was proposed as part of the THERP methodology (Swain & Guttmann, 1983) for quantifying human reliability. They prepared an extensive list of so-called Performance Shaping Factors (PSFs) that was used to modify human error probabilities based on the status of the factors. The PSFs were divided into three main groups and subgroups:

- External
  - Situational – general to one or more jobs in a work situation, e.g. quality of work environment, working hours, and organizational structure
  - Job and task instructions, e.g., procedures and work methods
  - Task and equipment characteristics – specific to tasks in a job, e.g., complexity, task repetitiveness, and task criticality
- Stressor
  - Psychological – directly affect mental stress, e.g., task speed, task load, and distractions
  - Physiological – directly affect physical stress, e.g., fatigue, pain, and hunger/thirst
- Internal
  - Organismic factors – Characteristics of people, e.g., training, personality, and motivation

A more recent taxonomy is HFACS (Human Factors Analysis and Classification System) that was originally proposed by Shappell and Wiegmann (2000). HFACS was based on the theoretical framework of Reason (1990) and has gained widespread recognition and use. A special version adapted for the maritime industry has also been developed (Chen et al., 2013). This is illustrated in Figure 7.2. HFACS-MA is divided into five levels, with similarities to Rasmussens sociotechnical hierarchy (Rasmussen, 1997). From the bottom, the four levels are as follows:

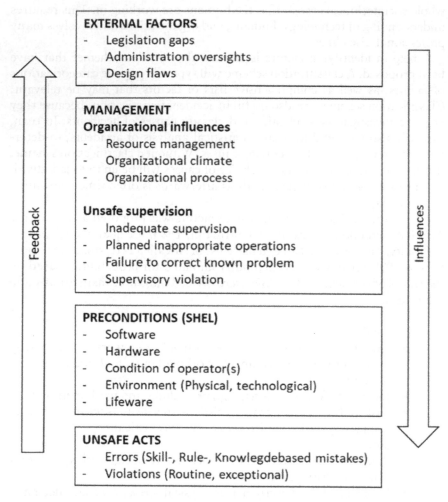

Figure 7.2 HFACS-MA (adapted from Chen et al., 2013).

- Unsafe Acts: This is split into Errors and Violations and is based on Reasons classification. Errors are split into Skill-based, Rule-based, and Knowledge-based errors, and Violations are split on Routine and Exceptional.
- Preconditions: This is concerned with the immediate context where unsafe acts occur, covering the technical systems (software and hardware), the environment, and other people, as well as the condition of the operator(s).
- Management: This level is split into two separate parts; Unsafe Supervision and Organizational Influences. Unsafe supervision is in

principle the same as unsafe acts, but errors and violations on the super-visory level rather than on the execution level. Organizational influences are split on resource management, organizational climate, and organiza-tional process.
- External factors: This is the top level and includes gaps in legislation, design flaws, and administration oversights.

HFACS-MA is based on the original HFACS framework but has been mod-ified to fit the maritime domain and in particular to match terminology and definitions used by IMO.

It may be noted that the list of PSFs and the classification scheme pro-posed in HFACS-MA have been developed for different purposes. HFACS is mainly for classifying errors and mistakes at different levels of the socio-technical hierarchy, while the PSFs influence the performance of humans and thus the probability of errors and mistakes being made. At the same time, the HFACS structure also implies that the higher levels in the model influence the lower levels. There is therefore some overlap between them. "Job supervision" is, e.g., a PSF, but we can also see that "Unsafe supervi-sion" is a level in HFACS-MA.

EMSA, the European Maritime Safety Agency, has developed a taxon-omy for reporting of accidents and incidents (EMSA, 2016) to the EMCIP (European Marine Casualty Information Platform). EMCIP is a database and contains information about all aspects of accidents in European waters and with European flagged ships, including information about causes.

7.4 HUMAN FACTORS INFLUENCING ACCIDENTS

The lack of a commonly accepted classification scheme causes problems for doing statistical analysis of the causes of accidents. Available databases often do not contain information about causes at all (and in particular not human causes) and the only way to find out more about this is to do a thor-ough review of detailed accident reports, when such reports are available. Serious accidents will be investigated by various national authorities, like NTSB (National Transportation Safety Board) in the USA, MAIB (Marine Accident Investigation Branch) in the UK, and NSIA (Norwegian Safety Investigation Authority) in Norway. Reports are publicly available and can be studied in detail. It will however typically require a lot of work to do this.

EMSA (2020) publishes an annual report providing an overview of acci-dents and incidents in the preceding five years. In this report, they also provide some information about causes of accidents. Figure 7.3 shows a breakdown on the most common contributing factors.

The largest proportion is "Shore management" with 27%, but this covers a whole range of factors like operations management, safety management,

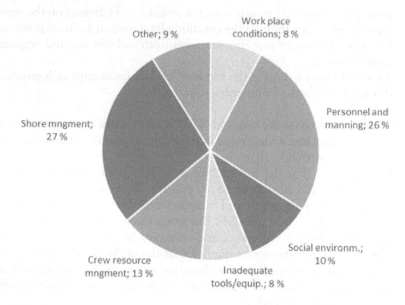

Figure 7.3 Contributing factors to accidents and incidents (EMSA, 2020).

design, regulatory activities and emergency preparedness. The second largest proportion is "Personnel and manning", with 26%. According to the EMCIP taxonomy (EMSA, 2016), this covers a range of more specific factors. Personnel includes factors related to mental, physical, and psychological state and capability, knowledge and skills, and motivation. Manning includes factors like manning level, too high or too low workload, job satisfaction, and long working hours. It is also noted that crew resource management has a relatively high contribution, with 13% of the total.

It may also be interesting to look at safety recommendations from investigations performed. This is also a fairly good indicator of causes, since recommendations typically will address the causes of accidents and incidents that have occurred. In the period of 2014–2019, EMSA issued a total of 2,206 safety recommendations resulting from investigations. A summary is shown in Table 7.3.

First, it is interesting to note that 88% were related to "Human Factors", "Other Procedures", and "Ship-Related Procedures". All of these would be within the scope of this chapter. If we look specifically at "Human Factors", the three most common subcategories can all be classified as organizational issues. It may be argued that "Training, skills and experience" is a personal factor, but at the same time, it is the responsibility of the organization to ensure that personnel are qualified for the positions they have, either by employing people with the right background or provide training to them.

Table 7.3 Safety recommendations issued by EMSA

Focus area	No of recomm.	%	Most common subcategories
Human factors	278	13	Training; skills; experience (48%), Management (22%), Company and organization (10%)
Other procedures	263	12	Compliance (37%), Other/unspecified (30%), Port and terminal facilities (14%)
Ship-related procedures	1,410	64	Operation (42%), Compliance (15%), Information dissemination (12%)
Ship structure and equipment	209	9	Ship equipment/system (51%), Other/ unspecified (24%), Safety of navigation (6%)
Shore and water equipment	46	2	Almost evenly distributed on Navigation aids, Study/review, VTS, and Other equipment
Total	2,206	100	

EMSA (2020).

7.5 FATIGUE

Fatigue can be both psychological and physiological. Psychological fatigue can also be termed mental fatigue while physiological fatigue is related to bodily fatigue and tiredness in muscles. Both will however tend to lead to the same effects on human performance, causing decline in alertness, concentration, and motivation (Akhtar & Utne, 2014). In practice, this will typically be expressed as general weariness, slower reactions, and reduced vigilance.

Rumawas (2016) looked at human factors on offshore supply vessels in the North Sea. This was not a study of accidents, but how various factors influenced the performance of the crew. He found that slamming, noise, and ship motion were the factors that lead to disturbed sleep. Slamming potentially leads to both noise, sharp motions, and vibrations. It is thus not surprising that this is at the top of the list. Further, slamming and motions also disturbed watches while noise was less of a problem.

Fatigue is a significant problem for seafarers, and this has been studied by Smith et al. (2006). They found that there were a number of factors that partly are specific for the maritime industry that contributed to fatigue:

- Poor quality sleep
- Negative environmental factors
- High job demands
- Stress
- Physical work hazards
- Long working hours
- Low job support
- Operational patterns (port visits)

Operation of ships requires 24-hour operation and will involve shiftwork and work during nighttime. Humans have a natural need for sleeping during nighttime and means that the performance level is best during daytime. Lack of sleep or low-quality sleep during daytime is therefore a problem associated with shiftwork at sea. It is noted that these factors listed may interact with each other, e.g., poor quality sleep can be a result of negative environmental factors like noise, vibration, and motion.

Fatigue will typically increase significantly during the first week of a tour on the ship. It will typically also take 1–2 weeks to recover from fatigue after starting on a period of leave.

In the EU project HORIZON, fatigue was investigated, and some results are presented by Barnett et al. (2012). A particular focus in this project was the effect of watches on sleepiness. It was found that night watches are the largest problem and that the effect was particularly strong at the end of the night watch from midnight to 06:00 in the morning. It was also found that about 20% of the participants in the study had short period of sleep during the watch. There is also a difference in fatigue for different watch regimes. Working 6 hours followed by 6 hours off is more tiring than working 4 hours and then having 8 hours off. Importantly, the tiredness is more dependent on the quality of the sleep in the rest periods than the workload on watch.

Akhtar and Utne (2014) have studied the effect of fatigue on grounding. A Bayesian network was developed based on accident reports, and they found that if bridge personnel were subject to fatigue, the probability of grounding increased by 23%. The study also found that the probability of fatigue is strongly influenced by organizational and middle-management factors. In the best case, the probability that the bridge team would be fatigued was as low as 1% while in the worst case it was 77%.

IMO has published guidelines on fatigue that explain causes and effects in general and specifically for different groups of seafarers. They define fatigue as follows:

> A reduction in physical and/or mental capability as the result of physical, mental or emotional exertion which may impair nearly all physical abilities including strength, speed, reaction time, coordination, decision making or balance.

It may be argued that this is not really a definition of fatigue as such but a description of what causes fatigue and what the effects are. However, it may be useful for our purpose in highlighting the importance of this topic.

The guidelines also point out some of the unique features of the maritime industry that may contribute to fatigue as a greater problem compared to other industries. Among these are the fact that the worker is away from home for extended periods of time, unable to get away from the workplace sometimes for weeks. This makes it difficult to separate work and recreation. Further, this means that workers from different nationalities and with different background not just have to work together but also live together. The workplace is also moving and is exposed to changing environmental conditions. All of these features, in particular when combined, can strengthen fatigue in shipping compared to most other industries.

The guidelines distinguish between four categories of factors that influence fatigue. Only some examples of factors within each category are shown in the following summary:

- Crew-specific factors – These are factors that are individual, from one person to another. One person may be strongly affected by these factors while another is not affected at all. Example are sleep and rest, biological clock/circadian rhythms, psychological and emotional factors such as stress, health, use of alcohol/drugs/caffeine, age, etc.
- Management factors – These are partly what we have called organizational factors and partly operational factors. Among the organizational factors are staffing policies, paperwork requirements, economics, schedules, overtime, company culture, rules and regulations, etc. Operational factors include frequency of port calls, time between ports, weather and traffic along the route, and nature of work while in port.
- Ship-specific factors – These are factors that are related to the ship design and that may vary significantly from one ship to another. These factors can affect workload (e.g., automation, equipment reliability, age of ship), some affect sleep (e.g., noise, vibration, accommodation spaces) while a third group affect stress levels (partly the same factors that affect sleep).
- Environmental factors – These can be external to the ship (associated with weather conditions or also port conditions) or internal to the ship. The internal factors will partly be related to ship design, but also ship operations.

The effects of fatigue are in general to reduce the performance of the person that is fatigued. One of the problems is that an individual is not good at judging whether one is fatigued or not. Some concrete examples of how our performance is affected are described in the following.

- Our attention is reduced, and we are more likely to miss or forget important information. A typical example is that one or more steps are omitted when we are performing a sequence of tasks.
- We are more likely to select the strategy that requires the least work, instead of the strategy that is most likely to succeed or be safe. In other words, our willingness to take risk increases.
- The process of maintaining and updating our situational awareness can be slowed down, because we are slower at perceiving, interpreting, and understanding signals and information from the surroundings.

– Similarly, fatigue will also affect our ability to solve problems.

Grech et al. (2008) provide some recommendations for how to organize shiftwork based on findings of Kroemer and Grandjean (1997):

- Have short shift rotations.
- As far as possible use personnel between 25 and 50 years.
- Employ primarily healthy and emotional stable persons.
- Avoid continuous shiftwork.
- Ensure adequate breaks per shift for meals and rest.

7.6 PHYSICAL WORKING ENVIRONMENT

The following physical work climate factors are thought to be relevant for the performance of the crew:

- Thermal climate
- Noise
- Vibration
- Illumination

These factors were studied in depth by Ivergård (1978) on Swedish vessels. The results shall be summarized in the following section but it should be kept in mind that the findings may not necessarily be fully representative for the conditions today.

7.6.1 Thermal climate

The thermal climate is a function of a number of factors: Temperature, humidity, air speed (circulation), and heat radiation. The subjectively experienced climate must further be seen in relation to the workload and clothing of the individual. The thermal climate represents a challenge on vessel as:

- They operate in all climate zones
- The conditions are varying highly in the different sections of the vessel
- The crew members are exposed to different thermal stress
- The crew can be subject to extreme thermal loads in emergency situations

In discussing thermal stress, it is necessary to make a distinction between the different sections of the vessel. Table 7.4 gives a summary of the problems related to thermal control in the various sections of a vessel.

Table 7.4 Thermal climate factors on vessels

Section	Situation	Solution
Living quarters	Reasonable control.	Ventilation and air condition is standard.
Navigation bridge	Thermal control up to 22–28 °C. Heat sources: Large windowpanes, electronic equipment, and open doors.	Ventilation and air condition is standard.
Galley	A number of heat sources: Stove etc. Long periods above thermal comfort criteria.	Improved isolation of heat sources.
Engine rooms	Temperature often 10°–20° higher than outside temperature. Heat radiation from hot surfaces. Heavy repair and maintenance work. High relative humidity in smaller, confined spaces. Work in engine rooms will in general take place well outside thermal comfort criteria. Under winter conditions large temperature variations within the same space.	Air cooling of engines.
Cargo tanks	Inspections and final cleaning operations may be stressing in hot outside climate. Protective clothing may contribute to the stress.	Cleaning by permanently installed equipment.
On deck	Both high and low temperatures depending on time of the year and geographical latitude. Sun radiation. Effect of wind (air speed) under winter conditions.	Clothing and protection.

Seppanen et al. (2006) studied the effect of temperature on task performance in an office environment. This is not directly comparable to tasks on a ship, although bridge tasks can be compared to some extent. They found that optimal performance was observed at 22°C, with no significant effect observed in the range of 21–25°C. For lower or higher temperatures, a decrease in performance could be observed. At 30°C, a reduction in performance of around 9% was found.

Pilcher et al. (2002) did a meta-study of earlier investigations and found similar results. Temperatures above 32°C and below 10°C resulted in performance decrease of around 15%. The effect was however dependent on the time of exposure and the duration of the task.

The experience of well-being and task performance is a function of the climate. The body will be in thermal balance if the heat produced by

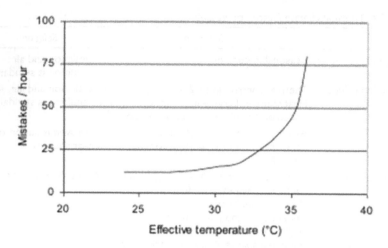

Figure 7.4 Monitoring mistakes as a function of effective temperature (Mackworth, 1946).

metabolism equates the heat lost through evaporation, radiation, convection, and by work accomplished. As evaporative heat loss is a key source during most conditions, humidity also plays a key role. In practice, we therefore apply the *Effective Temperature* concept (ET) that is a function of both dry-bulb temperature and relative humidity. The body is in balance with respect to heat loss at a dry-bulb temperature of 25°C and a relative humidity of 50% (McCormick, 1976).

Mackworth (1946) made one of the early studies on the temperature effect on monitoring tasks as shown in Figure 7.4. It can be seen that the number of monitoring mistakes starts to increase sharply above an effective temperature of 30°C. This corresponds to the upper limit that is experienced as comfortable.

Obviously, the question of acceptable temperature is also related to the duration of exposure. As shown in Figure 7.5, the duration of unimpaired mental performance is decreasing quickly as the effective temperature rises above 30°C (Wing, 1965).

Finally, it should be mentioned that there seems to be a unique combined effect of warmth and sleep loss (Pepler, 1959). Observed phenomena are reduced performance in tracking tasks and gaps in serial responses. Simply stated the following mechanisms are seen: "Warmth reduces accuracy", whereas "sleep loss reduces activity".

7.6.2 Noise

The primary noise sources onboard are: main engine, propeller, auxiliary engines, and engine ventilation. The noise (or unwanted sound) is measured

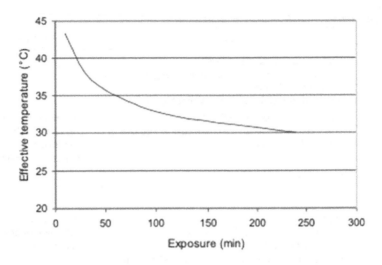

Figure 7.5 Upper limit for unimpaired mental performance (Wing, 1965).

in terms of sound intensity and expressed by dB(A). The level of noise varies for each section of a vessel as indicated in Table 7.5.

A particular kind of noise is infrasound. Infrasound is acoustic waves with frequency below 20 Hz. Infrasound is not audible by the human ear and is as such not experienced as noise. However, it has been found that for high intensities, infrasound has negative effects on the human in terms of reduced well-being, tiredness, and increased reaction time.

The relation between infrasound and possible negative effect is still only partly understood. It has been suspected that even higher intensities (above 100 dB) may have adverse effects on control of balance, disturbance of vision, and choking. Typical ranges for infrasound measured on vessels are: 4 Hz–55 dB to 16 Hz–95 dB. The fact that infrasound is present both on the bridge and in the engine control room may pose a problem for safe operation.

Szalma and Hancock (2011) have done a meta-study of the effect of noise on human performance. Some of their findings are that intermittent noise is more disruptive than continuous noise. On ships, we may have both (e.g., waves vs engine). The disruption is strongest for cognitive tasks and accuracy in performing tasks is more affected than the speed of performing the task. The probability of making errors can in other words increase. Somewhat surprising is perhaps that the intensity of the noise is not a very important factor.

7.6.3 Vibration

Vibration is a phenomenon that can cause both discomfort and be a health hazard and can thus also affect human performance. Vibration can be defined by the magnitude (the oscillation or amplitude) and by the frequency

Table 7.5 Noise on vessels

Section	Situation	Solution
Living quarters	Highest values: dB(A) = 58–70 Depends mainly on the distance to engine rooms. Variation with ship type. 28% experienced the noise as troublesome.	
Navigation bridge	Highest values: dB(A) = 65–73 Some effect on direct communication and use of internal communication equipment.	Lowest noise levels on vessels with bridge in the fore part (passenger, ro-ro)
Galley	Mean value: dB(A) = 71–77 Recommended value 65 dB(A) is exceeded due to the background noise.	
Engine rooms	Highest values: dB(A) = 93–113 Recommended value: 100 dB(A). Factors: Power, engine type. Risk of physiological damage (reduced hearing). 50% of engine personnel report the noise as troublesome.	Wear ear protection. Sectioning of engine room

(the repetition rate). On a ship, both of these factors can vary considerable depending on the source. The magnitude can be measured by displacement or velocity, but in practice it is often acceleration that is used, and standards also use acceleration to specify requirements. Vibration can further be characterized as being stochastic (e.g., from waves) or constant (e.g., from engine). Studies have shown that discomfort will grow with increasing amplitude, and it will also depend on the frequency (Griffin & Erdreich, 1991).

The main sources of vibration are the main engine and the propeller. The induced vibrations are further transmitted through the steel hull and deck houses. High superstructures are further subject to resonance phenomena. The same is experienced in the aft part as an effect of hogging-sagging movement of the hull girder and giving very low frequency resonance (0.5 Hz).

Vibrations are seen by a significant part of the crew as one of the most disturbing environmental factors onboard. This is partly explained by the fact that it is experienced both in work and off-duty. This means that persons affected are given no opportunity for restitution during free hours. Vibration is expressed by acceleration (m/s^2) in selected frequency bands (Hz). Vibrations are usually measured for a range from 0.5 to 125 Hz.

Measurements of vibrations by Ivergård (1978) can be summarized as follows:

- Navigation bridge:
  Most vessels have a vibration level below recommended values. Exception is tankers. However, crew still reports some uncomfort.

- Engine room:
  Most observations are laying between the comfort and performance curves. Although the comfort limit is not exceeded personnel reports problems due to the reduced opportunity for restitution during the free watch.
- Cabin area:
  Vibration level coincides with comfort line in critical area: 8–16 Hz. Confirms that restitution becomes a problem.

Grether (1971) looked at the effect of vibration on human performance in general (not specifically for maritime applications) and found that vibration influences visual acuity (i.e. our ability to discern shapes and details of what we see). Not surprising, tasks requiring precise movements also become more difficult. On the other hand, reaction time, monitoring, and pattern recognition are not affected.

A more recent study is a meta-analysis performed by Conway et al. (2006). First and foremost, they concluded that there were relatively few studies that provide information about the effect of vibration, and they partly attributed this to the fact that "everybody knows" that vibration will influence human performance. What they found was that "perceptual tasks suffered the largest decrement, followed by continuous and discrete fine motor tasks." This confirms what was noted earlier, that vision and precise movements will be most affected.

There is an ISO standard (ISO 9996, 1996) that describes the range of effects that can influence human activity and performance when exposed to vibration. The standard does not indicate the strength of the influence but provides a qualitative description. There is also an ISO standard (ISO 2631-1, 1997) that provides a method for quantifying whole-body vibration with respect to among others comfort. However, it does not specify any limits for acceptable vibration.

7.7 MOTION

One of the classical torments of the life at sea is the so-called *seasickness* that may affect both passengers and crew. The more correct term is motion sickness as it is present in different transport modes and certain other situations. Motion sickness may be experienced during anything from riding a camel, operating a microfiche reader to viewing an IMAX movie. How strongly one may be affected is differing, but the effects usually include stomach discomfort, nausea, drowsiness, and vomiting to mention some.

The problem has been surveyed by Stevens and Parsons (2002) in a maritime context. As the authors point out, one of the consequences of seasickness is less motivation and concern for the safety critical tasks onboard. However, for the majority of persons there will be an adaption to the sea motion over time so that the sickness fades away after a few days. There is also some comfort in the fact that the susceptibility is usually decreasing with increasing age.

The effect of motion is sometimes appearing in the form of the *sopite syndrome* that is less dramatic and usually only experienced as drowsiness and mental depression. This effect may however also be safety critical. It will have an effect on performance and is not always evident to the subject that experiences it. One may therefore continue to work and believe that he or she is performing as usual.

Various explanations of motion sickness have been proposed, but contemporary theory points to the conflict or mismatch between the organs that sense motion:

- *Vestibular* system in the inner ear: semicircular canals that respond to rotation and otholits that detect translational forces.
- The *eyes* (vision system) that detect relative motion between the head and the environment as the result of motion of either or both.
- The *proprioceptive* (somato sensory) system involves sensors in body joints and muscles that detect movement or forces.

The theory assumes that under normal situations the three systems detect the movements in an unambiguous way. However, under certain conditions the senses give conflicting signals that lead to motion sickness. A typical maritime scenario is experiencing sea motion inside the vessel without visual reference to sea, horizon, or land masses. In this case the vestibular detects motion in the *absence* of visual reference. Watching the waves from the ship may give rise to *conflict* between vision and vestibular system. The opposite conflict can be observed in ship simulators, in particular when simulating maneuvering of high-speed craft. In this situation, the eyes will observe movements, but the vestibular system and the body will not experience these movements. The conflict can cause rapid onset of seasickness.

Extensive experiments in motion simulators found that subjects were primarily sensitive to vertical motion (heave) and that maximum sensitivity occurred at a frequency of 0.167 Hz (Griffin, 1990). Given the fact that the principal vertical frequency in the sea motion specter is 0.2 Hz, the occurrence of seasickness is understandable. There are two methods for estimation of motion sickness: Motion Sickness Incidence (MSI) and Vomiting Incidence (VI). Both methods are outlined by Stevens and Parsons (2002).

There are a number of effects of seasickness:

- Drowsiness and apathy, leading to reduced motivation to perform assigned tasks

- Motion-Induced Fatigue (MIF), leading to reduced mental capacity and performance
- Reduced physical capacity
- Added energy expenditure to counterbalance motion
- Sliding, stumbling, and loss of balance
- Some interference with fine motor control tasks

The effect on cognitive tasks is more inconclusive as it has been difficult to isolate the effect of physical stress. Another problem is the fact that bridge tasks are involving a number of cognitive processes and skills which may be influenced differently by the sea motion (Wertheim, 1998).

In order to minimize the risk of seasickness it is necessary to establish operational criteria. Baitis et al. (1995) point out that it is not the roll angle in itself that limits the operation but rather the vertical motions and accelerations associated with them. Table 7.6 shows the criteria proposed by NATO (NATO STANAG 4154, 1997) which are based on both earlier and recent principles. An alternative set of criteria make a distinction between different ship types (NORDFORSK, 1987) as summarized in Table 7.7. It can however be concluded that the two sets of criteria agree fairly well.

Motion also has another effect that influences human performance or motion-induced interruptions (MII) (Rumawas, 2016). MIIs are all the interruptions that occur during the performance of a task and that are caused by motions of the vessel. This can be relevant for manual work, but

Table 7.6 Sea motion operability criteria

Motion sickness incidence (MSI)	20% of crew in 4 hours
Motion-induced interruption (MII)	I tip per minute
Roll amplitude	4.0° RMS
Pitch amplitude	1.5° RMS
Vertical acceleration	0.2 g RMS
Lateral acceleration	0.1 g RMS

NATO STANAG 4154 (1997).
RMS: Root mean square; g: acceleration of gravity.

Table 7.7 General bridge operability criteria for ships

	Merchant ships	Naval vessels	Fast small craft
Vertical acceleration (RMS)	0.15 g	0.2 g	0.275 g
Lateral acceleration (RMS)	0.12 g	0.1 g	0.1 g
Roll (RMS)	6.0°	4.0°	4.0°

NORDFORSK (1987).

also for other types of work, e.g., radar observation and reading instruments. Criteria for MII are expressed as number per minute and according to NATO (2000), recommended limits are 1 per minute.

7.8 VISION

A key task during navigation is to observe the surroundings to avoid hazardous situations. This task can be demanding on the officer on watch and the lookout, in particular during periods of darkness or reduced visibility. Key sources of information are as follows:

- The seaway, landscape, sea marks, and marine traffic (marine environment)
- The visual and aural displays on the bridge
- The behavior of the vessel (movement, acceleration, etc.)

7.8.1 Lookout

Although all vessels are equipped with radar, visual lookout is still an important method for assessing the position of the vessel in a restricted seaway and detecting and monitoring fixed objects and other traffic. Lookout is often a challenging task for various reasons. One factor is that at open sea there are few objects to observe. Prolonged observation of a uniform field such as calm sea, open sky, or fog can have both a psychological and a physiological cost.

The lack of changes in scenery and events occurring makes it hard to maintain vigilance. The lookout can easily be distracted by other activities/tasks and may also fall asleep due to boredom.

Physiologically, one may also experience *blanking out* after 10–20 minutes. This results in reduced motor coordination and ability to maintain balance. Another related phenomenon is *empty field myopia* that means that the eye accommodation is in a constant state of fluctuation due to the lack of objects to focus on. This problem may be solved by looking away on objects like masts or other objects on the vessel every 5 minutes.

There have also been extensive discussions about what is the best lookout strategy (Madison, 1974), either a slow rowing gaze or repeated fixations at widely separated positions. It seems that systematic search is most efficient for low contrast targets (during darkness or low visibility, objects that do not stand out clearly from the surroundings), while free search is best for high visibility targets.

There are also various illusions that can affect observations. Some well-known factors are as follows:

- Refraction – This is the phenomenon where the direction of the light is broken by passing through different media. This can easily be observed by looking at an object placed in water. During periods of rain, water may build up on the windows of the bridge and result in refraction. This results in a misinterpretation of the relative direction (bearing) of other vessels or objects.

- Fog and haze – In conditions of reduced visibility, objects appear to be smaller than they are, and this can be interpreted as if they are further away than they actually are. This may put the vessel at risk in approaching situations by reducing the time and distance to stop or change heading. A compounding factor is also the tendency to under-estimate the relative speed of other objects under marginal visual conditions.
- Texture – The texture of an object may be a cue about its distance from the observer. The more details we can observe, the closer we will assume that they are. Unusual objects with very few detailed features (as can be seen e.g. on modern naval vessels designed to minimize radar returns) may therefore appear to be further away than they are.
- Auto kinesis – This is a phenomenon that is relevant during night and may occur when looking continuously at a single light source against an otherwise dark background. The light will apparently move and sometimes in an oscillating fashion. Both light markers and lanterns of other vessels might be the source of this phenomenon. This may lead to misinterpretation of what the light source, e.g., if a fixed light appears to be moving.

7.8.2 Night vision

The ability to observe under night (dark) conditions is especially critical. Madison (1974) states that it takes in the order of 25 minutes to fully adapt to the darkness following exposure to sunlight or artificial light. Other sources mention even longer periods, up to 40 minutes (Lamb, 1990). This slow adaptation can be explained by the fact that the eye has two types of photoreceptors namely *rods* and *cones*:

- Cones react to higher light levels, observation of finer details, and perception of colors.
- Rods are associated with night vision and low-intensity light. They are relatively insensitive to red light which is in the lowest frequency of the visual spectrum. Regardless of frequency of stimulus objects are seen in shades of black and white.

It is hard to explain why adaptation to low light takes so long, but one possible explanation is that light exposure naturally changes with the onset of dusk which is a slow process. Before the invention of artificial lighting, slow adaptation was therefore acceptable (Lamb, 1990).

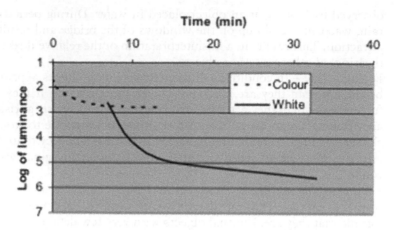

Figure 7.6 Time for night vision to be fully developed (Jayle et al., 1959).

Figure 7.6 shows that the dark adaption curve is discontinuous. The cones adapt in relatively few minutes, whereas the sensitivity of the rods is rapidly lost in high-level illumination and therefore needs longer, of the order 25 minutes, to adapt. This is of course crucial to our ability to detect other objects during night. After a shift changeover, it will take time for the new bridge personnel to adapt, but the same also applies if they have to leave the bridge to go to the toilet or if they need to turn on lights to, e.g., look at charts, make entries in logbook, or other tasks.

It is also a well-known fact that one should not stare directly on a dim spot under dark conditions in order to avoid the interference of the blind spot in the eye. Night light sensitivity is in fact greatest 15°–20° away from the center of vision.

The lookout function may also be degraded by *night blindness*. The existence or degree of night blindness may be unknown to the person itself as it is not apparent during daytime conditions. The source may be either psychological or physiological. Physiological factors that result in night blindness are dietary deficiencies (vitamin A), diabetes, glaucoma, or congenital (born with) night blindness. It has also been shown that aging is an important factor. It will develop from roughly the age of 40 and becomes pronounced after 50.

Another phenomenon that interferes with the lookout function is the so-called night myopia (nearsightedness) that is maintained as long as the eyes are night adapted. Farsighted persons may for practical reasons therefore see better without corrective glasses, whereas one with normal sight will improve night vision with a corrective lens of roughly –1.5 diopters. Jayle (1959) found that when night myopia was corrected the night vision threshold was improved by 50%.

Under low level illumination at night, we are dependent on the rods. The rods are insensitive to red light, and this has some advantages. "Pre-adaptation", e.g., before going on duty, is possible, by using red goggles before moving into a dark space.

The effect can also be used on the bridge. Instruments to be used during darkness can use red illumination, thus avoiding that the night vision is affected. However, there are certain disadvantages with this also. Firstly, the visual acuity is poorer which means that the ability to discriminate finer details is reduced. Secondly, the visual fatigue will increase. In certain instances where visual acuity is critical other colors are therefore applied (e.g., blue), as for instance on radars and CRT screens. This on the other side will reduce the night adaptation.

Both the use of red and blue lighting may also affect the reading of charts and should therefore also been taken into consideration. Colors are perceived by the cones and not the rods. This means that distinguishing different colors on e.g. charts will be difficult. Further, red, orange, and buff colors have a tendency to disappear under red lighting.

A particular effect of adaptation to darkness is that we become sensitive to light, but much less so to spatial variables (Powers & Green, 1990). This among other means that it is much more difficult to judge distance in darkness than in daylight. This can influence our situation awareness.

7.8.3 Radar operation and vigilance

Radar operation may be a quite demanding task for various reasons. One situation is navigation in coastal waters with heavy traffic where the operator has to monitor many targets and assess collision risk. However, similar to lookout, maintaining vigilance during radar watch in open waters with little or no traffic for longer periods can be quite difficult. Radar watch and detection of targets have of course been made much easier with the introduction of ARPA (Automatic Radar Plotting Aid).

It has been well established that the detection of radar targets may fail for a number of reasons:

- The signal may be weak
- The radar target is veiled by signal noise
- No warning
- Increased boredom due to the monotony of the task
- Lack of rest pauses

The fact that performance deteriorates with time is expressed in the so-called "Mariners Law": "Maximum vigilance can be maintained for a period of about 30 minutes – after this the performance deteriorates sharply" (Elliott, 1960). This observation has later been stated in a more precise form by

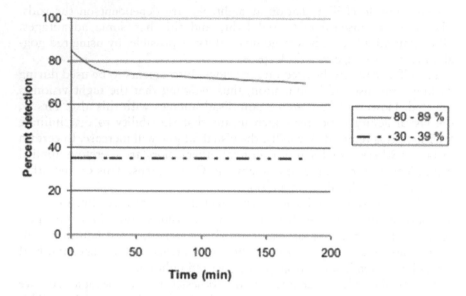

Figure 7.7 Detection probability versus time on watch for different initial detection probabilities (Teichner, 1972).

Teichner (1972) who found that the probability of detecting a visual signal is a function of:

- The initial probability of detection
- The duration of the watch
- Whether the detection demands continuing adjustment of eye focus or not.

Two of these factors are illustrated in Figure 7.7. It shows that given a low initial detection frequency there is insignificant deterioration with time whereas a high initial detection deteriorates sharply during the first 30 minutes. This phenomenon is ascribed to the fact that the first situation is reflecting a demanding task requiring high concentration, in contrast to the other that is less demanding one and thereby soon results in less concentration.

7.9 SITUATION AWARENESS

Situation awareness is an important concept that is often mentioned in accident investigations. Situation awareness can be understood as "the ability of an individual to possess a mental model of what is going on at any one time and also to make projections as to how the situation will develop" (Hetherington et al., 2006). In simple terms, we can say that situation awareness is our understanding of what is going on and our ability to understand how that situation may develop. It is important to note the last part of the sentence – that situation awareness also includes ability to predict what may happen in the future, although usually only short-term.

Situation awareness

Situation awareness is dependent on a number of factors, not least our experience. A novice who was suddenly placed on the bridge of a ship in an area with a lot of other traffic would have great difficulties comprehending the situation. Firstly, it takes experience to understand where ships are heading and what speed they have, secondly, you need to know COLREG and other rules influencing navigation, and thirdly you also need a picture of traffic patterns to quickly understand the situation. A novice would be able to observe the status at a given point in time but would have difficulties projecting how the situation might develop.

An example of lack of situation awareness is from the collision between the Norwegian frigate Helge Ingstad and the tanker TS Sola in 2018. This occurred in darkness and the navigation crew on Helge Ingstad believed that the lights on the tanker were fixed lights on shore. Their situation awareness was therefore wrong, and this also meant that they were not able to predict that the tanker could move into their planned path.

Boeing 737 MAX

An example from the aviation industry may serve to illustrate challenges associated with increasing automation also in the maritime industry. In 2018 and 2019, two Boeing 737 MAX aircraft crashed under similar circumstances, leading to all aircraft of this type being put on the ground for an extended period. The reason for these accidents was that the aircraft had been fitted with automatic systems that would compensate for certain design deficiencies, but when input from a specific sensor failed, the aircraft would behave in a way that the pilots did not expect, because they had not been trained to understand this behavior. Because of this, they started fighting against the automatic system, eventually leading to the aircraft crashing.

In this situation, the pilots did not fully understand how the technology worked, thus they did not have a full understanding of why the aircraft behaved as it did, and they were unable to project what would happen when they took certain actions. Complete situation awareness was in other words missing. However, the pilots could hardly be blamed for this since they reacted in the way they normally should if the aircraft had behaved "normally".

This is clearly also a pitfall for ship technology, as more and more tasks are automated, and we move toward increasing autonomy. Presently, the technology that is being tested still relies on human operators' supervision of the operation, but if they do not understand the technology properly, they may well misjudge the situation and take the wrong action if they have to intervene.

Situation awareness, as defined above, can be divided into three main steps (Flin et al., 2003):

- Gathering information – for this step we use our senses, vision, hearing, etc., to gather information about the present situation. This can be gathered directly from the environment or indirectly through displays, computer screens, etc. This is a step that anyone who has their senses intact can achieve. We can, e.g., see various lights around us if we are on a ship at night.
- Interpreting information (perception) – here we need training and experience to understand what the information means. An experienced mariner will, e.g., know that a combination of white lights at different heights and a red or green light is a ship, and based on the position and color, he can determine the heading of the ship. In this way, a picture of the present situation is built that someone with no nautical experience cannot achieve.
- Anticipating future states – finally, the experienced mariner will also be able to predict future states. Based on the interpretation above, the heading of another ship, he is also able to say whether the other ship is moving away or getting closer and whether action needs to be taken to avoid collision. Again, because the novice is unable to interpret the information, it is also hard to predict future states.

Lack of (or loss of) situation awareness is obviously critical, because we plan our actions based among others on what we believe or expect will happen in our surroundings. If these expectations are incorrect, it is obviously also far more likely that the wrong decisions will be made and the wrong actions taken.

Failure of situation awareness can be divided into three levels (see, e.g., Endsley, 1995):

- Level 1: Failure to correctly perceive situation – this may be because we don't have the information, it is difficult to access, or we fail to observe it.
- Level 2: Failure to comprehend the situation – this is the next step, moving from gathering information to interpreting the information we have. If we are unable to make sense of what we have observed, we have a level 2 failure. The Helge Ingstad example can be classified as level 2 – the crew observed the lights but were not able to understand what those lights meant. This can also be a problem with highly automated systems because we as humans do not necessarily understand

the full functionality of the system and thus have problems under-standing what is happening. This also leads to the next level.

- Level 3: Failure to project situation into future – This is the final step and may occur even if we understand the situation. This can be par-ticularly important in emergencies, because the situation is unusual, and we may not be able to understand what will happen next, even if we have interpreted the present situation correctly.

The three levels are also sometimes described as "What?", "So what?", and "Now what?". We can see that this is a sequential process (although typi-cally with iterations back and forth), and if a level 1 failure occurs, we are of course also unlikely to succeed with levels 2 and 3.

There are studies indicating that the most common failure is on Level 1. In a study by Jones and Endsley (1996), 78% of situation awareness failures were found to be associated with level 1.

Collision between USS John S McCain and tanker Alnic MC in 2017

The following description is based on the NTSB accident investigation report (NTSB, 2019). USS John S McCain hit Alnic MC and 10 people on the navy vessel were killed and 48 injured. No one was injured on Alnic MC.

On US Navy vessels, it is common to have a helm position that con-trols the heading of the vessel (steering) and a lee helm that controls speed (power). Control can be transferred between the two stations if necessary. Before the accident occurred, the helmsman was going to transfer control of speed to the lee helm, but unintentionally transferred control of both speed and heading, without being aware of this. For the helmsman, it therefore appeared as if control of heading had been lost. There was an emergency override, but the function of this was not fully understood by the crew and this was therefore not activated until too late. There was no malfunction of any of the technical systems.

This is a typical example of a situation where situation awareness is lost. The helmsman was not aware of the status of the system and could not control the heading of the vessel from his position anymore. If we consider the three levels in the process described above, we can describe it like this:

- Level 1 Perception: The helmsman perceived that he could not control the heading of the vessel anymore from his position at the helm. However, he had not perceived the reason for this, being unaware that the control had been transferred to the lee helm. There was thus a partial perception failure.
- Level 2 Interpretation: Because the helmsman had not perceived that control had been transferred, his interpretation was based only on the fact that he could not control the vessel anymore. A logical interpretation of this is that there is a problem with the steering. The status of the system is thus not fully understood.

- Level 3 Projection: When the status is not understood, the choice of actions is also likely to be wrong. The simple solution in this case would have been either to move to the lee helm or transfer control back again, but this was not a relevant action with the understanding the helmsman had of the situation. In addition, the problem was worsened by the fact that the crew did not fully understand how the emergency override functioned.

These types of situations have also been associated with pilots taking over active control of a vessel in restricted fairways. The pilot may not be fully familiar with the rudder control system and this may result in wrong operation or unable to deal with unexpected response of the vessel.

Hetherington et al. (2006) looked at several studies of accidents, from the maritime and aviation industries, and found that around 70% of human errors are related to inadequate situational awareness. This is therefore obviously an important issue.

Humans rely very much on pattern matching when we are observing and perceiving situations. In simple terms, we can say that when we are in a new situation, we do not consciously perceive all the information that we are presented with, especially through our eyes. Instead, we start searching our huge "library" in the brain and try to find earlier experiences that "match" the new situation. This is not based on all the information, but typically just a selection of "markers" that we can use to identify a situation as being similar to something we have seen before. A simple example is when we are driving a car and approaching a crossroad with a traffic light. We will then focus on the color of the traffic light and also the other cars that are on the road. We will not really notice the surroundings, e.g., what color there are on the houses around the crossroad. The reason for this is that our brain knows that the color of the houses is completely irrelevant in this situation, so this information is effectively "shut out" from the brain.

There are several well-known examples illustrating this effect. The most well-known one is probably the video where the viewer is asked to count the number of times a basketball is passed between players. During the passing, when the viewer is concentrated on counting passes, a gorilla enters the view and stays in clear view for a few seconds before leaving again. It turns out that only about half the viewers actually see the gorilla. This is a typical example of how we are very selective in what we actually observe, even when something very unusual happens (Simons & Chabris, 1999).

The advantage of this is that pattern matching is efficient and quick. We save time and free up resources in our brain for solving the next problem or addressing the next situation that we face. The disadvantage is that the pattern matching can miss important information that in most cases is not relevant for the interpretation of the situation, e.g., because it is irrelevant or has a low probability of occurring. However, in certain circumstances it may be relevant. If crucial information is missed, we may match the situation at hand with the wrong pattern and we end up with a (sometimes completely) wrong situation awareness.

7.10 PERCEPTION AND DECISION MAKING

Perception is basically the task of making sense of what you detect and observe. As for observation there are a number of sources of error. Typical questions are: Do I see a vessel or a stationary object? Is the light from a marker or another vessel? or Is the object moving away or toward me? Perception involves:

- Understand the signal, apply meaningful concepts
- Relate visual input to known pictures
- Assess movement relative to mental, dynamic models
- Giving priority to alternative information sources

Problems may arise due to the fact that the human apply sequential processing (one thing at the time), input in short-term memory is forgotten or mental overload. The performance of human information processing rests very much also on correct decision making: How to assess the situation on the basis of the perceived information and how to respond in light of the mission plan and the detected obstacles or constraints. Both perception and decision making may be flawed by what we call false hypothesis and habits.

7.10.1 False hypothesis

A rational perception of what we sense can be influenced by the so-called false hypothesis phenomenon or what we may term anticipation or illusions. It appears in different ways or by different mechanisms:

- A tendency to take in limited information and assume the rest
- High expectancy: Long experience that things happen in a certain pattern makes you "see" things regardless of the actual stimulus
- Hypothesis as a defense: Interprets things in way that reduce stress or anxiety (not facing reality)
- After a demanding work period the concentration or vigilance drops and makes you more vulnerable to error

7.10.2 Habit

Safe operation of a vessel is to a considerable degree based on sound habits or essentially behaviors. The objective of training is to establish habitual actions or skills. Through acquired experience the repertoire of skills is further developed. A negative aspect of time or age is that physical fitness reaches a peak in early years. A classic question is therefore whether physical superiority of the young is beaten be the greater experience of the mature crew member. It may however be a certain risk with habits under changing conditions:

- Habitual responses may be inappropriate due to change of the dynamic characteristics of the vessel, another propulsive system, and so on.
- "Cannot make myself forget": New response patterns have to be trained, but under stress or with focus elsewhere one return to earlier habits.
- The problem may be overcome by "over-learning" of critical responses.
- But as new responses and improved skills are developed the ability to stay calm is also improved.

7.10.3 End-spurt effect

It has been shown that for vigilance demanding tasks the knowledge about the remaining duration of the assignment has an effect on the performance of the operator. In experiments subject knowing the how long time they had to stay vigil performed superior to subjects not knowing.

7.11 COMMUNICATION

A central factor in many high hazard industries is communication between individuals involved in operations. This is also the case within the maritime industry, both internal communication among crew members, communication with pilots, and external communication with other ships, vessel traffic services, etc. In a study by Acejo et al. (2018), 693 accident reports from six countries (the UK, Australia, the USA, New Zealand, German, and Denmark) in the period of 2002–2016 were analyzed. They concluded that of all accidents, failure of communication was an immediate or contributory cause in 24% of the cases. For grounding and collision, failure of communication was involved in 41% and 36%, respectively. For grounding, it is likely that it is failure of internal communication on the ship which is the main cause, while for collisions, failure of communication between ships is likely to be important.

Another example of the importance of this is quoted by Hetherington et al. (2006), based on a review of 273 incidents by the Canadian Transportation Safety Board. They found that 42% of accidents involved misunderstandings or lack of communication between the pilot and the crew. The relationship between the pilot and the captain/officer of the watch is a complex issue that involves more than communication. When the pilot takes over navigation, the normal roles and responsibilities on the ship are changed and different pilots may have somewhat different views on their role either as advisor or active navigator. This may create uncertainty and misunderstandings that are not just related to communication issues.

One problem with communication in the maritime domain is that seafarers come from all over the world and have different languages. This can create problems with internal communication on the ship, but also external communication with other ships. External communication can also be made difficult by technical issues with the equipment being used and the transmission.

IMO have addressed this problem in several ways. Firstly, regarding external communication, SOLAS specifies that:

> English shall be used on the bridge as the working language for bridge-to-bridge and bridge-to-shore safety communication as well as for communications on board between the pilot and bridge watchkeeping personnel.
>
> *(IMO, 2004)*

Further, STCW (IMO, 2010) has requirements for language proficiency among the crew.

A simple way of reducing the probability of failure of communication is to request that the receiver responds with the same information that he or she has received rather than just confirming with an "OK" or "Received". This is commonly used in many situations and greatly increases the probability that the information that the sender intends to transmit is correctly received.

Collision between Mesabi Miner and Hollyhock

The description is based on the Marine Accident Brief from NTSB (2015). The US Coast Guard vessel *Hollyhock* was breaking ice for a convoy of six ships in the Straits of Mackinac that provides a link between Lake Michigan and Lake Huron. The vessel ran into thicker ice and could not maintain speed. The *Mesabi Miner* was the first in the convoy and was unable to slow down quickly enough to avoid hitting the stern of *Hollyhock*. No injuries or pollution resulted, but both vessels were extensively damaged.

Convoy operations in ice require very close communication between the vessels involved and in this case, it was concluded that this was not adequate. The bridge team on *Hollyhock* informed the convoy that they had stopped as soon as it occurred, but this was not acknowledged by *Mesabi Miner*. A second warning was also issued, but this was not heard by the *Mesabi Miner*.

7.12 BRIDGE RESOURCE MANAGEMENT

In 1977, there was a terrible accident at Tenerife Airport, where two Boeing 747 planes crashed, resulting in 583 fatalities. This is the worst accident that has occurred in aviation history. The investigation showed that there had been misinterpretations and false assumptions during the period leading up to the accident.

Among the outcomes of this accident was the introduction of what was called crew resource management (CRM), aimed at improving communication and working relationships in the cockpit. CRM covers a whole range of skills (often called non-technical skills), including communication, teamwork, situation awareness, leadership, assertiveness, decision making, and workload management (Hetherington et al., 2006). In the maritime industry, similar training programs have been developed, called Bridge Resource Management (BRM) (the terms Bridge Team Management (BTM) and Maritime Team Management are also being used).

Requirements for BRM have also been included in STCW as part of the Manila amendments in 2010 (IMO, 2010). According to Table AII/1 of Chapter II, officers in charge of navigational watch shall have knowledge of BRM principles, including:

- Allocation, assignment, and prioritization of resources
- Effective communication
- Assertiveness and leadership
- Obtaining and maintaining situational awareness
- Consideration of team experience

Similar knowledge is expected from masters and chief mates. Similar training for engine room crew is also required (commonly labeled Engine Room Resource Management, ERM).

The effect of BRM has proven difficult to measure. A study by O'Connor (2011) showed that BRM training had no significant effect on competence or attitudes of naval officers. Studies actually showing that this has an effect are hard to find.

7.13 HUMAN-MACHINE INTERFACE (HMI)

Navigation and control of vessels are based on information input for visual lookout and reading instruments on the bridge. The bridge system has controls for speed, heading, and setting of instruments. The bridge system may be seen as the interface between the navigator and the vessel systems, and it is paramount that this interface is designed in a manner to enhance user-friendliness and safety.

The design of the bridge system is subject to regulation by IMO (2000):

"Guidelines on Ergonomic Criteria for Bridge Equipment and Layout" for SOLAS certified vessels.

Another key document is IACS (2007):

> Recommendation for the Application of SOLAS Regulation V/15, Bridge Design, Equipment Arrangement and Procedures (BDEAP).

Design of HMI is subject to an ongoing development and innovation within most industries and not least all kinds of vehicles including sea-going vessels. Key technological elements are instrumentation, computerization, information integration, and application of visual display units (VDUs). The functionality is also dependent on a good understanding of how humans interact with machines or ergonomic design. Ergonomics is defined as follows by the International Ergonomics Association:

> The scientific discipline concerned with the understanding of interactions among humans and other elements of a system and the profession that applies theory, principles, data and methods to design in order to optimize human well-being and overall system performance.

It is important to consider the fact that humans and machines have different capabilities. Machines may work with consistent performance over time and are less influenced by external factors. Humans, on the other hand, are creative and in that sense able to deal with emergencies and unusual/unexpected situations through problem solving and improvisation.

ABS (2003) is offering guidance notes on "Ergonomic design of navigation bridges" and advocates the following eight basic principles:

1. Define the roles and responsibilities of bridge personnel:
   Characterize the main functions on the bridge (visual lookout, radar operation, rudder control, etc.) and specify requirements in terms of parameters to observe or control. Secondly, allocate bridge personnel to the functions. Specify procedures for dealing with loss of automatic functions.
2. Design for human limitations, capabilities, and expectations:
   It is important to assess the mental workload in all relevant operative situations. Control and display units shall mimic the spatial and functional relationships of the parameters they represent. Other important design requirements are: accommodate operator expectations, facilitate operator attention, limit perceptual and memory requirements, and focus on standardization.
3. Arrange bridge devices, controls, and displays to maximize access:
   The components and consoles shall be arranged on the basis of frequency, related task, or sequence of use. Important and frequently used displays and controls shall be located for easy reach and use by the operator. Safety and time critical tasks shall have high priority in the arrangement of components and consoles.

4. Design displays consistent with task requirements:

Coding formats for information shall be consistent across all displays. Design displays in a way to enhance situation awareness. Other important design principles are to provide precise information and avoid display complexity. The design of alarms should also have great consideration in order to make critical situations easily understood, avoiding nuisance alarms and ambiguous alarms.

5. Design simple, direct, and easy-to-use inputs and controls:

The bridge solution shall facilitate direct human control. Automated systems shall in a clear manner display present control mode and changes of control mode. When operator intervention in automated systems is necessary, guidance and immediate feedback should be provided.

6. Design for productive performance and to reduce human error:

Displays and controls shall be designed in a manner to avoid operator error. Communications systems shall be designed in a manner to be operated without having to move from the workstation. Controls may be operated from different workstations or locations but should not be possible to operate by two persons simultaneously.

7. Provide job aids and training:

The bridge system shall have necessary documentation to facilitate training and aid during operation: Documentation of training needs and provide system descriptions, procedures, and adequate labels and warnings.

8. Perform testing

The design process itself involves the following main steps (ABS, 2003):

- Designing overall layouts and component arrangement. The number of workstations should accommodate for alternative composition of the manning (number of persons and allocation of roles)
- Selecting devices to meet requirements
- Estimating and allocating workload on personnel
- Designing procedures and job aids
- Determining communication requirements
- Determining internal and external visibility requirements

The main functions to perform on the bridge are navigation, interfacing with deck operations, dealing with emergencies, and monitoring unattended engine rooms. An overview of these functions is presented in Table 7.8. The bridge also has certain maintenance and administrative functions but they are not discussed here.

The arrangement of the bridge shall be designed in a manner that gives maximum visual overview of the water around the vessel. The physical

Table 7.8 Overview of main bridge functions

Function area	Function
Navigation	Voyage planning and preparation, chart updating Maintain trim and ballast Visual lookout, area surveillance Surveillance of sea state, current, water depth, weather monitoring, visibility Position fixing, voyage track keeping Collision avoidance and traffic surveillance Helm control/conning, station keeping, berthing/unberthing Communication internal/external, signals and lights, tug interface
Interfacing with deck operations	Docking/undocking Anchoring, line handling, directing shore gangs Tug and towing operations Small boat deployment and recovery, evacuation training/operation Cargo planning, loading/unloading sequence Cargo management, bunkering, and storing
Emergency operation	Extreme weather and seas Fire – Large and small Flooding/ballast control Collision/Grounding/Stranding Internal security Loss of propulsion/steering Search and rescue/Man overboard and Vessel security Fuel spills/Environmental hazards Bridge – Housekeeping
Functions related to unattended engine rooms	Monitoring/control of propulsion systems (operation, start-up, emergency shutdown, fuel system, lube systems, cooling system, alarm conditions) Steering gear Monitoring electrical system Fire main control, fire suppression activation Bilge monitoring and management Pollution monitoring and control

Adapted from ABS (2003).

dimensions and location of the workstations shall be placed in a manner that gives unobstructed view through the bridge windows. There are further guidelines for design and dimensions of the windows in order to obtain a clear view from the workstations.

It is given guidelines for view ahead from the bridge. It is required that the view from the bridge should not be obscured by more than 500 meters or 2 ship's lengths whichever is less. See Figure 7.8.

The second aspect is external field of view around the vessel. The field of view requirement is dependent on the work station in question but for the centrally located conning position the requirements are as follows (Figure 7.9):

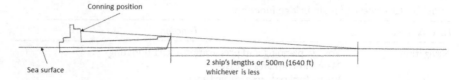

Conning position

Sea surface

2 ship's lengths or 500m (1640 ft)
whichever is less

Figure 7.8 **View of surface ahead from the bridge (ABS, 2003).**

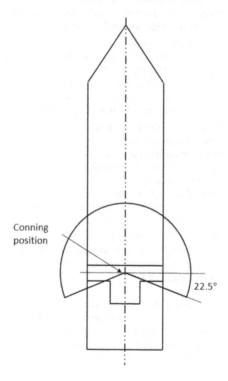

Conning
position

22.5°

Figure 7.9 **Field of view from conning position (ABS, 2003).**

The forward horizontal field of view from the navigating and maneu-vering workstations and from the conning position, should extend over an arc from 22.5° abaft the beam on one side of the vessel, forward through the bow centerline to 22.5° abaft the beam on the other side of the vessel at a minimum.

It is beyond the scope of this chapter to address the detailed design of the bridge in terms of individual workstations and their arrangement and detailed design of instruments. For illustrative purposes a general concept proposed by the guidance notes of ABS is shown in Figure 7.10.

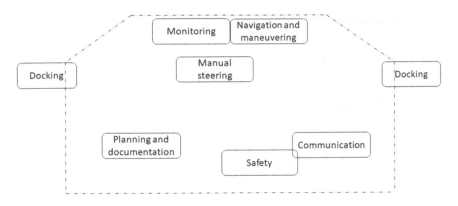

Figure 7.10 Typical bridge arrangement (ABS, 2003).

7.14 SAFETY CULTURE AND SAFETY CLIMATE

The concept of safety culture was first discussed after the Chernobyl accident in Ukraine (then part of the Soviet Union) in 1986. After this, it has been extensively described and studied and has become a factor that is very often mentioned as important for safety.

So what is safety culture? There are many definitions and explanations, but in broad terms we can say that it is the collection of the beliefs, perceptions, and values that employees share in relation to risks within an organization. If we analyze this definition, there are several things worth noting:

- First of all, a safety culture is something that is shared within an organization. Individuals may have their own values, beliefs, and perceptions, but these are the things that are shared, often as a result of adaptation when new members are introduced into the organization.
- Secondly, these are "beliefs, perceptions and values", i.e., work practices are not part of safety culture the way that it is defined here. However, in practice, beliefs, perceptions, and values will influence our work practices also and the manifestation of a safety culture is therefore primarily observed through work practices.
- Thirdly, this is "in relation to risks". However, beliefs, perceptions, and values will in many cases tend to be more wide-ranging than just being applicable to risks. Some will be specific, but not all, and this also means that we can regard safety culture as a subset of organizational culture in general.

Safety climate is another term that is being used and it may be difficult to distinguish between the two. The Queensland Government in Australia (worksafe.qld.gov.au) has described the difference as safety culture being

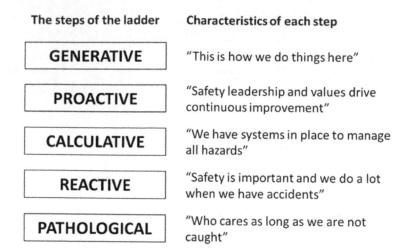

The steps of the ladder	Characteristics of each step
GENERATIVE	"This is how we do things here"
PROACTIVE	"Safety leadership and values drive continuous improvement"
CALCULATIVE	"We have systems in place to manage all hazards"
REACTIVE	"Safety is important and we do a lot when we have accidents"
PATHOLOGICAL	"Who cares as long as we are not caught"

Figure 7.11 The HSE culture ladder (adapted from Hudson, 2007).

the "personality" of an organization while the safety climate is the "mood" at a given time. This means that the safety culture will influence safety climate. Safety climate can be seen as a measurement or a snapshot of the situations at a specific point in time.

The HSE culture ladder (Hudson, 2007) is an illustration of different steps in the development of a safety culture. The ladder is shown in Figure 7.11. The idea is that organizations (including a small organization like the crew on a ship) can be classified according to where they are on the culture ladder. The top step is the "nirvana" of safety culture. The characteristics for each step should make this fairly self-explanatory, but some comments can be added.

The lowest step on the ladder, pathological, describes organizations that in practice do not care about safety at all. They will follow rules and regulations as long as there is a reasonable chance that it will be detected if they do not comply, but apart from this, they have no self-interest in ensuring the safety of workers or other people.

The next step, reactive, has been typical for the maritime industry for a long time. IMO have typically responded to serious accidents by introducing new regulations but have generally been slow at implementing improvements unless they have been triggered by accidents. The long list of conventions that can be tied to accidents (including key ones like SOLAS and MARPOL) is a testament to this.

The implementation of the ISM code may be regarded as one further step up the ladder, to the calculative step. The core of the ISM code is that shipowners and ships should have systems for managing safety, a typical characteristic of this step.

Where the maritime industry will go in the future is hard to say and there are of course large differences between different companies and different parts of the shipping industry. The top level, the generative level, can also be likened to the culture that one finds in High Reliability Organizations (HROs).

How can safety culture be improved? This rests on the view that safety culture is something that we can form as we like, provided we use the right tools. Hudson (2007) is an advocate of this view, and he has described how this has been done in a major international oil company.

Reason (1997) suggests that culture consists of certain "components" that can be shaped:

- Reporting culture – a culture where information about incidents, failures, and errors is reported, by anyone and everyone who observe it or make errors themselves.
- Just culture – a culture where apportioning blame is not important and where those who make unintentional mistakes and errors are not punished for this.
- Learning culture – a culture that can learn from failures and errors and improve the way things are being done to avoid the same mistakes from being made again.
- Flexible culture – a culture that can reorganize and adapt as necessary to new and unusual situations. This is related to the HRO theory.

Management can actively try to develop the culture in this direction, by using the right tools, eventually achieving a good safety culture.

However, there are also those who claim that safety culture is not something that can be managed and formed actively, although they acknowledge that it may be influenced over time (Haukelid, 2008). According to this view, it is the members of the organization that collectively contribute to the emergence of the culture. If this is the case, management will have very little influence over culture, except in the same way as any other member of the organization.

In the end, it is perhaps likely that the reality lies somewhere between the two views presented above. It is possible to influence and develop culture in a direction that we want to, but managing all aspects is likely to be difficult and maybe even impossible.

REFERENCES

ABS (2003). *Guidance Notes on Ergonomic Design of Navigation Bridges.* American Bureau of Shipping. Houston, TX.

Acejo, I., Sampson, H., Turgo, N., Ellis, N., & Tang, L. (2018). *The causes of maritime accidents in the period* 2002–2016. Seafarers International Research Centre (SIRC) Cardiff University, ISBN: 1-900174-51-0.

Akhtar, M. J., & Utne, I. B. (2014). Human fatigue's effect on the risk of maritime groundings–A Bayesian Network modeling approach. *Safety Science, 62*, 427–440.

Baitis, A. E., et al. (1995). 1991–1992 *Motion Induced Interruption (MII) and Motion Induced Fatigue (MIF) Experiments at the Naval Biodynamics Laboratory,* CRDKNSWC-HD-1423-01, Bethesda, MD: Naval Surface Warfare Center.

Barnett, M., Pekcan, C., & Gatfield, D. (2012, April). *The use of linked simulators in project "HORIZON": Research into seafarer fatigue.* MARSIM.

Chen, S. T., Wall, A., Davies, P., Yang, Z., Wang, J., & Chou, Y. H. (2013). A Human and Organisational Factors (HOFs) analysis method for marine casualties using HFACS-Maritime Accidents (HFACS-MA). *Safety Science, 60*, 105–114.

Conway, G. E., Szalma, J. L., Saxton, B. M., Ross, J. M., & Hancock, P. A. (2006). The effects of whole-body vibration on human performance: A meta-analytic examination. In *Proceedings of the human factors and ergonomics society annual meeting* (Vol. 50, No. 17, pp. 1741–1745). Sage, CA: Los Angeles, CA: SAGE Publications.

Coraddu, A., Oneto, L., de Maya, B. N., & Kurt, R. (2020). Determining the most influential human factors in maritime accidents: A data-driven approach. *Ocean Engineering, 211*, 107588

Elliott, E. (1960). Perception and alertness. *Ergonomics, 3*(October), 357–364.

EMSA (2016). EMCIP – *Glossary of reporting attributes,* Version 1.0, January 2016.

EMSA (2020). *Annual overview of marine casualties and incidents 2020.* https://www.confitarma.it/emsa-annual-overview-of-marine-casualties-and-incidents-2020/

Endsley, M. R. (1995). A taxonomy of situation awareness errors. In R. Fuller, N. Johnston, and N. McDonald (Eds.), *Human factors in aviation operations* (pp. 287–292). Aldershot: Avebury Aviation, Ashgate Publishing Ltd.

Flin, R., Martin, L., Goeters, K. M., & Hoermann, H. (2003). Development of the NOTECHS (non-technical skills) system for assessing pilots' CRM skills. *Human Factors and Aerospace Safety, 3*(2), 95–117.

Grech, M. R., Horberry, T. J., & Koester, T. (2008). *Human factors in the maritime domain.* Boca Raton: CRC Press, Taylor & Francis Group.

Grether, W. F. (1971). Vibration and human performance. *Human Factors, 13*(3), 203–216.

Griffin, M. J. (1990). Motion sickness. *In Handbook of human vibration* (pp. 271–330). New York: Academic Press.

Griffin, M. J., & Erdreich, J. (1991). *Handbook of human vibration.* London: Academic Press.

Haukelid, K. (2008). Theories of (safety) culture revisited—An anthropological approach. *Safety Science, 46*(3), 413–426.

Hetherington, C., Flin, R., & Mearns, K. (2006). Safety in shipping: The human element. *Journal of Safety Research, 37*(4), 401–411.

Hudson, P. (2007). Implementing a safety culture in a major multi-national. *Safety Science, 45*(6), 697–722.

IACS (2007). *Recommendation for the Application of SOLAS Regulation V/15, Bridge Design, Equipment Arrangement and Procedures (BDEAP)*. No. 95 (Corr. 2 July 2011). London.

IMO. (2000). *MSC/Circular 982 guidelines on ergonomic criteria for bridge equipment and layout*. London. https://www.imorules.com/MSCCIRC_982.html.

IMO (2001). *Guidance on Fatigue Mitigation and Management*. London: Inernational Maritime Organization. https://wwwcdn.imo.org/localresources/en/Our Work/HumanElement/Documents/1014.pdf

IMO (2004). *SOLAS consolidated edition 2004*. Chapter V, Regulation 14/3. London: International Maritime Organization.

IMO (2010). *International convention on standards of training, certification and watchkeeping for Seafarers (STCW)* (2017 Edition), London: International Maritime Organization.

ISO 9996 (1996). *Mechanical vibration and shock – Disturbance to human activity and performance – Classification*. https://standards.iteh.ai/catalog/standards/iso/d2cac7d2-f73b-4c3e-8b09-f7b188a6cf82/iso-9996-1996

ISO 2631-1 (1997). *Mechanical vibration and shock. Evaluation of human exposure to whole-body vibration*. Part 1: General requirements. https://www.iso.org/standard/7612.html.

Ivergård, T. et al. (1978). Work conditions in shipping – A survey. (Arbetsmiljö innom Sjöfarten – En kartläggning) (in Swedish). Stockholm: Sjöfartens Arbetarskyddsnämnd.

Jayle, G. E. et. al. (1959). *Night vision*. Springfield, MO: Charles C. Thomas.

Jones, D. G., & Endsley, M. R. (1996). Sources of situation awareness errors in aviation. *Aviation, Space, and Environmental Medicine*, 67, 507–512

Kroemer, K. H. E., & Grandjean, E. (1997). *Fitting the task to the human*. 5th ed. London: Taylor & Francis.

Lamb, T. D. (1990). Dark adaptation: A re-examination. In R. F. Hess, L. T. Sharpe, & K. Nordby (Eds.), *Night vision*, (pp. 177–222). Cambridge: Cambridge University Press.

Mackworth, N. H. (1946). Effects of heat on wireless telegraphy operators hearing and recording morse messages. *British Journal of Industrial Medicine*, 3, 143–158.

Madison, R. L. (1974). *Human Factors in Destroyer Operations* (NTIS: AD 775 011). M.Sc. Thesis. Naval Postgraduate School, Monterey, CA.

McCormick, E. J. & Sanders, M. S. (1983). *Human factors in engineering & design*. New York: McGraw-Hill.

NATO (2000). *Standardization Agreement (STANAG): Subject: Common procedures for seakeeping in the ship design process*. NATO, Military Agency for Standardization.https://iopscience.iop.org/article/10.1088/1757-899X/982/1/012041/pdf

NATO STANAG 4154. (1997). Chapter 7: Seakeeping criteria for general application. *In Common procedures for seakeeping in the ship design process*. Brussels: North Atlantic Treaty Organization (NATO).

NORDFORSK. (1987). *The Nordic Cooperative Project. Seakeeping performance of ships. Assessment of a ship performance in a seaway*. Trondheim: MARINTEK.

NTSB (2015). *Collision of Bulk Carrier Mesabi Miner and US Coast Guard Cutter Hollyhock*. Marine Accident Brief MAB-15-07.

NTSB (2019). *Collision between US Navy Destroyer John S McCain and Tanker Alnic MC Singapore Strait, 5 Miles Northeast of Horsburgh Lighthouse, August 21, 2017.* Marine Accident Report NTSB/MAR-19/01 PB2019-100970

O'Connor, P. (2011). Assessing the effectiveness of bridge resource management training. *The International Journal of Aviation Psychology*, 21(4), 357–374.

Pepler, R. D. (1959). Warmth and lack of sleep: Accuracy or activity reduced. *Journal of Comparative and Physiological Psychology*, 52, 446–450.

Pilcher, J. J., Nadler, E., & Busch, C. (2002). Effects of hot and cold temperature exposure on performance: A meta-analytic review. *Ergonomics*, 45(10), 682–698.

Powers, M. K., & Green, D. G. (1990). Physiological mechanisms of visual adaptation at low light levels. In R. F. Hess, L. T. Sharpe, & K. Nordby (Eds.), *Night vision*, (pp. 177–222). Cambridge: Cambridge University Press.

Rasmussen, J. (1997). Risk management in a dynamic society: a modelling problem. *Safety Science*, 27(2–3), 183–213.

Rumawas, V. (2016). Human factors in ship design and operation: Experiential learning, Doctoral thesis at NTNU, 2016: 11.

Schröder-Hinrichs, J. U., Hollnagel, E., Baldauf, M., Hofmann, S., & Kataria, A. (2013). Maritime human factors and IMO policy. *Maritime Policy & Management*, 40(3), 243–260.

Seppanen, O., Fisk, W. J., & Lei, Q. H. (2006). *Effect of temperature on task performance in office environment.* Berkeley, CA: Lawrence Berkeley National Laboratory.

Shappell, S. A., & Wiegmann, D. A. (2000). *The human factors analysis and classification system--HFACS.* Washington, DC: DOT/FAA/AM-00/7. Office of Aviation Medicine.

Simons, D. J., & Chabris, C. F. (1999). Gorillas in our midst: Sustained inattentional blindness for dynamic events. *Perception*, 28(9), 1059–1074.

Smith, A. P., Allen, P. H., & Wadsworth, E. J. K. (2006). *Seafarer fatigue: The Cardiff research programme.* Centre for Occupational and Health Psychology, Cardiff University.

Stevens, S. C. & Parsons, M. G. (2002). Effects of motion at sea on crew performance: A survey. *Marine Technology*, 39(1), 29–47.

Swain, A. D., & Guttmann, H. E. (1983). *Handbook of human-reliability analysis with emphasis on nuclear power plant applications.* Final report (No. NUREG/CR--1278). Sandia National Labs.

Szalma, J. L., & Hancock, P. A. (2011). Noise effects on human performance: A meta-analytic synthesis. *Psychological Bulletin*, 137(4), 682.

Teichner, W. H. (1972). *Predicting Human Performance III. The Detection of Simple Visual Signal as a Function of Time on Watch.* New Mexico State University, Department of Psychology Report NMSU-ONR-TR-72-1. New Mexico: Las Cruces.

Wertheim, A. H. (1998). *Mental Load Is Not Affected in a Moving Environment.* Report TNO-TM-98-A068. TNO Human Factors Research Institute. Soesterberg, Netherlands.

Wing, J. F. (1965). *A Review of the Effective Ambient Temperature on Mental Performance.* US Air Force Laboratory, TR 65-102. Dayton, OH: Wright-Patterson Air Force Base.

Chapter 8

Risk analysis methods

8.1 INTRODUCTION

In Chapter 3, the terms risk analysis and risk assessment were defined, but no further explanation of what they entail was given.

The main objective of this chapter is to describe a set of tools and techniques that can be used in the process of carrying out risk analysis. The first part of the chapter gives a brief introduction to risk analysis and risk assessment in general and what the main steps are. The second part of the chapter gives some useful basic theory related to system description and structures. Finally, the third and main part of this chapter deals directly with risk assessment techniques. The following techniques are described:

- Preliminary Hazard Analysis (PHA) – a general method that can be used for many different purposes and gives a quick overview of hazards and ranks them.
- Hazard and Operability Studies (HAZOP) – a detailed method, originally developed for process systems, but can be applied for all systems that have a flow in them, like a fuel system, ballast system, or hydraulic system.
- Failure Mode, Effect and Criticality Analysis (FMECA) – another detailed method, developed for technical systems in general, but used in particular for safety systems.
- Safe Job Analysis (SJA) – a simplified version of PHA, but specifically used for work operations, in particular when preparing to perform high-risk activities.
- System Theoretic Process Analysis (STPA) – also a detailed method, based on control theory.
- Fault Tree Analysis (FTA) – a graphical and quantitative method that can be used to analyze the causes of an event.
- Event Tree Analysis (ETA) – another graphical and quantitative method that can be used to analyze how an event can develop in different ways, depending on whether barriers are functioning or not.

DOI: 10.4324/9781003055464-8

- Bayesian Networks – this is a method that can be used in the same way as FTA, but which is more flexible.
- Risk Contribution Trees (RTC) – IMO describes this in the FSA guidelines (IMO, 2018) and this is in practice a combination of FTA and ETA.

All these techniques are used for risk analysis and as shall be seen, there are many similarities between several of the methods. However, they take different perspectives on the problem, they partly use different terminology and are tailored to analyze the system in different ways.

8.2 RISK ASSESSMENT AND RISK ANALYSIS

8.2.1 A General Risk Assessment Process

Risk assessment and risk analysis have been defined in Chapter 2, and Figure 8.1 illustrates the risk assessment process, including risk analysis

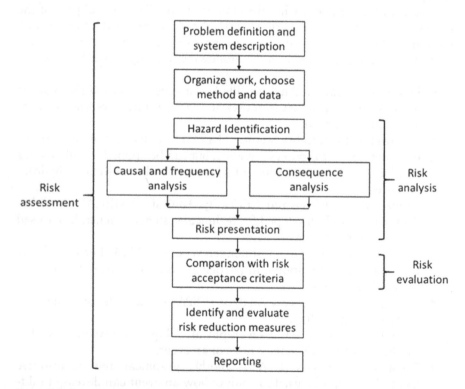

Figure 8.1 A general risk assessment process.

and risk evaluation in more detail. This is a general process description, and many such descriptions exist, with minor variations on how many steps are defined and what the exact content of the individual steps is. However, the main elements are the same in most descriptions.

The figure shows that risk analysis is the process of identifying hazards, determining causes and frequency, determining consequences, and summarizing this into a presentation of risk. Risk assessment is the complete process of planning and preparing the analysis, performing the risk analysis, and also evaluating the results and possible risk reduction measures. It may be noted that there is significant confusion about this terminology in literature and that risk analysis and risk assessment often is mixed. However, in standards, this terminology is becoming quite common, and we will stick to this as far as possible.

In the following, each of the steps is briefly described in separate subsections. Later in this chapter, specific methods for performing risk analysis will be described and the details of each step will then be more elaborated.

8.2.2 Problem definition and system description

The first step in the risk assessment process is to establish a problem definition and system description, i.e., to define and describe the vessel and/or the activity whose risks are to be studied and the limitations of the system.

It should be self-evident that we need to establish a problem definition before starting the work properly, but it is not always that this is done with sufficient clarity. Unless we understand what the purpose of the risk analysis is, we cannot expect to arrive at results that are useful.

Risk analysis is in most cases done to support one or more decisions and understanding these decisions is essential. To do this, we also need to understand the context that the decision is made. Some of the factors that we need to consider are:

- What are the decision criteria (risk acceptance criteria) relevant for the decision? Are they expressed as an FN curve, as a single risk value (e.g., individual risk per year), in a risk matrix, or qualitatively?
- Are there relevant regulations that influence the analysis and/or are there relevant standards that describe what should be done and how it should be done?
- What is the decision that the analysis should support? Is it verification of acceptable risk, is it input to improvement of design, is it to provide recommendations for reducing risk, etc.? Many other purposes can be envisaged, but it is essential to understand this in detail.
- Who are the key stakeholders that are going to use the results from the analysis and what is their knowledge/background in risk analysis?

When the problem is clear to the users and the analysts, a problem definition should be established. This should also include a description of the scope and limitations of the analysis.

The problem definition will in turn influence our system definition, including:

- How should the system breakdown be done? It is common to break down a system into smaller parts before analyzing the system. This can be done in different ways, where the most common methods are:
  - Functional breakdown, according to what functions are performed in the system. Functions on a ship may, e.g., be voyage, loading, unloading, and docking.
  - A system, subsystem, and component breakdown. For a ship, we may e.g., break it down into navigation system, loading system, and propulsion system.
  - A layout/geographical breakdown. For a cruise ship, this can e.g., cabins, restaurants, entertainment areas, engine and utility rooms, various tanks, and bridge.
- The level of detail that the system description should contain. Regardless of what method we choose for breakdown, we can go into more details. For the system breakdown, the propulsion system can be broken down into power system, power transmission system, and propellers/thrusters. The power transmission system can comprise shafts (to main propellers) and gears. This can be continued all the way to individual components if this is necessary or useful.
- What assumptions do we have to make? Since risk analysis is a prediction of possible future outcomes, we also have to make assumptions about what the future will look like. Also, we do not necessarily have all the information about the systems that we are analyzing, and we need to make assumptions about these aspects also. When designing a ship, we make assumptions about manning levels, maintenance levels, operational patterns, etc. In many cases, these assumptions are made implicitly, but we should try to describe them explicitly as far as possible.
- What simplifications do we make? A risk analysis can be seen as a model that we use to calculate/describe risk. Models never are "complete" representations of the phenomenon that we model since this will be too complex, time-consuming, and generally not possible in practice. Simplifications are therefore always necessary, and these simplifications will vary with the context and scope of the study. An example of a simplification could be for a bulk carrier. We will normally not specify precisely what cargo the ship will carry, only that it is bulk cargo and the maximum load it can take.

– What are the system limits? The limits of the system should be set to match the problem definition. To avoid unnecessary work, the limits should not be set too wide, but on the other hand, we have to make sure that the system includes everything of relevance for the risk analysis that is being done. An example of this may be an analysis of ship collision in a fairway. The system limits then have to be set so that the fairway and any other traffic are included. On the other hand, if the analysis is concerned with fire on a ship, external factors outside the ship can largely be ignored.

System description

For a ship, the system description may have to be much wider than just the ship itself, depending on the objective of the analysis to be performed. Some typical elements that may have to go into the system description are:

– Geographical area: Fairways, specific routes, navigation marks, particular hazards, etc.
– Ports and port facilities
– Environmental description: Sea conditions, meteorological relations, visibility, presence of ice, etc.
– Ship description: Number, capacities, sizes, safety critical systems, and technical descriptions
– Ship operation: Transport quantity, type of cargo, frequency/ scale of operations
– Other activities: Surrounding traffic and activities that may introduce hazardous situations

A system description may end up containing a lot of detail and it is important to remember that the focus should be on describing aspects that are important for safety, in a positive or a negative manner. It may be argued that it is only those aspects that will influence the risk assessment (that is considered in the risk assessment) that are necessary to describe.

When the problem definition, system limits, and analysis limitations are established, the type of information required to do the analysis can be established. If it turns out that information is not available, additional assumptions may have to be made.

The system description is important because it acts as a reference basis for those who are using the analysis. A good example is in projects, where the design of a system is still developing. If the description is not sufficient, it will be difficult to keep track of what the status of the design was at the point in time when the analysis was done. If the design has changed, the risk analysis may not be valid anymore, but we are not aware of it because the system description is inadequate (Figure 8.2).

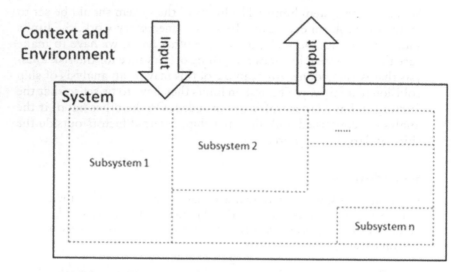

Figure 8.2 Conceptual illustration of a system.

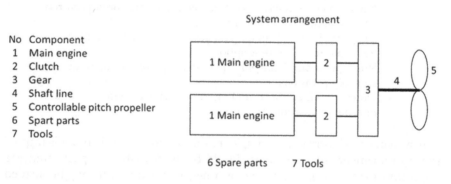

Figure 8.3 Propulsion system arrangement.

System description of a propulsion system

The propulsion system of an oil tanker is to be analyzed using different risk analysis techniques. To do this, a suitable system description must be established. The arrangement of the propulsion system is shown in Figure 8.3.

The main function of the propulsion system is to provide power to the vessel. It is used under normal operation and is to give the vessel the required speed and maneuverability in the whole lifetime of 24 years. The two main diesel engines (1) may be operated individually when maximum power is not required. The diesel engines are uncoupled from the gear using a clutch (2). When both diesel engines are in operation both clutches must be coupled, and the total propulsion power is transmitted to the propeller (5) by the gear (3) and shaft line (4). Spare parts (6) and tools (7) necessary to perform repairs and maintenance are also available.

Three days per year are allocated for maintenance (off-hire) and the ship spends 26 days in port. This leaves 336 days per year sailing time and 70% of this time (235 days) is at full power. In the last 30% (101 days), only one diesel engine is required in operation. The ship is to sail in ice-free sea conditions and ports. There are two engineers responsible for the operation and maintenance planning of the system.

Some comments are warranted to the above description. It may be noted that several of the components, e.g., the main diesel engine, could be described in much more detail, including subsystems or components and all the utility systems required for the main engines to operate. The level of detail in a description will depend on the purpose of the analysis. Further, it may be noted that the description contains technical information, but also operational information. To understand the risk associated with a system, we normally need both. It is also mentioned that the ship sails in ice-free sea conditions. This is an example of information about the external environment that may be relevant to include in many cases.

8.2.3 Work organization and choice of method and data

There are two parts of this step. The first part, work organization, is mainly about how we organize and conduct the risk analysis with respect to resources, time required, etc. This is significantly influenced by the method chosen, and we therefore discuss the choice of method first.

There are many risk analysis methods described in literature, with large differences in approach, degree of detail, workload required, and competence required to use them. Some aspects to consider when choosing method are as follows:

– How detailed should the analysis be? In general, we do not use a method that is more detailed than the scope and the problem requires.
– How much time (and resources) do we have available to perform the analysis? In many cases, it will be project schedule that determines when decisions are made, and it may be better to have coarse results available than no results (because the analysis is too time-consuming).
– How detailed information do we have available? If limited details are available, there is no point in specifying a very detailed method.
– How are the decision criteria formulated? This will often have strong implications for the method that is chosen.

Choice of data is an important issue in risk assessment and is often also one of the most difficult aspects. Data are often hard to find, may be old or irrelevant, and often are very limited. The choice of data can significantly influence the end result and needs to be done carefully. Data sources and criteria for evaluating data are described in Chapter 14.

In the organization of the work, there are several aspects that need to be considered:

- Who and how many should be involved in the preparation of the risk assessment? It is common that risk assessments are performed by a group of people rather than a single individual. This gives a broader input to the analysis, both in terms of knowledge, experience, and roles of the individuals involved. In the group, expertise in design and operation of the system being analyzed should be included, in addition to expertise in risk assessment.
- Is additional experts required for specific input/analyses? Sometimes, highly specialized expertise is required for specific parts of an analysis, e.g., for calculation of accidental loads (fire loads, explosion overpressure, impact loads, etc.) and load effects (effect of fires, explosions, and impact on structures, equipment, people, etc.).
- When should the results from the analysis be available? Risk assessment is normally performed to provide input to decision-making. These decisions often have far wider implications than just affecting risk and risk is just one input and one criterion to be considered. In such a context, it is important that the results are available in time for when the decision is to be made.
- What other resources are needed outside of the group involved in the preparation of the risk assessment? Experts have been mentioned already, but a lot of information is also required, and this has to be made available for the team in a timely and correct manner.

8.2.4 Hazard identification

The second step of the process is to perform a hazard identification.

The purpose of the hazard identification is to identify anything that may threaten the assets that we want to protect, such as humans, the environment, or economic values. This is usually done without much regard for the definitions that are presented in Chapter 3 of this book. It will therefore normally be a combination of hazards, hazardous events, influencing factors, etc. that is the outcome of a hazard identification.

The hazard identification is very often supported by some sort of checklist that acts as a reminder of things that may go wrong and also as a trigger for discussion. Many checklists have been prepared for different purposes. For the maritime industry, the list prepared by IMO (2018) is an example. Other lists that may be relevant also exist. In ISO 17776 (ISO, 2016), a list of hazards relevant for the oil and gas industry is included. This also covers many aspects that are relevant to consider in the maritime industry. The machinery safety standard, ISO 12100 (2010), contains a long list of example hazards in Appendix A to the standard. This also contains useful examples and goes into more detail than the IMO list.

It may be noted that these lists typically will contain both hazards, hazardous events, influencing factors, and triggering events, without any real distinction. Remembering that the purpose is to identify anything that can go wrong, this makes sense since more examples on these lists may contribute to trigger more discussion and ideas for what may be relevant.

However, when we are processing the results from a hazard identification, this is necessary to be aware of. Often, it can be found that what has been identified as "unique hazards" are contributors in the same accident scenario. If we are not aware of this, risk may be counted doubly and the risk associated with a system may be exaggerated. See Chapter 3 for an example of this.

It may also be noted that even if we call this step "hazard identification", the methods used in risk analysis in most cases assume that it is hazardous events that are identified. For the hazardous events we first the identify causes of the events and secondly how these events may develop into an accident.

The hazard identification is essential for the quality of any risk analysis process. If the hazard identification is not complete, there will be hazards and hazardous events that are not considered in the risk management process and that may come as a surprise if they are realized. The problem is of course that we have no guarantee that we identify everything in a process like this. The only thing we can do is try to do it in the best possible way, at least ensuring that we have done our best. Some key factors in improving hazard identification are:

- Apply systematic methods – in the rest of this chapter, some of the most used methods for systematic hazard identification are described.
- System description/breakdown – an important help in ensuring that the identification is done systematically is that the system is described and broken down in a suitable manner. Without a proper description (e.g., of what hazardous materials are used), it is harder to identify all relevant hazards. Similarly, a suitable breakdown will tend to focus the discussion in a better way and also makes systematic identification easier.
- Involve experienced personnel – personnel with experience from operation of the system (or similar systems) will normally also have gained experience with failures, near misses, incidents, and accidents. This may help to identify hazards and events that personnel without experience may disregard because they cannot see how it can happen or they disregard it for other reasons.

- Systematic use of past experience – In addition to involving experienced personnel, it is also helpful to systematically gather and use information from earlier accidents, experienced failures, etc. This can be from the company, from industry, or more in general.

Spending sufficient time on the hazard identification is essential: What is not identified will not be analyzed and will therefore not be considered in the management of risk!

The outcome of the hazard identification should be a structured list of hazardous events that can be analyzed further.

8.2.5 Causal and frequency analysis

For each of the identified hazardous events, the causes of the events are identified and described. Each hazardous event may have several causes. Further, we can often find that for the hazardous event to occur, a chain/combination of events and conditions may be required. As far as possible, all significant causes should be described, and what combinations of events and conditions are required.

Figure 8.4 illustrates this principle. From the hazards, there may be several causal chains composed of one or more events that lead to the hazardous event. Further, there may also be several consequence chains leading to the end consequences. The figure also illustrates how barriers can be introduced to stop development of the causal and consequence chains. Barriers are discussed in detail in Chapter 15.

How deep into details the analysis should go should be decided before the work is started. In many cases, it is possible to go into almost endless details. Some factors that may determine how detailed this should be done:

- What is the purpose of the analysis? For ranking of risk, less details are usually needed than for improving designs or operations.

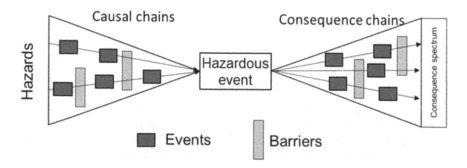

Figure 8.4 Bow-tie illustrating causal chains and consequence chains.

- How much effort is required to analyze causes? There should be a reasonable balance between effort and expected benefit.
- What causes can we influence and make changes to? Causes that are far beyond what can be influenced by the decision-makers are usually less useful to go into details on because it is not possible to reduce risk by changing them. Examples of such factors are regulations, market conditions, and external environment. We can identify that bad weather can be a cause of some types of accidents but going further into details on the causes of bad weather is in most cases not very relevant.

If the risk analysis is purely qualitative, no further activities are undertaken in this step, but if a quantitative analysis is performed, it is necessary to determine the probability or frequency of the hazardous event. This can be done in different ways, depending on the method applied and the data available.

Fire in a cabin on a cruise ship

Consider the hazardous event "Fire in a cabin on a cruise ship". The following is an illustration of how we can go deep into the causes of the event by repeatedly asking "How can this happen?" or "Why does this happen?"

- Direct causes of the event can be, e.g., smoking, electrical overheating, and open flame.
- We choose to explore the cause "electrical overheating" in more detail. Possible causes may be faulty wiring, faulty equipment connected to power or overloading.
- Looking into "faulty wiring" in more detail may, e.g., show that this can be due to faulty construction, failure of wiring, or faulty design.
- In turn faulty design can potentially be caused by mistake by the designer, inadequate competence of designer, inadequate standards for designing electrical systems.

8.2.6 Consequence analysis

For each of the hazardous events, the possible consequences are identified. This can often be divided into two parts:

- The hazardous event can often develop in many different ways, depending on specific conditions and events. These conditions and events must be identified, and we describe how the hazardous event can develop in different ways.
- When the chain of events has been developed, we need to determine what the effect of the event is. These effects are what we normally call the consequences.

In many cases, the end consequences may vary greatly and a "consequence spectrum" may be defined, describing possible consequences from worst case to least serious consequences (often zero). The different consequences are often associated with different scenarios ("paths") that the hazardous event may follow. These scenarios need to be described, including the events and conditions that lead to the consequences from the hazardous event.

The consequences may vary not only in severity and magnitude but also in type. The most common types of consequences that we distinguish between are humans (injury/loss of life), environment (for the maritime industry it is mainly spills to sea that are of concern), and economical values (damage to or loss of ship and cargo). For each consequence type, we may have a consequence spectrum.

As for the causal analysis, we need to decide on how detailed we want to do the consequence analysis and also what types of consequences we identify before starting the risk assessment. Many of the same factors will be relevant to consider also for the consequence analysis.

Fire in a cabin on a cruise ship

A fire may develop in many ways, depending on the situation. Some examples of events and conditions that may influence how it develops are:

— Are there people present in the cabin or not? If people are present, they may quickly detect and extinguish the fire, avoiding serious damage or injury. On the other hand, if there are people present and sleeping, they may be exposed to serious injury or even fatality.
— Does the fire detection system detect the fire? Normally, there will be smoke/heat detectors that detect the fire automatically, but these systems may fail. The development of the fire will depend on this.
— Similarly, there will be a sprinkler system that is triggered by a fire alarm. This may however also fail.

Based on the above, we can also see that the consequences can vary significantly depending on the specific chain of events:

— In the worst case, a fire in a cabin may spread outside the cabin to other cabins and eventually large parts of the ship. In such a case, many fatalities and many injuries may be the result.
— In the best case, the fire will be detected quickly and extinguished. No injuries will occur at all.
— In between these two extremes, there are a whole range of possible outcomes.
— In addition to injuries and fatalities, there may also be other consequences. Damage to the cabin or to larger parts of the ship is one possible consequence. Repairing the ship is a cost that has to

be covered by the ship owner/insurance company. Another possible consequence is that the ship may be out of operation for a shorter or longer period of time. This has cost implications for the ship owner.
 – If the ship is extensively damaged, there may also be potentially harmful spills of fuel, oil, and other liquids/chemicals.

8.2.7 Risk presentation

When both the frequency and the consequence of each hazard have been estimated, they are combined to form an expression of risk. Risk may be presented in many different and complementary forms (Chapter 9). The results of a quantitative analysis can be summarized in a single number, e.g., PLL, FAR, or IRPA. A more comprehensive result presentation is obtained through use of an FN curve. Similarly, for qualitative studies it is quite common to use a risk matrix to present the results.

In practice, the numbers that we arrive at will not give a sufficient description of risk. In order to understand the risk, and in particular when we want to make decisions regarding what to do with risk, we need more information than what a single number tells us (Figure 8.5). Some examples of additional information that is useful are as follows:

 – Identification of the main contributors to risk
 – Descriptions of important/typical accident scenarios
 – Descriptions of uncertainty associated with the risk analysis
 – Results from sensitivity analyses showing how results may change if key assumptions and input values are changed

Another aspect of this is that the way risk is presented can have a big impact on the decision that is made. Risk communication to relevant stakeholders

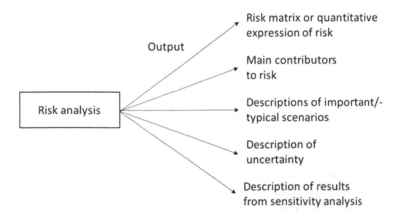

Figure 8.5 Typical output from a risk analysis.

has been a significant research topic, see, e.g., Fischhoff (1995). Care therefore must be taken when the way to present risk is decided. Among the factors that should be considered are:

- Who are the stakeholders and what expertise do they have in risk analysis and interpretation of results from risk analysis? Particularly important are of course the decision-makers and their competence.
- How are the decision criteria formulated? The results from the risk analysis must be presented such that it is possible to compare the results with the acceptance criteria, otherwise a decision will obviously be difficult.
- What is the effect of key assumptions and uncertainties? It is not sufficient to present the results from the risk analysis, but also to describe the uncertainties associated with the results. If the uncertainties are so large that there is uncertainty about whether we meet the decision criteria or not, that is obviously important information. Further, key assumptions that may significantly affect the results should also be highlighted, to give the decision-makers an opportunity to judge these assumptions and their impact on the results.
- Is there a need for presenting risk in alternative ways? Different ways of presenting risk have different strengths and weaknesses and it can often be useful to present risk in several ways, e.g., using an FN curve and an individual risk value. This will give the users a more nuanced picture of the results and gives a better basis for making decisions.

8.2.8 Comparison with risk acceptance criteria

When the results from the analysis have been summarized and presented, the next step is to compare these with the risk acceptance criteria. It is not the task of the risk analyst to decide whether it is necessary to reduce risk, only to make the comparison and present whether the results meet the criteria or not. To supplement this, information about sensitivity of the results and uncertainty should be provided. This is particularly important if the uncertainty may be so large that the conclusions also are uncertain. Typically, this will be the case if the results show that the risk is close to the risk acceptance criteria.

8.2.9 Identify and evaluate risk reduction measures

In most cases, the risk analysis will also include identification and evaluation of possible measures that can be introduced to reduce risk. Doing the risk analysis in itself will obviously not reduce risk; it is only when we implement improvements that a reduction will be achieved. To do this in the best way, with regard to risk reduction, cost, and other effects, a systematic identification and evaluation of risk reduction measures is essential.

Within the FSA framework, Risk Control Measures (RCM) and Risk Control Options (RCO) are two key terms. RCMs are means that can control a single element of risk. This can include means that contribute to remove or reduce hazards, reduce the probability of hazardous events occurring, or reduce the consequences.

RCM and RCO

One hazardous event that is potentially critical for ships is ship-ship collision. Some examples of possible RCMs are (without considering the practicality or the cost):

- RCM1: Traffic Separation Schemes (TSS) – external to the ship. Will increase distance between ships, thus reducing the probability of a collision.
- RCM2: Requirements for strengthened manning on the bridge in areas with high traffic density – will reduce the probability that a critical situation is not detected and interpreted correctly in time.
- RCM3: Requirements for simulator training for navigators sailing in areas with high traffic density and complex navigation – will reduce the probability the navigator misinterprets the situation or make other mistakes.
- RCM4: Traffic surveillance and warning to ships in potentially critical situations – external to the ship. Warns the navigator of critical situations, increasing the probability that it is detected and understood in time.
- RCM5: Double-sided hull – will make the ship more robust against penetration and flooding should a collision occur, thus reducing the consequences.

All of these contribute to reduce risk of ship-ship collision, but they work in different ways and are also implemented by different actors. A couple of possible RCOs are as follows:

- RCO1: Reducing the probability of collision through measures on the ship (combining RCM2 and RCM 3).
- RCO2: Reducing the probability of collision through measures external to the ship (combining RCM1 and RCM 4).

RCOs will typically be on a higher level than RCMs. The two RCOs could also have been combined, but in practice it is natural to separate measures related to the ship and its crew from measures that are external since there are very different stakeholders involved in the implementation.

Detailed methods for how to approach identification and evaluation of risk reduction measures can be found in Chapter 15. Cost-benefit analysis

(CBA) is often used to evaluate measures. In a CBA, a comparison is made of the costs involved in implementing a measure against the benefits that are achieved in terms of risk reduction. In order to do this comparison, the risk reduction also has to be expressed in monetary values. CBA is described in Chapter 15.

8.2.10 Reporting

There is no standard format for reporting risk analyses as this will depend on both the scope of the analysis, the level of detail, and the end users. The following is an example of a fairly comprehensive, quantitative analysis (Table 8.1). More information on content of reports can be found in e.g., ISO 31010 (2019) and Rausand and Haugen (2020).

8.2.11 Limitations of risk analysis

Risk analysis (and assessment) is a powerful tool in obtaining information and increased understanding of a system, its hazards, and the accident mechanisms. This information and understanding makes us able to implement risk control options and thus improve the system's safety. However,

Table 8.1 Typical content of a risk assessment report

Heading	Content
Executive summary	A brief summary of the study and the main results
Introduction	Objectives and scope, risk acceptance criteria
Work organization	Study team, timeline of study
System description	Description of the system and the context
Methods and data	What methods have been chosen and what data have been applied and why have these choices been made
Risk analysis	Overview of the risk analysis
Risk reduction measures	Identification and analysis of possible relevant risk reducing measures
Discussion of results	Discussion of results, focusing on main contributors and other important findings. Discussion of results compared to risk acceptance criteria. Should also include discussion of uncertainty and results from sensitivity analysis
Conclusions	Summary of the main conclusions and recommendations that result from the analysis
References	
Appendixes	The content of the appendixes can vary a lot, but typical information may be detailed system description including drawing, etc., data dossier containing details of all input values to the analysis, key assumptions that have been applied in the analysis work, detailed analysis models and calculations

one should be aware of the limitations of such analysis, especially in relation to quantitative analysis. The lack of good statistical data due to limited experience is probably the most significant and common limitation in quantitative analysis. This is particularly clear in a maritime context where the number of large-scale accidents is quite low. Lack of statistical data results in huge uncertainties in the analyses' outcomes, and one should therefore always evaluate these uncertainties and include this evaluation in the decision and recommendation process.

The complexity of most systems makes it necessary to make several simplifying assumptions in order to be capable to perform the analysis. These simplifications also create uncertainties.

A major limitation of traditional risk analyses is that human and organizational factors are usually not given adequate attention. During the last decades, it has become a well-established fact that human and organizational factors affect the safety of technically complex systems, conventional ships and other vessels being no exemption. These factors materialize themselves as active failures and latent conditions that breach the defenses that prevent hazards from becoming severe losses. In technical systems that interact with humans, active human failures are normally considered to be the largest single cause of accidents. Investigations suggest that approximately 60% of all accidents are caused directly by human errors. In addition, some accidents are more indirectly caused by human errors, being a result of so-called organizational factors, e.g., company policies and safety culture.

Risk analysis is not a prediction of whether or when an accident will occur. It can only provide a probability that an accident will occur, and this tells us something about how likely it is that an accident will occur. Even if that probability is high, say 0.9, this does not mean that an accident is certain to happen. Also, even if the frequency of an accident is small, this does not mean that it cannot happen tomorrow. It is too easy to make the mistake of assuming that if the frequency is 0.01 and the return period in other words is 100 years, the accident will not happen until after 100 years. However, this frequency only tells us that it will happen once in that 100-year period, not when it will happen.

After accidents have occurred, criticism is often raised against risk analysis because it did not specifically identify and analyze the accident that occurred. This will nearly always be the case, because we very seldom analyze all possible scenarios. This will be too time-consuming, and the value added will typically decrease the more detailed we analyze the scenarios. Instead, we choose representative scenarios that cover a whole range of more specific scenarios and analyze all together. This is far more efficient and adequate in most cases.

Remark: As a risk analyst, we can always claim that we are right. If, e.g., a ship collision occurs, we can always say that "I told you so, this was analyzed in the risk analysis, and we found that this could happen". On the other hand, if it does not occur, we can argue that "I told you that the probability of this happening was very low."

8.3 PRELIMINARY HAZARD ANALYSIS (PHA)

8.3.1 Introduction

A simple method for risk analysis that is used often is Preliminary Hazard Analysis (PHA). The method is qualitative but will still give a ranking of the identified hazardous events with respect to risk. It is simple to apply and does not require extensive resources. It is therefore commonly used to get a quick overview of risk.

PHA is a method that represents a group of methods with different names, using different terminology, and applicable to different types of problems. The principle of the approach is however the same in all methods. Several of the following methods described in this chapter belong to the same group of methods.

It should be noted that PHA also sometimes is used as an abbreviation for Process Hazard Analysis, which is a method mandated in the Process safety management regulations (OSHA, 2002).

Further reading about the method can be found in the following:

- Rausand, M. and Haugen, S. (2020). *Risk assessment: Theory, methods and applications*, Wiley, Hoboken, New Jersey, USA (Rausand and Haugen, 2020)
- Vincoli, J.W. (2006). *Basic guide to system safety* (Vol. 18). New York: Wiley-Interscience. (Vincoli, 2006)
- Kjellen, U. and Albrechtsen, E. (2017). *Prevention of accidents and unwanted occurrences: Theory, methods, and tools in safety management*. CRC Press. (Kjellen and Albrechtsen, 2017)

8.3.2 Objectives

The objectives of a PHA can be many, but it is commonly used to obtain an overview over the hazardous events and rank them according to their contribution to the total risk. As indicated by the name, it is often used as a preliminary analysis, forming the basis for more detailed analysis later. The highest contributors to risk are then typically analyzed more in detail. On the other hand, it also commonly used as a stand-alone analysis, in particular for simple systems (IMO, 2018).

8.3.3 Applications

PHA is a general and non-specific method that has not been developed for any specific application or industry. Accordingly, it is also being used very widely. Some examples where a PHA may be relevant to use in the maritime industry are as follows:

- Establish initial overview of hazardous events and associated risk levels for a new ship design, a port, an aquaculture facility, etc.
- Establish initial overview of hazardous events and risk levels for a technical system, e.g., power system, navigation system, and evacuation system.
- Establish initial overview of hazardous events and risk levels for an operation (e.g., a complex loading/offloading operation or a construction process).

8.3.4 Method description

In the method description, we focus on the four steps that comprise the risk analysis, i.e., hazard identification, causal and frequency analysis, consequence analysis, and risk presentation. For the other steps, consult the description in Section 8.2.1.

8.3.4.1 Preparations

If the system is large and composed of very different parts or activities, it is often useful to do a breakdown of the system before starting the process. This will normally help to make the process of identification more focused and systematic. The approach to the breakdown will depend on the problem and the system (see Section 8.2.2).

A PHA is usually performed by a team consisting of participants who know the system well and a team leader who is familiar with the method. The analysis is usually performed in one or a few work meetings.

A necessary part of the preparations is to define the categories to be used in the frequency and consequence analysis. The risk matrix is discussed in Chapter 9. In some cases, companies have standard risk matrixes that they use, or standards and guidelines may also specify this.

8.3.4.2 Hazard identification

In this step, the aim is to identify all relevant hazardous events. The process of identifying these often entails identification of a mix of hazards, hazardous events, accidents, and consequences. In principle, the longer the list of candidates is, the better in order to target all potential hazards. The identification is typically performed in a brainstorming session, usually supported by some form of checklist that can support the creative process.

The checklist that is used is then applied for one part at a time, before moving on to the next part of the system.

Hazard identification for a ship

A ship is considered to be a large system, with a range of hazardous events that may be possible. The hazardous events that are relevant to consider in the engine room are quite different from what is relevant for, e.g., the bridge or the cargo areas. Similarly, if activities are considered there will be large differences between, e.g., navigating in open seas, docking, and loading/unloading. This indicates that it may be useful to divide the ship into smaller parts.

The disadvantage with the breakdown is that the process will tend to be more time-consuming the more detail the system is broken down into. The objective of identifying all hazardous events may therefore have to be weighed against the effort required to achieve the objective.

Many standards and other documents contain checklists of hazards and hazardous events. In many cases, these are developed for specific purposes. Within the maritime industry, the IMO FSA Guidelines (IMO, 2018) contain an appendix called "Examples of hazards" that may be used. It may be noted that the list contains a mix of hazards and hazardous events. In addition, Appendix 1 also contains a table with "Examples of human-related hazards" which may be relevant to use. In addition to these, there are also other checklists that may be relevant, even if they are not specific for the maritime industry. ISO 12100 (2010) contains a comprehensive list of hazards that are relevant for all types of machinery. A lot of this may be directly relevant also for maritime applications. Further, ISO 17776 (2016) has a similar checklist for use in the offshore oil and gas industry. Significant parts of this list will also be relevant.

8.3.4.3 Causal and frequency analysis

The causal analysis can be divided into two steps:

- Identification and qualitative description of the causes of the hazardous events. Often, focus is on the direct causes, with less emphasis being given to indirect causes and detailed causal chains. This may vary however, depending on the scope and objectives of the study.
- Determining a frequency (or probability) ranking of the hazardous events. This is typically on a scale ranging from 1 to 4–7, depending on how detailed the ranking is done. More details about how to design risk matrixes are given in Chapter 9.

Causes of hazardous events

In the following, two examples of some possible causes of hazardous events (HE) are described:

HE1: Container dropped during handling. Causes: failure of crane, accidental release by operator, failure of container

HE2: Engine room fire. Causes: ignition of released flammable liquids, overheating of rotating equipment, overheating of electrical equipment, open flame (welding, cutting)

In the first example, the causes are partly failure specific and more detailed analysis of the causes is probably not required for the purpose of a PHA. It may be concluded, however, that "failure of the crane" needs to be investigated in more detail because it is likely that it can fail in many different ways, both critical and non-critical.

In the second example, several of the causes are broader and may be too generic to be of much use. "Release of flammable liquid" can be a number of different things (fuel, lube oil, hydraulic oil, others) and can come from a wide range of sources. We may therefore want to be more specific. Similar with "Overheating of rotating equipment" where it will be beneficial to identify more specifically what equipment may overheat. In this situation, it may therefore be useful to split HE2 into several more specific hazardous event, e.g., "HE2.1 Engine room fire due to release of flammable liquids" and "HE2.2 Engine room fire due to overheating of rotating equipment".

8.3.4.4 Consequence analysis

The consequence analysis is typically also done in two steps:

- Identification and qualitative description of the consequences of the hazardous events. In this step it is necessary to separately describe different types of consequences.
- Determine what consequence category the hazardous event should be placed in. This applies separately to each of the consequence types that are being considered. The results may be that a hazardous event falls in one consequence category for, e.g., personnel and in another for, e.g., environmental consequences.

As mentioned earlier, a hazardous event may have a wide spectrum of possible consequences, from nothing to catastrophic. In the description, it may be useful to indicate the whole spectrum, but when the consequence ranking is done, we have to choose one consequence category to represent the whole spectrum. This is clearly a weakness with PHA, and it may be difficult to choose the "correct" category. Different approaches are used for this:

- One approach is to select the category that represents the worst-case consequences. This may seem reasonable, based on the argument that it ensures that we do not underestimate risk. However, this will tend to overestimate the risk very much, sometimes by several orders of magnitude. In many cases, it is therefore not a good option.

- A second approach is to use an "average" consequence. This will give a risk that is closer to a correct value based on the definition of risk that is applied in this book.
- Defining what an "average" consequence is may however be difficult. An alternative may therefore be to say that it is the "most likely consequences" that should be considered.
- A final approach is to select a "credible worst-case" scenario. This is something in between the two first approaches and gives conservatives estimates of risk at the same time as not being overly conservative. In many cases, this is a reasonable approach to apply. It may be difficult to define precisely what "credible" means and it may be necessary to provide examples to illustrate the concept. This may be done by specifying a probability limit (e.g., "one in ten hazardous events of this type will have this consequence or worse").

Choosing consequence category when using a risk matrix

In an earlier example, fire in a passenger cabin was considered and a range of possible consequences were described.

If we apply the "worst-case" approach, the consequences would be classified as causing many fatalities, extensive damage to the ship and environmental damage due to spills. Instead, if we consider a "credible" worst case, it is probably more reasonable to say that the fire is primarily limited to one cabin, but that persons in that cabin may be severely injured. The consequences for the ship will be limited and no environmental consequences are foreseen.

8.3.4.5 Result presentation

The main results from a PHA will typically comprise:

- A table summarizing the hazardous events, causes, and consequences. Many variants of tables exist but a typical example would contain the following columns:
  - Identificator (numbering of the identified hazards/events)
  - Part of the system
  - Hazard or hazardous event
  - Description of causes
  - Description of consequences
  - Ranking of probability/frequency (according to categories)
  - Ranking of consequences (according to categories)
- A risk matrix with the individual hazardous events illustrated.

This can also be supplemented with other types of information:

- Ranking of the individual hazardous events, based on a risk index or simply the position of the event in the risk matrix.
- More detailed description of the highest-ranking hazardous events, to the extent that this is available.

Extract of a PHA for a combined passenger/car ferry

In Table 8.2, a few examples of events that may occur on a ferry have been illustrated. This is by no means a complete analysis, but it illustrates the range of events that may be included and the information that may be included in a PHA. Note that this is a simplified example of how the reporting format may look like.

The definition of Frequency (F) and Consequence (C) are not included here and fictious numbers are included in the table for illustration only.

In this case, events are listed rather than hazards. This means that everything listed here are something that occurs. We could also have listed hazards or a combination of events and hazards. The hazard associated with the event "Fire in engine" could be "Flammable material present".

Table 8.2 Extract of events analyzed in a PHA

Id	Event	Causes	Consequences	F	C
1	Grounding	Loss of power Navigator distracted Low visibility	Minor damage, denting Penetration, flooding Sinking	1	4
2	Fire in engine	Release of fuel oil Hot surface Welding Overheating	Minor fire Extensive fire in engine room Fire spreads outside engine room	1	4
3	Fire in galley	Electrical short circuit Overheating of equipment	Local fire Fire spreads outside galley	3	3
4	Passenger falling into the sea	Leaning too far over side Wet/slippery deck Movement of ferry Passenger unsteady	Hypothermia Drowning	2	4
5	Blackout	Engine failure Electrical failure	Ferry stars drifting Grounding	2	2
6	Car hitting passenger/crew	Too high speed Driver not observant Passenger walking in front of car	Light injury Serious injury Fatality	3	3
7	Passenger falling in stairs	Movement of ferry Slippery/wet stairs Passenger dizzy	Light injury Serious injury Fatality	4	3

This is not an event, but a state or a condition. This could be turned into an event by calling it "Release of flammable material" instead.

Under causes and consequences, only some examples are listed. In many cases, it is possible to identify a large number of causes and also a wide range of possible consequences. Normally, we will limit ourselves to perhaps a handful of the most likely causes.

One possible consequence of blackout is listed as "grounding". Grounding is also listed as an event. This is something that is observed quite often in PHAs, that events that can belong in the same event chain are identified as separate event. As long as the results are used purely qualitatively and to identify risk reduction measures, this is not a problem but if we later use the results for quantification, we need to make sure that we do not double count risk.

8.4 SAFE JOB ANALYSIS (SJA)

8.4.1 Introduction

Safe Job Analysis (SJA) is a simplified version of a PHA, where the identified hazardous events not necessarily are categorized and plotted in a risk matrix. Instead, the analysis may be purely qualitative. Several names are used for this method, including Job Safety Analysis (JSA), Task Hazard Analysis (THA), and Job Hazard Analysis (JHA). In principle, they are identical although small variations in terminology and reporting may occur.

Some references that provide more detailed descriptions of the method are as follows:

- Roughton, J. and Crutchfield, N. (2011). *Job hazard analysis: A guide for voluntary compliance and beyond.* Butterworth-Heinemann (Roughton and Crutchfield, 2011)
- Rausand, M. and Haugen, S. (2020). *Risk assessment: Theory, methods and applications.* Wiley, Hoboken, New Jersey, USA (Rausand and Haugen, 2020)
- NOROG (2017). *090 Norwegian Oil and Gas recommended guidelines on a common model for safe job analysis*, Rev. 4 (NOROG, 2017)
- OSHA (2002). *Job hazard analysis* (OSHA, 2002)

8.4.2 Objectives

The objective of an SJA is to contribute to ensuring that work operations can be completed safely, without injury or fatalities. Further, the SJA will also normally identify needs for modification to the planned work to make it safer and identification of needs for risk reduction measures (usually personal protective equipment, PPE).

8.4.3 Applications

SJA is primarily aimed at analysis of work operations that have the potential to cause injury or fatality to personnel. Non-routine operations that are unfamiliar to the people involved and operations with high loss potential are the main target of analysis. The method can be applied to:

- establish a safe work procedure for one-off operations.
- introduce personnel involved in the work to a work operation.
- prepare personnel in the work before starting the work operation.
- improve existing work procedures where accidents or incidents have occurred.

The approach is largely the same in all cases although the preparations before starting the analysis may be different.

8.4.4 Method description

8.4.4.1 Preparations

The preparations before starting the SJA consist of:

- Collecting information about the work operation and breaking the operation down into suitable steps for analysis
- Deciding who should take part in the analysis. In most cases, this will be the personnel that is going to perform the work operation
- Deciding on a reporting format and a risk matrix if this is going to be applied.

8.4.4.2 Hazard identification

The hazard identification is based on a breakdown of the work operation into distinct steps or activities and each step is examined systematically to identify what may go wrong. A checklist can be useful in this process, in the same way as with PHA. In many cases, checklists for work in general (not necessarily specific for maritime applications) will be adequate to use. Examples are the checklist in the OSHA Job Hazard Analysis booklet (OSHA, 2002) and the checklist in the NOROG guidelines for Safe Job Analysis (NOROG, 2017).

The focus in the hazard identification will tend to be on the workers involved and the hazardous events that can lead to injury or fatality among them. Other consequence types, such as environmental damage and economic loss, are more seldomly covered.

8.4.4.3 Causal and frequency analysis

There are different variations of safe job analysis, with different approaches to the causal and frequency analysis. In the simplest versions, the causal analysis is dropped completely, instead the focus is purely on what the possible consequences are, and the measures recommended to prevent and protect the workers. In more comprehensive versions, the analysis is principally very similar to what was described for PHA, with listing of causes and a ranking of the frequency based on a set of predefined categories.

8.4.4.4 Consequence analysis

The consequence analysis is also in most cases quite simple, listing the possible injuries that can occur to the workers involved. This can be, e.g., burns, crushing, falls, and chemical exposure. As for the causal analysis, ranking of consequences similar to a PHA can also be used in some cases.

8.4.4.5 Result presentation

The results are usually presented in a tabular form, using a predefined format. A simple example is shown in Table 8.3. An important part of the

Table 8.3 Illustration of SJA for entry of a tank

Operation	Hazard	Risk reduction measures
Opening hatch	Overpressure	
	Toxic/flammable gas escapes	
Securing hatch	Hatch falls and injures person	
Ventilate space	Insufficient oxygen	Check air in space prior to entry. Wear breathing equipment if necessary
	Toxic gas	Check for toxic gases prior to entry. Wear breathing equipment if necessary
Enter space	Being overcome by gas	Inform duty officer prior to entry Carry oxygen analyzer Carry lifeline Standby watch outside space Rescue and first aid equipment available
	Falls	Wear harness
Perform work	Hazards depend on what work is taking place, e.g., inspection, welding, and sandblasting	
Leave space	Falls	Wear harness
Close hatch	Crushing, etc.	Keep body parts clear of hatch Use lifting equipment as necessary

safe job analysis is to identify measures that can contribute to reduce risk. This will primarily be operational measures that are possible to implement immediately before starting the work. Examples are modifying the work procedure and introducing personal protective equipment (PPE).

Safe job analysis of a work operation

A common operation on ships that can be hazardous is entry of an enclosed space, e.g., a tank. This operation can be hazardous for several reasons, in particular because of potential lack of oxygen or presence of toxic fumes, but also for other reasons. In Table 8.3, a simple illustration of an SJA for entry of a tank is shown. The illustration is not complete, neither with respect to the steps in the operation, hazards nor risk reduction measures.

The format used here is very simple and many other variants of form can be found in different applications. In practical use, this is however adequate in many situations.

8.5 FMECA

8.5.1 Introduction

Failure Modes, Effect and Criticality Analysis (FMECA) is a systematic method for identification of failures of technical systems and what their effect may be on system operation. FMECA was one of the first risk analysis methods that was developed, for military applications in the late 1940s. Today, it is one of the most commonly used methods.

FMECA is sometimes also called FMEA (Failure Modes and Effect Analysis). In most applications, there is no distinction between these two and the term FMECA will therefore be used in this book.

More details of FMECA can be found in, e.g.:

- ABS (2003). *Guide for risk evaluations for the classification of marine-related facilities*, June 2003, Houston, TX, USA
- IEC 60812 (2018). *Analysis techniques for system reliability - Procedures for failure mode and effect analysis (FMEA)*, IEC, Geneva (IEC 61802, 2018)
- Rausand, M. and Haugen, S (2020). *Risk assessment: Theory, methods and applications*, 2nd Ed, Wiley, Hoboken, NJ, USA

8.5.2 Objectives

In an FMECA, we identify all failure modes of technical systems and analyze the effect of these failures on the system function. The objectives of this may be several:

- To identify needs for improving the design, to improve reliability of the system. Failure modes that have a critical impact on the functioning of the system should be eliminated as far as possible through design improvements, e.g., redundancy or high reliability components.
- To form a basis for maintenance planning. The criticality of the failure modes can also be used to prioritize maintenance activities. Failure modes with high criticality will typically require preventive maintenance actions while for those with lower criticality, corrective maintenance may be acceptable.
- To form a basis for troubleshooting system failures. By observing the way that the system has failed, this can be compared to the overview over effect of failure modes from the FMECA. This can point toward what failure modes have occurred and help the repair of the system.

8.5.3 Applications

FMECA is developed for use on technical systems, and in particular safety critical technical systems. Examples can be fire detection systems, fire extinguishing systems, ballast systems, and safety critical utility systems. It may also be used for production critical systems, where the need for minimal downtime of the system is important.

8.5.4 Method description

8.5.4.1 Preparations

As for all types of analysis, the first part of the preparation consists of establishing a system description and breakdown of the system, either a functional breakdown or a subsystem and component breakdown. The level of detail of the analysis should also be determined and the level of detail in the breakdown is determined by this.

Let us consider a simple ballast system that can be described as shown in Figure 8.6. This is by no means a complete system description but illustrates how we can break down the system into subsystems and component groups.

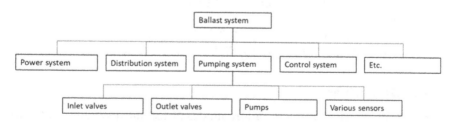

Figure 8.6 System breakdown for ballast system.

This can also be further broken down into specific components, e.g., the individual valves that are part of the system.

In addition to this, the operational modes of the system also need to be defined. System will often function differently and be in different states in different operational modes and this will have to be considered in the analysis.

Operational modes for ballast pumps

The ballast pumps will have at least two operational modes that must be considered in the FMCEA:

– A standby mode, where the pumps are not active but waiting to be activated when there is a need to pump ballast from or to a tank.
– An operational mode, where the pumps are running and pumping water from/to tanks.

The failure modes will clearly be different. For the pumps, a relevant failure mode in standby mode is "pump does not start". In the operational mode, this is not relevant since the pumps already are running, but "pump stops" is relevant to consider. Other modes may also be relevant to consider, e.g., by specifying more precisely what the pumps are doing, e.g., filling tanks, emptying tanks, and moving water between tanks. The failure modes and the consequences are not necessarily the same in all these situations.

FMECA is often performed by a team, with design engineers from different disciplines taking part together with a team leader with FMECA experience.

It is also necessary to decide the categories to be used in the risk matrix.

8.5.4.2 Hazard identification

In FMECA, failure mode is a key term, and it is necessary some terms related to this (adapted from Rausand and Haugen, 2020):

– Fault: A state where a component or a system is not able to perform as required
– Failure: An event where a component or a system loses its ability to perform as required
– Failure mode: The manner in which the failure occurs, independent of the cause of the failure

In other words, a system that experiences a failure event will go into a fault state and the way that the failure occurs is the failure mode. In FMECA, it is the failure modes that we identify. Examples of failure modes for a pump could be "does not start", "does not stop", or "does not deliver the required amount of liquid". Note that a failure does not necessarily mean complete loss of a function of a system or a component. A partial loss may also be regarded as a failure, if this means that the system or component does not do what we require it to do anymore.

System failure vs component failure

Failure of a component in a system does not necessarily mean that the system experiences a failure. If we have a system with two pumps in parallel, and where each pump can supply 100% of the requirement, failure of one of the pumps will not lead to failure of the system, as long as the second pump still is running.

8.5.4.3 Causal and frequency analysis

As for other methods, identification of the causes of the failure modes is the next step. Focus will normally be on the direct causes of failure. The frequency analysis consists of determining which frequency category the failure mode belongs to.

In the example in Table 8.4, six categories are used, but it is also common in some applications to use a higher number, up to ten categories. Other ways of classifying can be where categories are defined based on the probability that a failure mode will occur in the lifetime of the system, e.g., "Failure mode very unlikely to occur in the lifetime of the system (Prob<0.01)".

An additional aspect that is often included in FMECA is the probability of detection of a failure mode (Table 8.5). This is particularly relevant for many safety systems, because they often are in a standby mode in normal operation, being activated only when an emergency arises when they are required. This means that there may be "hidden" failures in the system that are not easily detected before the system is required. Failure modes that have serious consequence and at the same time cannot be easily detected are

Table 8.4 Example of categories of frequency (occurrence) in FMECA

Occurrence (O)	Description
1	Occurs once in 10,000 years
2	Occurs once in 1,000 years
3	Occurs once in 100 years
4	Occurs once in 10 years
5	Occurs once every year
6	Occurs once a month

Table 8.5 Example categories of detectability in FMECA

Detectability (D)	Description
I	Failure mode will always be detected (Prob > 0.99)
2	Failure mode will normally be discovered (Prob = 0.80)
3	Failure mode may be discovered (Prob = 0.50)
4	Failure mode unlikely to be discovered (Prob = 0.20)
5	Failure mode cannot be detected (Prob < 0.01)

Note: The ranking should be defined such that the rank (D) decreases with increasing probability of detection.

more critical than those that can be detected immediately. The probability of detection is therefore often also categorized in a similar manner as the frequency, although usually on a separate scale.

It is noted that the probability of detection also will have to be related to time. An example can be for rank 1, where we can state "Failure mode will always be detected before the component is required to function". Most failure modes will be detected eventually, but the interesting aspect is whether it can be detected early enough to have any serious consequences. How the time to detection is defined will depend on the situation.

8.5.4.4 Consequence analysis

The consequence analysis in an FMECA focuses on the consequences for the system function and not the consequences to vulnerable assets. Failure of a safety system may clearly be important for, e.g., consequences to people, but since it is the failure mode and not the hazardous event that is being considered in the analysis, it is not possible to specify what the effects will be outside the system. An example may be a fire water system that is intended to extinguish fires. FMECA considers failure of this system but does not look at why the system is needed. If the event requiring the fire water system to be operational is not known, we cannot say what the consequences will be.

The consequences (or effects which is the term more commonly used in FMECA) are usually considered on two levels:

- Local effects – effects locally to where the component is located in the system. The local effects of the failure mode "fire water pump does not start when required" is that there is no water supplied from the pump.
- Global effects – effects for the whole system. Continuing the example with the fire water system, the global effect of "fire water pump does not start when required" is nothing, because we assume that the second pump operates as normal.

Table 8.6 Categories of consequences in FMECA

Detectability (D)	Description
1	Catastrophic – death, permanent total disability
2	Critical – permanent partial disability, injuries, or occupational illness that may result in hospital treatment for at least three people
...	Marginal – lost time injury
10	Negligible – minor injury/illness

Based on MIL-STD-882E (2012).

The last item points to an important aspect of FMECA: it is always assumed that just one failure occurs at a time. Simultaneous failures are therefore not considered, even if these may be very critical in many cases. This means that in systems with redundant components, the global effects of failure modes will normally be none.

The global effects are normally classified in the same way as is done in PHA. The definition of the categories will typically be very different, however. In Table 8.6, a simple example of possible categories is shown.

This example considers only risk to people, but other consequences can also be considered, in the same way as in a PHA.

8.5.4.5 Result presentation

The main results from an FMECA are reported in a table that may have different formats. The most common columns to include in the table are as follows:

- Id
- Operational mode
- Component
- Failure mode
- Failure causes
- Local effect
- Global effect
- Detection
- Occurrence ranking (O)
- Severity ranking (S)
- Detectability ranking (D)
- Risk Priority Number (RPN – see below)

The FMECA table may also act as an action plan, where the person responsible for resolving the issues identified is recorded, together with a deadline for completing the work.

It is quite common to calculate an RPN (Risk Priority Number) in FMECA. In FMECA, RPN is not defined in the same way as in a PHA (or in risk matrixes in general). Instead, it is common to express RPN as follows:

$$RPN = O \cdot S \cdot D$$

where
O = Occurrence ranking (frequency of occurrence)
S = Severity ranking (the global effect of the failure mode)
D = Probability of detecting the failure mode

RPN is an expression of the criticality of the identified failure mode, where a high RPN, implies high criticality. It is noted that criticality is not the same as risk, since we multiply with a third factor compared to our definition of risk.

FMECA of a propulsion system

The loss of propulsion power directly results in a loss of the controlled mobility of the vessel. In the HAZOP of the propeller (see example in Section 8.6) it was assumed that the controllable pitch propeller (CPP) was a critical subsystem for the propulsion system. The criticality is, however, dependent of the failure consequence and the failure likelihood. Hence an FMECA of the whole propulsion system may be appropriate.

A description of the propulsion system is given in Section 8.6.4. The resulting FMECA form is shown in Figure 8.7. This is by no means a complete FMECA of the system but gives an indication of what the outcomes may look like.

8.6 HAZOP

8.6.1 Introduction

HAZOP (HAZard and OPerability study) was developed for the process industry, aimed at analyzing hazards and operational problems in process plants. The origin of the method is not very clear, but the chemical company ICI described the method in the 1960s. The method was aimed at being "a lantern in the bow rather than a light at the stern". This points precisely to one of the main features of modern safety management, namely that we should try to identify what may happen in the future (what lies ahead) rather than just looking at what has happened in the past.

The method is very popular, and several variations of the method have been developed:

System: **PROPULSION**
Ref Drawing no.:

Performed by:
Date:

Page:

Ref. No	Function	Operational mode	Failure Mode	Failure cause or Mechanism	Detection of failure	On sub-systems	On system function	Resulting state	Failure rate	Severity Ranking	Risk Reducing Measures	Comments
						Effect of Failure						
1	Propulsion	Normal operation	Stop	No fuel feed	On watch (sound)	Gear over-load	Half effect	Reduced speed	Remote	Minor - Minor	Reducing pitch, cut off	Engine 2 is functioning Onboard repair
			-"-	Crankshaft Failure	-"-	-"-	-"-	-"-	Occasional	-"-	-"-	Engine 2 is functioning Harbour repair
			Reduced Function	Piston running hot	-"-	-"-	Reduced effect	-"-	Very Unlikely	-"-	-"-	Engine 2 is functioning Onboard repair
2	Gear - reduce number of revolutions (i.e. RPM) transmit power	Normal Operation	No power transmittal	Broken cog	-"-	Main Engine Over-Load	No propulsion	Loss of manoeuvr-ability	Remote - Occasional	Critical	Stop main engine, lock shaft line	Both primary and secondary gearwheels are failed Harbour repair
3	Shaft line- Transmit Power	Normal Operation	No power transmittal	Broken shaft	-"-	-"-	-"-	-"-	Remote	Critical	-"-	Repair in dock
4	Controllable pitch propeller (CPP) – transmit power	Normal operation	Reduced Function	Broken blade	-"-	-"-	Reduced effect	Reduced Speed	Occasional	Minor	Reduce main engine power, reduce propeller pitch	Damage to only one propeller blade Repair in dock

Figure 8.7 FMECA form for propulsion system.

- Procedure HAZOP – developed to analyze work procedures (a detailed SJA)
- Software HAZOP
- Computer HAZOP (CHAZOP)
- Human HAZOP

As the names indicate, they are aimed at specific purposes, where the main modifications compared to the original HAZOP method is that the system breakdown and the guidewords applied to support the hazard identification are different.

Further reading about HAZOP can be found, e.g., in:

- Crawley, R. (2015). *HAZOP: Guide to best practice*, Elsevier
- IEC 61882 (2016). *Hazard and operability studies (HAZOP studies) - Application guide*, IEC, Geneva
- Kletz, T.A. (1999). *HAZOP and HAZAN: identifying and assessing process industry hazards*, 4th Ed. IChemE.

IEC 61882 describes a generic HAZOP method, without reference to any particular application.

8.6.2 Objectives

The objective of HAZOP is to identify all deviations that affect safety or operation of the system. This is mainly used to improve the design of systems, both from a safety and operational point of view.

8.6.3 Applications

HAZOP was originally developed for the process industry but has gained wide recognition in many other industries. The oil and gas industry is one important example. Within the maritime industry, examples of systems where it can be useful are fuel systems, ballast systems, and lubrication systems. With new fuels such as LNG and hydrogen, it is also becoming increasingly relevant.

For other variants than the original process HAZOP, the use is much wider and is not limited to any particular industry or application.

HAZOP can be used in various stages of development of a system, from conceptual design to detailed design, operation, and modifications. The level of detail in the study will depend on the progress of the design details.

8.6.4 Method description

8.6.4.1 Preparations

In a HAZOP, the system is divided into parts (sometimes called nodes) before starting the analysis. How this is done will depend on the purpose of the HAZOP and also the level of detail in the study. In a process HAZOP, the system will be divided according to the flow through the system and where the flow changes (two flows merge, flows separate, pressure increases or drops, tanks, etc.). In a procedure HAZOP, the procedure will be studied step by step.

A HAZOP is typically performed by a team, in structured workshop meetings. The team will typically comprise:

- A HAZOP leader that is familiar with methodology and leads the discussion in the meeting
- A HAZOP secretary that records the discussion
- Relevant discipline engineers, e.g., process, mechanical, instrument, safety, and other specialists as needed
- Operators

The size of the team can vary significantly, from 6–8 people up to more than 20. The most efficient meetings and discussions are usually achieved if the number is limited to a maximum around ten people.

8.6.4.2 Hazard identification

The hazard identification is based on the use of guidewords combined with process parameters or properties of the system. The guidewords act as a checklist that is used to guide the identification. In a process HAZOP, there originally was a standardized set of guidewords, but for other purposes, other guidewords may need to be applied or they may get a different interpretation. Example guidewords are shown below (adapted from IEC 61882, 2016).

For a specific application, other guidewords may also be applied, and it is recommended that a list of guidewords suitable for the system is established before starting the analysis. In general, the set of guidewords described above can be used for many applications although one may choose to reformulate some of them (e.g., OTHER THAN can be replaced by WRONG ACTION if it is a procedure HAZOP).

In a process HAZOP, the guidewords are supplemented with process parameters to establish what we call deviations. In the two first lines of Table 8.7, the guidewords are combined with flow, to make up a meaningful deviation. "Deviations" in HAZOP terminology are the equivalent of the hazardous events that are identified in PHA.

Examples of process parameters may be flow, pressure, temperature, level, viscosity, etc. These will also have to be adapted to the system being considered and may therefore be very different for different systems. In, e.g., Computer HAZOP, the parameters are called "attributes" and can be, e.g., data flow, data rate, and timing of events.

Table 8.7 Examples of HAZOP guidewords, interpretation and use

Guideword	Interpretation	Example of use for ballast system
NO/NOT	Failure to achieve the intended outcome	NO flow of ballast water through the pumps
MORE/LESS	An increase/decrease occurs	LESS pressure of ballast water after the pumps
AS WELL AS	Something occurs/is done in addition to what was intended	AS WELL AS opening valve A, valve B is also opened
PART OF	Only part of the intention is achieved	PART OF the operation is executed – valve A is opened, but not valve B
REVERSE	The flow through a system is reversed compared to the intention	REVERSE flow through the ballast pumps
OTHER THAN	Something else than the intention is achieved	A ballast tank OTHER THAN intended is flooded
EARLY/LATE	A step in a procedure is performed earlier/later than intended	Ballast pumps stopped EARLY, before required level is reached
BEFORE/AFTER	A step in a procedure is performed before/after another step in the procedure	Ballast pumps started BEFORE valves have been configured correctly

The analysis is done in practice by selecting one part of the system at a time, then systematically applying all the combinations of guidewords and parameters by asking whether this deviation can happen in the system. Only those that are relevant are normally recorded.

8.6.4.3 Causal and frequency analysis

When a deviation has been identified, the next step is to describe the possible causes of this deviation. If we take "NO FLOW of ballast water" as an example, some possible causes may be no supply of water, valves not opened, or pumps not working. Normally, focus will be on the direct causes, without going into details on underlying and indirect causes.

In a HAZOP, categorization of frequencies may also be applied similarly to PHA and FMECA, although it is not very common in practical applications. IEC 16882 does not specify this as part of the HAZOP procedure.

8.6.4.4 Consequence analysis

For each deviation, the consequences of the deviation are described, in the same way as the causes. Categories may also be used here.

8.6.4.5 Result presentation

The output from a HAZOP study will mainly be a set of tables. There is no standard format, but a fairly common approach is to include the following columns:

1. Id – unique identifier for each deviation
2. Deviation (in some cases, this may be split into three columns: Guideword, parameter, and deviation)
3. Causes of deviation
4. Consequences of deviation
5. Existing risk controls/barriers
6. Recommendations/actions

The report from a HAZOP will often also act as an action plan, where the recommendations and actions are assigned to one of the HAZOP team members for follow-up and with a deadline for the action to be completed. In such cases, it may be the responsibility of the HAZOP team leader to verify that the actions are completed on time.

HAZOP of propeller

The mobility of a vessel is highly dependent on the propeller. If the propeller for some reason fails, the whole propulsion system and navigation system are put out of operation and the ship's movement is out of control. It is therefore clear that the propeller is a critical component. As part of a HAZOP procedure the controllable pitch propeller (CPP) in Figure 8.8 is identified as a part with individual intention. We want to perform a single loop in the HAZOP procedure for the CPP.

It is assumed that the analysis should emphasize loss of propeller function. The case of degraded operation is not considered.

Definition of CPP intention: The propeller is to transform rotational energy, transmitted through the propeller shaft, into a pressure

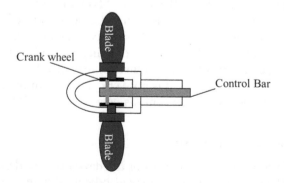

Figure 8.8 Controllable pitch propeller (CPP).

Table 8.8 Identification of deviations

No.	Guideword and parameter (Deviation)	Description
1	NO pitch	No rotational energy is transformed
2	NO blade	No rotational energy is transformed
3	NO control bar	All blades on random pitch, loss of operational control
4	NO crank wheel	One or all blades have independent pitch
5	NOT enough material strength	Parts of the propeller break down
5	MORE pitch than optimal	Too heavy load on propulsion system. Cavitation
6	LESS pitch than optimal	Too little load on propulsion system. Cavitation
7	LESS draft than allowed	Propeller is not sufficiently submerged. Loss of thrust
8	LESS depth than necessary	Propeller hits the ground and is damaged

Table 8.9 Identification of causes and safety measures

No.	Causes	Safety measures
1	Operation failure or control mechanism failure, alignment mechanism defect	See 2–5
2	Object in the water breaks the blade	Implementation of propeller protection like gratings or water jet. Sail in ice-free waters. See 7 and 8
3	Material weakness	Improve design and construction
4	Material weakness	Improve design and construction
5	Wrong design, corrosion or cavitation, alignment mechanism is defect and causes different pitch on the blades which again causes extra load on bearings and shaft line	Validate propeller design, cathodic protection, appropriate propeller material, test the propeller against cavitation, periodic alignment adjustment
5	Operation failure	Surveillance, increase operator competence
6	Operation failure	Surveillance, increase operator competence
7	Operation failure	Surveillance, increase operator competence
8	Operation failure	Technical equipment, operator competence and surveillance

difference over the propeller blades. It is this pressure difference that accelerates and maintains the speed of the vessel. The controllable pitch's intention is to optimize this energy transformation for various operational conditions.

Tables 8.8 and 8.9 provide a simplified illustration of the output from a HAZOP, split on two tables and including only some of the columns normally found in a HAZOP table.

These lists/tables are not exhaustive. Other deviations are possible, and to find these one must be creative and have good system understanding. Further, to do a complete HAZOP of the propulsion and navigation system, other subsystems should also be covered.

8.7 STPA

8.7.1 Introduction

STPA (System Theoretic Process Analysis) is a more recent method that was developed around 2000 (Leveson, 2004). Underlying the method is an accident theory that is based on control theory, called STAMP (System Theoretic Accident Models and Processes). A strength of the method is that it is suitable for systems that combine software and hardware, a feature that is relevant in more and more advanced systems.

STPA is probably the method that is most difficult to use in practice and it requires considerable experience before the method is mastered fully. We will not go into much detail, but provide an overview and interested readers can look at some of the references.

Information about STPA can be found at https://psas.scripts.mit.edu/home/. A couple of references that give an overview of STAMP and STPA are:

- Leveson, N.G. and Thomas, J. (2018). *STPA Handbook* (available from https://psas.scripts.mit.edu/home/)
- Leveson, N.G. (2016). *Engineering a safer world: Systems thinking applied to safety*. The MIT Press.

This paper gives an example of application of STPA for a maritime application:

- Rokseth, B., Utne, I.B. and Vinnem, J.E. (2017). A systems approach to risk analysis in maritime operations. *Journal of Risk and Reliability*, 231 (1), 53–68.

8.7.2 Method description

The underlying assumption in STPA is that maintaining safety is seen as a control problem, and that we can describe the system as a hierarchy of control loops, see Figure 8.9.

Technical control loops can usually easily be described in this way, but the same thinking is also applied to human and organizational activities to control risk. A high-level example can be training of personnel. The action is the decision to provide training (control action), the "actuator" is the training course, the controlled process is the performance of the tasks that personnel have been trained in and a feedback loop can be established, e.g., through observation of the performance of the work, where the observer is the "sensor". A hierarchy can be established, e.g., if we assume that the

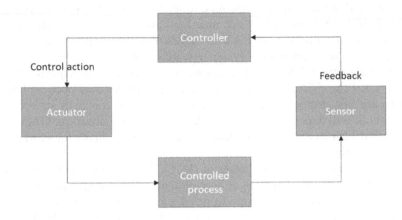

Figure 8.9 A generic control loop.

training that is provided is related to operation of a ballast system. The operator then issues "orders" to the ballast system by operations on the ballast panel. These orders can be to open or close valves, start or stop pumps, etc., and the operator will get feedback in various forms, through indicator lights on the control panel, through measurements of flow through pumps, levels in tanks, etc.

A key part of the preparations for doing an STPA is to establish the control hierarchy that is in place (or should be in place) to control hazardous events. Identifying high-level hazardous events is also part of the preparations, since we need this to describe what control structures are in place to avoid/control these hazardous events.

When the preparations are in place, the detailed analysis is done by considering all the control loops and identifying Unsafe Control Actions (UCAs). UCAs are control actions that can lead to negative consequences and four generic UCAs are defined:

– Control action not provided (i.e., not performed)
– Control action is unsafe when provided
– Control action started too early or too late
– Control action stopped too early or too late

If we assume that a control action is to stop the ballast pumps, we can identify four UCAs based on this:

– Ballast pumps not stopped
– Ballast pumps stopped
– Ballast pumps stopped too early or too late

We can see that use of generic UCAs can lead to a lot of repetition and also a lot of UCAs that are in fact not unsafe. This is a similar problem as we have described both for FMECA and HAZOP and is an issue that we have to live with if we want to do a systematic analysis.

For each of the UCAs, we then consider if there are any negative consequences. Of the UCAs identify, negative consequences may occur if ballast pumps are not stopped and also if they are stopped too early or too late. However, it will depend on the particular context and situation. Only those UCAs that have negative consequences are relevant to bring forward to the final step in the analysis.

This step is to identify the causes of the UCAs. This could of course also have been done before looking at consequences but doing it in this way is more efficient since we save time in not looking at those that do not have negative consequences. When we identify the causes, we go back to the control loop and consider failure in each part of the loop:

- Can the command fail?
- Can the actuator fail?
- Can the action fail?
- Can the feedback fail?

Often, several failure modes can be identified for each part of the loop.

8.7.3 Comparison with FMECA

FMECA and STPA will eventually end up with partly the same results, but the system perspectives are different, and this will also tend to mean that STPA may identify more potential safety critical problems than an FMECA will do. In FMECA, we look at the component hierarchy while in STPA we look at the control hierarchy. Since the controls often are implemented by components, there will be significant overlap. However, looking at controls will give a better understanding of the functions of the system than an FMECA, and in particular so when the functions are provided by software and humans, i.e., not technical components.

On the other hand, analyzing a system in the way that is done in STPA is far more demanding and requires more experience than with FMECA. This needs to be kept in mind when choosing which method to use.

8.8 FAULT TREE ANALYSIS

Fault tree analysis is a method to identify, illustrate, and analyze the causes of an event. For risk analysis purposes, it is most commonly used to analyze causes of hazardous events or failures.

The method is a graphical technique that can visualize the relationship between causes and events, and at the same it is also a set of numerical methods that can be used to quantify the probability or frequency of an event or a failure. In the following, we will first describe how the graphical visualization is developed. This is the qualitative part of the fault tree analysis. This is followed by the numerical approach.

More detailed descriptions of the method can be found in among others:

- Rausand, M., Barros, A., and Høyland, A. (2021). *System reliability theory – Models, statistical methods and applications*, Wiley
- NASA (2002). *Fault tree handbook with aerospace applications*, Washington DC, USA
- ISO 61025 (2006). Fault tree analysis

8.8.1 Constructing fault trees

A fault tree is based on identifying and drawing the event that we want to model and the causes of this event and then linking the causes and the events by using logical gates. Standardized symbols are normally used to illustrate events, causes, and the logical gates. Some of the most commonly used symbols are shown in Figure 8.10.

Symbol	Gate Name Casual Relation
	AND gate - Output event occurs if all input events occur simultaneously
	OR gate - Output event occurs if any one of the input events occurs
	Basic event with sufficient data
	Undeveloped event
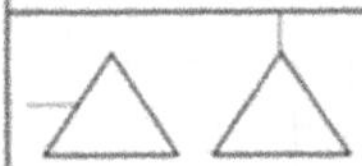	Transfer in and Transfer out symbols

Figure 8.10 Fault tree symbols.

The first task of a fault tree analysis is to describe the system and its components/subsystems down to a sufficient level of detail. How detailed will depend on how detailed the analysis should be, what data are available, as well as availability of time and resources for doing the analysis. The next task is to construct the fault tree for a particular unwanted system failure using this system description. It is important that all the failures in the fault tree are given precise definitions. The unwanted event or accident target for the analysis is referred to as the TOP event of the fault tree. The description of the TOP event should give answer to what the event is, where it occurs, and when it occurs.

The occurrence of the TOP event is always dependent on two or more conditions or failures on a more detailed, i.e., lower, level. The main task in the FTA approach is to systematically define and structure the conditions or causes that lead to the top event. These events should be defined in such a way that only a limited number of causes lead to the top event. Some recommend only defining two causes on the lower level at a time, but for some complex system failures this may not be realistic. The causes directly leading to the top event are at the second level in the fault tree.

When they have been identified, the next task is to assess the logical relationship between the causes. In most cases, the TOP event is either dependent on a simultaneous occurrence of two or causes on the second level or only one of the causes alone may lead to the TOP event. In the first case an AND-gate is used and in the last case an OR-gate is used.

AND-gates symbolize that all the causes pointing into the gate need to occur for the TOP event to occur. Assume that we have two pumps in parallel that supply water to a safety critical system. As long as one of the pumps is working, water will be supplied. This corresponds to a parallel structure as shown in Figure 8.11. If we then say that the TOP event is "No water from pumps", we use an AND-gate to connect "Pump 1 not working" AND

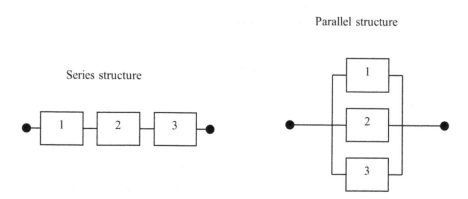

Figure 8.11 General system structures.

"Pump 2 not working" to the TOP event, to illustrate that both pumps must fail at the same time. An AND-gate may have many events pointing into the gate, but there are always at least two.

OR-gates symbolize that as long as at least one of the events pointing into the gate occur, the TOP event will occur. If we have a system with a pump with a closed valve located just after the pump and the pump is going to supply fuel to the engine, we can see that if "Pumps fails to start" OR "Valve fails to open", the TOP event will occur. This corresponds to a series structure, Figure 8.11. An OR-gate may also have any number of events pointing into it. In theory we could also have just one, but normal practice is to have at least two.

One thing to be aware of when applying the gates is that it is very easy to mix up the gates. In particular, this is easy when we are working with component failures. The pumps that we used to illustrate the AND-gate may be an example. The AND-gate symbolizes that Pump 1 and Pump 2 have failed. Experience has shown that it is very easy to think that "as long as Pump 1 OR Pump 2 is working, the top event will not occur". This is of course the opposite logic of what the fault tree is aimed to illustrate, but it is a very easy mistake to make.

This procedure is then repeated to establish the logical relations between the causes on the third level of the fault tree, and so on. When the causes are described to the level that we want, the fault tree construction is finished.

Figure 8.10 also contains some other symbols, of which the most important is the symbol for Basic Events. The basic events in fault trees are at the "final level" of the fault tree and represent the most specific causes shown in the fault tree. What we define as basic events can vary a lot depending on the level of detail that we want to go into and the information available for the analysis. They may represent failure of whole systems, subsystems, or components and also other types of events, e.g., human errors or external events. If we are looking at, e.g., the reliability of a ballast system, we will typically model failure of the ballast pumps as basic events. On the other hand, if we are looking specifically at the reliability of the ballast pumps, we will break the pump down into its individual components and model the components as basic events.

What all basic events have in common is that they are binary events, in the sense that they either occur or do not occur. We cannot define basic events that have three states, e.g., "Working", "Failed", and "Partially failed". This is important to understand and represents a limitation in what fault trees can be used for. Human and organizational factors influencing the performance of technical systems or humans are often hard to model properly in a fault tree because of this. Typically, we cannot say that "Inadequate maintenance" will automatically lead to failure of a component, although it can increase the probability of failure. For situations like this, Bayesian Networks (Section 8.10) is an alternative method.

In some cases, we may end up stopping the event tree development before we have reached the level of detail we had planned or hoped to reach. This may be due to lack of information or insufficient time available. The symbol for Undeveloped events in Figure 8.10 may then be used to signify that this is the case.

Finally, Figure 8.10 also shows transfer symbols. These are used when fault trees are too big to illustrate on one page or screen. We can then split the fault tree into smaller parts, using the transfer symbols to illustrate that the fault tree is continued on another page ("Transfer out"-symbol) and that a fault tree is a continuation from another page ("Transfer in"-symbol).

8.8.2 Minimal cut sets

One objective of fault tree analysis is to establish a general view and understanding of what single causes or combination of causes can lead to the TOP event. A good understanding of this can also help us to improve system.

An important tool for this understanding is what we call Minimal Cuts Sets in fault trees. We can start by explaining what a cut set is. A cut set is a set of basic events (causes) that will cause the TOP event to occur. This means that if all the causes are present at the same time, the TOP event will occur. The simplest example of a cut set is all the basic events that we have identified in the fault tree. If all occur at the same time, it is obvious that the TOP event will occur.

However, this is not very useful, because it will normally be extremely unlikely that all events will occur at the same time. It is more interesting to find out if it is sufficient that fewer events occur, for the TOP event to occur.

Minimal cut sets are those cut sets that are such that they cannot be reduced without losing their status as a cut set. Normally, not all causes need to occur for the TOP event to occur, therefore the cut set containing all events will normally not be a minimal cut set.

In the fault tree in the earlier example, illustrated in Figure 8.12, there are four basic events, E_1 to E_4. We can define a cut set as $\{E_1, E_2, E_3, E_4\}$. However, we see that this is not a minimal cut set, because it contains both "Combustible gas present" and "Combustible substance present". It is sufficient for one of these to occur for the TOP event to occur. We can therefore define two minimal cut sets for this fault tree, $\{E_1, E_2, E_4\}$ and $\{E_1, E_3, E_4\}$. In both cases, we need to have heat and oxygen present, but it is sufficient that either a combustible substance or combustible gas is present.

A minimal cut set can then be defined as a set of basic events that together are sufficient and necessary to cause the TOP event to occur. A minimal cut set can contain just one basic event.

This is a very simple example where it is easy to see what the minimal cut sets are simply by inspection of the fault tree. For larger systems, it quickly becomes difficult to identify the minimal cuts, and systematic methods

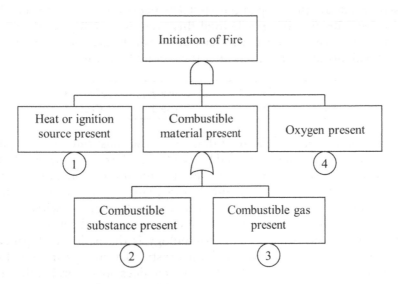

Figure 8.12 Simplified fault tree for a fire.

are required. One method that is commonly used is MOCUS – Method of Obtaining Cut Sets. The MOCUS algorithm can be described in the following steps:

1. Start with the Top event.
2. Replace the Top event with the events on the second level according to the following criteria: If the events on the lower level are connected through an OR-gate they are written in separate rows. If they are connected through an AND-gate they are written in separate columns.
3. Repeat step 2 successively for all events that are not basic events.
4. When all events are basic events the events in each row constitute a cut set (NB! Not a minimal cut set).

The fault tree in Figure 8.12 can be used to illustrate the use of the MOCUS algorithm. The starting point of the algorithm is the top event according to step 1. In the fault tree in Figure 8.12, this is {Initiation of fire}.

This event is then replaced by the events on the lower level according to step 2. Because the events on the second level of the fault tree are connected through an AND-gate they replace the top event in three columns:

Cause 1	Combustible material present	Cause 4

Causes 1 and 4 are basic events and not treated any further according to step 2 in the MOCUS algorithm. However, the event {Combustible material present} needs another loop of the MOCUS algorithm in order to complete the cut sets. Because the gate beyond this event is an OR-gate, the causes on the third level are written in separate rows. Hence, according to the MOCUS algorithm, the cut sets after the second loop are:

K_1	Cause 1	Cause 2	Cause 4
K_2	Cause 1	Cause 3	Cause 4

According to step 4 of the algorithm, each row constitutes a cut set, and hence there are two cut sets, K_1 and K_2, for this fault tree. Consequently, the general conditions for a fire, i.e., the event {Initiation of fire}, are satisfied when, for example, Cause 1, Cause 2, and Cause 4 occur simultaneously. Because none of the causes in the two cut sets can be removed without them losing their status as cut sets, both K_1 and K_2 are minimal cut sets.

Another term in the fault tree terminology is the path set. A path set assembles a set of causes with the characteristic that non-occurrence of the causes in the path sets ensure that the top event does not occur. For the same fault tree, the non-occurrence of Cause 1 {Heat or ignition source present} ensures that the top event does not occur. Hence Cause 1 is a path set.

Both the minimal path sets and the minimal cut sets give important information about the properties of the system. The number of elements in the minimal cut sets should be as large as possible to avoid triggering off the top event due to a few causes. Barriers may be built into the system to achieve this. The number of path sets should be large because this implies that the system is designed to have multiple ways of avoiding the top event.

Minimal cut sets are also commonly used in the calculation of the TOP event probability in the fault tree. Any fault tree can be redrawn based on the minimal cut sets:

- The TOP event is followed by an OR-gate, with all the minimal cut sets pointing into the OR-gate.
- Each of the minimal cut sets is represented by an AND-gate, with all the basic events in the cut set pointing into the AND-gate.

This gives a fault tree with just two levels, an OR-gate at the top and AND-gates below and this is convenient for quantification.

8.8.3 Quantification of fault trees

Quantification of fault trees is a two-stage process, involving first assignment of probabilities/frequencies to all the basic events and second calculation of the TOP event probability/frequency using these data.

The first step can be quite time-consuming since data in most cases are not readily available and information from a variety of sources usually is required. The types of data that may be relevant to include can vary, depending on the basic events. Some examples of data are:

- Probability of a basic event being in a failed state at a random point in time. This can be, e.g., a sensor that we expect to function when needed, but may be in a failed state.
- Probability of failure on demand. This can be, e.g., a pump that has to start in a certain situation or it can be a person who is expected to perform a certain task.
- Frequencies of events. Input here is how often events occur, e.g., natural events such as storms.

More information about data sources can be found in Chapter 14.

For quantification, it may be useful to start by showing how we can calculate the probability of functioning (reliability) of a series structure and a parallel structure (ref Figure 8.11). The equations are based on the assumption that R_i, $i = 1,...,n$ is independent.

Series structure:

$$R_s = R_1 \cdot R_2 \cdot .... \cdot R_n = \prod_{i=1}^{n} R_i \qquad (8.1)$$

Parallel structure:

$$R_s = 1 - (1 - R_1) \cdot (1 - R_2) \cdot .... \cdot (1 - R_n) = 1 - \prod_{i=1}^{n} (1 - R_i) \qquad (8.2)$$

where
    Reliability of structure $= R_s$
    Probability of failure of structure $= U_s = 1 - R_s$
    Reliability of element $i = R_i$
    Probability of failure of element $i = U_i = 1 - R_i$

It is noted that failure of a series structure can be represented by an OR-gate in a fault tree and correspondingly that failure of a parallel structure can be represented by an AND-gate. If we have identified all the minimal cut sets in a fault tree, we can use these two equations to calculate the TOP event probability.

We use $P_{MCS,i}$ to signify the probability of occurrence of minimal cut set i, P_{TOP} to signify the probability of the TOP event and $P_{i,j}$ to signify the probability of occurrence of basic event j in minimal cut set i. We can then calculate P_{TOP} as follows:

$$P_{TOP} = \prod_{i=1}^{n} (1 - P_{MCS,i}) = \prod_{i=1}^{n} \left(1 - \prod_{j=1}^{n} P_{i,j} \right) \qquad (8.3)$$

This utilizes the fact that any fault tree can be drawn with the minimal cut sets pointing into an OR-gate to the TOP event.

It is important to remember that these equations assume that the basic events are independent and that the minimal cut sets are independent. This is not the case in many applications, because some of the cut sets may have one or more basic events in common. In these cases, the calculation will not be accurate and instead we will calculate and upper bound for the TOP event failure probability. This is called the Upper Bound Approximation, and we write it as follows:

$$P_{\text{TOP}} \leq \prod_{i=1}^{n} (1 - P_{\text{MCS},i}) = \prod_{i=1}^{n} \left(1 - \prod_{j=1}^{n} P_{i,j} \right) \tag{8.4}$$

In most cases, this approximation is sufficiently close to the exact result to be used. When we consider the inherent uncertainty in the input data in most cases, this is more than adequate. It is also important to note that the exact failure probability will be smaller than what we calculate using the approximation. The value will therefore be conservative.

When using fault tree analysis to improve the reliability of systems, it is important to understand which of the basic events contributes the most to the TOP event probability. For this purpose, we use various importance measures, to rank the basic events. Two of the most commonly used measures are Birnbaum's measure and Fussel-Vesely's measure.

Birnbaum's measure is based on the sensitivity of the TOP event probability to changes in the basic event probabilities. Formally, it can be expressed as follows:

$$I^{B}(i) = \frac{\partial P_{\text{TOP}}}{\partial P_i} = P_{\text{TOP } P_i=1} - P_{\text{TOP } P_i=0}$$

where $P_{\text{TOP } P_i=1}$ is the probability of the TOP event given that basic event always occurs, while $P_{\text{TOP } P_i=0}$ is the probability of the TOP event given that basic event does not occur. The larger $I^{B}(i)$ is, the higher the importance of basic event i.

Fussel-Vesely's measure, $I^{\text{FV}}(i)$, is calculated from the minimal cut sets, based on the assumption that the more minimal cut sets a basic event is represented in, the more likely is that it will be of high importance for the reliability of the system. $I^{\text{FV}}(i)$ is the probability that at least one minimal cut set that contains basic event i is failed when the TOP event has occurred. It can be approximated by the following equation:

$$I^{\text{FV}}(i) = \frac{1 - \prod_{j=1}^{m} (1 - P_{\text{MCS},j}^{i})}{P_{\text{TOP}}}$$

where $P^i_{\mathrm{MCS},j}$ is the probability of the jth minimal cut set that contains basic event I and m is the number of minimal cut sets containing basic event i.

More details about importance measures can be found in, e.g., Rausand and Haugen (2020).

Fault tree analysis of propulsion system

The failure modes of a tanker's main propulsion system have been established earlier in the chapter using an FMECA analysis. The connections and relations between the failures are unknown, and must therefore be modeled in a fault tree.

It is assumed that the following information is commonly available and known:

The top event is already defined as "loss of propulsion for the tanker". A simple way to break down the propulsion system is to emphasize on power transition in the main propulsion system. There are three independent events that may result in the top event. These are the "loss of propulsion power transmission" in the shaft lines or gear, "loss of propulsion power generation" from the engines, and "loss of propulsion power consumption" due to propeller failure. Only one of these events has to occur in order to trigger the top event. Hence these three events have to be combined by an OR-gate. The fault tree can be structured as shown in Figure 8.13. The "loss of propulsion power transmission" event can be caused by gear failure and/or shaft line failure (see FMECA) and must therefore be combined through the use of an OR-gate. The "loss of propulsion power consumption" event only includes the event of controllable pitch propeller (CPP) failure. In terms of the event of "loss of propulsion power generation", both the starboard and port engines must fail to deliver power to the gear. An AND-gate must therefore be used for these two events. There are two ways each engine can fail to deliver power to the gear; by failure of the clutch and by failure of the engine itself. An OR-gate must be used for these events because one is sufficient for the engine to fail to deliver power to the gear. The events of main engine failure (both starboard and port engines) in Figure 8.13 need to be treated in further detail. According to the FMECA, the causes or basic failure events 1–3 are all gathered in the "main engine failure" event and these have to be combined through the use of an OR-gate since one of the causes is enough for the main engine to fail. The main engine failure modes can be arranged/modeled in a fault tree as shown in Figure 8.14.

The MOCUS algorithm is applied to find the minimal cut sets (subscript S=Starboard, subscript P=Port):

MOCUS step 1:

"Loss of main propulsion power for a specified tanker under one year of normal operation"

MOCUS step 2:

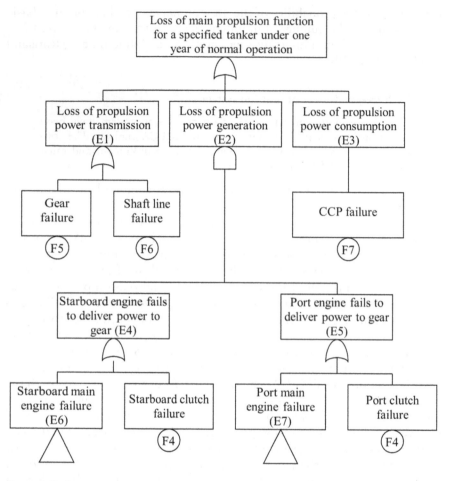

Figure 8.13 Fault tree for the top event of "loss of propulsion for the tanker".

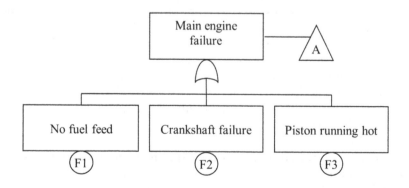

Figure 8.14 Main engine failure modes.

E1
E2
E3

MOCUS step 3.1 – for the "loss of propulsion power transmission" event (i.e., E1 in the fault tree):

F5
F6
E2
E3

MOCUS step 3.2 – for the "loss of propulsion power generation" event (i.e., E2 in the fault tree):

F5	
F6	
E4	E5
E3	

MOCUS step 3.3 – for the event that "starboard engine fails to deliver power to gear" (i.e., E4 in the fault tree):

F5	
F6	
A_s	E5
$F4_s$	E5
E3	

MOCUS step 3.4 – for the event of "starboard main engine failure" (i.e., E6):

F5	
F6	
$F1_s$	E5
$F2_s$	E5
$F3_s$	E5
$F4_s$	E5
E3	

MOCUS step 3.5 – for the event of "port main engine failure" (i.e., E7 in the fault):

K_1	F5	
K_2	F6	
K_3	FI_s	FI_p
K_4	$F2_s$	FI_p
K_5	$F3_s$	FI_p
K_6	$F4_s$	FI_p
K_7	FI_s	$F2_p$
K_8	$F2_s$	$F2_p$
K_9	$F3_s$	$F2_p$
K_{10}	$F4_s$	$F2_p$
K_{11}	FI_s	$F3_p$
K_{12}	$F2_s$	$F3_p$
K_{13}	$F3_s$	$F3_p$
K_{14}	$F4_s$	$F3_p$
K_{15}	FI_s	$F4_p$
K_{16}	$F2_s$	$F4_p$
K_{17}	$F3_s$	$F4_p$
K_{18}	$F4_s$	$F4_p$
K_{19}	F7	

MOCUS step 4:

There are 19 possible combinations of basic causes (or basic event failures) for the propulsion system (each row). There are mostly two basic causes in each cut set. It is advantageous to have as many basic causes in each cut set as possible, and one and two basic causes in each cut set are not much. The cut sets K_1, K_2, and K_{19} include only one basic cause. Hence the top event occurs when one of these basic events occur. It would therefore be advantageous to implement redundancy or other reliability improving measures for these cut sets. For example would the use of two independent propeller systems create redundancy and hence reduce the risk for top event occurrence. This may, however, not be practicable.

There are several interesting calculations that should be performed. The probability of the top event Q_0 is certainly of particular interest. The probabilities for each cut sets are also of interest. Normally some computerized calculation program, such as a spreadsheet, would be applied to calculate the top event probability. Here, on the other hand, the events are calculated manually using the series- and parallel structure equations presented in the Basic Theory section of this chapter. The series structure equation is used to calculate OR-gates and the parallel structure is used to

Table 8.10 Failure data calculated for a sailing operation of one year (336 days)

Failure	Failure description	Reliability probability p	Failure probability q
F1	No fuel feed	0.730	0.270
F2	Crankshaft failure	0.973	0.027
F3	Piston running hot	0.984	0.016
F4	Clutch failure	0.948	0.052
F5	Gear failure	0.764	0.236
F6	Shaft line failure	0.971	0.029
F7	CCP failure	0.813	0.187

calculate AND-gates (it must be remembered that the reliability $p_i=1-q_i$, where q_i is the failure probability). Failure data are given in Table 8.10.

As shown in Table 8.11 the probability for the top event of "loss of main propulsion function for a specified tanker under one year of normal operation" is 0.465. This means that there is a 46.5% chance that this particular unwanted, and potentially very dangerous, event will occur.

To assess the importance of the different basic causes the cut sets' failure probability is calculated using the given failure probability data in Table 8.12:

According to Fussel-Vesely's measure of component importance, the following importance ranking of the basic causes (or failures) is established (Table 8.13).

The ranking of the components is the "repairman's" ranking. If propulsion is lost, the most likely failure is related to the gear, i.e., basic cause/failure event F5, and so on. Other measures of importance should be applied at the design stage.

Table 8.11 Calculation of top event failure probability Q_0

Q_{E7}	$=1-P_{E7}=1-[p_{F1}\cdot p_{F2}\cdot p_{F3}]=1-[(1-q_{F1})\cdot(1-q_{F2})\cdot(1-q_{F3})]$	0.301
Q_{E6}	$1-[(1-q_{F1})\cdot(1-q_{F2})\cdot(1-q_{F3})]$	0.301
Q_{E5}	$1-[(1-Q_{E7})\cdot(1-q_{F4})]$	0.337
Q_{E4}	$1-[(1-Q_{E7})\cdot(1-q_{F4})]$	0.337
Q_{E2}	$=1-P_{E2}=1-[1-(1-P_{E4})\cdot(1-P_{E5})]=Q_{E5}\cdot Q_{E4}$	0.114
Q_{E3}	q_{F7}	0.187
Q_{E1}	$1-[(1-q_{F5})\cdot(1-q_{F6})]$	0.258
Q_0	$1-[(1-Q_{E1})\cdot(1-Q_{E2})\cdot(1-Q_{E3})]$	0.465

Table 8.12 Calculation of cut sets' failure probabilities

K_1	F5		$Q_{K1}=0.236$
K_2	F6		$Q_{K2}=0.029$
K_3	FI_s	FI_p	$Q_{K3}=0.073$
K_4	$F2_s$	FI_p	$Q_{K4}=0.0073$
K_5	$F3_s$	FI_p	$Q_{K5}=0.0043$
K_6	$F4_s$	FI_p	$Q_{K6}=0.014$
K_7	FI_s	$F2_p$	$Q_{K7}=0.0073$
K_8	$F2_s$	$F2_p$	$Q_{K8}=0.00073$
K_9	$F3_s$	$F2_p$	$Q_{K8}=0.00043$
K_{10}	$F4_s$	$F2_p$	$Q_{K10}=0.0014$
K_{11}	FI_s	$F3_p$	$Q_{K11}=0.0043$
K_{12}	$F2_s$	$F3_p$	$Q_{K12}=0.00043$
K_{13}	$F3_s$	$F3_p$	$Q_{K13}=0.00026$
K_{14}	$F4_s$	$F3_p$	$Q_{K14}=0.00083$
K_{15}	FI_s	$F4_p$	$Q_{K15}=0.014$
K_{16}	$F2_s$	$F4_p$	$Q_{K16}=0.0014$
K_{17}	$F3_s$	$F4_p$	$Q_{K17}=0.00083$
K_{18}	$F4_s$	$F4_p$	$Q_{K18}=0.056$
K_{19}	F7		$Q_{K19}=0.187$

Table 8.13 Importance ranking based on Fussel-Vesely's measure of importance

	Relevant cut sets	$1-\Pi(1-Q_{Ki})$	I^{VF}	Ranking
FI	K_3, K_7, K_{11}, K_{15}	0.0966	0.208	3
F2	K_4, K_8, K_{12}, K_{16}	0.0098	0.021	6
F3	K_5, K_9, K_{13}, K_{17}	0.0061	0.013	7
F4	$K_6, K_{10}, K_{14}, K_{18}$	0.071	0.150	4
F5	K_1	0.236	0.507	1
F6	K_2	0.029	0.062	5
F7	K_{19}	0.187	0.402	2

8.9 EVENT TREE ANALYSIS

8.9.1 Principles

If the consequences of an event or incident are to be analyzed, event tree analysis (ETA) may be applied. The event tree is good at visualizing different ways an event can develop, from initiating event. The event tree does this by splitting up a given initiating event forwardly and is therefore an inductive method.

Similarly to fault trees, event trees are also a graphical illustration of an event sequence, combined with rules for quantification that can be used to calculate probabilities and frequencies of various outcomes from an initiating event.

The logical diagram used in an ETA describes the relation between an initiating event and the events that describe the possible consequences.

The principle of an ETA can perhaps best be illustrated by an example.

Event tree for blackout on a ship

Assume that we have a situation where a ship has a sudden blackout close to shore. One possible scenario that we can envisage is then as follows:

- Ship starts drifting toward shore
- Ship is unable to regain power in time
- Hull penetrated when drifting aground
- Ship sinks rapidly

We can see that in all these steps, alternative outcomes are also possible. In the first step, the ship may avoid drifting toward shore, either because it can release the anchor and remain in position or because wind, waves and current makes it drift toward open sea. It may also be able to regain power quickly, and thus avoid hitting shore. There are thus many possible outcomes from the initiating event, and we can illustrate this using an event tree, as shown in Figure 8.15.

It is underlined that many other factors also could be included in the event tree.

In the example in Figure 8.15, the branching points are not used for all outcomes. On the first branch, the tree is divided into two branches, but when we get to the second branching point, we can see that it is only the first branch that is further divided into two. The reason for doing this is to simplify the event tree. We could have included all the branching points on all branches, but this would mean that the event tree would have had 16 end events instead of the five shown in the figure. However, the analysis

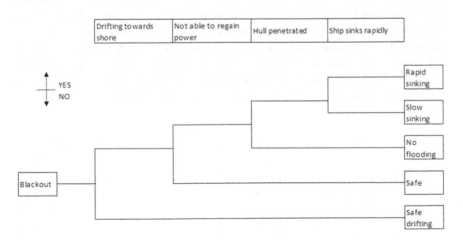

Figure 8.15 **Example event tree – blackout close to shore.**

would not really change. This can be seen if we consider the second branch from the first branching point, where we have answered "No" to "Drifting towards shore". As long as the ship is not drifting toward shore, it will not run aground, the hull will not be penetrated, and sinking will not occur. Therefore, the consequences are the same for all outcomes and we can simplify it by representing this with just one end event.

This should always be kept in mind when constructing event trees, because they quickly tend to become big, with many end events. This will usually make it more difficult to interpret and use in practice. We therefore always try to keep it as simple as possible, obviously without losing important information that should have been included in the event tree.

8.9.2 Constructing event trees

In the event tree, the points where the tree splits are sometimes called branching points. In the example above, we see that all branching points split into two new branches or outcomes. However, we can also have branching points with more than two outcomes and in principle it can be as many as we like. The advantage of this can in some cases be that the event tree becomes more compact.

To illustrate this, consider a situation where we want the final outcomes of the event tree to be split on events with "No fatalities", "1–5 fatalities", and "More than 5 fatalities". This can be modeled in two ways as shown in Figure 8.16. On the left-hand side, (a) shows how this can be modeled using just two outcomes in each branch, while (b) shows how we can use three outcomes instead. Logically they are equivalent and mathematically they will also give the same result.

When fault trees were discussed, standard symbols for drawing fault trees were presented. For event trees, there is no similar standardization. The initiating event may be drawn at the top, with the events developing underneath, or to the left, with events to the right. Further, in Figure 8.16, the description of the branches is provided in the event tree itself, next to the branch. Some also place the names on top, as illustrated in Figure 8.15.

The first step that we need to go through before constructing an event tree is to select and define the initiating event that we are going to analyze. For this purpose, it may be useful to remember the three questions that

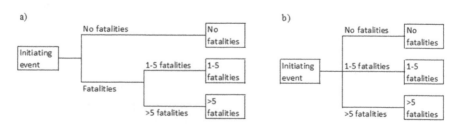

Figure 8.16 Alternative ways of drawing event trees.

were used for defining TOP events in fault trees, i.e., What, Where, and When. If we use the blackout example, we can formulate this as follows:

What: Blackout
Where: Close to shore
When: During voyage

The initiating event can then be formulated as "Blackout close to shore during a voyage".

Next, we have to decide what factors that should be included in the event tree. There are some questions to consider in connection with that:

- What types of factors should be included in the event tree? This can be a variety of factors, most common is that we include various safety systems and whether they function or not. However, other types of factors may also be relevant to include, such as Ignition/Not ignition, environmental conditions, etc.
- Have all factors that have a significant influence on the outcome been included? A key purpose of the event tree is to illustrate how different chains of events lead to different consequences. All factors that will influence the outcome should therefore be included, although this has to be balanced against the size and complexity of the tree. Often, we therefore end up including only the most important factors in the event tree.
- In what order should the factors be included? The common approach is to include the factors chronologically, meaning that when we follow a chain of events through the event tree, it is also a description of the chronology of the event. However, this is not always practical because the event tree will then sometimes become much larger than necessary and also because it is not always possible to say that the events will occur in one specific order.

8.9.3 Quantification of event trees

Event tree analysis is also a quantitative method and not just used for graphical illustration of event chains. We can use the event tree to calculate the probability or frequency of specific consequences. We can also use it to determine which factors will have the largest influence on the consequences and this can in turn give guidance on how to reduce risk.

If we are going to use the event tree quantitatively, there are two criteria that must be met:

- All the possible outcomes from a branching point in the event tree must be mutually exclusive, i.e. they cannot overlap. In most cases this is not a problem (we can have e.g., "System fails/System works" or "Event occurs/Event does not occur" as possible outcomes), but not always. In theory, we could e.g., say that one outcome is "Speed<8 knots" and another outcome "Speed>6 knots". In this situation, there is overlap between the two and we should reformulate to, e.g., "Speed<8 knots" and "Speed>8 knots". We therefore have to be careful in the formulation of the branches.

– Second, the complete set of outcomes from a branching point must be complete, in the sense that they jointly describe all possible outcomes. Again, for the examples mentioned for the first criterion, we can see that this applies. However, in other cases it is not necessarily like this. An example can be if we want to model the speed of a vessel and say that one outcome is "Speed<5 knots" and the second outcome is "Speed 5–10 knots". In this situation, we have not covered the possibility that the speed is more than 10 knots, i.e., it is not complete.

The quantification is done by first assigning a frequency to the initiating event, and subsequently assigning probabilities to all the branching points. Because of the criteria just discussed, we know that the sum of the probabilities in a branching point always becomes 1. If we have defined "System fails" and "System works" as the two outcomes, and we know that the probability of "System fails" is 0.05, we also know that the probability of "System works" is $1-0.05=0.95$. Similarly, if we have three outcomes, the third probability can be calculated once we know the probabilities of the two first ones.

One very important fact to be aware of is the conditional dependency between the probabilities. An event tree represents a chain of events. When we have reached a certain point in the event tree, we then know that a set of events has led up to this point. The probabilities that we assign are thus conditional on the previous events that have occurred. In many cases, this has no impact on the probabilities that we use, but in other cases, this may have a considerable impact.

Conditional probabilities in event trees

Consider the initiating event "Fire in engine room". The first important factor is whether the automatic fire detection system works or not, followed by whether the fire extinguishing system works or not. In this case, it is assumed that the fire extinguishing system normally is initiated upon a confirmed fire detection, but it can also be initiated manually by one of the crew. This is illustrated in Figure 8.17. The figure also shows probabilities assigned to the event tree.

Here we see that the probability of the fire extinguishing system working is lower if the fire detection has failed. This is because manual initiation of the system is less likely than automatic. We can also see that the differences in probability of failure are very large, with the probability being 30 times higher in the second case. These numbers are not from

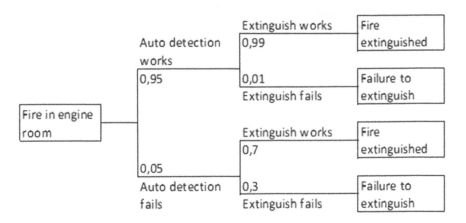

Figure 8.17 Illustrating conditional probabilities in event trees.

a real case, but differences like this are not unrealistic. Being aware of these dependencies is therefore very important for the quantification.

The calculations are simple for event trees and can be illustrated by considering the event tree shown in Figure 8.18.

The probability of the end events (EE1 to EE6) given that IE has occurred can be calculated as follows:

$$P(\text{EE1}) = P_{F1} \cdot P_{F2} \cdot P_{F3} \cdot P_{F4}$$

$$P(\text{EE2}) = P_{F1} \cdot P_{F2} \cdot P_{F3} \cdot (1 - P_{F4})$$

$$P(\text{EE3}) = P_{F1} \cdot P_{F2} \cdot (1 - P_{F3})$$

$$P(\text{EE4}) = P_{F1} \cdot (1 - P_{F2}) \cdot P_{F4}$$

$$P(\text{EE5}) = P_{F1} \cdot (1 - P_{F2}) \cdot (1 - P_{F4})$$

$$P(\text{EE6}) = 1 - P_{F1}$$

In simple terms, we can say that we follow the path through the event tree and multiply together all the probabilities that we "pass" on the way to the end event. We can also calculate the frequency of the end events, by multiplying with the frequency of the initiating event:

$$F_{\text{EE}i} = F_{\text{IE}} \cdot P(\text{EE}i), \quad i = 1, 2, \ldots, 6$$

where F_{IE} is the frequency of the initiating event.

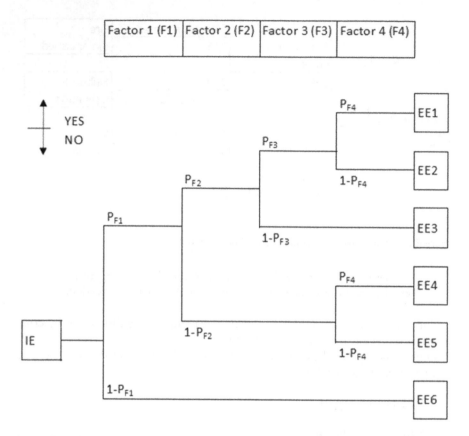

Figure 8.18 Quantification of event trees.

Another thing that may be noted is that if we add together the probabilities of all the end events, 1 to 6, we arrive at a total probability of 1. This follows from the two criteria that we initially established, that the branches are mutually exclusive and complete. With no overlap and complete coverage of all outcomes, the sum of probabilities must be 1.

Quantification

Using the event tree in Figure 8.17, and assuming that the frequency of the initiating event is 0.01 per year, we can calculate the probabilities and frequencies of the end events in this event tree. This is shown in Table 8.14.

This also shows that the sum of probabilities becomes 1 and the sum of frequencies is the same as the initiating event frequency. This property is handy to use to verify that all input data and calculations are correct.

Table 8.14 End event probabilities and frequencies calculation

End event	Probability	Frequency
1	$0.95 \cdot 0.99 = 0.9405$	$0.9405 \cdot 0.01 = 0.009405 = 9.4 \cdot 10^{-3}$
2	$0.95 \cdot 0.01 = 0.0095$	$0.0095 \cdot 0.01 = 0.000095 = 9.5 \cdot 10^{-5}$
3	$0.05 \cdot 0.7 = 0.035$	$0.035 \cdot 0.01 = 0.00035 = 3.5 \cdot 10^{-4}$
4	$0.05 \cdot 0.3 = 0.015$	$0.015 \cdot 0.01 = 0.00015 = 1.5 \cdot 10^{-4}$
SUM	1	0.01

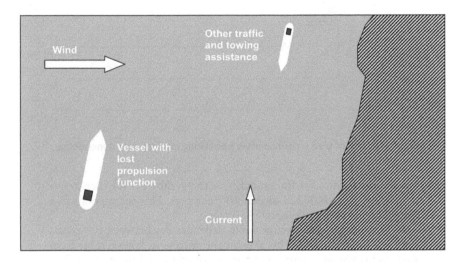

Figure 8.19 System description.

Loss of propulsion power

In the following, we will look at a more comprehensive example. The loss of propulsion power results in loss of controlled mobility for an oil tanker. The event of loss of propulsion power has been examined earlier in this chapter using the fault tree analysis (FTA) approach. The potential consequences for the loss of propulsion power have, however, not been analyzed in detail. Because the oil tanker has large oil spill potential, with devastating effects on the environment, it is of interest to estimate the likelihood of an oil spill if propulsion is lost. The accident scenario is presented in Figure 8.19. The aim of the analysis is to investigate possibilities of oil spill.

It is assumed that the probability of loss of propulsion power is 0.465 per year.

The following initial event tree is designed (Figure 8.20).

With this event tree structure, we can see that there is only one end event that leads to "Oil spill" and the frequency of this event can be found simply by multiplying the initiating event frequencies and all the

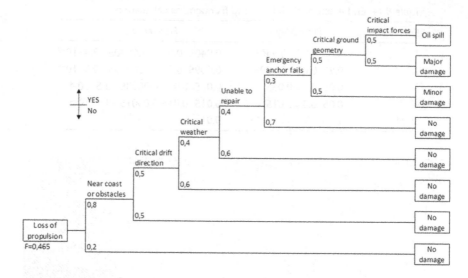

Figure 8.20 Initial event tree – probabilities not taking into account dependence.

"Yes" probabilities for the branches in the event tree. We then arrive at a frequency of 0.00223 or $2.2 \cdot 10^{-3}$ per year. This corresponds to a return period of approximately 450 years.

In this initial event tree, possible dependencies between the events have not been assessed. This assessment process is, however, far from easy and is normally carried out at the discretion of the analyst. The problems involved in assessing dependencies between events exist for all quantitative methodologies.

A couple of the dependencies that can be identified in this event tree are as follows:

- "Critical impact forces" will depend on "Critical weather". This is because the weather conditions will influence the drifting speed which in turn influences impact forces. It is thus more likely that the impact forces will be "critical".
- "Emergency anchor fails" will also depend on "Critical weather". Weather conditions will influence the forces on the anchor and more severe weather will increase the probability that the anchor will not hold.

A simple way of systematically identifying dependencies is to use a format as shown in Table 8.15.

A table like this will only tell us which factors will influence others, and the "direction" of the influence, but does not specify the strength of the influence. In some cases, we may be able to find data that can help us determine the conditional probabilities that should be applied in the event tree, but often we have to rely on expert judgment, based

Table 8.15 Identifying dependencies between factors in the event tree

	Critical drift direction	Critical weather forces	Unable to repair	Anchoring fails	Critical stranding geometry	Critical impact forces
Near coast/obstacles	+		+		+	
Critical drift direction						
Critical weather forces			+	+		++
Unable to repair						
Emergency anchoring fails						+
Critical stranding geometry						++

Increasing probability of occurrence: +; Decreasing probability of occurrence: –

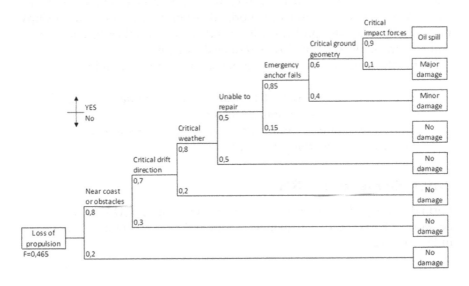

Figure 8.21 Updated event tree probabilities.

on an understanding of the scenario. Updated probabilities are found in the event tree shown in Figure 8.21.

If we now recalculate the frequency with the updated probabilities, we arrive at a frequency of 0.0478 or $4.8 \cdot 10^{-2}$ per year. This corresponds to a return period of approximately 21 years. The frequency is thus considerably higher when dependencies are taken into account. In this case the numbers are not "real" numbers, but they are not unrealistic, and we often see that when dependencies are taken into account, they can have a very strong impact on the final results. We should therefore always be very careful in considering these when doing quantitative analysis of event trees.

8.10 BAYESIAN NETWORKS

Bayesian Networks (BN) have become increasingly popular in risk analysis in recent years. They are also commonly called Bayesian Belief Networks (BBN) and sometimes the term influence diagram is being used to mean more or less the same thing.

Bayesian networks can, in the same way as fault trees, be used for two purposes:

- To graphically illustrate the relationship between an event and the causes of that event
- To calculate the probability or frequency of an event

The main advantage of a BN compared to a fault tree is that it is more flexible and can be used to model causal relationships more generally than fault trees. Where fault trees only can model deterministic causal relationships, BN can also deal with probabilistic causality. BNs are therefore more general and any fault tree can also be modeled using BN.

More information about Bayesian Networks can be found in, e.g.:

- Jensen, F.V. and Nielsen, T.D. (2007). *Bayesian networks and decision graphs*, Second edition, Springer
- Pourret, O., Naim, P. and Marcot, B. (2008). *Bayesian networks – A practical guide to applications*, Wiley

8.10.1 Elements of a BN

Formally, a Bayesian network is a directed acyclic graph (DAG). A Bayesian network is based on two elements: Nodes and arcs. A simple BN is shown in Figure 8.22.

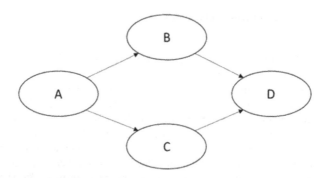

Figure 8.22 Bayesian network.

Nodes are normally a representation of a stochastic variable and they are commonly illustrated by an oval with text inside providing a name for the node. A, B, C, and D are all nodes. An example of a node can be "Wind speed". Wind speed can take on any value from zero up to as strong as the wind can conceivable be in the location we are considering and there would be a statistical distribution associated with this. In reality, this is therefore a continuous stochastic variable. Another node can be "Main engine" and we may choose to define this as a discrete stochastic variable with just two states: "Working" and "Failed". We can see that this is exactly the same way that we would define a Basic Event in a fault tree.

In practice, we will normally define all the stochastic variables as discrete variables, primarily to simplify the quantification of the Bayesian network. For the node "Wind speed" that was mentioned earlier, this could be solved by defining, e.g., four states: "Wind <5 m/s", "Wind 5–20 m/s", "Wind 20–32 m/s", and "Wind >32 m/s".

Bayesian networks are widely used for many applications, but for risk analysis purposes, we will often define the initiating event (or TOP event) as a node in the Bayesian network, with two states: "Initiating event occurs" and "Initiating event does not occur". When we quantify the BN, we can then determine the probability that the initiating event occurs.

Arcs are the arrows connecting the nodes (symbolizing that this is directed graph). When an arc points from Node A to Node B, this means that Node A has a direct influence on Node B. In practical terms, this means that the state of Node A, will have an influence on what the state of Node B is. In Figure 8.22, we can also see that Node A influences Node D through both Node B and C. Node A thus has an indirect influence on Node D.

An example that can illustrate influence and at the same time illustrate the difference between BN and fault trees is the following. Assume that we have an electric motor. Two factors that influence whether this motor will work or not are supply of electricity and maintenance. We could model this with "Electric motor" as a child node and "Supply of power" and "Maintenance" as two parent nodes. If supply of power is lost, we know with certainty that the motor will not work. However, even if maintenance is inadequate or completely lacking, this does not mean that the motor automatically stops working. We would expect though, that the probability of failure is higher than if it is properly maintained.

"Supply of power" could also be modeled in a fault tree, because the causation is deterministic (loss of power automatically leads to failure of the motor), but "Maintenance" cannot be modeled in a fault tree. The reason is that the causation here is probabilistic rather than deterministic.

Since BNs are acyclic graphs, the arcs cannot form closed loops. This means that it should not be possible to follow the arcs from a node and find a path that points back to the node that we started from. From a practical point of view, this is also meaningless since it means that a node indirectly will influence itself.

Some specific terms are being used to describe nodes and these can be compared to how we describe family relations. A child node is a node that has at least one arc pointing into the node. In Figure 8.22, Node D is a child node, but so is also Node B and C. Parent nodes are correspondingly nodes that have an arc pointing to at least one other node. Nodes A, B, and C are thus all parent nodes, while Node D is not. More generally, we may also use the terms ancestors and descendants. Node A, B, and C are also ancestors of Node D, and Node B, C, and D are all descendants of Node A.

Finally, we also use the term root node to describe nodes that have no arcs pointing into them. Node A is the only root node in Figure 8.22.

8.10.2 Constructing a BN

Construction of a BN has many similarities with the construction of a fault tree. The first thing we need to do is define the event that we want to analyze the causes of. The What/Where/When-questions that were discussed under fault trees can be applied also here.

Next, we start asking questions about the direct causes of the event. Since BNs can model both deterministic and probabilistic causality, it may be useful to reformulate the question and ask "What are the factors that directly influence the probability that this event will occur?"

The factors that we identify then become the parent nodes of the event that we started with. The factors should be formulated as nodes. Often it is beneficial to define them in neutral terms instead of using negative formulations like "Failure of radar" or "Inadequate competence". Alternative formulations could be "State of radar" and "Competence level" or similar. If we are going to use the BN for quantification, we should also define discrete states for the identified nodes. Preferably, we should try to keep the number of states to a maximum three or four, to avoid a combinatorial explosion of required input data.

The process is then repeated for each of the parent nodes that were identified, by asking again what influences the state that the node will be in. This continues until we have developed the model to the stage where we want it to be.

During the process of identifying factors, it is highly likely that the same factors will appear several times. In a fault tree, we can have the same basic event appearing in several places in the fault tree, but in a BN each node should only appear once, with arcs going from the node to all child nodes that are influenced. When the identification process is going on, it may be easy to use slightly different names on the nodes so a careful review should be undertaken to ensure that all cases of identical nodes have been identified.

On the other hand, there may also be situations where we use the same name for what in reality are different nodes. A typical example is "Competence". An example can be the event "Grounding" which may be caused either by inadequate navigation or loss of power. Navigation is likely to be influenced by the competence of the navigator. Loss of power can have different causes, but one may be wrong operation, which can be influenced by the competence of the engine crew. In both cases, competence can be defined as a node. However, in the first case, it is the competence of the navigators, whereas in the second case it is the competence of the engine crew. We may well have a situation where one is good and the other is poor. It is therefore necessary to define two nodes and call them, e.g., "Competence navigation crew" and "Competence engine crew".

Experience has shown that building a consistent and logical BN requires in-depth expertise on causality, technical systems, operations, and organization. It also requires time and usually many iterations to arrive at the final result. Discussing the BN in a group is in most cases very helpful and will contribute to improve the final model.

8.10.3 Quantification of BN

For quantification of BNs, it is first of all necessary to comply with the requirements that the network has to be acyclic, as discussed earlier. In addition, we also make the following assumptions:

- When we know the state of all the parents of a node, information about the state of the other ancestors will not add any more information. Looking at Figure 8.22, this means that if we know the state of B and C, collecting information about A will not tell us more about D. This property is useful when quantifying BNs.
- If we know the state of all parents of a node, we assume that the node is conditionally independent. B and C are dependent because they are both influenced by A. However, if we know the state of A, we assume that they are independent of each other.
- Two nodes are assumed to be conditionally independent when there are no arcs connecting them, directly or indirectly.

In a fault tree, we can calculate the probability of the TOP event occurring (or that the TOP event is in what we call a "failed state"). When we are using a BN, we can do exactly the same thing, i.e., define that the event that we are analyzing has two states: "Failed" (event has occurred) or "Not failed" (event has not occurred). The states of the ancestors of this event will then determine the probability that the event will "Fail". If we do not know the state of the ancestors but know the probability distributions, we can also calculate the probability.

The following BN is a very small extract of a larger BN for grounding, with just two direct nodes that influence grounding, lookout, and

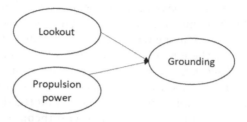

Figure 8.23 Extract of a BN for grounding.

propulsion power. We define the node Grounding to have two states: "Grounding" and "Not grounding". Further, we define three states for "Propulsion power", i.e., "Full power", "Reduced power", and "No power". Finally, for lookout, we define two states: "Adequate lookout" and "Inadequate lookout" (Figure 8.23).

Let us now make the following assumptions (all probabilities are assumed):

- If we have full power, grounding will not occur as long as the lookout is adequate.
- If we have full power, but the lookout is inadequate, grounding will occur with a probability of 0.3.
- If we have reduced power, grounding will occur with a probability of 0.01 as long as the lookout is adequate.
- If we have reduced power, grounding will occur with a probability of 0.30 if the lookout is inadequate.
- If we have no power, grounding will occur with a probability of 0.2 as long as the lookout is adequate.
- If we have no power, grounding will occur with a probability of 0.35 if the lookout is inadequate.

We can see how the probability of grounding will depend on the states of the parent nodes of grounding. Another thing that also may be noted is that we could have included other nodes in the BN. First of all, a necessary condition for grounding is that the ship is in coastal waters. This could therefore have been included as node. Secondly, reduced power is not necessarily a problem apart from the fact that the speed is reduced, but in bad weather conditions, the ship may not be able to keep track if power is reduced and may thus drift aground. Weather conditions could therefore also have been included as a node.

In the example, probabilities were presented for various combinations of states of the parent nodes. These assumptions can also be presented in a more compact format in a Conditional Probability Table (CPT).

In the first two columns of this table, the two parent nodes are listed and the rows represent all possible combinations of states of the nodes.

The third column represents the probability of one of the states of the grounding node. Since there are only two states, we know that the probability of the other state ("Not Grounding") will be 1 minus the probability of the "Grounding" state and it is therefore not shown in the table. If the "Grounding" node had three or more states, we would have to include probabilities for all states except the last one (again because it would be 1 minus the sum of the other states).

One observation that can be made straight away is that the need for data is much higher here than in a fault tree. In a fault tree, we need the probabilities of the basic events and then we can calculate the TOP event probability. Here, many more probabilities are required. Even in this extremely simple example, six values were required. If we had a slightly more complex case, with four parent nodes, each with four states, the number of probabilities required would increase to 256, i.e., combinatorial explosion!

We are now touching upon probably the biggest problem with using BNs quantitatively. By nature, BNs are good at modeling complex causalities where the nodes may have many different states. However, in practical use, it becomes almost impossible to populate the model with all the probabilities that we need. In most cases, we end up with the majority of the data being based on expert judgment. Using BN quantitatively will therefore always be a question of balancing modeling accuracy against what is realistic to quantify.

In Table 8.16, the probability of grounding was provided for six different combinations of states of the parent nodes. In risk analysis, we normally want to calculate an "average" probability, not for a specific combination of states. To do this, we also need the probabilities that each individual state of the parent nodes will occur. This is unconditional probabilities and to illustrate the calculation, we can assume the following:

P(Inadequate lookout)=0.02
P(Adequate lookout)=0.98
P(No power)=0.001
P(Reduced power)=0.01
P(Full power)=0.989

Table 8.16 Example conditional probability table

State of "Lookout"	State of "Propulsion power"	P(grounding|states)
Adequate	Full power	0.0
Adequate	Reduced power	0.01
Adequate	No power	0.2
Inadequate	Full power	0.3
Inadequate	Reduced power	0.3
Inadequate	No power	0.35

Table 8.17 Calculation of the total probability of grounding

"Lookout"	"Prop. power"	P(combination of states)	P(grounding|states)	P(grounding)
Adequate	Full power	$0.98 \cdot 0.989 = 0.969$	0.0	0
Adequate	Reduced power	$0.98 \cdot 0.01 = 9.8 \cdot 10^{-3}$	0.01	$9.8 \cdot 10^{-5}$
Adequate	No power	$0.98 \cdot 0.001 = 9.8 \cdot 10^{-4}$	0.2	$2.0 \cdot 10^{-4}$
Inadequate	Full power	$0.02 \cdot 0.989 = 2.0 \cdot 10^{-2}$	0.3	$5.9 \cdot 10^{-3}$
Inadequate	Reduced power	$0.02 \cdot 0.01 = 2.0 \cdot 10^{-4}$	0.3	$6.0 \cdot 10^{-5}$
Inadequate	No power	$0.02 \cdot 0.001 = 2.0 \cdot 10^{-5}$	0.35	$7.0 \cdot 10^{-6}$
Total probability of grounding				$6.3 \cdot 10^{-3}$

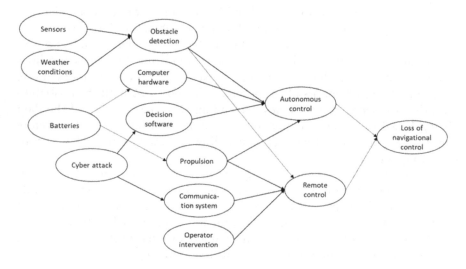

Figure 8.24 Extract of BN for loss of navigational control for an autonomous ship.

As long as "Lookout" and "Propulsion power" are independent (no arcs connecting them), it is now simple to calculate the total probability of grounding. The calculation is shown in Table 8.17.

8.10.4 Example

For autonomous ships, collision with ships or other objects is one of the events where the causes change very much compared to conventional ships with a manned bridge. Figure 8.24 illustrates a part of a BN for loss of navigational control for an autonomous ship. This is based on a paper by Guo et al. (2021). The ship normally performs all navigation autonomously, but there is a remote operator in an onshore control room that can take over control of the ship if needed.

The event in this case has been called "Loss of navigational control" and this node has two states: "Loss of control" and "Control" (event occurs or not). Since there are two modes of operation, autonomous and remote, the first "level" in the BN is simply the two alternative control modes. These can also be defined in the same way as the first node, with just two states (functioning/not functioning).

In the autonomous mode, there are four parent nodes: Obstacle detection, computer hardware, decision software, and propulsion system. These represent first the ability to detect the hazard, next the hardware and software to process the information and make decisions, and finally the propulsion system to execute the decision. For the remote control, it is partly the same nodes: obstacle detection, communication system, operator intervention, and propulsion system. The operator replaces the software and hardware in the autonomous mode, but in addition we need a communication system to link the ship to the remote control center.

The part of the BN that has been described so far could also have been described by a fault tree. The control modes could have been connected to the event with an AND-gate (both modes have to fail to lose control) while the nodes connected to the two nodes could have been modeled using two OR-gates.

For the rest of the BN, only a few nodes are included. Obstacle detection will be dependent on a variety of sensors (mainly radar, lidar, and cameras). In addition, weather conditions will influence the ability to detect obstacles. Heavy precipitation and fog will impair the sensors to a varying degree and reduce their reliability. However, this is not a node that could easily have been modeled in a fault tree because the sensors will not work perfectly up to a certain weather condition and then stop working. Instead, it will be a slow degradation of performance as the weather gets worse.

The ship has batteries for power and the batteries also provide power to the electronic systems on board. In the BN, arcs from batteries to computer hardware and propulsion system are included. This could have been modeled using a fault tree, although it depends on the system. Reduced performance of the batteries may also cause reduced performance of the propulsion system, but not necessarily to a complete failure. If that is the case, the BN approach is better suited than the fault tree.

The final node that is included is cyber-attack. Cyber-attacks can primarily be done through the communication system but can potentially attack all software-driven systems onboard. In the figure, it is shown to influence the decision system (the software) and the communication system. All control systems (not shown in the BN) are however potentially vulnerable.

In this example, the majority of the nodes could have been modeled using a fault tree, with some exceptions. In view of the difficulty in finding data for BNs, it may be questioned whether it is not better to accept that the model is simplified somewhat and use a fault tree instead? It is then important to understand that if we convert a fault tree to a BN, the data needs will not increase. It is only when we introduce probabilistic causality and nodes that have more than two states that data for additional probabilities start to increase. Because of this, it can be argued that we should always use BN instead of fault tree, since there are no major disadvantages. However, fault trees have been used for decades and are still far better known than BNs and this may be a reason for continuing to use them.

The BN that is shown is not the only possible way to model this event. This may seem to be a weakness (although it applies to all risk analysis methods) but is primarily due to the complexity of the problem. Accidents can have many causes and can develop in many ways and when we are modeling this in a risk analysis, we are always making a lot of simplifications. It will be up to the expertise and experience of the analyst to make these simplifications and it is likely that this will lead to differences in modeling. This is not something that is specific for risk analysis; to a smaller or larger degree it will apply when we are developing models of any real-world phenomenon. However, this should not be used as an argument against using risk analysis – the models that we develop can still be very useful even if they are not perfect.

8.11 RISK CONTRIBUTION TREES

In the IMO FSA Guidelines (2018), it is recommended to use Risk Contribution Trees (RCTs) in quantitative risk analysis. RCT is not a method in itself but is rather a combination of fault trees and event trees into a complete risk model for a specific event. In Figure 8.25, an RCT has been illustrated, where the left part is the fault tree and the right part is the event tree.

In practice, the RCT is not analyzed differently from the individual parts that it is comprised of. The only difference is that the initiating event frequency in the event tree is provided by the fault tree calculations rather than from another source.

It may be noted that although IMO recommends this approach, experience from FSAs that have been conducted is that the left-hand side of the RCT (the fault tree analysis) is very often omitted and that the initiating event frequency instead is based purely on statistics for the initiating events. This is a clear disadvantage in at least two respects:

- Without a detailed understanding of the causes of initiating events, it is more difficult to propose risk reduction measures that address the left-hand side of the bow-tie. In view of the fact that it is often recommended to prioritize these measures over measures to reduce consequences (Chapter 14), this is an obvious weakness.

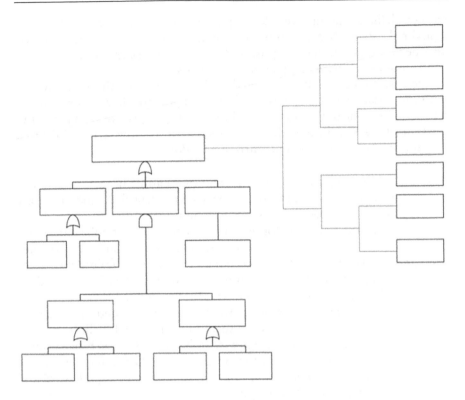

Figure 8.25 Risk contribution tree.

- If the causes of the initiating events are different or that the importance of causes is different for the ship being analyzed, it is hard to
  reflect this fact in the analysis. We may thus miss out on important
  differences that can lead to a less accurate risk picture.

8.12 CONCLUDING REMARKS

In this chapter, a number of methods for hazard identification and risk
analysis have been introduced. When we have a system and a problem that
have to be analyzed, choosing the best method can be quite difficult. Some
comments can be given to this.

The first consideration to make is whether we need to do a quantitative risk
analysis or not. Quantitative analyses are generally more time-consuming
and require more advanced expertise than qualitative methods. In general,
we should therefore choose a qualitative method, unless quantification is
necessary. An obvious situation when we need quantitative analysis is when

we have defined quantitative risk acceptance criteria, e.g., expressed as individual risk or an FN curve. If we are going to use cost-benefit analysis, we also need a quantitative analysis, and it is clearly better for ranking risk contributions compared to qualitative methods.

The first methods that are covered in this chapter are all qualitative (or semi-quantitative, using a coarse ranking) and probably have more similarities than differences. However, they differ in the perspective with respect to the type of system, the intended application area, and the terminology that is used. Briefly, this can be summarized as follows:

- PHA is a completely general method that can be used for many different applications. Hazardous events are identified, and these are ranked with respect to consequence and probability.
- FMECA is mainly used for technical systems, in particular safety systems. The system is broken down into subsystems and components, and the hazard identification is done by identifying failure modes for the components. Ranking is often also used in FMECA, based on probability, consequence, and often also detectability.
- HAZOP is originally developed for process systems and can be generally used for systems with some sort of flow involved (ballast systems, fuel systems, hydraulic oil systems, etc.). The system is broken down based on the flow through the system rather than components. The hazards are termed deviations and are composed of a guideword and a process parameter that is combined.
- STPA has a control theory perspective and describes the system as a control structure. The hazards in STPA are unsafe control actions, based on a predefined set of generic unsafe control actions.

The differences in terminology can be confusing at first and may give the impression that the methods are more different than they actually are.

Another observation is related to the way the hazard identification is performed where alternative methods focus on different aspects of the accident scenarios. If we consider the bow-tie model, the hazardous events (that are identified in the bow-tie) are at the center of the bow-tie, with causes to the left and consequences to the right. In an FMECA, the failure modes that are identified may also be identified in a (detailed) PHA, but they will then typically be labeled as causes of the hazardous events. One reason for this is that an FMECA will limit the analysis of consequences only to effects on the technical system being analyzed while a PHA will focus on the consequences to a wider system consisting of people, environment, and assets.

The differences in system perspective (components in FMECA, flow in HAZOP, and control structure in STAP) also mean that several methods can be applied to the same system and give different results. An example is a ballast system where we can use FMECA to inspect the individual components and how they can fail. Alternatively, we can use HAZOP to look at the flow of water through the system and how that can fail. But we can also use STPA to look at how the flow is controlled and how it can fail. In practice, it will be the same physical events and failures that are being considered, but different perspectives applied, and different methods can give different insights which may be valuable for improving the design and operation of the systems.

For a comprehensive review of methods for hazard identification (including the methods described herein), the report "Review of hazard identification techniques" (HSL, 2000) may be useful.

REFERENCES

ABS (2003). *Guide for risk evaluations for the classification of marine-related facilities*, June 2003, Houston, TX: American Bureau of Shipping.

Crawley, R. (2015). *HAZOP: Guide to best practice*, Amsterdam: Elsevier.

Fischhoff, B. (1995). Risk perception and communication unplugged: Twenty years of process 1. *Risk Analysis*, 15(2), 137–145.

Guo, C., Haugen, S. and Utne, I.B. (2021). Risk assessment of collisions of an autonomous passenger ferry. *Proceedings of the Institution of Mechanical Engineers, Part O: Journal of Risk and Reliability*, 1748006X211050714.

HSL (2000). *Review of hazard identification techniques*. Health and Safety Laboratory. Derbyshire: High Peak.

IEC 60812 (2018). *Analysis techniques for system reliability - Procedures for failure mode and effect analysis (FMEA)*. Geneva: IEC.

IEC 61882 (2016). *Hazard and operability studies (HAZOP studies) - Application guide*. Geneva: IEC.

IMO (2018). *Revised guidelines for Formal Safety Assessment (FSA) for use in the IMO rule-making process*. MSC-MEPC.2/Circ.12/Rev.2, 9 April 2018

ISO (2010). ISO12100:2010 *Safety of machinery — General principles for design — Risk assessment and risk reduction*. Geneva: International Organization for Standardization.

ISO (2016). ISO17776:2016 *Petroleum and natural gas industries — Offshore production installations — Major Accident hazard management during the design of new installations*. Geneva: International Organization for Standardization.

ISO 61025 (2006). *Fault tree analysis (FTA)*. Geneva: International Organization for Standardization.

ISO 31010 (2019). *Risk management – Risk assessment techniques*. Geneva: International Organization for Standardization.

Jensen, F.V. and Nielsen, T.D. (2007). *Bayesian networks and decision graphs*, Second edition, Berlin: Springer Verlag.

Kjellen, U. and Albrechtsen, E. (2017). *Prevention of accidents and unwanted occurrences: Theory, methods, and tools in safety management*. Boca Raton, Florida: CRC Press. ISBN 9781498736596.

Kletz, T.A. (1999). *HAZOP and HAZAN: Identifying and assessing process industry hazards*, Fourth edition. Boca Raton, Florida: CRC Press.

Leveson, N.G. (2004). A new accident model for engineering safer systems, *Safety Science*, 42(4), 237–270.

Leveson, N.G. (2016). *Engineering a safer world: Systems thinking applied to safety.* Cambridge, MA: The MIT Press.

Leveson, N.G. and Thomas, J. (2018). STPA Handbook (available from https:// psas.scripts.mit.edu/home/)

MIL-STD-882E (2012). *System Safety.* Washington DC: Department of Defense.

NASA (2002). *Fault tree handbook with aerospace applications.* Washington, DC: NASA Office of Safety and Mission Assurance.

NOROG (2017). *090 Norwegian oil and gas recommended guidelines on a common model for safe job analysis*, Rev. 4. Stavanger, Noorway: Norwegian Oil and Gas Association.

OSHA (2002). *Job hazard analysis.* Washington DC: Department of Labor, Occupational Safety and Health Administration.

Pourret, O., Naim, P. and Marcot, B. (2008). *Bayesian networks – A practical guide to applications.* Hoboken, NJ: Wiley.

Rausand, M., Barros, A. and Høyland, A. (2021). *System reliability theory – Models, statistical methods and applications.* Hoboken, NJ: Wiley.

Rausand, M. and Haugen, S. (2020). *Risk assessment: Theory, methods and applications.* Hoboken, NJ: Wiley.

Rokseth, B., Utne, I.B. and Vinnem, J.E. (2017). A systems approach to risk analysis in maritime operations. *Journal of Risk and Reliability*, 231 (1), 53–68.

Roughton, J. and Crutchfield, N. (2011). *Job hazard analysis: A guide for voluntary compliance and beyond.* Amsterdam: Butterworth-Heinemann.

Vincoli, J.W. (2006). *Basic guide to system safety* (Vol. 18). New York: Wiley-Interscience.

Chapter 9

Measuring risk

9.1 INTRODUCTION

In Chapter 3, risk was defined, and it was shown that risk commonly is expressed as the product of probability and consequence. This is simple and straightforward, but in practice, it is possible to develop numerous ways of expressing risk, based on this product.

In this chapter, some of the most common ways of expressing risk will be described. The main focus is on risk to people, but we will also look briefly at ways of expressing risk to the environment. One important point that may be mentioned straight away is that the different measures have different properties, leading to the somewhat surprising result that we may draw different conclusions regarding how to reduce risk, depending on what risk measure we are using. This is obviously important to understand when we are making decisions about risk.

The chapter starts by looking at the risk matrix. The risk matrix is not a quantitative measure of risk but is so extensively used to illustrate risk that it is worth discussing separately. The risk matrix can be used for all types of consequences, not just risk to people. Next follows various ways of measuring risk to people, related to injuries and fatalities. A brief introduction to measuring risk to the environment is also provided. This is mainly focused on accidental spills and the effect of these. Finally, a very brief introduction to measuring risk to other assets is included.

Only a few risk measures are mentioned in this book, with primary focus on those mentioned in IMO documents. A more comprehensive review of various risk measures can be found in, e.g., Rausand and Haugen (2020).

9.2 RISK MATRIX

The risk matrix is probably the expression of risk that is most well known. Most people that have been involved in risk management and risk analysis will recognize this. It is very widely used in many contexts, simple to use, easy to understand, and does not require quantitative analysis in most cases.

DOI: 10.4324/9781003055464-9

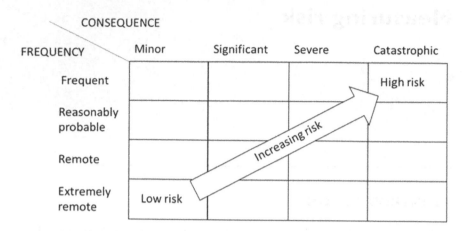

Figure 9.1 Example of risk matrix.

The risk matrix is based on combining frequency (or probability) and consequence in a table or a matrix, as shown in Figure 9.1. For both frequency and consequence, a set of categories or classes has been defined. All the events/hazards that have been identified are classified according to the relevant frequency and consequence categories and can thus be plotted in the risk matrix. An event that is "Reasonably probable" and with "Catastrophic" consequence will then end up in the rightmost column, second line from the top.

Since risk increases with increasing frequency and/or increasing consequence, the risk will increase diagonally from bottom left to top right in the risk matrix in Figure 9.1. It may be noted that there is no standard for drawing the risk matrix. Therefore, the frequency and consequence scales can both be reversed, such that, e.g., the frequency decreases from bottom to top (in Figure 9.1, it is opposite). If the frequency scale is reversed, the lowest risk will be in the upper left-hand corner and the highest risk in the lower right-hand corner. The most common way of presenting a risk matrix is however to have the highest risk in the upper right corner.

The first impression is that the risk matrix is straightforward to construct and use. However, experience has shown that there are several aspects that need to be considered carefully, both in relation to designing the matrix and using it.

Either frequency or probability can be used in the matrix (in the same way as when risk is defined). However, we need to stick to one of these and ensure that categories are defined accordingly.

There are no fixed rules for the number of categories. In Figure 9.1, four categories are used both for frequency and consequence. The number of categories can also be different for frequency and consequence. A high number of categories means that it is easier to distinguish between different hazards when they are plotted in the matrix. If there are few categories, many hazards will tend to fall in the same categories and ranking and prioritizing will be more difficult. On the other hand, the more categories we have and the "narrower" each category is, the harder it becomes to classify the hazards without using quantitative methods and data. From experience, four to six categories seem to work well in most cases, although it is noted that IMO uses seven frequency categories in their FSA guidelines (IMO, 2018).

Each category, for both frequency and consequence, must have a definition associated with it. These need to be unambiguous and as clear as possible. It may be tempting to use categories like "often"/"seldom" or "probable"/"unlikely" for frequency and probability and terms like "minor damage" or "catastrophic" for consequence. However, these qualitative terms are ambiguous because different people may have different interpretations of what the terms mean. Is "often" something that happens every day, on every voyage, or every year?

The categories should be defined such that everyone who is using the risk matrix (as far as possible) has the same understanding of the categories. In Figure 9.1, names of the categories were provided, but no further explanation. Table 9.1 provides examples of how categories can be defined in a more precise manner.

The table gives examples, but these need not be relevant for all situations. When defining the "boundaries" (i.e., the upper and lower categories), we should consider what is the highest and lowest likely frequency and consequence that we need to be able to distinguish between hazards. If we are considering a ship with a crew of say 5, it is of limited interest to define a consequence category with more than 5 fatalities because this will never be used. This may mean that a risk matrix defined for one purpose not necessarily is equally useful for another purpose. This should be considered as part of the preparations for the study where the risk matrix is planned to be used.

In Table 9.1, categories for consequences only to people are shown. If other types of consequences are considered, e.g., environment, categories also need to be developed to describe these. One way of doing this can, e.g., be to define categories based on spill size.

Table 9.1 Example categories for frequency and consequence

	Frequency		Consequence
5	>1 per year	5	10 or more fatalities
4	0.1–1 per year	4	1–10 fatalities
3	0.01–0.1 per year	3	1–10 major injuries
2	0.001–0.01 per year	2	1–10 minor injuries
1	<0.001 per year	1	Negligible consequences

Risk matrix used by IMO

In the FSA guidelines from IMO (2018), example categories and an example risk matrix are included in Annex 4. Consequence categories are shown for injuries/loss of life and for material damage (damage to ship). Proposed frequency categories are shown in Table 9.2.

This table contains only four categories, but it is noted that FI (frequency index) goes from 1 to 7. In effect, seven categories could therefore have been defined. The corresponding frequencies for the categories also indicate that "intermediate" categories could have been defined (by using one order of magnitude instead of two between the categories).

One comment related to the frequency categories is that the description is related to the "lifetime of a ship" for several of the categories. This can be a more useful approach because mariners can more easily relate to their own experience. However, the values may also be measured in relation to other exposure units, e.g., all ships owned by the shipowner or all ships in a specific category globally. Clearly, this has significant impact on how we classify hazards with respect to frequency. This therefore must be clearly specified as part of the definition of the risk matrix. IMO proposed four categories for severity as shown in Table 9.3.

Based on these categories, a risk matrix can be constructed. See Table 9.4.

The example from the FSA guidelines describes the frequencies in four ways: By a frequency index (from 1 to 7), by a word (e.g., Frequent), with a description, and with a frequency (per year). This minimizes the chances of misunderstandings or conflicting interpretations.

The final aspect of how we define the categories are the "steps" between the categories or relative magnitude of neighboring categories. It is often recommended that when we move from one frequency category to the next,

Table 9.2 Definition of frequency categories

Frequency index			
FI	Frequency	Definition	F (per ship year)
7	Frequent	Likely to occur one per month on one ship	10
5	Reasonably probable	Likely to occur once per year in a fleet of 10 ships or likely to occur a few times during a ship's life	0.1
3	Remote	Likely to occur one per year in a fleet of 1,000 ships or likely to occur once in the total life of a group of similar ships	10^{-3}
1	Extremely remote	Likely to occur once in a lifetime (20 years) of a world fleet of 5,000 ships.	10^{-5}

IMO (2018).

Table 9.3 Definition of consequence categories

		Severity Index		
SI	Severity	Effects on human safety	Effects on ship	S (equivalent fatalities)
1	Minor	Single or minor injuries	Local equipment damage	0.01
2	Significant	Multiple or severe injuries	Non-sever ship damage	0.1
3	Severe	Single fatality or multiple severe injuries	Severe damage	1
4	Catastrophic	Multiple fatalities	Total loss	10

IMO (2018).

Table 9.4 Risk matrix

		Risk Index (RI)			
		Severity (SI)			
		1	2	3	4
FI		Minor	Significant	Severe	Catastrophic
7	Frequent	8	9	10	11
6		7	8	9	10
5	Reasonably probable	6	7	8	9
4		5	6	7	8
3	Remote	4	5	6	7
2		3	4	5	6
1	Extremely remote	2	3	4	5

IMO (2018).

the step should be an order of one magnitude, or a factor 10. An example can be that one category is defined as 0.1–0.01 events per year and the next category is 0.01–0.001 events per year. This approach is applied by IMO (2018). This is not necessary, but it is recommended that the steps are identical, and not vary from one step to the next. In Table 9.5, two examples are shown.

The reason why varying steps should be avoided is that the ranking (based on combining frequency and probability categories) can turn out wrong if the steps are not the same.

When we calculate risk, we normally multiply frequency and consequence to arrive at a risk number. In a risk matrix, a risk number (sometimes called Risk Priority Number, RPN, or Risk Index, RI) may be calculated. However, rather than using the actual frequency or consequence, we use the category numbers instead. It may then be argued that the numbers should

Table 9.5 Example frequency categories

Frequency categories that should **not** be used	Frequency categories that can be used
>1 per year	1 per year
0.1–1 per year	0.1–1 per year
0.05–0.1 per year	0.01–0.1 per year
0.001–0.05 per year	0.001–0.01 per year
<0.001 per year	<0.001 per year

be added instead of multiplied, and that the numbers that we arrive at should be interpreted as logarithms of risk rather than actual risk numbers. The following example illustrates this.

Illustration of calculating risk numbers from a risk matrix

Assume that we have a risk matrix with five frequency categories, from 1 to 5, with 2 corresponding to frequency 0.001 and 3 corresponding to 0.01. Further we have defined five consequence categories, where 4 corresponds to 1 fatality and 5 corresponds to 10 fatalities.

Further we assume that we have identified two events and have analyzed them and plotted them in the risk matrix:

Event 1: $F = 2$ and $C = 5$
Event 2: $F = 3$ and $C = 4$

First, we can calculate the risk based on the definitions of the categories:

$$RE1 = 0.001 \cdot 10 = 0.01 \tag{9.1}$$

$$RE2 = 0.01 \cdot 1 = 0.01 \tag{9.2}$$

That is, the risk associated with these two events is equal.

We can now calculate risk from the risk matrix in two ways:

$$\text{Addition: } RE1 = 2 + 5 = 7, RE2 = 3 + 4 = 7 \tag{9.3}$$

$$\text{Multiplication: } RE1 = 2 \cdot 5 = 10, RE2 = 3 \cdot 4 = 12 \tag{9.4}$$

The result is not the same for the two approaches. If we add the consequence and frequency, the resulting risk number is the same. However, if we multiply, they are different.

The reason for this is that the category numbers can be regarded as logarithms, and we know that if we want to find the logarithm of a product, we can add the logarithms together. When we add together the frequency and consequence category, we can thus view the sum as a logarithm of the risk level.

If we use the earlier example, we can also see that this is correct. Say that we have the following risk categorizations instead:

Event 1: F=2 and C=5
Event 2: F=3 and C=5

If we calculate the risk now, based on the definitions of the categories, we find that the risk associated with Event 2 is ten times higher than the risk associated with Event 1. If we add together F and C for the two events, we find that Event 1 gets a value of 7 while Event 2 gets a value of 8. A logarithm of 8 corresponds to a value ten times higher than a logarithm of 7, therefore the addition produces logical comparisons between the two events. If we were to use multiplication approach instead, we would get risk numbers of 10 and 15, respectively.

The example illustrates how we can interpret the RPN (as long as we add the frequency and consequence categories). In practice however, the RPN should only be used for ranking, without assigning any specific meaning to the RPN.

The risk matrix usually incorporates risk acceptance criteria, although this is not always clearly recognized. In most cases, the risk matrix will be divided into three areas that usually are color coded red, yellow, and green. Normally, the decision criteria that follow with a risk matrix will specify that if the risk is in the red region, risk must be reduced, while if it is in the green region, there is no need for risk reduction. This corresponds to the ALARP principle (Chapter 6), with the yellow region corresponding to the ALARP region.

However, there are some aspects of this that are important to be aware of. The ALARP principle is normally applied in relation to the total risk level, i.e., the sum of risks associated with different events. In a risk matrix, the comparison is made for individual events that have been identified, not the total risk. Each event is plotted separately in the risk matrix, and it is the position of that event in the risk matrix that determines whether it is acceptable or not.

This clearly has some implications for how we set the acceptance limits in a risk matrix compared to the total risk.

The "coloring" of the risk matrix is often done based on the visual impression of the matrix. A suitable part of the matrix is colored red, a band in the middle is yellow and the rest becomes green. The size of the red part is fairly "standardized". However, if we consider that the categories can be defined freely and the colors represent acceptance criteria, we expect that the size of the red area would vary.

A final comment related to the use of the risk matrix is that this is intended as a tool for coarse ranking. This means of course that small differences between events can most likely not be identified. The way that we recommend that the risk matrix is defined, it can typically only distinguish between orders of magnitude. A side effect of this is that the effect of risk reduction measures can be difficult to visualize in the risk matrix. Only measures that have a significant effect on risk will lead to an event moving from one category in the risk matrix to another.

This coarseness may be seen as a weakness of the method, but the advantage compared to more detailed methods is that it is far more efficient and less resource consuming. It is therefore often useful to start with a coarse method and then go into more details if the coarse method does not give us the answers that we need to make decisions.

9.3 MEASURING RISK TO PEOPLE

9.3.1 Societal and individual risk

There are numerous ways of measuring and expressing risk to people and we may start by distinguishing between two groups of measures that have somewhat different properties.

The first is societal or group risk measures. As indicated by the name, these are measures that say something about the risk to a group of people or to society at large (which also can be considered as a group). They can express both injury risk and fatality risk but are most commonly used to express fatality risk. Societal risk measures will depend on the size of the group – the larger the group, the larger the calculated risk.

Individual risk measures will say something about the risk to an individual, often an average individual in a group. These risk measures are thus independent of group size.

9.3.2 Injury risk

Risk analysis seldom deals with injuries, but injuries are commonly used to produce statistics on safety performance. One of the most common measures is the LTI-rate (Lost Time Injury), also abbreviated LTIR. Sometimes, LTIF (Lost Time Injury Frequency) and LTIFR (Lost Time Injury Frequency Rate) are also used. An LTI accident is an accident that causes the person involved to have to take time off from work to recover. This will also include fatalities although the number of fatalities normally will be so low compared to the number of injuries that this makes little difference to the calculated LTI-rate. The minimum time lost to be counted as an LTI accident is usually defined as one shift. LTI-rate is commonly defined as follows:

$$\text{LTIR} = \frac{\text{No of LTI accidents}}{\text{Hours of exposure}} \cdot 2 \cdot 10^5 \text{ hours} \tag{9.5}$$

If we assume that a working year is 2,000 hours, the normalizing factor $2 \cdot 10^5$ hours corresponds to 100 working years. The LTI-rate can then be interpreted as the number of LTIs among a population of 100 persons working for 1 year. An LTI-rate of 2 means that among those 100 persons, two injuries occur during a year.

One issue to be aware of when comparing LTI-rates, is that they are not always calculated in this way. An example is IMCA (2020) that uses a multiplication factor of $1 \cdot 10^6$ instead of $2 \cdot 10^5$ hours. This will of course mean that the LTI-rate will be five times higher since the exposure time is five times higher. We should therefore ascertain how the calculation has been done before concluding on whether an LTI-rate is high or low.

In Figure 9.2, an example of LTI-rates from IMCA (International Marine Contractors Association) is shown. These are taken from their annual safety statistics (IMCA, 2020). A linear trendline has also been added in the figure, to indicate how the LTI-rate has declined over the period shown. The value in recent years has been below 0.5. Considering that IMCA uses a higher multiplication factor, the value calculated using the equation above would have been less than 0.1. This is a fairly low number, in particular considering the industry that this is taken from.

Another quite common measure is Total Recordable Injury Rate, TRIR (or Total Recordable Injury Frequency, TRIF). This is calculated in the same way as LTI-rate, but the calculation includes all recordable injuries, i.e., also those that do not lead to lost time as defined in the calculation of the LTI-rate. Minor cuts are typical examples of recordable injuries that do not necessarily lead to time off from work. This means of course that TRIR always is greater than or equal to LTIR.

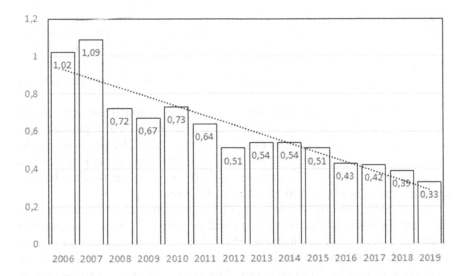

Figure 9.2 Example LTI-rates (IMCA, 2020).

A final injury measure that is used is LWD-rate (Lost Workdays Rate). This is also called DART (Days Away, Restricted or Transferred). Rather than just counting the number of lost time injuries, the number of days off work (or when you are not able to do your normal work) is counted.

$$\text{LWDR} = \frac{\text{No of Lost Workdays}}{\text{Hours of exposure}} \cdot 2 \cdot 10^5 \text{ hours} \tag{9.6}$$

This means that the more severe the injury, the more it will "weigh" in the calculation because the number of days off work will be higher. Again, the LWD-rate will as a minimum be the same as the LTI-rate. Fatalities are also included. The number of days used to include fatalities may vary, but 7,500 days is being used in some applications.

9.3.3 Potential Loss of Life (PLL)

PLL is one of the most common group risk measures. PLL has slightly different meanings in different texts, but most common is Potential Loss of Life. PLL can be defined as follows:

> The (statistically) expected number of fatalities within a specified population due to a specified activity or within a specified area per annum.

> *(Rausand & Haugen, 2020)*

This definition specifies that PLL is tied to a group, an activity/area, and a period (year). An example can be that we can calculate PLL for the crew on a container ship operating worldwide for one year. Another example can be to calculate PLL for all employees in a company for a period of one year.

PLL is defined as "the expected number of fatalities". This means that PLL can be smaller than one, if the risk is low (or the group is small). For the crew on a container ship, one would expect many years between each time a fatal accident occurs. This means that PLL will be correspondingly low.

PLL can be calculated in different ways. Often, statistics are used to calculate PLL, and then the statistics are assumed to be applicable also to describe the future. PLL is then calculated simply by considering the group in question and counting the number of fatalities in that group over a period of time. The period can be more than one year, but we then have to divide by number of years to arrive at a PLL value.

One comment to this way of using PLL is that fatal accidents are seldom and that the statistics because of this will be very uncertain. The example with the container ship illustrates this. If, e.g., a 10-year period is considered, there will most likely be no fatalities. It is thus not possible to estimate a PLL value from statistics. The same with a company – even in large companies there may be years between each time someone is killed.

Table 9.6 Input data to calculation of PLL

Event	f(E)	P(killedE)	$N_{fat,E}$
Collision	0.02 per year	0.2	2
Fire	0.08 per year	0.1	1

Calculating PLL based on statistics is therefore in practice only feasible for large groups, e.g., all seafarers globally or for all seafarers within a (large) shipping company.

The alternative is to calculate PLL from risk analysis. In practice this is often done as follows, assuming that E_i is an event that has a certain risk associated with it:

$$PLL_E = f(E) \cdot P(\text{killed}|E) \cdot N_{fat,E} \qquad (9.7)$$

where $f(E)$ is the frequency of event E occurring, $P(\text{killed} | E)$ is the probability that E causes one or more fatalities, and $N_{fat,E}$ is the number of people killed in the event. If there are several events, the contributions from all events are added together to arrive at a total PLL.

Calculating PLL

Let us assume that two events are being considered, collision and fire and that we have the information shown in Table 9.6.

PLL can then be calculated as follows:

$$PLL = 0.02 \cdot 0.2 \cdot 2 + 0.08 \cdot 0.1 \cdot 1 = 0.016 \qquad (9.8)$$

This can be interpreted as 1 fatality per approximately 60 years (~1/0.016) or mean time between fatalities.

9.3.4 FN curve

A very different way of measuring societal risk is to use an FN curve. Compared to PLL, the FN curve provides more information about risk because it illustrates both frequency and consequence and does not combine it into a single number. An example of an FN curve is shown in Figure 9.3.

The FN curve illustrates the frequency of accidents (F) and the consequence of accidents (N). N is the number of fatalities per accident, i.e., it is an expression of the severity of individual accidents and not the total risk. N = 5 then means accident with exactly five fatalities. F is the frequency of accident with N or more fatalities, i.e., it can be seen as the frequency of exceedance of a certain consequence level. The F for N = 5 is then the frequency of all accidents that cause 5 or more fatalities.

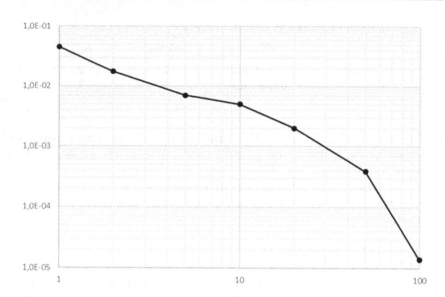

Figure 9.3 Example FN curve.

Because of the way the frequency is defined, the FN curve will always be flat or falling with increasing N. The reason is that $F(N) \geq F(N+1)$ for all values of $N \geq 1$. This is because the frequency of $F(N)$ is equal to $F(N+1)$ plus the frequency of accident with exactly N fatalities. Since the frequency cannot be negative, $F(N)$ cannot be smaller than $F(N+1)$.

The N-axis will start at 1, not at zero, because it is fatal accidents that are illustrated in an FN curve. Accidents with no fatalities are therefore not relevant. N can in other words take on values from 1 and upwards.

It is common to use logarithmic scales both for the frequency and number of fatalities. The main reason for this is that the range of both values can be large. The frequency will typically span several orders of magnitude and it will then be hard to interpret the curve unless logarithmic scales are used.

The FN curve is also used in the FSA guidelines (IMO, 2018), and Figure 9.4 shows FN curves for some ship types.

Figure 9.4 also illustrates how risk acceptance criteria can be introduced as part of the FN curve. The two diagonal dotted lines represent the upper and lower limits for the ALARP region, and the figure shows that all the ship types considered fall within this region.

Acceptance criteria in the FN diagram can be defined by two parameters:

- An anchor point, often defined by the frequency F that corresponds to $N=1$. The anchor point is not an expression of the total acceptable risk, only what the acceptable frequency of fatal accidents (accidents with 1 or more fatalities) is.

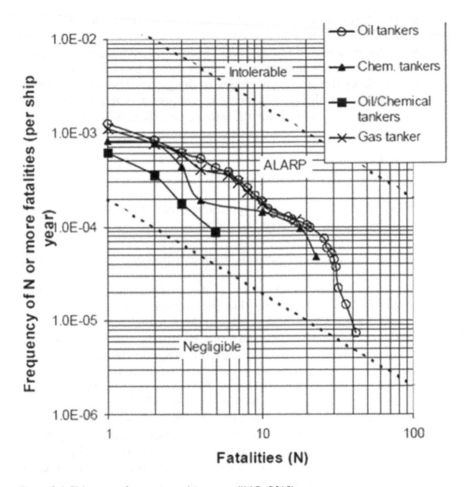

Figure 9.4 FN curves for various ship types (IMO, 2018).

- The slope of the line. The slope of the line can be interpreted as our aversion to accidents with very large consequences. If the slope is 1, we often say that we are "risk neutral", i.e., we accept the same risk for accidents with one or few fatalities as we do for accidents with many fatalities. If the slope is >1, we are "risk averse".

The acceptance criterion can also be expressed using a line with two different slopes. Often the slope will then be higher for the right part of the line.

9.3.5 Individual Risk (IR)

Probably the most common individual risk measure has also the most obvious name, being called simply Individual Risk (IR). The term IRPA (individual risk per annum) is also often used. IR is simply the probability that an individual will be killed related to the system or the activity being considered during a period of one year. This can be calculated in different ways, but most common approach is to calculate the total risk for a group and then divide by the number of people in the group. The result will then be an average IR for all the individuals in the group. The total risk for a group is the same as the PLL for the group, i.e., IR can be calculated as follows:

$$\text{IR} = \frac{\text{PLL}}{N_E} \qquad (9.9)$$

where N_E is the number of people exposed (number of people in the group). IR can also be calculated directly, as follows:

$$\text{IR} = P(\text{event}) \cdot P(\text{exposed} \mid \text{event}) \cdot P(\text{killed} \mid \text{exposed}) \qquad (9.10)$$

The first factor is the probability of an event occurring that can cause fatalities, the second is the probability that the individual being considered is exposed to the event when it happens, and the third is the probability that the individual is killed when exposed to the event.

This equation highlights that exposure is a factor in the calculation and this is relevant to consider for all work-related accidents, including maritime accidents. When we are calculating PLL for a ship, we take the total size of crew that is onboard at any time and use that in the calculation. However, since none of the crew members will be on board the ship for one whole year, we must take into account the proportion of time they spend on board when calculating IR. If we use the equation based on PLL, we must use the total number of people working on the ship over a year to represent N_E, not just the average manning.

Calculating PLL and IR

Consider the following situation:

- A ship has a normal manning of 20 persons.
- There are three crew shifts that are keeping the ship in operation on all year basis. In total, there are therefore 60 individuals that work on the ship.
- Each individual among these will spend one-third of the year on the ship.
- We want to calculate PLL and IR due to one specific event. This event has an annual frequency of 0.0001 and if the event occurs, 2 people will be killed.
- The probability of being killed if you are exposed to the event (are on board the ship when the event occurs) is 0.1.

First, we can calculate PLL. This is done by taking the frequency of the event and multiplying by the consequence:

$$PLL = 0.0001 \text{ per year} \cdot 2 \text{ fatalities} = 0.0002$$
$$= 2 \cdot 10^{-4} \text{ fatalities/per year} \tag{9.11}$$

Now, we can use the equation above to calculate IR based on PLL:

$$IR = \frac{PLL}{N_E} = \frac{2 \cdot 10^{-4}}{60} = 3.3 \cdot 10^{-6} \text{ per year} \tag{9.12}$$

Alternatively, we can also calculate IR as shown in Eq. 9.10:

$$IR = 0.0001 \text{ per year} \cdot \frac{1}{3} \cdot 0.1 = 3.3 \cdot 10^{-6} \text{ per year} \tag{9.13}$$

In both cases, we arrive at the same result.

9.3.6 Fatal Accident Rate (FAR)

Another risk measure that is extensively used is FAR, or Fatal Accident Rate. The FAR value is also an individual risk measure but does not explicitly express the probability of fatality as IR does. FAR can be defined as follows.

The expected number of fatalities in a defined population per 100 million hours of exposure (Rausand & Haugen, 2020).

Formally, this can be expressed as follows:

$$FAR = \frac{\text{Expected no of fatalities}}{\text{No of exposed hours}} \cdot 10^8 \text{ hours} = \frac{PLL}{H_E} 10^8 \tag{9.14}$$

At first glance, this may look like a group risk measure since it is the number of fatalities in a defined population. However, since we consider risk for a given exposure (100 million hours), this means that changes to the size of the group also will lead to a corresponding change in exposure. The effect of the group size is therefore canceled out in the calculation.

It may also be noted that the equation for FAR has the same form as the equation for LTI-rate. The differences are that "expected no of fatalities" is replaced by "no of LTI accidents" and the multiplication factor 10^8 is replaced by $2 \cdot 10^5$.

The multiplication with 10^8 hours is in practice a "scaling factor". To give an indication of what this value means in practice, we can use the following reasoning. If we assume that a working life is 50 years and a working year is 2,000 hours, this corresponds to 100,000 or 10^5 hours. This means that 10^8 hours corresponds to 1,000 working lives. A FAR value can then be interpreted as the number of fatalities in a group of 1,000 people working their entire working life in a specific industry, company, or activity.

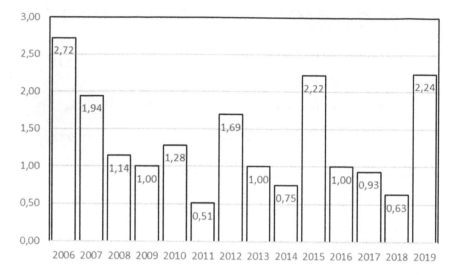

Figure 9.5 FAR values from IMCA (2020).

Figure 9.5 illustrates examples of FAR values from IMCA (International Marine Contractors Association). The values range from 2.7 down to about 0.5 in the period shown in the diagram. There is only a weak downward trend mainly because there were high peaks both in 2015 and 2019. This is a good illustration of how major accidents can influence statistics like this quite strongly. Both in 2015 and in 2019 there were severe accidents with 9 and 11 fatalities respectively. The total number of fatalities in 2019 was 18, meaning that if the one accident with 11 fatalities had not occurred, the FAR value would have been 0.87 instead, i.e., more or less in line with the average for the period of 2016–2018. This is not an argument for not including very serious accidents like this in the statistics but rather a warning that accidents like that will influence the statistics quite strongly.

It may also be noted that the values shown can be regarded as typical for medium to high-risk industries.

9.3.7 Combining injuries and fatalities

In quantitative risk analysis, it is common to focus on fatalities rather than injuries. Most risk measures therefore consider only fatality risk (or only injury risk). However, in some instances, we may also have a need for combining injuries and fatalities into one risk value.

One approach to this is to assume that a fatality corresponds to a certain number of injuries. Values that are sometimes used are as follows:

1 fatality $= 10$ major/serious/permanent injuries

$\qquad = 100$ minor injuries

(9.15)

In the FSA guidelines (IMO, 2018), "equivalence ratios" like the above are also proposed to be used.

A more comprehensive approach is to use QALY or DALY (also mentioned in IMO, 2018). These measures have been developed by WHO (World Health Organization) and is originally aimed at measuring the effect of diseases. However, they can be applied also to accident risk. QALY (Quality Adjusted Life Years) assumes that one year of perfect health has a value of 1. If your health is less than perfect, the value is less than 1. This means that a permanent injury due to a work accident will give a QUALY value of less than 1 for each of the remaining years the person is living. Shortened life expectancy can also be taken into account, by using QUALY$=0$ for the years of reduced life expectancy. The same will apply to fatal accidents. DALY (Disability Adjusted Life Years) is essentially the same, but the scale is turned around, meaning that perfect health is assigned a value of 0 and less than perfect a value larger than 0. WHO have proposed values for some diseases and injuries. Some examples are blindness (0.195), fractured femur (0.308), and amputation of finger (0.030) (WHO, 2013).

Traditionally, life expectancy has been based on Japanese statistics since Japan has had the longest life expectancy. Alternative approaches, based on national life expectancy statistics, have also been used. This last approach means that the DALY/QALY values will be different for a country with a short life expectancy compared to a country with long life expectancy, even if the injury is the same and it occurs to persons at the same age.

Calculating QALY

Let us make the following assumptions. We will use an average life expectancy of 80 years in the calculation. A crew member is involved in an accident at age 28, when he loses a finger. Further, he is involved in a far more serious accident at age 45, when he is killed.

With this information we can calculate QALY. With a life expectancy of 80 years, the maximum QALY is 80, if he has no accidents and remains healthy his entire life. The accidents however reduce the value:

$$\text{QUALY} = 80 - (80 - 45) \cdot 1 - (45 - 28) \cdot 0.030$$

$$= 80 - 35 - 17 \cdot 0.030 = 44.49$$

(9.16)

The person lives until he is 45, i.e., 45 life years, but the quality adjustment reduces this to 44.49 taking into account that he lost a finger at the age of 28.

9.4 MEASURING RISK TO THE ENVIRONMENT

Measuring the risk to the environment can also be done in many ways, as we have seen for risk to people. In both cases it is a matter of combining frequency and consequence, but it is primarily the consequence that can be expressed in different ways. We will here look only at risk associated with oil spills.

Consequences of oil spills can be measured in (at least) four different ways:

- Spill size – this is the easiest way of measuring the consequences since it can be related directly to the direct consequence of the accident, not having to analyze what happens to the oil after it has been released. If we use this, we can calculate risk as an expected spilled volume per year (equivalent to how we calculate PLL). This is also a value that is easy to compare with statistics from oil spills in the past, published by ITOPF (2021).
- Effect on wildlife, ecosystem, etc. This is harder to measure and will require firstly that we define precisely what effect we are interested in (is it effect on seabirds in general, specific species of seabirds, fish stocks in general, specific species of fish, etc.). Next, we need to consider what happens to the oil after it has been released and determine when and how it exposes the resources we are concerned about. This requires competence in modeling of dispersion of oil in sea, under varying environmental conditions and for varying types of oil. Finally, the actual effect on the relevant resources of being exposed to the oil needs to be considered. This requires competence on the specific species being considered. Precisely how the consequences are measured can vary, but one possibility is to express it in terms of reduction in the size of a population, e.g., 50% reduction of the population of cod in the affected area because of a spill.
- Effect on coastline. This is a variation of the previous measure, where the length of coastline being affected is considered. A precise definition of what we mean when we say that the coastline is "affected" is necessary. This is somewhat simpler than the previous measure and requires competence on oil spill dispersion only. Consequences can be expressed in terms of km of coastline affected.
- Duration of effect of the spill. The idea here is to look at how long the environment is affected by a spill or to phrase it differently: How long will it take until the environment has recovered to the state it was before the spill occurred. This can be related to various aspects, e.g., how long until the population of relevant species is back to normal or time until clean-up and natural processes have removed all effects of oil from a shoreline or a beach. This is perhaps the most difficult way to measure the consequences because it also requires competence about how nature recovers from such negative effects.

In general, it is not easy to choose the best way of expressing consequences of spills. By far the easiest choice is to use spill size only, but this will not reflect the major differences in consequence if we have a spill in the middle of the ocean compared to if it occurs close to or in a highly sensitive environmental area. It may therefore be argued that this is too simplistic.

What is done in some cases though is that separate environmental studies are performed for specific areas (vulnerable areas), where the effect of a "generic" spill is determined. These results can then be combined with calculations of spill size from all ships coming in the vicinity of these areas.

In a study for the European Commission (EC, 2006), it was concluded that agreed methods for calculating environmental effects are lacking and that this also has made it difficult to measure risk and specify risk acceptance criteria.

9.5 MEASURING RISK TO OTHER ASSETS

9.5.1 Economic risk

Economic risk can also be calculated in the same way as risk to personnel, by multiplying the frequency of events with the consequence, where the consequence is expressed in terms of money rather than injuries or loss of life. This calculation will give us the statistically expected loss per year. If an event has a frequency of $1 \cdot 10^{-6}$ per year and causes a loss of 100 million USD, this gives a statistically expected loss of just 100 USD per year. This is a low number that most will consider to be an acceptable risk. However, this number does not reflect the true situation, either we lose nothing, or we lose 100 million USD. The "average" value of 100 USD therefore only gives us limited information about risk (as all risk measures based on a single number will do). This loss may be catastrophic to even a large business, potentially causing bankruptcy. It may therefore be necessary not just to consider the calculated risk but also the actual loss that will occur if the event occurs. If this loss is too high, we will typically secure ourselves by buying insurance. The insurance premium can then be seen as the annual "loss" (or more correctly cost) that we are willing to accept to avoid a catastrophic loss.

Macondo/Deepwater Horizon oil spill

In 2010, a blowout occurred at the BP-operated Macondo field in the Gulf of Mexico, from the drilling rig Deepwater Horizon. This caused 11 fatalities and the largest offshore oil spill in US history was a fact. The costs associated with this spill were considerable and in 2018 BP stated that the total costs to the company up to that point were 65 billion USD. BP survived, but in the first period after the disaster, there was discussion about whether even a mega-company like BP could go bankrupt because of this catastrophic accident. Another illustration of the economic effects for the shareholders was the drop in share price after the accident. The share price dropped by more than 50% in the first 40 days after the accident and the total value of the company dropped by 105 billion USD. By 2018, BP had dropped from being the second largest oil company in the world to being the fourth largest (Wikipedia, 2021).

9.5.2 Reputation

A final type of consequence that is becoming increasingly important to consider is the reputation of the company. For many companies, it is important that they are trusted by clients, passengers, and society in general. Without this, they will lose business and may eventually be driven out of business completely.

In the maritime domain, the cruise business is a good example of this. With the competition among cruise companies, negative news about a company due to accidents can be very damaging. In fact, it can even be damaging to the whole industry. The grounding of the Costa Concordia in 2012 is an example of this. Obviously, Costa Cruises (the owner) suffered the largest losses, but there was industrywide concern about the accident and one of the results was that the Cruise Lines International Association (CLIA) was formed, among others to improve safety image of cruise line operators.

Measuring reputation is in general not easy although it may be possible to use surveys about consumer confidence in a company and similar tools. For public listed companies, effect on share price of adverse events may partly be seen as an indirect measure of effect on reputation. When using risk matrixes, effect on reputation may be expressed in ordinal categories as already discussed in this chapter. From experience the categories are however often quite vague, such as "significant effect on reputation" and "minor effect on reputation".

REFERENCES

EC. (2006). *Land use planning guidelines in the context of article 12 of the Seveso II directive 96/82/EC as amended by directive 105/2003/EC* (Report No. EUR 22634 EN). https://minerva.jrc.ec.europa.eu/EN/content/minerva/3b0cfe29-cf09-4b74-b41e-4a64949e95ae/lupguideart12pdf

IMCA. (2020). *2019 Safety Statistics.* https://imcaweb.blob.core.windows.net/wp-uploads/2020/06/IMCA-InformationNote-1513-STS137.pdf

IMO. (2018). *Revised guidelines for Formal Safety Assessment (FSA) for use in the IMO Rule-Making Process.* https://wwwcdn.imo.org/localresources/en/OurWork/HumanElement/Documents/MSC-MEPC.2-Circ.12-Rev.2%20-%20Revised

%20Guidelines%20For%20Formal%20Safety%20Assessment%20
(Fsa)For%20Use%20In%20The%20Imo%20Rule-Making%20Proces...%20
(Secretariat).pdf

ITOPF. (2021). *Oil tanker spill statistics 2020*. https://www.itopf.org/fileadmin/
uploads/itopf/data/Documents/Company_Lit/Oil_Spill_Statspublication_
2020.pdf

Rausand, M., & Haugen, S. (2020). *Risk Assessment: Theory, Methods and
Applications (2nd ed.)*. Wiley.

WHO. (2013). *WHO methods and data sources for global burden of disease estimates
2000–2011*. https://www.who.int/healthinfo/statistics/GlobalDALYmethods_
2000_2011.pdf

Wikipedia. (2021). *Economic effects of the Deepwater Horizon oil spill*. https://en.
wikipedia.org/wiki/Economic_effects_of_the_Deepwater_Horizon_oil_spill

Chapter 10

Methods for navigational risk analysis

10.1 INTRODUCTION

Traffic-related accidents represent a major part of ship accidents and losses. By traffic-related accidents, we mean the following accident types:

- Grounding or stranding: The ship hits a shoal or the shore seafloor.
- Allision or ramming: Impact with an object above the waterline.
- Collision: Hitting or being hit by another vessel.

Since the early 1990s, there has been increasing focus on understanding why these accidents occur and how we can estimate their frequencies. The accident phenomenon may be analyzed from different perspectives and with a large number of factors influencing risk:

- Technical: Functional standard of the vessel and reliability of navigation and control systems.
- Human factors: Organization of the crew, competence and motivation of crew members, and human error probabilities.
- The seaway: Physical dimensions, complexity in terms of heading changes, water depth, and current.
- Traffic pattern: Meeting, crossing and overtaking vessels, traffic density, rules of the road, and conflicting vessels.
- Environment: Weather conditions, such as precipitation (rain, snow), fog (visibility), wind, and waves.
- Aids to navigation: Pilot service, seamarks, Traffic Separation Schemes (TSS), and Vessel Traffic Systems (VTS).

It is a characteristic of marine traffic accidents that the dominant causal factors are human and organizational factors (HOF). Different studies

DOI: 10.4324/9781003055464-10

indicate relative frequencies from 70% to 90%. This is a strong motivation for developing models that take into account human error mechanisms and the interaction between humans, technical systems, and the environment. Methods incorporating these factors may be applied to estimate the effect of alternative risk control measures. A second motivation is risk analysis of waterways focusing on improvement of sailing conditions and aids to navigation (AtoN).

The start-of-the-art today is characterized by different modeling approaches putting different weight on the perspectives outlined above. It is fair to say that there is still no analytical approach that captures all these factors. Qualitative models can typically rank the relative performance of, e.g., alternative waterways. Quantitative models may on the other hand give numerical estimates of the accident frequency. In the following sections, the following will be discussed:

- Probabilistic models with focus on waterway characteristics and traffic flow
- Causation models that estimate the probability of HOF
- Qualitative models

10.2 GROUNDINGS

10.2.1 Early studies of powered groundings

Early probabilistic models are primarily focusing on the geometric properties of the fairway and were inspired by the pioneering work of Macduff (1974), Fujii and Takanaka (1971), and Fujii (1982). Macduff defined a model for the probability of grounding (P_{RG}) as follows:

$$P_{RG} = P_G \cdot P_C \tag{10.1}$$

The probability is given by the product of the geometrical probability (P_G) and the causation probability (P_C).

The geometrical probability expresses the probability of grounding being on a random course in the fairway with a width W (see Figure 10.1) and a ship with stopping distance T:

$$P_G = \frac{4T}{\pi W} \tag{10.2}$$

The causation probability simply expresses the probability of being on a random course and not taking any evasive action.

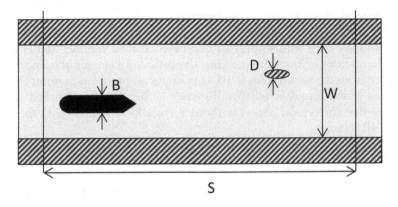

Figure 10.1 A simple model of a grounding accident.

Fujii (1982) has given a slightly different model for the grounding frequency:

$$N = P_C \, (D + B) \rho V \tag{10.3}$$

where

D = linear cross section of the obstacle in the fairway
B = breadth of vessel
ρ = density of traffic flow (ships/area-unit)
V = average speed of vessels

An alternative formulation of the model proposed by Fujii (1982) is:

$$N = Q \cdot P_{RG} = Q \cdot \frac{(D + B)}{W} \cdot P_C \tag{10.4}$$

where Q is the traffic frequency (ships/time unit). Estimation of P_C may be based on observations of the number of accidents and traffic for a given fairway over a period:

$$P_C = \frac{N \cdot W}{Q(D + B)} \tag{10.5}$$

The expression $W/(D+B)$ is given by the dimension of the seaway and mean ship-breadth. Note that the underlying assumption in this expression is that the traffic is uniformly distributed across the width of the fairway.

Investigations of three fairways by Fujii (1982) gave estimates for P_C in the range of $0.8–3.3 \cdot 10^{-4}$. As an average, it was decided to apply $2.0 \cdot 10^{-4}$.

Later studies (Friis-Hansen, 2008) have indicated estimates in the same range: $1.0-6.3 \cdot 10^{-4}$. This variation simply reflects the fact that the studied fairways may vary with respect to geometry, traffic volume, and environmental conditions. Roughly speaking the ratio between highest and lowest estimate from the two studies is 10. It is important to keep in mind that the model is based on given fairway distance S. Various unpublished studies indicate that the typical period without navigational control is in the order of 30–40 minutes.

Grounding frequency

The grounding accident frequency is to be estimated for the following straight fairway with one obstruction.
  Fairway data: $S=9$ nautical miles$=16,668$ m
  $W=1,000$ m
  $D=100$ m
  Vessel traffic: $Q=10$/hour$=87,600$/year
  $V=15$ knots$=7.7$ m/s
  $B=30$ m
  Assuming time without control to be conservatively 40 minutes, the vessel will sail 18,480 m which is of the same order as the length of the fairway. The grounding frequency is:

$$N = 2.0 \cdot 10^{-4} \cdot 87,600 \frac{(100+30)}{1,000} = 0.23 \text{ groundings/year}$$

This means that the mean time between groundings (the return period) corresponds to 4.3 years (1/0.23).

10.2.2 Developments of powered grounding models

A very obvious simplification in the early models was the assumption that traffic was uniformly distributed across the fairway. Pedersen (1995), Simonsen (1997), and Friis-Hansen (2008) proposed models that take the lateral distribution of the ship traffic into consideration and also different ship type or size (Figure 10.2). They also made a distinction between the following two accident scenarios:

1. Ships following the ordinary direct route at normal speed (category I). Accidents in this category are mainly due to human error but may include ships subject to unexpected problems with the propulsion/ steering system that occur in the vicinity of a fixed marine structure or the ground.
2. Ships that fail to change course at a given turning point near the obstacle (category II).

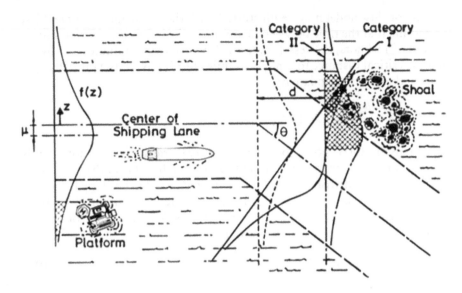

Figure 10.2 Model for predicting grounding frequency for ship traffic with lateral distribution (Pedersen, 1995).

The number of groundings was estimated as follows:

$$N_1 = \sum_i P_{C,i} Q_i \int_{z_{min}}^{z_{max}} f_i(z)dz \qquad (10.6)$$

$$N_2 = \sum_i P_{C,i} Q_i \exp\left(-d/a_i\right) \int_{z_{min}}^{z_{max}} f_i(z)dz \qquad (10.7)$$

where

i=index for ship class
$P_{C,i}$=causation probability for ship class i
Q_i=number of ships in category i
$f_i(z)$=probability density function for the ship traffic
z_{min} and z_{max}=transverse coordinates for an obstacle
d=distance from waypoint to the obstacle
a_i=average distance sailed between position checks

Model 2 incorporates two errors, namely, (1) missing the waypoint and (2)
not correcting the heading in due time. It was assumed that the probability
of not checking the position from the waypoint to the obstacle could be
modeled by a Poisson process and is given by:

$$P_{NT} = e^{-d/a_i} \qquad (10.8)$$

The use of this model may be illustrated with the following example. First, we assume that the time between position checks is 3 minutes. By assuming a speed of 15 knots, the vessel travels 1,389 meters in 3 minutes (a_i). By setting the critical distance to 3 nautical miles ($d=5,556\,\text{m}$), the probability of missing the position update is:

$$P_{NT} = e^{-5,556/1,389} = 1.8 \cdot 10^{-2}$$

Friis-Hansen (2008) points out that the lateral traffic distribution may be expressed by different statistical models and further shows that by assuming the Gaussian distribution a definite integral can be developed.

COWI (2008) has developed the first grounding model (imprecise navigation) further by incorporating two extra factors. The frequency of grounding due to imprecise navigation is given by:

$$N = Q \cdot P_G \cdot P_C \cdot k_{DC} \cdot k_{RR} \tag{10.9}$$

where k_{DC} is a distance correction factor and k_{RR} is risk reduction factor.

Based on AIS observations, the distribution of ship headings in a given fairway is estimated based on a normal distribution $N(\mu, \sigma)$. The geometrical probability then becomes:

$$P_G = F(a_1) - F(a_2) \tag{10.10}$$

This expresses the relative frequency for the critical headings leading to a grounding (Figure 10.3). They chose a slightly higher causation probability, $P_C = 3 \cdot 10^{-4}$, based on own studies.

A critical factor in assessing the grounding risk is the distance S between the traffic observation point and the obstacle. The nearer the observation point is to the obstacle, the greater the fraction of observed vessels that will

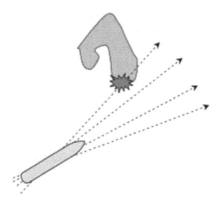

Figure 10.3 Grounding scenario (COWI, 2008).

have changed their heading away from the obstacle. A distance correction factor is therefore incorporated in the model:

$$k_{DC} = \frac{10}{S}$$

(10.11)

Based on traffic observation experience, the mean distance is set to 10 nm reflecting observation periods in the interval half an hour to one hour.

Finally, the model made a distinction between vessel having and not having a pilot. With pilot onboard, a risk reduction factor $k_{RR} = 0.5$ was applied.

10.2.3 Drift grounding

A vessel may lose control due to electric power loss (black out), steering system failure, or loss of propulsion. Factors that influence the probability of grounding will include:

- Distance to obstacle (depends on drifting direction)
- Vessel orientation: transverse, longitudinal, or something in between
- Effect of wind and waves
- Drifting speed

The majority of drift groundings take place in restricted waterways like ports, rivers, and canals. Mølsted et al. (2012) proposed a circular motion as shown in Figure 10.4 without any attempt to model the factors above.

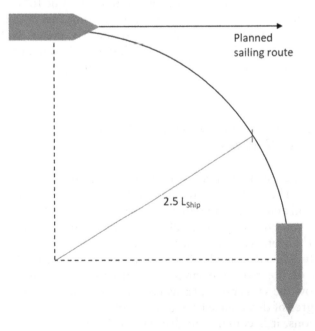

Figure 10.4 Ship track after steering machine failure (Mølsted et al., 2012).

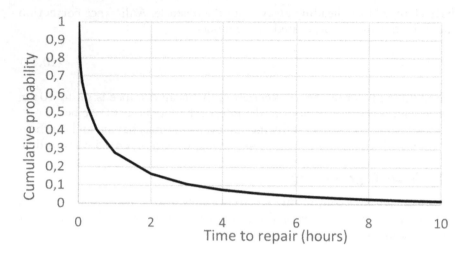

Figure 10.5 Cumulative distribution for time to repair machinery failure. Weibull distribution with scale parameter $\eta=0.605$ and shape parameter $\beta=0.5$.

The authors proposed failure rates per hour of $1.5 \cdot 10^{-4}$ for loss of propulsion and $6.3 \cdot 10^{-5}$ for loss of steering. A mean drift speed of 1 knot was further proposed.

The authors recognize that even if a failure occurs, this can in most cases also be repaired by the crew. They propose that the time to repair a failure in the machinery system can be expressed by a truncated Weibull distribution with scale parameter $\eta=0.605$ and shape parameter $\beta=0.5$. This is illustrated in Figure 10.5 and shows, e.g., that after 2 hours, the probability that the machinery still is failed is about 0.15, dropping to less than 0.05 after 6 hours.

10.3 ALLISION WITH FIXED OFFSHORE INSTALLATIONS

There is an increasing number of fixed installations in areas where ships are traveling. The first installations were lighthouses and buoys, followed by offshore oil and gas installations and more recently offshore wind farms. The largest consequences are usually seen for offshore oil and gas installations. In the worst case, the allision can lead to a total loss as seen at the Mumbai High North Complex Platform in July 2005 with 22 fatalities.

Analysis of the risk associated with allision with these installations largely follow the same principles as powered groundings, but models with varying degree of detail have been developed.

For allisions, it is common to distinguish between passing and visiting vessels. Visiting vessels are various work vessels that are approaching the installation on purpose, being it a supply vessel, a service vessel, or an oil shuttle tanker. Statistics show that visiting vessels represent roughly 75% of the accidents. For offshore installations, exploration units are more exposed to accidents than permanent or fixed installations.

It is noted that many of the models that have been developed have been aimed at analyzing the risk to the installation that is being hit and not the risk to the ship. The models will therefore typically calculate the frequency for an installation and not for a ship.

Before looking at modeling, it may also be mentioned that frequencies also can be estimated from statistics. Estimation of ramming frequency is based on different statistical sources. Accident data may be found in sources like *World Offshore Accident Database* (WOAD) maintained by DNV-GL and for the *United Kingdom Continental Shelf* (UKCS) by the UK Health and Safety Executive (HSE). Similar databases were developed for marine traffic in exposed waters but since the introduction of AIS, it has been possible to track vessels and produce up-to-date information of traffic volume, density, and routes. See Haugen (1998) and Hassel et al. (2017).

In practice statistics will have limited use since the risk for a specific installation will be highly dependent on the traffic volume and type of traffic in the area. Statistics can therefore only give average frequencies for an area.

10.3.1 Powered passing vessels

COLLIDE II (Haugen & Vollen, 1998) was one of the pioneering models. The model elements were developed on the basis of an elaborate analysis of how vessels were operating in the vicinity of platforms. Separate models were developed for passing traffic, nearby fishing vessels, visiting offshore vessels, drifting floating units, and drifting vessels. We will here present the passing vessel scenario, where the collision frequency was expressed as follows:

$$P = \sum_i \sum_j \sum_k Q_{ijk} \sum_l P_{G,jkl} \cdot P_{CS,jkl} \cdot P_{CP,jkl} \tag{10.12}$$

The model is based on the assumption that the traffic can be sorted in traffic lanes, where vessels follow similar courses and within a fairly well-defined range from a centerline. The traffic is thus segmented into lanes (i), vessel types (j), and vessel size categories (k). Further, the vessels are also divided into four traffic groups, l, reflecting their navigational "behavior". The geometric allision probability ($P_{G,jkl}$) is estimated for each combination of the traffic segments. The causation probability is split into two probabilities, covering the fact that the ship may take action to avoid allision on its own (P_{CS}), but also that the platform may initiate actions that can lead to the

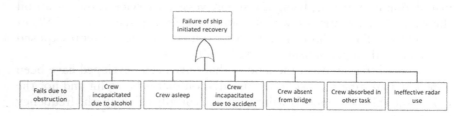

Figure 10.6 Causes of the ship not taking action to avoid allision.

ship taking action (P_{CP}). Figure 10.6 illustrates the causes of the ship failing to take action to recover from a situation where it is on course toward an installation.

It should be kept in mind that this model was developed before the introduction of satellite navigation, electronic charts, and AIS. Some of the assumptions that formed the basis for the model have been changed by these technological developments. The strength was however a thorough discussion of factors that influence the allision probability.

A basic assumption in the model is that a vessel at risk must be on an allision course before it reaches 12 nm from the installation. This was assumed to be a distance where the installation typically could be observed by radar or visually. It was further assumed that the ship traffic followed more or less fixed routes (lanes).

Navigational behavior was influenced by three main factors:

- Whether the installation was known to the vessel or not before it could be observed. With paper charts and manual updates, it was not given that all installations, especially new installations, were known to all vessels. Voyage planning could therefore be based on inadequate information and the course could be set straight for the platform.
- Normally, if vessels knew about the platform and it was located in the middle of where the ships were planning to pass, avoidance planning would be applied. This meant that the route was planned to the side of the installation, at what the navigator considered a safe distance.
- Alternatively, some would also use the installations as convenient points for fixing their position. With GPS, this is less relevant, but when the model was developed this was not common. Some ships would therefore plan their course closer to the platform than the shortest course line would indicate, simply to get a confirmed fix of their position.

The navigational behavior would in turn influence both the geometrical probability of allision and also the probability of avoiding impact.

The sub-models took a number of factors into consideration:

- Type of platform: Permanent or temporary
- Time since installation
- Probability that the platform is known (updating of charts)
- Availability and use of navigation systems
- Lateral distribution of traffic for planning and non-planning vessels
- Distribution of heading when approaching the platform

In retrospect, it may be questioned whether the fairly detailed models could be defended given the scarcity of data.

10.3.2 Allision with wind farms

With the introduction of wind farms, a number of different models were introduced for estimation of the allision probability. Most models were based on the same approach as for offshore installations but with different modifications of both the geometrical and causation probability. Wawruch and Stupak (2011) discussed the various approaches and explained that the different results could be explained by:

- Lack of accident data or use of different data sources
- The estimate of the causation probability
- Adjustment of causation probability
- Lack of human error probability estimates
- Lateral distribution of ship traffic (alternative statistical distributions)
- Different segmentation of the ship traffic

The authors cite a comparison of the results from models developed by MARIN and Germanischer Lloyd (GL). Table 10.1 shows that the estimates differ with more than one order of magnitude.

SSPA (Forsman et al., 2008) proposed a model with a few modifications:

- The traffic was modeled with both lateral distribution and heading distribution along the critical route segment leading up to the wind turbine.

Table 10.1 Probability of allision between ship and wind turbine

Model	Distance between center line of seaway and wind turbine		
	0 nm	0.5 nm	1 nm
GL	0.148	0.0481	0.0137
Marin	0.0060	0.0024	0.0009

Wawruch and Stupak (2011).

- The causation probability combined the following factors:
  - Error in passage planning and execution
  - Navigation error due to technical and human error
  - Failure of installation or standby vessel to warn passing vessel
  - Failure of vessel to react and correct heading

The model is expressed as follows:

$$F_{CP} = \sum_x \sum_{offset} \sum_{course} N \cdot P_x \cdot P_{offset} \cdot P_{course} \cdot P_{C1} \cdot P_{C2} \cdot P_{C3} \cdot P_{react}(x) \tag{10.13}$$

where

F_{CP} = Frequency of powered passing ship allision (per year)
N = Traffic in the shipping lane (vessels per year)
x = Position on the shipping lane
P_x = Probability of being in position x on the shipping lane
P_{offset} = Probability of an offset from the current x-position
P_{course} = Probability of a certain heading toward the object
P_{C1} = Probability of human error during planning and during voyage
P_{C2} = Probability of failure of navigational equipment or failure of watch keeping
P_{C3} = Probability of failure of wind farm safety measures
$P_{react}(x)$ = Probability of crew not reacting in time to correct error.

The probability of not being able to react in time is modeled with the following expression:

$$P_{react}(x) = e^{-0.2\, x^{1.5}} \tag{10.14}$$

where x denotes the distance sailed in nm. The mean distance is 2.25 nm, and the probability of reacting after 5 nm is 0.9.

10.3.3 Korean study

A recent approach has been proposed by Mujeeb-Ahmed et al. (2018). A basic assumption is that traffic observed by AIS is structured in separate routes with lateral distribution and heading deviations.

The apparent cross section of the platform is given by (Figure 10.7):

$$W_A = L_P \sin\theta + W_P \cos\theta \tag{10.15}$$

where θ is the angle of orientation of the platform relative to the traffic.

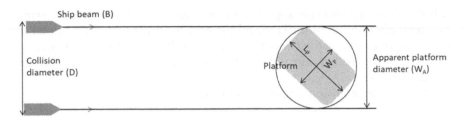

Figure 10.7 Collision scenario (Mujeeb-Ahmed et al., 2018).

The collision diameter is then given by $D = WA + B$ where B is the ship breadth. By setting the width of the lane to two times the standard deviation (2σ) and assuming a normal distribution, 68.2% of the traffic in the lane is captured. The proportion of traffic in a lane that is heading for the platform is $P_{G1} = D \cdot f(A)$. The normal probability density function at distance A between the center of the lane and the platform is given by:

$$f(A) = \frac{1}{\sqrt{2\pi}\sigma} \, e^{-\frac{k^2}{2}} \tag{10.16}$$

where $k = A/\sigma$ is the number of standard deviations between the center of the lane and the platform. Based on observation of the AIS tracks the distribution heading error is estimated and expressed by a normal distribution P_{G2}. The geometrical collision probability is then given by $P_G = P_{G1} + P_{G2}$. Table 10.2 shows the analysis of the traffic-related probabilities for an offshore installation near the port of Busan in South Korea. The traffic in each lane differs considerably, whereas the separation (A) and breadth of the lanes (σ) vary less. The geometrical probability (P_G) ranges from 0.0002 to 0.001.

Table 10.2 Geometrical collision probability of lanes

Route	Average heading (degrees)	Annual traffic (Q)	Separation A (nm)	σ (nm)	P_{G1}	P_{G2}	P_G
A	7	151	−2.51	3.03	0.019	0.012	0.0002
B	24	773	−2.55	2.77	0.012	0.035	0.0004
C	45	1,422	−2.04	2.59	0.024	0.023	0.0005
D	68	516	−1.67	2.53	0.027	0.036	0.001
E	120	777	−1.73	−2.72	0.025	0.028	0.0007
F	170	89	−1.5	2.19	0.030	0.010	0.0003

Mujeeb-Ahmed et al. (2018).

It is worth noting that this approach of describing the traffic both in terms of lateral distribution of the lane and the heading error is slightly similar to the approach devised by Forsman et al. (2008). The difference is that the latter performed an integration of the joint effect along the distance up to the platform.

The expected collision frequency is given by:

$$N = Q \cdot P_G \cdot P_{C1} \cdot P_{C2} \cdot P_{C4} \cdot M_1 \cdot M_2 \cdot M_3 \tag{10.17}$$

where

P_{C1} = Probability of failure of pre-voyage planning
P_{C2} = Probability of failure of vessel-initiated recovery
P_{C3} = Probability of failure of platform-initiated recovery
P_{C4} = Probability of failure to correct navigational error
M_1 = Mitigating effect of ARPA, VTS, and radar early warning (REWS)
M_2 = Mitigating effect of platform ability to rotate and minimize exposure

The P_{C1} value is determined for ship category and platform type. The highest values are seen for new installations and mobile platforms in connection with passing vessels where the knowledge of the presence of the platforms is less likely. Supporting data was taken from Spouge & CMPT (1999). Typical value at the outset is 0.95 and is reduced over time down to 0.1. For supply and standby vessels, the value is in the range of 0.05–0.1.

P_{C2} is estimated on the basis of expert judgment and FTA. Key determining factors are human errors like being absent from the bridge, distracted, asleep, incapacitated, influenced by alcohol and drugs, or due to reduced visibility. The probability of platform recovery (P_{C3}) is judged to be low due to the short time available and communication problem with the incoming vessel. The study chose to assess P_{C2} and P_{C3} jointly. The values offered were from $2.6 \cdot 10^{-3}$ for fishing vessels in good visibility to $8.2 \cdot 10^{-2}$ for standby vessels in bad visibility.

The probability of correcting the navigational error (P_{C4}) is seen as a function of distance (x in nm) between the platform and the position where the vessel lost control. The study applies the same model as Forsman et al. (2008), i.e., $P_{C4} = e^{-0.2x^{1.5}}$

The study discusses alternative radar-based mitigation measures and concludes that VTS monitoring is most effective by reducing the collision probability by 87% which means a mitigation factor of $M_1 = 1 - 0.87 = 0.13$. It is assumed that certain platform types like turret-moored FPSOs and FSUs may be able to rotate to minimize their exposed cross section and thereby reduce the collision probability. This factor is estimated to $M_2 = 0.28$.

10.4 SHIP COLLISION

10.4.1 Basic approach

Modeling of collisions is somewhat more complicated than for groundings and allisions because it involves two vessels in motion. Secondly, vessels may approach each other under different situations: head-on, overtaking, and crossing under different relative angles. Theoretically these situations may be expressed in a general model for the geometrical probability (P_G) but in practice separate models are normally used. Besides, the different meeting situations are also judged to place different demands on the navigator and may therefore result in different estimates for the causation probability (P_C).

One of the pioneering models was proposed by Macduff (1974). The geometrical probability was formulated as follows:

$$P_G = \frac{SL\sin(\theta)}{D^2\,925} \tag{10.18}$$

where the distance traveled is S (nm), the length of the ship is L (m) and D (nm) is the average distance between the ships. The vessels meet at a relative angle θ. The assessment of Li et al. (2014) is that the model overestimates the geometrical probability.

Fujii and Tanaka (1971) suggested the following model for vessels traveling through an area:

$$P_G = \int_{\text{entrance}}^{\text{exit}} \frac{\rho\,D\,V_{\text{rel}}}{V}\,dx \tag{10.19}$$

where ρ is the ship density (number of ships per unit area), D is the diameter of evasion (passing distance), V_{rel} the is relative speed, and V is the speed of the passing vessel. The passing distance D was set to 9.5–16.3 times the ship length. Li et al. (2014) point out that also this model overestimates the geometrical probability since they found that the passing distance may be as low as three times the ship length in congested waters.

10.4.2 Head-on collisions

The basic modeling approach may be demonstrated with a simple head-on collision case as shown in Figure 10.8.

It is assumed that a straight fairway has traffic in one direction with arrival frequency Q_1. In a given time period T, $Q_1 \cdot T$ ships will enter the fairway. This traffic will occupy an area $V_1 \cdot T \cdot W$ of the seaway. The traffic density is therefore

$$r = \frac{Q_1}{V_1 \cdot W} \tag{10.20}$$

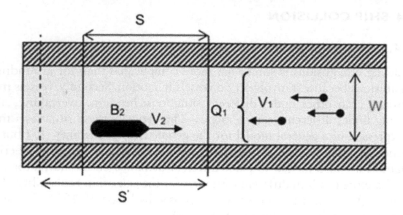

Figure 10.8 Head-on collision situation.

A ship meeting this traffic spends time $T_2 = S/V_2$ to pass the fairway segment of distance S. The ship is however exposed to traffic over a distance S' due to the relative speed of the traffic in the opposite direction $V = V_1 + V_2$. The distance is given by

$$S' = v \cdot T_2 = \frac{(V_1 + V_2)S}{V_2} \tag{10.21}$$

The collision cross section is determined by the breadth of the vessels $B = B_1 + B_2$ and the exposed area to a collision is given by $A = B \times S'$. This gives:

$$A = (B_1 + B_2)(V_1 + V_2)S/V_2 \tag{10.22}$$

The number of meeting situations for vessel 2 is $N_2 = A \times \rho$. Assuming an arrival frequency of Q_2 for vessel 2 the total meeting frequency is:

$$N_G = \frac{(B_1 + B_2)}{W} \frac{(V_1 + V_2)}{V_1\,V_2} S\,Q_1\,Q_2 \tag{10.23}$$

A numerical example gives:

$$N_G = \frac{(20 + 30)}{1000} \frac{(6 + 9)}{6 \cdot 9} 36{,}000 \cdot \frac{10{,}000 \cdot 10{,}000}{3{,}600 \cdot 24 \cdot 360} = 1{,}608 \text{ meetings/year}$$

By assuming a head-on collision causation probability $P_C = 4.9 \cdot 10^{-5}$ the annual collision frequency is estimated to:

$$N = N_G \cdot P_C = 1.61 \cdot 10^3 \cdot 4.9 \cdot 10^{-5} = 7.9 \text{ collisions/year}$$

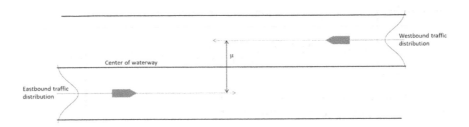

Figure 10.9 Head-on collision in straight fairway with lateral distribution of traffic.

This model was applied by COWI (2008). The present approach assumes that the traffic is distributed evenly laterally across the seaway which is not realistic. A vessel will have greater or less tendency to keep to starboard (right) in the seaway depending on the ship type and size. A more realistic approach may be to assume a normal distribution of the traffic (Figure 10.9).

The number of meeting situations can be expressed as follows:

$$N_G = S \sum_{i,j} P_{G\,i,j} \frac{V_{i,j}}{V_i^{(1)} V_j^{(2)}} Q_i^{(1)} Q_j^{(2)} \qquad (10.24)$$

where the index i expresses the segment of type and size for vessel 1 and index j similarly for vessel 2. This expression is not suitable for hand-calculation but is implemented in the IWRAP model (Friis-Hansen, 2008). By assuming a normal distribution for the traffic geometric probability can be expressed by the standard normal distribution (Friis-Hansen, 2008):

$$P_{G\,i,j} = \phi\left(\frac{\bar{B}_{i,j} - \mu_{i,j}}{\sigma_{i,j}}\right) - \phi\left(-\frac{\bar{B}_{i,j} + \mu_{i,j}}{\sigma_{i,j}}\right) \qquad (10.25)$$

where the average ship breadth is given by: $\bar{B}_{i,j} = \dfrac{B_i^{(1)} + B_j^{(2)}}{2}$, the mean separation distance is μ_{ij} and the standard deviation of the joint distribution is $\sigma_{ij} = \sqrt{(\sigma_i^{(1)})^2 + (\sigma_j^{(2)})^2}$. Let us make a comparison with the previous model. The mean ship breadth is 25 m (= (20 +30)/2), the mean separation $\mu_{ij} = 500$ m, and the standard deviation $\sigma_{ij} = (200^2 + 200^2)^{1/2} = 283$ m. The geometric probability is:

$$P_{G\,i,j} = \phi\left(\frac{25 - 500}{283}\right) - \phi\left(\frac{25 + 500}{283}\right) = 0.016$$

The number of meetings assuming similar vessels across the fairway is:

$$N_G = 36,000 \cdot 0.016 \cdot \frac{(6+9)}{6 \cdot 9} \cdot \frac{10,000 \cdot 10,000}{3,600 \cdot 24 \cdot 360} = 51.4 \text{ meetings/year}$$

By assuming a head-on collision causation probability $P_C = 4.9 \cdot 10^{-5}$ the annual collision frequency is estimated to:

$$N = N_G \cdot P_C = 51.4 \cdot 4.9 \cdot 10^{-5} = 2.5 \times 10^{-3} \text{ collisions/year}$$

By taking the lateral traffic distribution into consideration the collision risk was reduced by more than three orders of magnitude ($7.9/2.5 \cdot 10^{-3}$).

10.4.3 Crossing collision

Let us first assume a normal crossing situation ($\theta = 90°$) with linear distribution of the traffic laterally. It can be shown that the number of crossing situations is:

$$N_G = \frac{Q_1 Q_2}{V_1 V_2} \left[(B_1 + L_2) V_1 + (B_2 + L_1) V_2 \right] \tag{10.26}$$

where Q_1 and Q_2 are the arrival frequency in the respective directions and similarly the main ship dimensions are L (length) and B (breadth). The expressions in the brackets represent the collision diameter when ship 1 hits ship 2 and vice versa. Assuming the same data set as for head-on collisions, the number of crossing situations is:

$$N_G = \frac{10,000 \cdot 10,000}{6 \cdot 9} \cdot \frac{\left[(20+150)6 + (30+100)9 \right]}{360 \cdot 24 \cdot 3,600} = 130 \text{ crossing/year}$$

Assuming a crossing causation probability $P_C = 1.2 \cdot 10^{-4}$ the collision frequency becomes:

$$N = N_G \cdot P_C = 130 \cdot 1.2 \cdot 10^{-4} = 0.016 \text{ collisions/year}$$

The general expression for the number of crossing situations with crossing angle θ can be expressed as follows (Friis-Hansen, 2008):

$$N_G = \sum_{i,j} \frac{Q_i^{(1)} Q_j^{(2)}}{V_i^{(1)} V_j^{(2)}} D_{ij} V_{ij} \frac{1}{\sin \theta} \quad \text{for } 10° < |\theta| < 170° \tag{10.27}$$

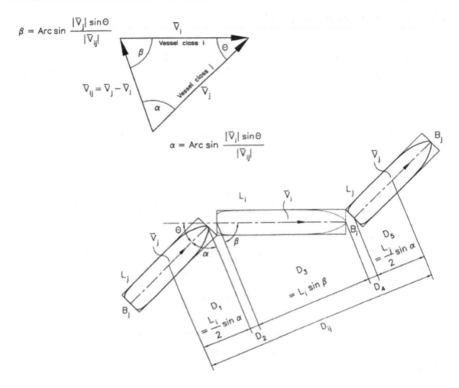

Figure 10.10 Geometrical collision diameter (IALA, 2022).

The collision diameter is based on the scenario shown in Figure 10.10 and is given by:

$$D_{ij} = \frac{L_i^{(1)}V_j^{(2)} + L_j^{(2)}V_i^{(1)}}{V_{ij}} \sin\theta + B_j^{(2)}\left\{1 - \left(\sin\theta \frac{V_i^{(1)}}{V_{ij}}\right)^2\right\}^{1/2}$$

$$+ B_i^{(2)}\left\{1 - \left(\sin\theta \frac{V_j^{(2)}}{V_{ij}}\right)^2\right\}^{1/2}$$

(10.28)

The relative speed is: $V_{ij} = \sqrt{(V_i^{(1)})^2 + (V_j^{(2)})^2 - 2\,V_i^{(1)}\,V_j^{(2)}\cos\theta}$ (10.29)

Let us calculate a case where the vessels cross at an angle of 45° and else apply the vessel data in the previous case and assuming only one vessel category. The relative speed is:

$$V_{ij} = (6^2 + 9^2 - 2 \times 6 \times 9 \times \cos 45)^{1/2} = 6.3 \text{ m/s}$$

The collision diameter is given by:

$$D_{ij} = \frac{100 \cdot 9 + 150 \cdot 6}{6.3} \sin 45 + 30 \left\{ 1 - \left(\sin 45 \frac{6}{6.3} \right)^2 \right\}^{\frac{1}{2}} + 20 \left\{ 1 - \left(\sin 45 \frac{9}{6.3} \right)^2 \right\}^{\frac{1}{2}}$$

$D_{ij} = 229$ m.

Note that the value within the root expression may become negative and in that case one shall apply the absolute value.

The annual number of meeting situations is:

$$N_G = \frac{10,000 \cdot 10,000}{6 \cdot 9} 229 \cdot 6.3 \cdot \frac{1}{0.71} \cdot \frac{1}{360 \cdot 24 \cdot 3,600} = 121 \text{ crossings/year}$$

Assuming a crossing causation probability $P_C = 1.2 \cdot 10^{-4}$ the collision frequency becomes:

$$N = N_G \cdot P_C = 121 \cdot 1.2 \cdot 10^{-4} = 0.015 \text{ collisions/year}$$

This result indicates that advanced modeling of the lateral distribution of the traffic will only marginally improve the estimation of the collision probability.

10.4.4 Traffic modeling based on AIS observation

In pioneering projects, the sea traffic was observed and analyzed by radar-based data. One had among other options access to data from VTS (Vessel Traffic Systems) that monitor and advice vessels in congested seaways. The analysis work became less demanding by the introduction the ARPA radar. One of the studies was undertaken by Lewison (1978) for the British Channel. One of the objectives of the study was to assess the effect of the introduction of a TSS (Traffic Separation Scheme) in this very busy waterway. A by-product was the estimation of the effect of visibility on the collision frequency as shown in Figure 10.11.

Since the introduction of AIS (Automatic Identification System) for vessels an improved method for the estimation of the lateral traffic distribution is possible. By tracking the traffic by means of AIS position signals over longer periods, one may get a more correct picture. And these data lend themselves to more efficient computer analysis. In the majority of risk studies, it has been common to apply the normal distribution to describe how the traffic varies across the seaway. Yoo and Kim (2019) have demonstrated how the traffic distribution may be estimated with greater precision. They

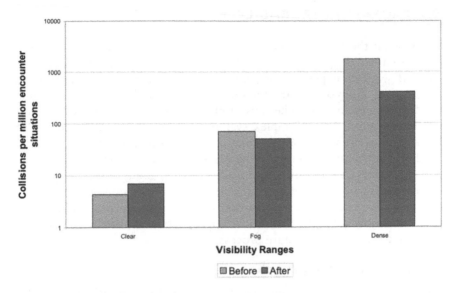

Figure 10.11 Collision frequency in the English Channel before and after the introduction of TSS. Visibility ranges: Clear: greater than 4km, Fog: 0.2–4km, Thick/dense: Less than 0.2km (Lewison, 1978).

have studied the traffic in the port of Ulsan in South Korea. Based on the observation of the inbound and outbound traffic at five "gates" and by application of the best-fit PDF, it was possible to obtain improved estimates of the lateral distribution. Following alternative distributions were tested: Wakeby, Log-Logistics, and Cauchy. As shown in Table 10.3, the normal distribution gave consistently higher estimates of the collision frequency than the Best-Fit distribution. The ratio was from 1.5 to 2.4 which is a significant finding.

Table 10.3 Estimated head-on collision frequency (number per year) at five locations in the port of Ulsan. Comparison of estimates based on Best-Fit PDF and Normal PDF for the geometrical probability

Gate	Best-Fit PDF (10^{-4})	Normal PDF (10^{-4})	Ratio (Normal/Best-Fit)
A	0.2900	0.5723	1.97
B	0.2898	0.7043	2.43
C	0.7061	1.0826	1.53
D	0.3475	0.4991	1.47
E	0.2915	0.5079	1.74

Yoo and Kim (2019).

10.5 CAUSATION PROBABILITY

Let us recall the basic expression for the probability of an traffic-related accident namely $P_{RG}=P_G \cdot P_C$. It has been shown that the geometrical probability (P_G) which expresses the probability of having a conflict, primarily is a function of physical characteristics like vessel and fairway dimensions, speed, and traffic density. The causation probability (P_C) on the other hand is reflecting much more complex systems and processes. The probability that a conflict situation leads to an accident is a function of the navigation standard and technical factors of the vessel, fairway complexity, and external factors like oceanographic and meteorological conditions.

Since the introduction of maritime traffic analysis a number of methods have been proposed for the estimation of the causation probability. They represent highly different theoretical approaches and different degrees of ambition to emulate this failure process.

10.5.1 Empirical approach

In the pioneering work by Fujii (1982), the traffic volume (Q) and number of grounding accidents (N) were observed for longer periods. It was thereby possible to estimate the accident probability:

$$P_{RG} = Q/N \tag{10.30}$$

The geometrical probability (P_G) was computed on the basis waterway and ship dimensions and it was therefore possible to estimate the causation probability:

$$P_C = P_{RG}/P_G \tag{10.31}$$

Fujii (1982) studied different Japanese waterways which gave different estimates as shown in Table 10.4. The ratio between highest and lowest estimate was 4.1. This fact may simply reflect the fact that vessel and traffic factors

Table 10.4 Estimates of causation probability in Japanese waters

Fairway	Vessel category	P_C
Uraga Strait	Greater than 300 RT	$1.1 \cdot 10^{-4}$
Bisanseto Fairway	> 100 RT	$2.8 \cdot 10^{-4}$
	100–500 RT	$3.3 \cdot 10^{-4}$
	500 RT <	$2.0 \cdot 10^{-4}$
Naruto Strait		$0.8 \cdot 10^{-4}$
Akashi Strait		$1.4 \cdot 10^{-4}$

Fujii (1982).

Table 10.5 Estimates of causation probability

Accident type	$P_c (10^{-4})$ Min Max		$P_c (10^{-4})$ Mean	Median
Grounding	1.0–6.3		2.3	1.6
Collision – Head on	0.49–5.18		2.3	1.8
Collision – Crossing	0.95–1.23		1.1	1.1
Collision – Overtaking	1.1		1.1	1.1
Bridge ramming	0.4–5.4		1.8	1.3

Friis-Hansen (2008).

may vary significantly. It was however decided to propose $P_C = 2.0 \cdot 10^{-4}$ as a general value for future risk studies.

Friis-Hansen (2008) gives an even broader summary of proposed estimates for the causation probability for groundings, collisions, and ship-bridge rammings (Table 10.5). It shows that the estimates vary with a factor of 10 and again indicates that the conditions may be quite different. However, it also shows that the mean or median value lies in the range from $1 \cdot 10^{-4}$ to $2 \cdot 10^{-4}$. Another lesson is that an estimate from one study should be applied with caution in another study unless the conditions are viewed as similar.

10.5.2 Fault tree analysis (FTA)

The introduction of this chapter pointed to the fact that it is a number of factors that may lead to a traffic-related accident. Mazaheri et al. (2016) see maritime transportation as a complex socio-technical system where the following elements interact:

- State of the system
- Environment of the system
- Interface of the system
- Operator of the system

These elements must be broken down further to make the analysis more concrete. However, greater detail raises the question of capturing the interactions between these factors. This can be illustrated with FTA approach proposed by Fowler and Sørgård (2000) as shown in Figure 10.12. The model is reflecting the interaction between the ships and captures some technical and human failures. However, the latter factors are described in a very general form. A more basic weakness is the inability of FTA as a model to express the interaction between these factors.

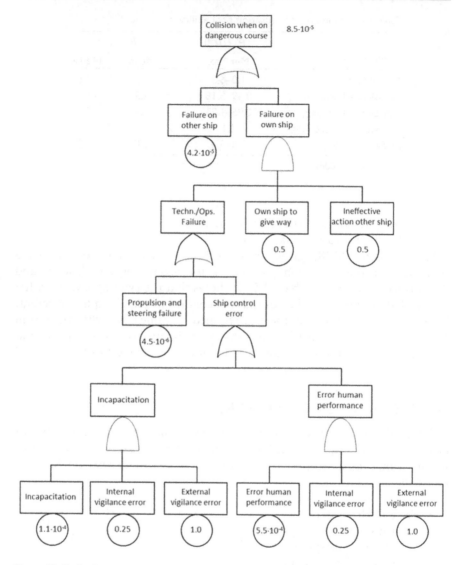

Figure 10.12 Fault tree analysis of collision in good visibility (Fowler & Sørgård, 2000).

10.5.3 Bayesian Belief Network

A Bayesian Belief Network (BBN) is a directed acyclic graph and similar to influence models which have been applied for a long period in risk analysis (NTSB, 1990). The BBN method has also a numerical form that makes it a powerful tool for analysis of accidents. The BBN consists of nodes and arcs and offers a way describing states and interdependencies between states. It quite flexible as nodes may be discrete or continuous and binary or multi-state. Another key property is that it may handle probabilistic data. Interdependencies or causal mechanisms in discrete models are handled by means of Conditional Probability Tables (CPT). Zhang and Thai (2016) gave the following reasons for application of BBN in accident analysis:

- Explicit presentation of causal relationships
- Making both forward and backward inferences
- Combination of expert's knowledge and empirical data
- Power to deal with uncertainty
- Making updates with new information/observation

Antao et al. (2009) have made a pioneering study of accident characteristics (Figure 10.13). The analysis focused primarily on situational factors related to accident types. It gave little insight into why accident happens and where limited to a simple factor that what was defined as probable cause. The approach reflected the fact that little systematic accident data was available at that time.

One of the early BBN studies of causal mechanisms was undertaken by Hu et al. (2007). Although addressing only six factors, it had a broad view by focusing both on external factors, fairway and pilotage, and how these factors were related to accident type and severity (Figure 10.14).

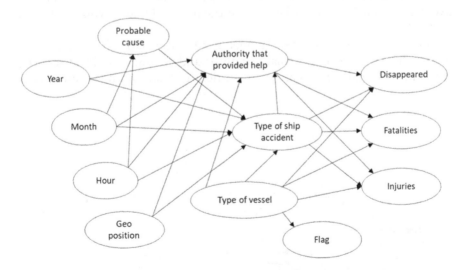

Figure 10.13 Analysis of accident characteristics with BBN (Antao et al., 2009).

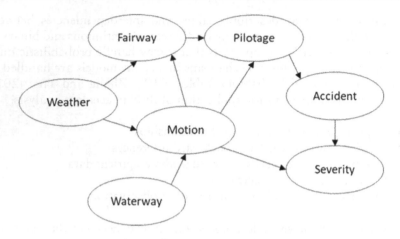

Figure 10.14 BBN analysis of ship navigation accidents (Hu et al., 2007).

Itoh et al. (2007) developed a slightly more advanced model for collisions (Figure 10.15). The upper right part of the figure shows the estimated
unconditional probabilities in the network and also termed "parent" nodes.
The lower part of the figure shows the conditional probabilities for the
so-called "child" nodes. The model demonstrates clearly how fast the size
of the CPT increases with the number parents. "Situational awareness" and
"decision making" have three parents both and thereby require the estimation of eight probabilities. The fact that "Physical restriction" has three
states also contributes to the increasing size of the CPT of the "Collision
avoidance result". The model estimated the probability of having a collision
to $4 \cdot 10^{-4}$.

Building of a ship accident model needs systematic data. A normal
approach is to perform systematic analysis and coding of accident reports.
It is still no broad agreement on an accident taxonomy among researchers

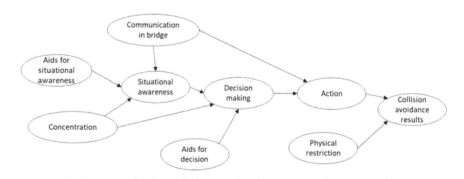

Figure 10.15 BBN analysis of collision (Itoh et al., 2007).

and investigators. The different maritime administrations offer accident statistics with different focus and terminology on accident events, states, and causal factors. A system that has got increased use is the Human Factors Analysis and Classification System (HFACS) developed initially by US researchers working with aircraft accidents (Wiegmann & Shappel, 2017). HFACS has gradually found application within the maritime sector and is known as the HFACS-MA system (Schröder-Hinrichs et al., 2011).

Slightly different versions have later been proposed by Mazaheri et al. (2015) and Yıldırım et al. (2019). The taxonomy is structured in five hierarchical levels as shown in Table 10.6. The top level is external factors that

Table 10.6 Marine-based HFACS (M-HFACS)

Level	1st layer	2nd layer	3rd layer
5	Outside factors	Regulatory factors Other factors	
4	Organizational influence	Resource management Organizational climate Organizational process	
3	Supervisional influence	Inadequate supervision Planned inappropriate operations Failed to correct know problems Supervisory violation	
2	Preconditions	Environmental factors	Physical environment Technological environment Infrastructures
		Condition of operators	Cognitive factors Psychobehavioral factors Adverse physiological states Physical/mental limitations Perceptual factors
		Personal factors	Coord./Communic. planning Personal readiness Bridge Resource Management
1	Unsafe acts	Errors	Skill-based errors Judgment/decision errors Perceptional errors
		Violations	Routine Exceptional

Mazaheri et al. (2015).

influence the shipping activity: Maritime administration and shareholders such as ports and shipbuilders. The framing of the safety-related conditions is determined by the top-level organizational influences: asset management, organization, and management process. The role of middle management is grouped in unsafe management and typical flaws like inadequate work planning, failure to fix known problems and other management violations. The preconditions for unsafe acts or risk influencing factors (RIFs) are broadly speaking workplace conditions, work organization, and individual conditions both mental and physiological. The so-called unsafe acts are either errors or violations. An immediate question is whether this classification system is too general for coding purposes.

A Finnish research group has discussed the methodological aspects of the development of BBN-based ship accident models (Mazaheri et al., 2015, 2016). The following slightly different steps are identified:

1. Define the system with its characteristics and boundaries.
2. Specify relevant factors in the model.
3. Develop the structure model's structure (qualitative version) based on expert assessments.
4. Collect data and code data from accident and incident reports.
5. Parameterize the model.
6. Validate the model.

Based on expert experience and preliminary studies, the model and its factors are defined. Implicitly, it means that the borders of the model are determined. The main source of data is accident reports primarily from national accident investigation agencies. Some agencies produce reports of good quality and following a systematic and multidisciplinary approach.

By accepting the fact that traffic-related accidents have a number of causal factors has resulted in larger and more complex BBN models. A grounding accident model proposed by Mazaheri et al. (2016) had 33 nodes and up to 7 parent nodes for one node (Situational awareness) as shown in Figure 10.16. It should be observed that this model is based on analysis of accidents only and therefore only have relative probabilities in contrast to the previous model (Itoh et al., 2007) that was based on absolute probabilities. An important aspect of the analysis is to identify which factors have the strongest impact on the accident probability. The authors propose a so-called Qualitative Feature test where each node is evaluated as follows: The difference in grounding probability for the best and worst value of a factor is determined. The factors with the highest differences are viewed as most important causal factors.

The sensitivity analysis showed that two of the causal factors had significantly higher values than the other factors, namely, "Being off course" and "Loss of control" (Figure 10.17).

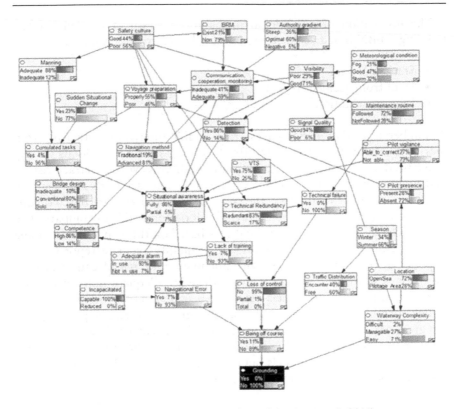

Figure 10.16 A BBN model of grounding accidents (Mazaheri et al., 2016).

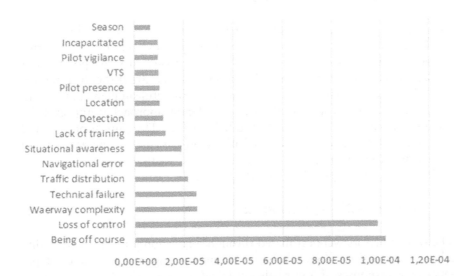

Figure 10.17 Sensitivity analysis based on a Qualitative Feature test (Mazaheri et al. 2016).

The population of conditional probabilities is still a challenge when building a BBN-based accident model. It is basically three sources:

- Statistical data
- Data from accident reports
- Expert elicitation

A number of statistical databases are available today and primarily from maritime administrations. They offer large sets of data but focus mainly on static data like place, time, external conditions, vessel characteristics, and consequences. Causal factors, on the other side, are fairly rudimentary described. Another weakness is the fact that the databases are based on different coding schemes and therefore difficult or impossible to compare. The model offered by Antao et al. (2009) was based on this type of data. Some maritime investigation agencies offer high-quality accident reports and it is possible to produce datasets with adequate coverage of causal mechanisms. It is however a challenge to produce an adequate number of data sets. Typically, the BBN models may have been based on around 100 cases. It may obviously be questioned whether that gives an adequate picture given the great variability in factors leading to a ship accident. The third option is to base estimation of CPTs on expert elicitation. The approach has been adopted in a number of models but also raises many methodological questions (Brooker, 2011; Zhang & Thai, 2016).

10.6 QUALITATIVE METHODS

The risk analysis methods discussed so far in this chapter may be characterized as qualitative methods. It has been pointed out that one of the critical aspects is the estimation of the causation probability. As pointed out causation is determined by a number of factors that are difficult to model. So-called qualitative models may be seen as a powerful supplement to quantitative tools. Kim et al. (2019) discuss different approaches and give the theoretical basis for a Korean method, NURI-C. The philosophy behind the method is to rank the risk of a fairway on the basis of the natural and human-made environment as outlined in Figure 10.18.

The risk is the weighted sum of 16 factors:

$$R = R_{f1} \cdot w_i + R_{f2} \cdot w_2 + \ldots + R_{fn} \cdot w_n \qquad (10.32)$$

The weight factors were estimated through a questionnaire study involving a broad group of expertise with nautical, engineering, and academic competence. The responses (R_{fi}) were also weighed based on the degree of experience. At the outset for the expert evaluation certain criteria were set where

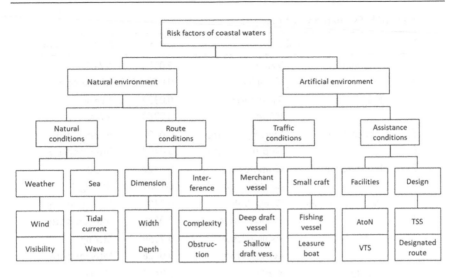

Figure 10.18 Maritime traffic risk factors of NURI-C (Kim et al., 2019).

Table 10.7 Risk criteria

Factor	Risk factor
Wind	Days with daily wind speed over 8 m/s divided with 365
Visibility	Days with visibility less than 1 km divided with 365
Tidal current	Days with more than 2 knots current divided by 365
Wave	Days with more than 2 m significant wave height divided by 365
Width of seaway	Area less than 6.4 L divided by the target area
Depth of seaway	Area less than 30 m depth divided by the target area. Shallow water effect when Depth/T < 2
Complexity	Complex area divided with target area. Complexity: More than 30° turn and less than four times L turning radius
Obstruction	Sum of obstructions in the target area. A ship is reluctant to pass obstructions with less than 1 nm distance
Vessel traffic	Number of vessels divided by the maximum capacity
Assistance conditions	1 – safe area/target area. Safe area has either Assistance to navigation (AtoN), Vessel Traffic System (VTS), Traffic Separation Scheme (TSS), or designated route

it was judged that a factor would have a negative effect on the human operation. The model ship was a container ship of 12,000 TEU with following
dimensions L, B, and T: 398, 55, and 15 m. The criteria are summarized in
Table 10.7. Based on the evaluation of the experts the relative weighting factors were estimated as shown in Table 10.8. It can be concluded that strongest weight was put on fishing vessels (20.59%) and obstructions (10.01%).

Table 10.8 Estimated weight factors

No	Risk factors		Weight	Relative ranking
1	Weather	Wind	0.0335	Weak
2		Visibility	0.0708	Intermediate
3	Sea	Tidal current	0.0332	Weak
4		Wave	0.0263	Weak
5	Dimensions	Width of route	0.0500	Intermediate
6		Depth of route	0.0703	Intermediate
7	Interference	Complexity	0.0427	Weak
8		Obstruction	0.1001	Strong
9	Merchant vessel	Deep draft	0.0700	Intermediate
10		Shallow draft	0.0705	Intermediate
11	Small craft	Fishing vessel	0.2059	Strong
12		Leisure craft	0.0728	Intermediate
13	Facility	AtoN	0.0434	Weak
14		VTS	0.0568	Intermediate
15	Design	TSS	0.0349	Weak
16		Designated route	0.0187	Weak

Kim et al. (2019).

The methodology was applied on a number of fairways along the South Korean coast that is characterized by a number of islands and high local fishing activity. The riskiest fairway was the Ongdo area. A summary of the risk evaluation is shown in Table 10.9. It was viewed that 58% of the area was affected by fishing vessels and thereby gave a risk rate of 11.94 which represents 38% of the total risk (31.19). It was established that neither VTS, TSS, nor designated route was implemented in the area. The next highest risk rate was found to be absence of VTS but still found to be of relative low criticality with a risk rate of 5.68 or 18% of the total risk.

As pointed out this method may be viewed as a useful supplement to the quantitative methods especially when focus is on improvement of the seaway.

10.7 A FINAL COMMENT

A general formulation of a traffic-related accident frequency is:

$$N = Q \times P_G \times P_C \qquad (10.33)$$

where N is the number of accidents per time unit, Q is the number of ship movements per time unit, P_G is the probability of impact given loss of control, and P_C is the probability of loss of control. The least challenge is to

Table 10.9 Risk assessment of the Ongdo area

No.	Risk factors		Incidence	Weight (%)	Risk
1	Weather	Wind	4.11	3.35	0.14
2		Visibility	10.41	7.08	0.74
3	Sea	Tidal current	65.75	3.32	2.19
4		Wave	9.86	2.63	0.26
5	Dimension of route	Width	6.50	5.00	0.33
6		Depth	29.25	7.03	2.06
7	Interference of route	Complexity	6.50	4.27	0.28
8		Obstructions	19.50	10.01	1.95
9	Merchant vessel	Deep draft vessel	4.08	7.00	0.29
10		Shallow draft vessel	3.11	7.05	0.22
11	Small craft	Fishing vessel	58.00	20.59	11.94
12		Leisure boat	0.00	7.28	0.00
13	Facility	AtoN	0.00	4.34	0.00
14		VTS	100.00	5.68	5.68
15	Design	TSS	93.50	3.49	3.27
16		Designated route	100.00	1.87	1.87
Total				100	31.19

Kim et al. (2019).

assess the ship traffic Q which simply may be based on observation or counting of ship movements. The probability of having an impact P_G is estimated on the basis of the physical characteristics related to ship characteristics, waterway dimensions, and exposed traffic. Since the introduction of AIS it has been possible to give fairly realistic estimates of the impact probability. It is however important to keep in mind that it is questionable to apply data from one fairway to another. The impact probability is in other words scenario specific. The weakest element in the model is the loss of control probability P_C. The key challenge is to analyze and model the complex causal mechanisms leading to an accident. It has been shown how the approach has developed from solving P_C as the unknown in the equation above to elaborate BBN modeling. But it is still important work to be done in order to produce valid parameters in these models.

REFERENCES

Antao, P., Guedes Soares, C., Grande, O., & Trucco, P. (2009). Analysis of maritime accident data with BBN models. In S. Martorell et al. (Eds.), *Safety, Reliability and Risk Analyis: Theory, Methods and Applications* (pp. 3265–3274). Milton Park, Oxfordshire: Taylor & Francis Group.

Brooker, P. (2011). Experts, Bayesian Belief Networks, rare events and aviation risk estimates. *Safety Science, 49*(8–9), 1142–1155.

COWI. (2008). *Risk analysis of sea traffic in the area around Bornholm* (Report No.: P-65775-002).

Forsman, B. et al. (2008). *Methodology for assessing risks to ship traffic from off-shore farms* (Report No. 2005 4028).

Fowler, T. G., & Sørgård, E. (2000). Modeling ship transportation risk. *Risk Analysis, 20*(2), 225–244.

Friis-Hansen, P. (2008). *Basic modelling principles for prediction of collision and grounding frequencies.* https://www.iala-aism.org/wiki/iwrap/images/2/2b/IWRAP_Theory.pdf

Fujii, Y. (1982). Recent trends in traffic accidents in Japanese waters. *The Journal of Navigation, 35*(1), 90–99.

Fujii, Y., & Tanaka, K. (1971). Traffic capacity. *The Journal of Navigation, 24*(4), 543–552.

Hassel, M., Utne, I. B., & Vinnem, J. E. (2017). Allision risk analysis of offshore petroleum installations on the Norwegian Continental Shelf—an empirical study of vessel traffic patterns. *WMU Journal of Maritime Affairs, 16*(2), 175–195.

Haugen, S., & Vollen, F. (1989). COLLIDE Reference manual. Trondheim: Dovre Safetec.

Haugen, S. (1998). An overview of ship-platform collision risk modelling. *Risk and Reliability in Marine Technology.* Rotterdam: A. A. Balkema Publishers.

Hu, S., Cai, C., & Fang, Q., 2007, Risk assessment of ship navigation using Bayesian learning. In *2007 IEEE International Conference on Industrial Engineering and Engineering Management* (pp. 1878–1882). IEEE.

IALA. (2022). *IWRAP theory* (https://www.iala-aism.org/wiki/iwrap/index.php/Theory). Accessed 21. April 2022.

Itoh, H., Kaneko, F., Mitomo, N., & Tamura, K. (2007). A probabilistic model for the consequences of collision casualties. *ICCGS; Proc., Hamburg, 9–12.*

Kim, I., Lee, H. H., & Lee, D. (2019). Development of a new tool for objective risk assessment and comparative analysis at coastal waters. *Journal of International Maritime Safety, Environmental Affairs, and Shipping, 2*(2), 58–66.

Lewison, G. R. G. (1978). The risk of encounter leading to a collision. *Journal of Navigation, 31*(3), 384–407.

Li, K. X., Yin, J., Bang, H. S., Yang, Z., & Wang, J. (2014). Bayesian network with quantitative input for maritime risk analysis. *Transportmetrica A: Transport Science, 10*(2), 89–118.

Macduff, T. (1974). The probability of vessel collisions. *Ocean Industry, 9*(9), 144–148.

Mazaheri, A., Montewka, J., Nisula, J., & Kujala, P. (2015). Usability of accident and incident reports for evidence-based risk modelling – A case study on ship grounding reports. *Safety Science, 76,* 202–214.

Mazaheri, A., Montewka, J., & Kujala, P. (2016). Towards an evidence-based probabilistic risk model for ship-grounding accidents. *Safety Science, 86,* 195–210.

Mujeeb-Ahmed, M. P., Seo, J. K., & Paik, J. K. (2018). Probabilistic approach for collision risk analysis of powered vessel with offshore platforms. *Ocean Engineering, 151,* 206–221.

Mølsted, R. F., Kürstein, G. K. A., Kristina, M., Gamborg, H. M., Koldborg, J. T., Tue, L. S., & Søren, R. T. (2012). Quantitative assessment of risk to ship traffic in the Fehmarnbelt Fixed Link project. *Journal of Polish Safety and Reliability Association, 3*(1–2), 1–12.

NTSB. (1990). *Grounding of the US tankship Exxon Valdez on Bligh Reef, Prince William Sound near Valdez, Alaska* (Report No. PB 90–916405). National Transportation Safety Board. https://www.hsdl.org/?view&did=746707

Pedersen, P. T. (1995). Collision and grounding mechanics. *Proceedings of WEMT, 95*(1995), 125–157.

Schröder-Hinrichs, J. U., Baldauf, M., & Ghirxi, K. T. (2011). Accident investigation reporting deficiencies related to organizational factors in machinery space fires and explosions. *Accident Analysis & Prevention, 43*(3), 1187–1196.

Simonsen, B. C. (1997). Mechanics of ship grounding. PhD thesis. *Department of Naval Architecture and Offshore Engineering.* Lyngby: University of Denmark.

Spouge, J., & CMPT. (1999). *A guide to quantitative risk assessment of offshore installations.* Publication 99/100a. London: DNV Technica.

Wawruch, R., & Stupak, T. (2011). Modelling of safety distance between ships route and wind farm. *Archives of Transport, 23*, 413–420.

Wiegmann, D. A., & Shappell, S. A. (2017). *A human error approach to aviation accident analysis: The human factors analysis and classification system.* Milton Park, Oxfordshire: Routledge.

Yıldırım, U., Başar, E., & Uğurlu, Ö. (2019). Assessment of collisions and grounding accidents with human factors analysis and classification system (HFACS) and statistical methods. *Safety Science, 119*, 412–425.

Yoo, Y., & Kim, T. G. (2019). An improved ship collision risk evaluation method for Korea Maritime Safety Audit considering traffic flow characteristics. *Journal of Marine Science and Engineering, 7*(12), 448.

Zhang, G., & Thai, V. V. (2016). Expert elicitation and Bayesian Network modeling for shipping accidents: A literature review. *Safety Science, 87*, 53–62.

Chapter 11

Human reliability analysis

11.1 INTRODUCTION

Modern economic activities like industrial production, transport, and power generation are based on an increasing degree of automation and computer control. Advances in materials, technology, and engineering have resulted in systems with decreasing use of manpower or operators compared to the output. This is also a fact for maritime transport. Ships are built with greater and greater capacity, utilizing advanced technical subsystems and are based on smaller and smaller crews. Autonomous ships are high on the agenda of IMO, national authorities, and shipowners and will reduce crews even further. Despite these developments, the discussion about how to reduce human error in accidents is still high on the agenda.

Various safety studies indicate that human error represents somewhere from 60% to 80% of the causes leading to accidents and losses of ships. At first glance, one may argue that modern technology should reduce the role of the human operator in the system. This may be true to a certain degree in the sense that the navigator or engineer to a lower degree initiate processes that lead to accidents. However, even the most advanced systems are planned, designed, built, and maintained by humans and thereby subject to flawed decisions that make the system vulnerable. At the same time, it is also clear that modern systems based on advanced control systems and automation have led to higher reliability and safety.

Humans and machines show some quite important differences in the discussion of reliability. Machines can perform a function over longer periods with relatively small deviations from specified requirements but are unable to repair themselves. Humans, on the other hand, show less stable performance over time, but are able to detect and correct errors and adjust to changing requirements. Machines are designed by well-developed engineering methods, including application of reliability methods, and the aim is to develop systems that can operate with few failures. For people, quite different approaches have to be used to secure the reliability, including education, training, and personnel selection and supervision.

DOI: 10.4324/9781003055464-11

Quantitative risk analysis (QRA) is used to quantify the risk in a system and in human-operated systems *Human Reliability Analysis* (HRA) will be a critical element for assessing the human contribution to the risk. HRA is a systematic approach based on both qualitative and quantitative methods. Two simple arguments for HRA are as follows:

- Most systems involve a number of interactions between the system and the operator and may be the source of pathways to mission failure.
- In safety critical operations, the human may be the dominant source of risk both by initiating events and aggravating accident processes.

Human Error (HE) or *Human Failure Events* (HFE) may be defined as follows (Gertman et al., 2005):

> An out-of-tolerance action, or deviation from the norm, where the limits of acceptable performance are defined by the system. These situations can arise from problems in sequencing, timing, knowledge, interfaces, procedures or other sources.

An important part of this definition is "defined by the system". This means that it is the system that determines what the acceptable performance is. When systems are designed, it is obviously necessary to be aware of what the capacities and capabilities of humans are and design the systems in accordance with this. Otherwise, it is only to be expected that many human errors will occur.

Related to this definition, *Human Error Probability* (HEP) is defined as follows:

> The probability that an error will occur when a given task is performed, or simply as the number of errors relative to the number of opportunities.

Two main theoretical works on human errors have influenced the development of HRA methods. Rasmussen et al. (1981) defined three forms of human behavior that may be viewed hierarchically in terms of their mental demand (the SRK model):

- *Skill-based behavior*: Familiar or well-practiced tasks requiring little conscious control or an almost automatic response. Changing heading or speed of a ship could be an example of this.
- *Rule-based behavior*: The operator applies a rule or learned procedure to a situation or deviation from the normal operation. Driving in traffic may be an example of rule-based behavior, where we have to follow traffic signs and traffic lights plus a range of other rules that regulate our behavior.

- *Knowledge-based behavior*: The most demanding form of behavior requiring both perception, diagnosis, and problem-solving. An example can be solving a problem that we have never come across before, e.g., when writing an exam.

The types of errors that are made are different in the three different modes of behavior. Further, it is noted that what is skill-based, rule-based, and knowledge-based will vary from one individual to another. Typically, as we gain experience with a task, we will move "up" one step in behavior, e.g., from rule-based to skill-based.

Reason (1990) proposed a somewhat different approach for classifying error that also gives important insight into unsafe acts:

- Unintended actions: or execution errors manifesting itself in two forms:
  - *Slips* or attentional failures
  - *Lapses* or memory failures
- Intended actions:
  - *Mistakes* related to rule-based or knowledge-based behavior
  - *Violations*

11.2 HUMAN RELIABILITY ANALYSIS

The development of HRA methodologies was initially driven by the need for better approaches to understand and mitigate human error in the operation of military systems and nuclear power plants in the USA. Two events that demonstrated the criticality of the human element in nuclear plants were the *Three Mile Island* accident in Pennsylvania in 1979 and the *Chernobyl* accident in 1986 in Ukraine. Today HRA is applied in all technological sectors.

HRA involves three main phases where the two first may be defined as qualitative HRA and the third phase as quantitative HRA (Figure 11.1).

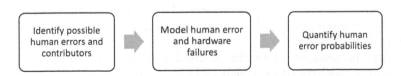

Figure 11.1 **Phases in HRA.**

Boring and Gertman (2016) have defined the two approaches as follows:

Qualitative HRA

- Focused on identification and modeling of the human failure event (HFE)
- Commonly employs some form of task analysis to identify potential human errors (HEs)
- Commonly looks at performance-shaping factors (PSFs)

Quantitative (Probabilistic) HRA

- Focused on producing human error probability (HEP)
  - Screening analysis performed for all HFEs
  - Detailed quantitative analysis for subset of all HFEs

Qualitative and quantitative HRA are complementary, and a qualitative study will always be performed before quantifying probabilities of human error.

Advanced technological systems are planned and designed by means of QRA to control reliability and safety. The methods were initially focusing on the technical systems and the related failures. The objective of HRA is to have a broader scope incorporating the human-machine subsystems. The early HRA methods had a kind of "machine-like" view on the human element with main focus on observable actions manifesting themselves as skilled-based and rule-based behavior and not related specifically to cognitive processes. Human error probability (HEP) is modified with *Performance-Influencing Factors* (PIFs) (also called Performance-Shaping Factors – PSFs). The PIFs address work conditions, personal states, and organizational aspects. The second-generation HRA methods had the ambition to also model and estimate the effect of error in cognitive functions.

In the following discussion of HRA methods we will mainly discuss two well-known methods, namely THERP (Technique for Human Error Rate Prediction) (Swain & Guttman, 1983) and CREAM (*Cognitive Reliability and Error Analysis Method*) (Hollnagel, 1998). They are often characterized as representing first- and second-generation HRA methods, respectively. CREAM offers two versions for HEP estimation: the Basic and the Extended method. They differ mainly in the degree of task detailing and modeling of PIFs.

A brief presentation of another method, HEART (Williams, 1986), and an example with SPAR-H (Gertman et al., 2005) are also included. The HEART method was also developed with focus on the nuclear industry but is slightly more generic and thereby somewhat easier to apply to other

Figure 11.2 Basic steps in the HRA process. Adapted from Sgobba (2017).

systems. The SPAR-H method is discussed in connection with a study that emphasized benchmarking of HEP estimates with accident data.

Figure 11.2 illustrates the main steps in the HRA process. The individual steps are discussed in the following subsections.

11.2.1 Problem definition

The first step in the HRA process is problem definition or identification of the human actions that will be studied (Figure 11.2). The tasks that are studied may either be behaviors related to the normal operation of the system or emergency tasks. The latter addresses the process of bringing a system that is out of control back to a normal state. The HRA can be based on familiarization with the system, documentation, operational logs, or accident reports. Given limited time and resources the problem definition phase may also serve as a way of prioritizing which parts that will be subject to HRA. Since HRA normally is a time- and resource-consuming process, only safety critical actions will usually be studied.

11.2.2 Task analysis

Task analysis or task decomposition is the second step in the process. The problem or operation that has been identified may involve a number of tasks and these should be broken down in subtasks and human actions. In certain instances, the task analysis may be started with a simplified approach or a so-called screening analysis in order to identify the most critical functions or when time or resources are limited. The method is demonstrated here for the task to prepare to leave port for a vessel (Figure 11.3). The task is broken down into four subtasks that are further decomposed in varying detail. The subtask to check traffic conditions actually required break down of actions in four levels. Rausand and Haugen (2020) give a more complete presentation of alternative approaches for task analysis.

11.2.3 Human error identification

Based on the task analysis the next step is to identify potential human error forms for the actions targeted. Human error identification (HEI) must be based on different approaches supported by system understanding,

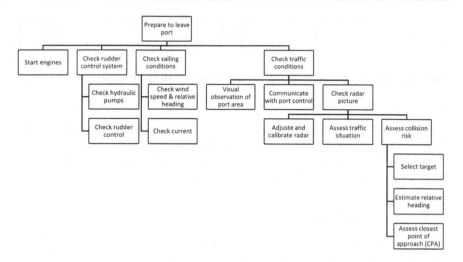

Figure 11.3 Task analysis of preparation to leave port.

operation experience, and human factors understanding. The HEI process may be supported by different methods and types of checklists like HAZOP (human hazard and operability) (see Chapter 8) and AEMA (action error mode analysis) (Øien & Rosness, 1998).

11.2.4 Error modeling

The causal mechanisms behind human error must be understood to be able to estimate the HEP. Key concepts in this relation are performance-influencing factors (PIFs) or performance-shaping factors (PSFs). It is common to make a distinction between internal and external PSFs:

- **Internal PSF:** Factors related to the individual that performs the action. Key categories are physical, physiological, and mental factors. The factors may be either temporary or permanent.
- **External PSF:** Factors that are outside the control of the individual and related to design of the workplace, the work organization, information systems, and other supporting systems.

A PSF may influence an action both positively and negatively. In the context of the HRA we are primarily interested in understanding how PSFs affect the behavior in a negative way. Different classifications of PSFs have been proposed for different HRA methods.

11.2.5 Human error probability HEP

The quantitative analysis phase of HRA involves estimation of the HEP and involves the following methodological steps in the first-generation methods:

1. Estimate the nominal HEP for each error identified.
2. Estimate the effect of identified PSFs on the nominal HEP.
3. Combine related errors in a model for the accident scenario.
4. Compute the resulting HEP for the scenario.

In complex tasks the situation may be that a number of actions are tightly coupled. Steps 3 and 4 are therefore necessary to model the interaction or effect of the potential errors. A common approach is to model the interaction between errors in a kind of event tree analysis (ETA). Apart from the computation of HEPs this is a straightforward process in first-generation methods.

In second-generation HRA the estimation of HEP becomes more demanding by the fact that the HEP is seen as a function of three sets of parameters, namely, human error type, PSF, and cognitive process. And secondly, it becomes even more demanding by the ambition to model the joint interaction both within each parameter set and between the parameter sets. Another problem associated with increasing model complexity is the challenge to find empirical data to prove the validity of the HEP model. This will be discussed in a later section.

11.3 THE THERP METHOD

11.3.1 Human error identification by means of PHEA

The starting point for THERP as for other HRA methods is the identification of human errors. The origin of the method used in THERP was the SHERPA method, initially developed by Embrey (1986). The method was subsequently developed and refined and is today known as Predictive Human Error Analysis (PHEA) technique (Embrey, 1993). Like the other methods for human error identification, the identification of errors is based on a taxonomy or checklist divided into six behavior types: Planning, operation, checking, information retrieval, communication, and selection. The nature of the error types is about timing, sequencing, omission, error of commission, wrong execution, and mismatch of action and object. It has been pointed out by critics that the approach like many other so-called first-generation HRA techniques was only focusing on skill-based and rule-based behavior (Table 11.1).

Table 11.1 The PHEA classification of erroneous actions

Planning errors	• Incorrect plan executed • Correct but inappropriate plan executed	• Correct plan executed, but too soon or too late • Correct plan but in wrong order
Operation errors	• Operation too long/short • Operation mistimed • Operation in wrong direction • Operation too little/much • Misalignment error	• Right operation on wrong object • Wrong operation on right object • Operation omitted • Operation incomplete • Wrong operation on wrong object
Checking errors	• Check omitted • Check incomplete • Right check on wrong object	• Wrong check on right object • Check mistimed • Wrong check on wrong object
Retrieval errors	• Information not retrieved • Wrong information retrieved	• Incomplete information retrieval
Communication errors	• Information not communicated • Wrong information communicated	• Incomplete information communicated
Selection errors	• Selection omitted	• Wrong selection

Hollnagel (1998).

11.3.2 Error modeling by means of THERP

The THERP method (Swain & Guttman, 1983) makes a distinction between external and internal PSFs as presented in Table 11.2.

Table 11.2 PSFs according to THERP

External PSFs *Situational factors*	**Internal PSFs** *Psychological stressors*	**Internal PSFs** *Organizational factors*
• Control room architecture • Quality of the working environment • Work hours and work breaks • Shift rotation and night work • Availability/adequacy of tools and supplies • Manning parameters	• Suddenness of onset • Duration of stress • Task speed • Task load • High jeopardy risk • Threat of failure, loss of job • Monotonous, degrading, or meaningless work	• Previous training/experience • Current practice or skill • Personality and attitudes • Motivation and attitudes • Knowledge of required performance standards • Sex differences • Physical condition

(continued)

Table 11.2 (Continued)   PSFs according to THERP

• Organizational structure and actions by others • Rewards, recognition, and benefits *Task and equipment characteristics* • Perceptual requirements • Motor requirements • Control-display relationships • Anticipatory requirements • Interpretation • Decision making • Complexity/information load • Frequency and repetitiveness • Task criticality • Long- and short-term memory • Calculation requirements • Feedback • Dynamic versus step-by-step activities • Team structure • Man-machine interface factors *Job and task instructions* • Operating procedures • Oral instructions	• Long, uneventful vigilance periods • Conflicts of motives about job performance • Reinforcement absent or negative • Sensory deprivation • Distraction (noise, glare, movement, flicker) • Inconsistent *Physiological stressors* • Duration of stress • Fatigue • Pain, discomfort • Hunger, thirst • Temperature • Radiation • G-force • Atmosphere • Vibration • Movement constriction • Physical exercise • Circadian rhythm	• Attitudes based on influence of family and other outside persons or agencies • Group identification

Chandler et al. (2006).

11.3.3 HEP estimation in THERP

The THERP method was originally developed for military systems and later modified and adapted for nuclear power systems. It will later be discussed how it has found its application in other sectors. THERP was at the outset primarily used to analyze systems heavily based on operational procedures both for maintenance tasks and emergency operation. A key element in the modeling process was therefore use of event trees to model the combined effect of error dependencies.

The THERP handbook (Swain & Guttman, 1983) provides information to estimate the nominal HEP for both normal operations such as calibration tasks, tests, and other maintenance tasks, but also for restoration tasks to bring a system out of control back to acceptable tolerance. The THERP handbook gives nominal HEP values (HEP_n) for tabulated error types. The uncertainty in the values is described as error bounds, representing the

90% confidence interval ($HEP_{0.05}$ and $HEP_{0.95}$). The interval is expressed through an error factor, given by:

$$EF = \sqrt{\frac{HEP_{0.95}}{HEP_{0.05}}} \qquad (11.1)$$

For screening purposes, the handbook offers some indicative values of HEPs for both rule-based action and diagnosis behavior in abnormal situations. As shown in Table 11.3, control room personnel fail to diagnose the situation with a probability of 0.05, i.e., one out of twenty attempts, based on rule-based behavior and written procedures. Taking into account the possibility of recovery, the HEP is reduced to half this value, at 0.025. It is further indicated that with lack of written procedures the diagnosis is certain to fail.

The success of diagnosis (knowledge-based behavior) is strongly related to the available time. Swain and Guttman (1983) have shown the probability of failure (HEP) as a function of time in minutes from the operator becomes aware of the situation. The probability is obviously highly dependent on the complexity of the situation and the experience of the person facing the situation. In this context, it may be noted that the probabilities are based on diagnosis of situations in the control room of a nuclear power plant. This is a highly complex technological system, where experience has shown that it may be difficult to correctly diagnose abnormal situations. It should also be noted that the HEPs mainly are based on expert judgment and not empirical data. According to the information provided, it will in practice take minimum 10 minutes before the probability of successfully diagnosing the situation starts increasing significantly. After 20 minutes, it has increased to 0.9, to 0.99 after 20 minutes, and to 0.999 after 60 minutes.

The nominal error probability may be influenced by one or more PSFs. The set of PSFs were described in Table 11.2. The THERP handbook gives advice on how to choose PSFs and estimate their influence on the HEP. However, even here it is critical that the analyst carefully assesses the case

Table 11.3 Human error probability and error factor for rule-based diagnosis by control room personnel of an abnormal event

Procedures availability	Error type	HEP_n	EF
Failure to perform correctly when written procedures are available	Error per critical step without recovery factors	0.05	10
	Error per critical step with recovery factors	0.025	10
Failure to perform correctly when written procedures are not available or used	Error per critical step with or without recovery factors	1.0	---

Swain and Guttman (1983).

based on available information and own experience. The effect of a PSF on the basic HEP is given as follows:

$$HEP_b = HEP_n PSF \tag{11.2}$$

In principle, the basic HEP may be influenced by more than one PSF but Rausand and Haugen (2020) warn that incorporating the effect of more than three is unrealistic. The THERP method does not give any advice on how to model the joint effect of PSFs and simply applies a multiplication model:

$$HEP_b = HEP_n \prod_{i=1}^{k} PSF_i, \tag{11.3}$$

$$\text{if } HEP_n \prod_{i=1}^{k} PSF_i > 1 \text{ then } HEP_b = 1 \tag{11.4}$$

where k is the number of PSFs influencing the HEP.

It is important to note that this equation may result in predicting HEPB > 1. This is obviously not possible in practice and may be regarded as a weakness of the model.

In HRA, we often face situations where the activities can be broken down into several more specific tasks and where it is necessary to quantify several HEPs. In such situations, it is important to be aware of possible dependencies between the HEPs. An example could be a situation where an operator must open two identical valves. If an error is made in opening the first valve, the probability of an error also in opening the second valve is likely to be dependent on the fact that an error already has occurred, implying that the second probability is higher than the first. If dependencies like this are not taken into account, the total probability of a human error in an activity can be underestimated severely, in extreme cases by several orders of magnitude.

11.3.4 Data on HEPs and PSFs in THERP

As already mentioned, the THERP method was initially developed for the nuclear industry and this is reflected in the data offered for both HEPs and PSFs in chapter 20 of the handbook (Swain & Guttman, 1983). The data is organized in 27 tables, and it is clear that the factors are primarily related to use of written material for procedures and administrative control, operation of displays, and process equipment like valves. This is quite far from many of the functions and conditions experienced on bridge of a vessel. A vessel is operated in different phases like leaving/entering a port, sailing in coastal waters and on the high seas. Secondly, the external factors have an important role: sea traffic, weather condition, and sea state, to mention a few. This means that applying THERP in this context put a demanding task on HRA specialist. The THERP database is organized as shown in Table 11.4.

11.4 THE CREAM METHOD

11.4.1 Human error identification by means of CREAM

Hollnagel (1998) developed the *Cognitive Reliability and Error Analysis Method* (CREAM) that also incorporated knowledge-based behavior and thereby focused on the cognitive aspects and the context of human error. The

Table 11.4 Data included in the THERP database

Category	Factor
Screening	1. Diagnosis
	2. Rule-based actions
Diagnosis	3. Nominal diagnosis
	4. Post-event control room staffing
Errors of omission	Written materials mandated:
	5. Preparation
	6. Administrative control
	7. Procedural items
	No written materials:
	8. Administrative control
	9. Oral instruction items
Errors of commission	Displays:
	10. Display selection
	11. Read/record quantitative
	12. Check-read quantitative
	13. Control & MOV selection & Use
	Locally operated valves:
	14. Valve selection
	15. Stuck valve detection
PSFs	16. Tagging levels
	17. Stress/experience
	18. Dependence
Uncertainty bounds	19. Estimate UCBs
	20. Conditional HEPs and UCBs
Recovery factors	21. Errors by checker
	22. Annunciated cues
	23. Control room scanning
	24. Basic walk-around inspection

Swain and Guttman (1983).

role of the operator is related to four modes of behavior, namely observe, interpret, plan, and act. The basic error modes or *Phenotypes* characterize errors related to physical actions in operation of technical systems (Table 11.5). A set of 15 specific cognitive activities is defined in CREAM (Table 11.6) and is the basis for the specification of person-related error for the three cognitive behaviors, namely, observe, interpret, and plan as outlined in Table 11.7.

Table 11.5 Basic error modes (basic phenotypes)

Error mode	Specific effect
Timing	Too early, too late, omission
Duration	Too long, too short
Sequence	Reversal, repetition, omission, commission, jump forward/backward
Object	Wrong action, wrong object
Force	Too much, too little
Direction	Wrong direction, wrong movement type
Speed	Too fast, too slow
Distance	Too far, too near

Hollnagel (1998).

Table 11.6 Cognitive activities in CREAM

Cognitive activity	General definition
Coordinate	Bring system states and/or control configurations into the specific relation required to carry out a task or task step. Allocate or select resources in preparation for a task/job, calibrate equipment, etc.
Communicate	Pass on or receive person-to-person information needed for system operation by verbal, electronic, or mechanical means. Communication is an essential part of management.
Compare	Examine the qualities of two or more entities (measurements) with the aim of discovering similarities or differences. The comparison may require calculation.
Diagnosis	Recognize or determine the nature or cause of a condition by means of reasoning about signs or symptoms or by the performance of appropriate tests. "Diagnosis" is more thorough than "identify."
Evaluate	Appraise or assess an actual or hypothetical situation, based on available information without requiring special operations. Related terms are "inspect" and "check."
Execute	Perform a previously specified action or plan. Execution comprises actions such as open/close, start/stop, and fill/drain.
Identify	Establish the identity of a plant state or subsystem (component) state. This may involve specific operations to retrieve information and investigate details. "Identify" is more thorough than "evaluate."
Maintain	Sustain a specific operational state. This is different from maintenance that is generally an offline activity.

(continued)

Table 11.6 (Continued)   Cognitive activities in CREAM

Cognitive activity	General definition
Monitor	Keep track of system states over time or follow the development of a set of parameters.
Observe	Look for or read specific measurement values of system indications.
Plan	Formulate or organize a set of actions by which a goal will be successfully achieved. Plan may be short term or long term.
Record	Write down or log system events, measurements, etc.
Regulate	Alter speed or direction of a control (system) in order to attain a goal. Adjust or position components or subsystems to reach a target state.
Scan	Quick or speedy review of displays or other information sources to obtain a general impression of the state of a system/subsystem.
Verify	Confirm the correctness of a system condition or measurement, either by inspection or test. This also includes the feedback from prior operations.

Hollnagel (1998).

Table 11.7 Cognitive error types in CREAM

Cognitive function		Potential cognitive function failure
Observation	O1	Observation of wrong object. Response is given to the wrong stimulus or event.
	O2	Wrong identification is made due to a mistaken cue or partial identification.
	O3	Observation not made (omission), overlooking a signal or measurement.
Interpretation	I1	Faulty diagnosis, a wrong diagnosis, or an incomplete diagnosis.
	I2	Decision error: not making a decision or making a wrong or incomplete decision.
	I3	Delayed interpretation or not made in time.
Planning	P1	Priority error, as in selecting the wrong goal (intention).
	P2	Inadequate plan formulated when the plan is either incomplete or directly wrong.
Execution	E1	Execution of wrong type with regard to force, distance, speed, or direction.
	E2	Action performed at wrong time, either too early or too late.
	E3	Action on wrong object (neighbor, similar or unrelated).
	E4	Action performed out of sequence, such as repetitions, jumps, and reversals.
	E5	Action missed, not performed, and omission of actions in a sequence ("undershoot")

Hollnagel (1998).

11.4.2 Error modeling by means of CREAM

In second-generation methods such as the CREAM model, emphasis is placed on the cognitive aspects of error causation. Hollnagel (1998) defined a set of *Genotypes* rather than PSFs to characterize the causal mechanisms. The structure of the genotypes is outlined in Table 11.8.

Table 11.8 Genotypes

Subgroup	Specific genotype
Temporary person-related functions	• Memory failure: Forgotten, incorrect recall, incomplete recall • Fear: Random action (trial-and-error), person paralyzed • Distraction: Task suspended, task not completed, goal forgotten • Fatigue: Delayed response, no response • Performance variability: Lack of precision, increasing number of misses • Inattention: Signal missed • Physiological stress: Conditions resulting in many effects • Psychological stress: Conditions resulting in many effects
Permanent person-related functions	• Functional impairment: Deafness, bad eyesight, color blindness • Cognitive style: Simultaneous and successive scanning, conservative focusing • Cognitive bias: Focus gambling, incorrect revision of probabilities, hindsight bias, attribution error, illusion of control, confirmation bias, hypothesis fixation
Equipment failure	• Equipment failure: Actuator failure, blocking, breakage, uncontrolled release, speed-up, or slow-down • Software fault: Performance slow-down, information delays, command queues, information not available
Procedures	• Inadequate procedure: Ambiguous text, incomplete text, incorrect text, mismatch to actual equipment
Temporary interface problems	• Access limitations: Item cannot be reached or cannot be found • Ambiguous information: Item difficult to find, inappropriately labeled • Incomplete information: Information provided by the interface is incomplete, e.g., messages, directions, and warnings
Permanent interface problems	• Access problems: Item cannot be reached or cannot be found • Mislabeling: Incorrect information, ambiguous identification, labeling is incorrectly formulated or written in foreign language
Communication	• Communication failure: Message not received or not understood • Missing information: No information, not available when it was needed, incorrect information, lack of feedback, misunderstanding between sender and receiver
Organization	• Maintenance failure: Equipment not operational, indicators are not working • Inadequate quality control: Inadequate procedures, lack of resources or supplies • Management problem: Unclear roles, dilution of responsibility, unclear line of command

(continued)

Table 11.8 (Continued)   Genotypes

Subgroup	Specific genotype
	• Design failure: Anthropometric mismatch, inadequate HMI (Human-machine interface) • Inadequate task allocation: Inadequate managerial rule, task planning and procedure • Social pressure: Group thinking
Training	• Insufficient skills: Performance failure, equipment mishandling • Insufficient knowledge: Confusion about what to do, lost situation awareness
Ambient conditions	• Temperature • Sound • Humidity • Illumination
Working conditions	• Excessive demand. Excessive task demand or insufficient time or resources • Inadequate workplace layout: Narrow work space, dangerous place, or elevated work space • Inadequate team support: Unclear job description, inadequate communication, lack of team cohesiveness • Irregular working hours: Shift work leading to physiological and psychological effects, circadian rhythm disturbances

Hollnagel (1998).

11.4.3 HEP estimation in CREAM

The CREAM model offers two alternative approaches for estimating the HEP. The simplest approach is a kind of screening method (Basic method) for a task involving several actions. A more detailed approach (Extended method) can be used for analyzing each action separately.

11.4.3.1 Basic method

In the preceding section, we have defined and specified the main components in the CREAM model: basic error types, cognitive activities, cognitive error types, and PSFs (Genotypes). However, the basic method is based on a simplified parameter called *Common Performance Condition* (CPC) that combines the components of the abovementioned model. It should also be noted that this approach can be applied for the total task and not only for identified actions of the task.

Taking the task analysis as the starting point the main steps are:

1. Assess the performance reliability for each CPC
2. Adjust the performance reliability for dependence between CPCs
3. Add the performance reliabilities for the CPCs

4. Determine probable control mode
5. Determine the HEP (general action failure probability)

The basis for estimating the HEP is the assessment of the nine CPCs defined in Table 11.9. These parameters are used to characterize the nature of the task or set of actions. The initial study of the operation and the task analysis will, with supporting technical documentation and human factors input, serve as the basis for assessing the performance level for each CPC or Level descriptor. The assessed performance may have three alternative outcomes for reliability, namely, "Improved", "Not significant", or "Reduced".

Table 11.9 CPCs and performance reliability

CPC name	Level descriptor	Expected effect on performance reliability
Adequacy of organization	Very efficient	Improved
	Efficient	**Not significant**
	Inefficient	Reduced
	Deficient	Reduced
Working conditions	Advantageous	Improved
	Compatible	**Not significant**
	Incompatible	Reduced
Adequacy of HMI and operational support	Supportive	Improved
	Adequate	**Not significant**
	Tolerable	**Not significant**
	Inappropriate	Reduced
Availability of procedures and plans	Appropriate	Improved
	Acceptable	**Not significant**
	Inappropriate	Reduced
Number of simultaneous goals	Fewer than capacity	**Not significant**
	Matching current capacity	**Not significant**
	More than capacity	Reduced
Available time	Adequate	Improved
	Temporarily inadequate	**Not significant**
	Continuously inadequate	Reduced
Time of the day (circadian rhythm)	Day-time (adjusted)	**Not significant**
	Night-time/adjusted)	Reduced
Adequacy of training and experience	Adequate, high experience	Improved
	Adequate, limited experience	**Not significant**
	Inadequate	Reduced
Crew collaboration quality	Very efficient	Improved
	Efficient	**Not significant**
	Inefficient	**Not significant**
	Deficient	Reduced

Hollnagel (1998).

Hollnagel (1998) realized that four of these CPCs had synergetic effects on each other. Step 2 was therefore included. These four CPCs are shown in Table 11.10 together with the nine CPCs that are influenced. The modification of the effect of CPCs is done as follows:

1. If the performance level is *Improved* or *Reduced*, no adjustment is necessary.
2. If the performance level is *Not significant*, adjust the level as follows:
   a. If the majority of influencing CPCs are pointing in the same direction, the CPC is adjusted in the same direction: Either *Improved* or *Reduced*.
   b. If the influencing CPCs point in opposing directions, keep the initial assessment *Not significant*.

Step 3 is to add the assessed performance levels for the nine CPCs with a triplet comprising the number of "Reduced", "Not Significant", and "Improved" CPCs: $\{N_R R, N_{NS}, N_I\}$.

Moving on to step 4, Hollnagel (1998) defined four different control modes: Strategic, Tactical, Opportunistic, and Scrambled as they are described in Table 11.11. The modes are related to different HEP ranges as shown in Figure 11.4 determined by the combination of a number of Reduced and Improved performance reliabilities. The worst performance

Table 11.10 Rules for adjusting the CPCs

CPC name	Working conditions	Number of goals	Available time	Crew collaboration quality
Adequacy of organization	X			X
Working conditions		X		
Adequacy of HMI and operational support	X	X	X	X
Availability of procedures and plans		X	X	
Number of simultaneous goals			X	
Available time	X			
Time of the day (circadian rhythm)	X			
Adequacy of training and experience	X			
Crew collaboration quality				

Hollnagel (1998).

Table 11.11 Control modes in CREAM

Control mode	Description	Probability interval
Strategic	In strategic control, the person considers the global context, thus uses a wider time horizon and looks ahead at higher level goals. The strategic mode provides a more efficient and robust performance and may therefore seem the ideal to strive for.	0.000005–0.01
Tactical	In tactical control, performance is based on planning, hence more or less follows a known procedure or rule. The planning is, however, of limited scope and it needs to be taken into account that it may sometimes be ad hoc.	0.001–0.1
Opportunistic	In opportunistic control, the next action is determined by the salient features of the current context rather than by more stable intentions or goals. The person does very little planning or anticipation, perhaps because the context is not clearly understood or because time is too constrained.	0.01–0.5
Scrambled	In scrambled control, the choice of next action is in practice unpredictable or haphazard. Scrambled control characterizes a situation where there is little or no thinking involved in choosing what to do next.	0.1–1.0

Hollnagel (1998).

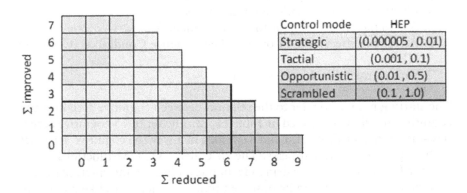

Figure 11.4 Control modes in CREAM and converting CPC values into control mode HEP. Adapted from Hollnagel (1998).

(Scrambled) is related to between 6 and 9 Reduced and zero Improved. The Strategic control mode is represented at the top of the diagram and is given by values of 4 to 7 for Improved combined with 1 to 3 for Reduced.

Table 11.11 also provides probability (HEP) intervals for the different control modes. It is noted that the intervals are very wide, in particular for the strategic control mode where the highest value is more than three orders of magnitude greater than the lower value in the interval. These values also form the basis for step 5.

A final comment to this simplified approach is worth mentioning. The basic method does not address HEP estimation directly. Instead, the concrete task is assessed relative to what is viewed as the normal performance. As shown in Table 11.9, Adequacy of organization rated as Efficient results in Not significant or in other words a normal performance. In the same manner, Compatible Working conditions result in Not significant. In a case where none of the CPCs are adjusted with respect to performance level, we would have a "normal task" characterized by zero Reduced and Improved. From Figure 11.4, we can see that control mode is tactical and that the HEP then falls in the range of 0.001–0.1.

11.4.3.2 Extended method

The alternative estimation method proposed by Hollnagel (1998) is based on the complete set of parameters outlined in the preceding sections and focuses on actions or subtasks in contrast to tasks as for the basic method. The main analytical steps are:

1. Identify the cognitive activities for the given subtask.
2. Assess the cognitive demands for each cognitive activity.
3. Specify the demand profile for the subtask.
4. Identify the likely cognitive failure types.
5. Assess the actual HEP for the failures identified.
6. Assess the effect of the Common Performance Conditions (CPCs) on HEPs.
7. Sum up the HEPs for the subtask.

The starting point is to refer to the relation between cognitive activities and cognitive functions as defined in Table 11.12. The set of 15 cognitive activities was defined in Table 11.6 and describes in a way the cognitive activities with respect to functional tasks. The so-called COCOM functions represent the four basic cognitive demands, namely, observe, interpret, plan, and execute (Table 11.6). It can be observed that some activities like communicate, compare, identify, plan, and scan are only involving a single demand, whereas the other activities engage two cognitive demands. Coordination, for instance, requires both planning and execution, and monitoring involves both observation and interpretation.

Let us consider the activity "Visual observation of port area" given in the task structure shown in Figure 11.3. The objective of this activity may be to scan the port area to evaluate whether the traffic may represent a too high risk to the vessel preparing to leave the port. The cognitive activities are *Monitor* and *Diagnose* and involve following COCOM functions, namely,

Table 11.12 A generic cognitive-activity-by-cognitive-demand matrix

	COCOM function			
Cognitive activity	Observation	Interpretation	Planning	Execution
Coordinate			•	•
Communicate				•
Compare		•		
Diagnosis		•	•	
Evaluate		•	•	
Execute				•
Identify		•		
Maintain			•	•
Monitor	•	•		
Observe	•			
Plan			•	
Record		•		•
Regulate	•			
Scan	•			
Verify	•	•		

Hollnagel (1998).

Observation and *Interpretation*, and *Interpretation* and *Planning*, respectively. The cognitive demand profile for the activity is simply assessed to be:

- Observation: 25%
- Interpretation: 50%
- Planning: 25%

This is simply counting the total number of COCOM functions involved in the cognitive activities and calculating the percentage of the total for each function.

The cognitive function failures have been defined in Table 11.7. The nominal HEPs and upper and lower bounds for the failure types are offered by Hollnagel in Table 11.13. The nominal failure probability ranges from $5 \cdot 10^{-4}$ for *Action on wrong object* to $2 \cdot 10^{-1}$ for *Faulty diagnosis*. The nominal HEP is 2,000 times higher for the most demanding function compared to the easiest function. The proposed values are supported by estimates from other similar studies or databases.

The next step is to assess which failure types are relevant or most probable for the identified cognitive functions. The findings in our case were:

- O2 – Wrong identification: $2 \cdot 10^{-2}$ (lower bound)
- I1 – Faulty diagnosis: $1 \cdot 10^{-1}$ (below basic value)
- P2 – Inadequate plan: $5 \cdot 10^{-3}$ (below basic value)

Table 11.13 Nominal values and uncertainty bounds for cognitive function failures

Cognitive function	Generic failure type	Lower bound (0.5)	Basic value	Upper bound (0.95)
Observation	O1 Wrong object observed	3.0 E-4	1.0 E-3	3.0 E-3
	O2 Wrong identification	2.0 E-2	7.0 E-2	1.7 E-1
	O3 Observation not made	2.0 E-2	7.0 E-2	1.7 E-1
Interpretation	I1 Faulty diagnosis	9.0 E-2	2.0 E-1	6.0 E-1
	I2 Decision error	1.0 E-3	1.0 E-2	1.0 E-1
	I3 Delayed interpretation	1.0 E-3	1.0 E-2	1.0 E-1
Planning	P1 Priority error	1.0 E-3	1.0 E-2	1.0 E-1
	P2 Inadequate plan	1.0 E-3	1.0 E-2	1.0 E-1
Execution	E1 Action of wrong type	1.0 E-3	3.0 E-3	9.0 E-3
	E2 Action at wrong time	1.0 E-3	3.0 E-3	9.0 E-3
		5.0 E-5	5.0 E-4	5.0 E-3
	E3 Action on wrong object	1.0 E-3	3.0 E-3	9.0 E-3
		2.0 E-2	3.0 E-2	4.0 E-2
	E4 Action out of sequence			
	E5 Missed action			

Hollnagel (1998).

These nominal HEPs shall then be modified with respect to the Common Performance Conditions (CPCs) that were defined in Table 11.9. A CPC may have differing effect on the COCOM functions which is documented in Table 11.14. The HEP may be reduced, kept unchanged, or increased depending on whether the CPC element is rated as above standard, on standard, or below standard. The weighting factors are ranging from 0.5 (very efficient) to 5.0 (deficient). The relevant weighting factors for the three nominal HEPs in our example are shown in Table 11.15. The aggregate effect is given by the product of the weighting factors (8.0, 8.0, 10.0). The modified HEPs are as follows:

- O2 – Wrong identification: $2 \cdot 10^{-2} \cdot 8 = 1.6 \cdot 10^{-1}$
- I1 – Faulty diagnosis: $1 \cdot 10^{-1} \cdot 8 = 8.0 \cdot 10^{-1}$
- P2 – Inadequate plan: $5 \cdot 10^{-3} \cdot 10 = 5 \cdot 10^{-2}$

The failure probability for the subtask is given by:

$$HEP = 1 - \left(1 - 1.6 10^{-1}\right)\left(1 - 8.0 10^{-1}\right)\left(1 - 5.0 10^{-2}\right) = 8.4 10^{-1}$$

The finding indicates that the subtask must be revised in order to improve the working conditions or CPCs.

Table 11.14 Weighting factors for CPCs

CPC name	Level descriptor	OBS	INTERPR	PLAN	EXE
		COCOM function			
Adequacy of organization	Very efficient	1.0	1.0	0.8	0.8
	Efficient	1.0	1.0	1.0	1.0
	Inefficient	1.0	1.0	1.2	1.2
	Deficient	1.0	1.0	2.0	2.0
Working conditions	Advantageous	0.8	0.8	1.0	0.8
	Compatible	1.0	1.0	1.0	1.0
	Incompatible	2.0	2.0	1.0	2.0
Adequacy of MMI and operational support	Supportive	0.5	1.0	1.0	0.5
	Adequate	1.0	1.0	1.0	1.0
	Tolerable	1.0	1.0	1.0	1.0
	Inappropriate	5.0	1.0	1.0	5.0
Availability of procedures and plans	Appropriate	0.8	1.0	0.5	0.8
	Acceptable	1.0	1.0	1.0	1.0
	Inappropriate	2.0	1.0	5.0	2.0
Number of simultaneous goals	Fewer than capacity	1.0	1.0	1.0	1.0
	Matching current capacity	1.0	1.0	1.0	1.0
	More than capacity	2.0	2.0	5.0	2.0
Available time	Adequate	0.5	0.5	0.5	0.5
	Temporarily inadequate	1.0	1.0	1.0	1.0
	Continuously inadequate	5.0	5.0	5.0	5.0
Time of the day (circadian rhythm)	Day-time (adjusted)	1.0	1.0	1.0	1.0
	Night-time (adjusted)	1.2	1.2	1.2	1.2
Adequacy of training and experience	Adequate, high experience	0.8	0.5	0.5	0.8
	Adequate, low experience	1.0	1.0	1.0	1.0
	Inadequate	2.0	5.0	5.0	2.0
Crew collaboration quality	Very efficient	0.5	0.5	0.5	0.5
	Efficient	1.0	1.0	1.0	1.0
	Inefficient	1.0	1.0	1.0	1.0
	Deficient	2.0	2.0	2.0	5.0

Hollnagel (1998).

Table 11.15 Adjustment factors (weights) of CPCs - example

CPC name	Assessed performance level	Weighting factors		
		O2	I1	P2
Adequacy of organization	Efficient	1.0	1.0	1.0
Working conditions	Incompatible	2.0	2.0	1.0

(continued)

Table 11.15 (Continued)   Adjustment factors (weights) of CPCs - example

CPC name	Assessed performance level	Weighting factors O2	II	P2
Adequacy of MMI and operational support	Tolerable	1.0	1.0	1.0
Availability of procedures and plans	Acceptable	1.0	1.0	1.0
Number of simultaneous goals	More than capacity	2.0	2.0	5.0
Available time	Temporarily inadequate	1.0	1.0	1.0
Time of the day (circadian rhythm)	Day-time	1.0	1.0	1.0
Adequacy of training and experience	Adequate, low experience	1.0	1.0	1.0
Crew collaboration quality	Deficient	2.0	2.0	2.0
Total influence of CPCs (product of)		8.0	8.0	10.0

11.5 HUMAN ERROR ASSESSMENT AND REDUCTION TECHNIQUE (HEART)

Another well-known first-generation method is the so-called HEART method (Williams, 1986). The initial qualitative phase is fairly similar to THERP. However, the estimation of the HEP is based on two slightly different concepts, namely the *Generic Task Type* (GTT) and the *Error-Producing Condition* (EPC). The method has been subject to later revisions and application in both nuclear power generation and other industries.

The first step in the HEP estimation process is to select one of the eight GTTs that best matches the analyzed task or subtask (Table 11.16). The generic task has a basic HEP value and the associated 5 and 95 percentiles (the 90% confidence interval). The HEP ranges from 0.0004 to 0.55 for the generic task types with the lowest and highest probability, respectively. So, even at this stage the analyst needs to judge whether to adjust the basic HEP for any discrepancy between the generic and actual task.

Table 11.16 Generic classification of HEP in HEART

Generic task	Nominal HEP (5th–95th percentile)
A Totally unfamiliar, performed at speed with no real idea of likely consequences	0.55 (0.35–0.97)
B Shift or restore system to a new or original state on a single attempt without supervision or procedures	0.26 (0.14–0.42)
C Complex task requiring high level of comprehension and skill	0.16 (0.12–0.28)

(continued)

Table 11.16 (Continued)   Generic classification of HEP in HEART

Generic task		Nominal HEP (5th–95th percentile)
D	Fairly simple task performed rapidly or given scant attention	0.09 (0.06–0.13)
E	Routine, highly practiced, rapid task involving relatively low level of skill	0.02 (0.007–0.045)
F	Restore or shift a system to original or new state following procedures, with some checking	0.003 (0.0008–0.007)
G	Completely familiar, well-designed, highly practiced, routine task occurring several times per hour, performed to the highest possible standards, by highly motivated, highly trained and experienced person, totally aware of implications of failure, with time to correct potential error, but without the benefit of significant job aids	0.0004 (0.00008–0.009)
H	Respond correctly to system command even when there is an augmented or automated supervisory system providing accurate interpretation of system stage	0.00002 (0.000006–0.0009)

Williams (1986).

The next step is to estimate the basic HEP based on the assessment of the operating conditions or what is termed *Error-producing conditions* (EPCs) in the HEART method. The first version of the method defined 26 EPCs as outlined in Table 11.17. The nominal HEP is adjusted in the following manner:

$$HEP_b = (EPC - 1) \cdot APA \cdot HEP_n \qquad (11.5)$$

The analyst must subjectively assess the relative contribution (APA) of the EPC on the nominal HEP. This is especially important because the method implies that HEP_n may be influenced by more than one EPC. The general expression is therefore:

$$HEP_b = \left\{ \prod_{i=1}^{k} (EPC_i - 1) \cdot APA_i \right\} \cdot HEP_n \qquad (11.6)$$

To give an indication of the output from the method, let us calculate the lowest and highest estimate of the basic HEP with some trivial assumptions:

Lowest HEP: $HEP = (1.4 - 1)1.0 \ 0 \cdot 00002 = 8.010^{-6}$

Highest HEP: $HEP = (17 - 1)1.0 \ 0 \cdot 55 = 8.8, \quad i.e., 1.0$

We can see that also this way of calculating HEP_b can give probability values greater than 1.0. These values correspond fairly well with the range of HEPs for the strategic and scrambled modes of behavior indicated for the CREAM method as shown in Figure 11.4.

Table 11.17 Error-producing conditions in HEART

	Error-producing condition	EPC
1	Unfamiliarity with a situation which is potentially important, but which only occurs infrequently, or which is novel	17
2	A shortage of time available for error detection and correction	11
3	Low signal-to-noise ratio	10
4	A means of suppressing or overriding information or features which is too easily accessible	9
5	No means of conveying spatial and functional information to operators in a form which they can readily assimilate	8
6	A mismatch between an operator's model of the world and that imagined by the designer	8
7	No obvious means of reversing an unintended action	8
8	A channel capacity overload, particularly one caused by simultaneous presentation of non-redundant information	6
9	A need to unlearn a technique and apply one which requires the application of an opposing philosophy	6
10	The need to transfer specific knowledge from task to task without loss	5.5
11	Ambiguity in the required performance standards	5
12	A mismatch between perceived and real risk	4
13	Poor, ambiguous, or ill-matched system feedback	4
14	No clear direct and timely confirmation of an intended action from the portion of the system over which control is to be exerted	4
15	Operator inexperience (e.g., a newly qualified tradesman, but not an "expert")	3
16	An impoverished quality of information conveyed by procedures and person/person interaction	3
17	Little or no independent checking or testing of output	3
18	A conflict between immediate and long-term objectives	2.5
19	No diversity of information input for veracity checks	2.5
20	A mismatch between the educational achievement level of an individual and the requirements of the task	2
21	An incentive to use other more dangerous procedures	2
22	Little opportunity to exercise mind and body outside the immediate confines of a job	1.8
23	Unreliable instrumentation (enough that it is noticed)	1.6
24	A need for absolute judgments which are beyond the capabilities or experience of an operator	1.6
25	Unclear allocation of function and responsibility	1.6
26	No obvious way to keep track of progress during an activity	1.4

Williams (1986).

11.6 APPLICATION OF THERP IN TRANSPORT

11.6.1 Crew error in aircraft takeoff

Yang et al. (2013) applied THERP for the analysis of the takeoff phase for an aircraft. The authors found it necessary to modify the terminology of the error categories somewhat (Table 11.18) compared to the PHEA classification used in THERP (Table 11.1). The error mode "Spatial illusion" is unique for an aircraft. The authors remarked that these error modes were the result of input from the O'Hare classification (O'Hare et al., 1994), Wickens's information processing model (Wickens & Flach, 1988), and the works of Rasmussen and Hollnagel.

The takeoff phase involves five tasks, namely, Before takeoff, Engine start, After start, Taxi, and Takeoff. The tasks were further broken down into 15 subtasks (Figure 11.5) and evaluated in terms of error mode, description of the error, and consequence. Table 11.19 gives a few examples of the task analysis format for the subtasks.

Table 11.18 Flight crew error modes

Human error modes	
1. Omit	6. Direction error
2. Lapse	7. Timing error
3. Spatial illusion	8. Speed error
4. Wrong pattern	9. Sequence error
5. Decision-making error	10. Force error

Yang et al. (2013).

Table 11.19 Examples of human error modes and consequences in the takeoff phase

Subtask	Human error mode	Description	Consequence
Check/adjust takeoff data	Omit	The flight crew forgets to calculate takeoff speed and neither checks takeoff configuration nor validates takeoff weight limitation	The appropriate takeoff speed cannot be determined
Close ground spoiler	Lapse	Ground spoiler is armed	The acceleration is abnormal, and the distance of takeoff needs to be longer
Rudder and pitch trim	Spatial illusion	Because of the different views, it is not trimmed indeed	The aircraft must be controlled manually
Nose up	Force error	Too much pitch angle	Tail strike
Thrust lever set to CL	Timing error	Set to CL too early or too late	Influence on climb performance. Influence on engine

Yang et al. (2013).

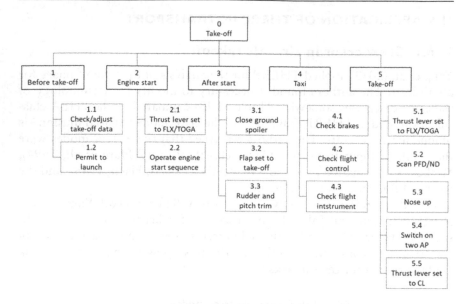

Figure 11.5 Task analysis of takeoff phase. Source: Yang et al. (2013).

By assuming that all 15 subtasks must be successful for a successful take-off to occur the authors imply that the outcome should be modeled with an event tree and that numerically the negative outcome (failure) is given by:

$$HEP = 1 - \prod_{i=1}^{15}(1 - HEP_i)$$

(11.7)

The estimation of the error probabilities in this study is done in a trivial manner by assuming the same values for all 15 subtasks:

Nominal HEP: $HEP_n = 0.0004$
Performance-shaping factor: $PSF = 2$
Basic HEP: $HEP_b = 0.0008$

The nominal HEP value was chosen by referring to the HEART method (Williams, 1986) and indicates that all subtasks were defined as a "*Completely familiar, well-designed, highly practiced routine task occurring several times per hour*". The main cause of determining the PSF was fatigue. The resulting human error probability for takeoff was:

$$HEP = 1 - (1 - 0.0008)^{15} = 0.0113$$

The analysis for estimating human failure in the takeoff phase can be criticized for the mismatch between a detailed task analysis and a simplified HEP estimation. The authors took the trouble to break down and analyze 15 subtasks in terms of error mode and detailed description. But on the other hand, this information is not applied in any manner to discriminate with respect to estimation of HEPs or PSFs.

11.6.2 Human contribution in marine traffic accidents

Martins and Maturana (2010) applied the THERP method in a Formal Safety Assessment of marine traffic accidents: collision, powered grounding, and drift grounding. The authors gave some reflections on the application of THERP for problems in maritime domain (cited):

1. *Human error: it is difficult to associate a probability to human error. Human beings, depending on the situation, can correct themselves before problems occur, bringing uncertainty to the available HEPs.*
2. *Impersonal error consideration: people do not have the same conduct as machines do.*
3. *Some factors are not considered (humor, disposition, courage, prudence, personality).*
4. *Lack of a specific database for the maritime sector.*
5. *The complete application of the THERP may require intensive application of resources due to the level of details required.*

In contrast to the conventional approach in HRA they model the accident scenarios or tasks by means of fault tree analysis (FTA) rather than event tree analysis (ETA). The collision model is shown in Figure 11.6.

Figure 11.6 shows that all basic events in the fault tree are human errors. Since specific data for the maritime industry does not exist, THERP data are used for the nominal HEPs. To demonstrate the method, the authors estimated the error probability for basic event no. 43 named *"Failure in drawing the route"* (Figure 11.6) or simply *Route planning*. The task analysis resulted in 11 activities outlined in Table 11.20. The next step was to select generic activities from the THERP database that best reflected the maritime activities. In this case the following generic activities were chosen:

- Use of checklist
- Read and recall information obtained from display
- Preparation of written material
- Checks made by another person

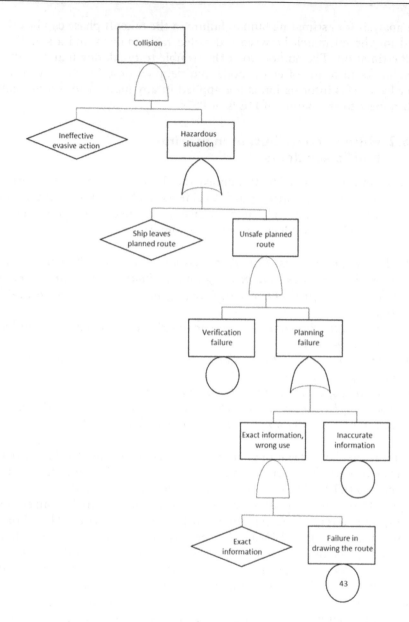

Figure 11.6 Collision fault tree (Extract from Martins & Maturana, 2010).

The estimated HEPs are shown in rightmost column in Table 11.20. The HEP estimates for similar generic activities vary depending on the amount of information that was handled by the navigator. The error probability for *Use of checklist* ranged from 0.001 to 0.01, or in other words one order of magnitude.

Table 11.20 Selection of Human error probabilities – Event 43 Route planning

Route planning activity	Relevant nuclear plant activity	HEP
1 Update the publication	Use of checklist	0.001
2 Reading information	Use of checklist	0.017
3 Establishment of minimal depth	Read and recall information obtained from display	0.01
4 Reading the depth in chart	Use of checklist	0.01
5 Identify restricted area	Read and recall information obtained from display	0.01
6 Drawing restricted area	Preparation of written material	0.001
7 Drawing successive courses	Preparation of written material	0.003
8 Identify waypoints	Preparation of written material	0.001
9 Choose reference points	Preparation of written material	0.001
10 Route revision by Officer	Checks made by another person	0.01
11 Route revision by Master	Checks made by another person	0.01

Martins and Maturana (2010).

The success of *Route planning* is dependent on the success of all the 11 activities and the total HEP can therefore be calculated by multiplying together the probabilities for all the activities. It is noted that this assumes independence between the failure probabilities for the individual activities:

$$\text{HEP}_{\text{Route planning}} = 1 - (1 - 0.001)^4 \cdot (1 - 0.017) \cdot (1 - 0.01)^5 \cdot (1 - 0.003) = 0.072$$

This process was applied for all the basic events in the fault tree and served as basis for computing the probability of the top event (collision) to $1.48 \cdot 10^{-4}$.

The next step was to perform a sensitivity analysis by studying the variation in top event probability as a function of the variation in each activity HEP within an event. The range of each activity HEP was given by the guidelines in the THERP database (Swain & Guttman, 1983). It was found that 16 activities gave a change in collision probability of more than 3% (Table 11.21). These activities were related to 9 of the 43 events in the fault tree.

Three *Performance-shaping factors* were identified:

1. Operator experience level: Experienced or inexperienced
2. Task demand or load: Routine or dynamic
3. Stress level: Low, excellent, middle, or high

It was analyzed how the 16 combinations of these PSFs influenced the most important events given in Table 11.21. By subjective assessment it was found

Table 11.21 Variation of top event (collision) probability as function of minimum and maximum estimated HEP for activity

Event		Activity	Variation of top event	
			Min.	*Max.*
6	Evasion pattern failure (COLREGs)	1 – Following established procedure	−98%	100%
10	Communication failure	1 – Deciding to make contact by radio	−85%	853%
		3 – Understanding the interlocutor	−1%	3%
		2 – To make contact	−1%	3%
		4 – Decide to make contact w/ other means	−2%	3%
11	Command failure	3 – Detection of error	−1%	7%
		1 – Understand correctly	−1%	7%
14	Helmsman error	1 – Manipulation of rudder	0%	4%
23	Command failure	3 – Detection of error	−57%	285%
		1 – Understand correctly	−57%	283%
		2 – Decision making	0%	3%
25	Helmsman wrong answer	1 – Manipulation of rudder	−16%	159%
29	Marking error	1 – Checks of equipment	−5%	13%
		6 – Captain detects error	−1%	5%
32	Captain failure	1 – Checks	−2%	11%
33	Nautical officer failure	1– Checks	−2%	11%

Martins and Maturana (2010).

that it was a potential for decreasing the HEP for all events, but four events were especially promising by indicating more than 10% improvement:

- 6 – Pattern proceeding failure (COLREGs)
- 10 – Communication failure
- 23 – Officer gives wrong order
- 25 – Helmsman gives wrong response

The authors discuss briefly how these PSFs might be modified with Risk Control Options (RCOs). Their assessment is outlined qualitatively in Table 11.22. They made a distinction between RCOs related to *Management and Organizational Factors* (MOFs) regulated by the ISM Code and those related to shipboard routines. The MOFs are the responsibility of the company management whereas the shipboard routines primarily must be kept up to date by the master and senior officers.

Table 11.22 RCOs relevant for important collision events

Event		RCO
6 – Pattern proceeding failure (COLREGs)	MOF type	Formalization Personnel selection
10 – Communication failure		Workload Formalization Coordination of work Physical resources Training process
23 – Officer gives wrong order	Shipboard routines	Supervision Shipboard training Indoctrination of bridge routines Coaching by senior officers
25 – Helmsman gives wrong response		

Martins and Maturana (2010).

11.7 APPLICATION OF CREAM IN TRANSPORT

Similar to the other methods that are described, CREAM (Hollnagel, 1998) was also originally developed for use by the nuclear industry. The method has later been applied and further developed for other industrial applications and also for marine engineering and operation projects. Akyuz and Celik (2015) applied CREAM in the analysis of the cargo loading process of LPG tankers. The method was also modified and applied for analysis of marine traffic-related accidents (Wu et al., 2017). Studies of how the CREAM method may be enhanced with fuzzy and Bayesian networks have also been demonstrated (Ung, 2018; Zhou et al., 2018).

Zhou et al. (2017) studied the general operator reliability in tanker shipping with the basic CREAM model and supported it with subjective assessment of data by sea-going personnel. The investigation did not focus on a specific task but rather the general work climate onboard and the associated HEPs. An important aspect is that the approach is feasible even without a supporting database. The model uses the nine *Common Performance Conditions* (CPCs) as a starting point for characterizing the work environment on a tanker (Table 11.9). The authors found it necessary to modify the set of CPCs in a manner that better reflected the marine working conditions (Table 11.23).

More weight was placed on physical environment and work characteristics (employment contract), and less on HMI and operational support, procedures/plans, and time of day. For all CPCs, three levels were specified: negative (decrease), neutral, and positive (increase). A distinction was also made between direct and indirect effects. The idea is that CPCs like physical environment and work characteristics influence the cognitive performance

Table 11.23 CPCs, levels, and effect on human reliability in tanker shipping

No.	CPC	Description	CPC level	Effect	Effect mode
C1	Training and experience	The quality and effect of training, the knowledge, skills, and experience of the crew	Adequate high experience Adequate but limited Inadequate	Increase Neutral Decrease	Indirectly
C2	Physical environment	Physical factors of the work environment, such as lighting, noise, vibration, temperature and humidity, sea state, port, and channel condition	Advantageous Acceptable Incompatible	Increase Neutral Decrease	Directly
C3	Organizational management	The process of organizing, planning, leading, and controlling when doing operations during shipping. Effect of organization, supervision, and administration	Very efficient Acceptable Deficient	Increase Neutral Decrease	Indirectly
C4	Work characteristics	Work stress, task characteristics during shipping	Satisfied Acceptable Unsatisfied	Increase Neutral Decrease	Directly
C5	Available time	The time available to carry out a task and corresponds to how well the task execution is synchronized to the process dynamics	Adequate Temporarily inadequate Continuously inadequate	Increase Neutral Decrease	Directly
C6	Contract information	Such as contract duration, the working time onboard	Satisfied Acceptable Unsatisfied	Increase Neutral Decrease	Directly
C7	Vessel facilities	The defenses, design, hardware, housekeeping, and maintenance of the vessel	Satisfied Acceptable Unsatisfied	Increase Neutral Decrease	Directly
C8	Crew collaboration quality	The quality of the collaboration between crew members, including the level of trust, the communication, and the general social climate among crew members	Efficient Inefficient Deficient	Increase Neutral Decrease	Both directly and indirectly

Zhou et al. (2017).

directly, whereas factors like experience and organizational factors develop over time and thereby have an indirect effect.

Depending on the assessed level of the CPC the so-called CPC score is given by:

$$\lambda_i = \begin{cases} +1 & \text{Positive} \\ 0 & \text{Neutral} \\ -1 & \text{Negative} \end{cases} \quad i = 1, 2, \ldots, n \quad (11.8)$$

Unlike Hollnagel (1998), this study introduced a weighting function to differentiate the relative effect of the CPCs. The reliability index is therefore expressed as follows:

$$R = \sum w_i \cdot \lambda_i \quad i = 1, 2, \ldots, n \quad (11.9)$$

The relative weight of the CPCs can be determined by pairwise comparison or what the authors calls *fuzzy analytic hierarchy process* (FAHP). The comparison of factor a_i and a_j is expressed by the term r_{ij}. Factors are compared on a scale from 0 to 1 as defined in Table 11.24.

The resulting set of comparisons defines the following fuzzy judgment matrix:

$$B = \left(r_{ij}\right)_{n \times n} = \begin{bmatrix} r_{11} & r_{12} & \cdots & r_{1n} \\ r_{21} & r_{22} & \cdots & r_{2n} \\ \vdots & \vdots & & \vdots \\ r_{n1} & r_{n2} & \cdots & r_{nn} \end{bmatrix} \quad (11.10)$$

Table 11.24 Criteria for fuzzy judgment matrix

Value	Definition	Description
0.5	Equally important	Equally important compared with one another
0.6	Slightly important	One factor is slightly more important than the other
0.7	Obviously important	One factor is obviously more important than the other
0.8	Much more important	One factor is much more important than the other
0.9	Extremely important	One factor is extremely more important than the other
0.1,0.2,0.3,0.4	Converse comparison	If a_i is r_{ij} compared with a_j, then a_j is $r_{ji}=1-r_{ij}$ compared with a_i

Zhou et al. (2017).

By computing the normalized eigenvector of the matrix, the relative weights are given by averaging across the rows of the eigenvector giving the set of weights:

$$T = \left(w_1, w_2, ..., w_n\right) \tag{11.11}$$

Given the eight CPCs for the present tanker case the reliability index R was found to be from −8 (all decrease) to 8 (all increase).

Recalling Hollnagel's definition of HEP values for alternative control modes (Figure 11.4), following extreme values are assumed: PHEP, min$=5 \cdot 10^{-6}$ and PHEP, max$=1.0$. By assuming a Naperian logarithm, the following expression describes the HEP as a function of R:

$$HEP = 0.002236 \cdot e^{-0.7629R} \tag{11.12}$$

By assuming that the number of errors is Poisson distributed, we know that the mean time between errors is given by an exponential distribution:

$$T_R = \frac{1}{HEP} \tag{11.13}$$

The described approach was tested with participation of five Masters with comparable experience. Their first task was to rank the CPCs from their individual experience. An example is given in Table 11.25.

The next step was to perform the pairwise comparison of the relative weights (w_i) and based on the fuzzy judgment matrix compute the weights for the CPCs. The Masters then had to assess for each CPC whether it was satisfied, acceptable, or inadequate and thereby setting the effect λ_i (+1, 0, −1). The reliability index R is then computed by combining the estimates of

Table 11.25 Example of the scores for CPCs

CPC	Inadequate/unsatisfied					Adequate/satisfied				
	1	2	3	4	5	6	7	8	9	10
Training and experience							x			
Physical environment						x				
Organizational management		x								
Work characteristics			x							
Available time		x								
Contract information					x					
Vessel facilities				x						
Crew collaboration quality		x								

Zhou et al. (2017).

Table 11.26 Estimates of HEP from five seafarers

Seafarer No.	Reliability index	Human error probability (HEP)	Control mode according to Hollnagel	Mean time to error
1	6.22	$1.94 \cdot 10^{-5}$	Strategic	5.9 years
2	−6.22	$2.58 \cdot 10^{-1}$	Scrambled	3.9 hours
3	−1.69	$8.11 \cdot 10^{-3}$	Tactical	5.1 days
4	−0.81	$4.15 \cdot 10^{-3}$	Tactical	10 days
5	−3.00	$2.21 \cdot 10^{-2}$	Opportunistic	1.9 days

Zhou et al. (2017).

weights and effects and summing over the eight CPCs. The resulting assessments of human error probability are summarized in Table 11.26.

The immediate observation is that there are large differences in the results, with the HEP varying by nearly four orders of magnitude. The HEP estimates correspond to variation from the *Strategic* to *Scrambled* mode. The large gap in estimates between the highest and lowest probability may partly be explained by differences in experience and differences between vessels where the two mariners have worked. On the other hand, the general view is that there is less variation in tanker shipping safety standards today. In total, the very large differences are therefore hard to explain. The study does not offer any comparison with empirical data.

11.8 CALIBRATION OF AN HRA MODEL WITH ACCIDENT DATA

So far, we have discussed models where error probabilities largely were based on expert opinion or adjustment of estimates from other systems (mainly nuclear industry). Burns and Bonaceto (2020) have formulated this inadequate situation with a quite critical comment:

> *An important limitation is that numerical inputs to HRA models, including nominal human error probabilities and performance shaping factors, are based on the subjective judgments of analysts and are seldom benchmarked against objective measures of human performance. This state of the art undermines the credibility of HRA in PRA, and encourages the continued proliferation of methods, models, and assumptions that are not supported by empirical evidence.*

Using general aviation (except scheduled airline service) as a case, they propose an alternative approach by combining the SPAR-H model (Gertman et al., 2005) with accident data from The National Transportation Safety Board (NTSB). The authors correctly emphasize the important differences between the nuclear industry (SPAR-H) and general aviation. Airplanes are usually operated by one pilot whereas nuclear plants are operated by crews. Aircraft accidents are primarily the result of pilots failing to perform routine actions in response to external events while in nuclear plants it is mainly equipment malfunction the operators must deal with. The time perspective is also different: pilot actions must normally be performed in seconds or minutes whereas the plant crew typically will have many minutes to hours.

The approach was to develop event trees (ETs) for eight sequences of the flight:

-Taxi out	-Descent
-Takeoff	-Approach
-Climb	-Landing
-En route	-Taxi in

Secondly, 30 fault trees (FT) were developed to model the critical actions in the ETs. The FTs had 38 pilot actions together with 12 external events as basic events.

The modeling of the PSFs was based on a slight modification of the SPAR-H model. Rasmussen et al.'s (1981) SRK model for three different behaviors was applied (skill, rule, and knowledge based). After an elaborate discussion it was found that the HEPs of the behaviors were related as follows: $HEP_K \gg HEP_R > HEP_S$. This contrasted with the original SPAR-H model that only distinguishes between diagnosis and action.

The SPAR-H model has eight PSFs:

-Stress	-Procedures
-Complexity	-Work processes
-Available time	-Experience/Training
-Ergonomics	-Fitness for duty

The investigators found that only the four PSFs in the left column were relevant for general aviation as the remaining ones not are not recorded in the historic accident data. The multipliers for the retained PSFs were identical

Table 11.27 PSF multipliers

PSF	PSF multipliers			
Stress/stressors	Nominal (1)	High (2)	Extreme (5)	
Complexity	Nominal (1)	Moderate (2)	High (5)	
Available time	Nominal (1)	Barely adequate (10)		
Ergonomics/HMI	Good (0.5)	Nominal (1)	Poor (10)	Missing misleading (50)

Burns and Bonaceto (2020).

Table 11.28 Pilot actions and associated PSFs (examples)

Event tree	Pilot action	SRK	Stress	Comp.	Available time	Ergonomics
Taxi out	4. Fail to manage systems and maintain control	S	I	I	I	I
Climb	11. Fail to maintain proper climb profile and heading	R	I	I	I	I
En route	16. Fail to avoid hazardous weather en route	K	2	2	I	0.5
Descent	22. Fail to avoid collision with hazard	K	5	I	10	I

Burns and Bonaceto (2020).

with the SPAR-H model except for some minor adjustments (Table 11.27). Using expert judgment, the PSF multipliers were set for the 38 pilot actions. Some examples are shown in Table 11.28.

The external event HEPs were estimated from the NTSB accident database and were related to collision hazard present (0.00001–0.1), departure/arrival airport controlled (0.40), and hazardous weather en route and at destination (0.05). The 30 fault trees gave 14 cut sets that gave a set of 14 equations that were the basis for estimating the control mode probabilities (HEP_K, HEP_R, HEP_S). The estimation process applied data on five-year average numbers from the accident database and a yearly traffic of $2.28 \cdot 10^7$ flights. The control mode probabilities were estimated to be:

$$HEP_K = 3.0 \cdot 10^{-4}$$

$$HEP_R = 1.5 \cdot 10^{-6}$$

$$HEP_S = 2.5 \cdot 10^{-7}$$

These estimates were less than the SPAR-H values with an order of magnitude and showed that application of data from the nuclear industry was not realistic in general aviation.

By applying these HEP values, the probability of annual number of accidents was estimated for the 14 accident scenarios ("Bins") as shown in Table 11.29 and Figure 11.7. The method gave good agreement with the data. The authors discuss the model in greater detail but it is evident that even by limiting the model to three HEP values and only four CPCs it shows

Table 11.29 Annual number of general aviation accidents in the USA

Bin	Type	Phase	Notes	Data	Model
1	K	Taxi	Collision with hazard	13	21
2	K	Takeoff	Collision with hazard	7	6
3	K	Takeoff	Insufficient runway	28	41
4	K	En route	Weather related	41	41
5	K	Approach	Collision with hazard	14	12
6	K	Approach	Failed approach/Go-around	18	12
7	K	Landing	Collision with hazard	7	6
8	K	Landing	Insufficient runway	40	41
9	R	Takeoff	Loss of control/Wake turbulence	134	137
10	R	En route	Fuel management	80	68
11	R	Approach	Loss of control/Wake turbulence	120	137
12	S	Taxi	Loss of control	13	11
13	S	Climb	Loss of control	1	6
14	S	Landing	Loss of control	218	228

Burns and Bonaceto (2020).

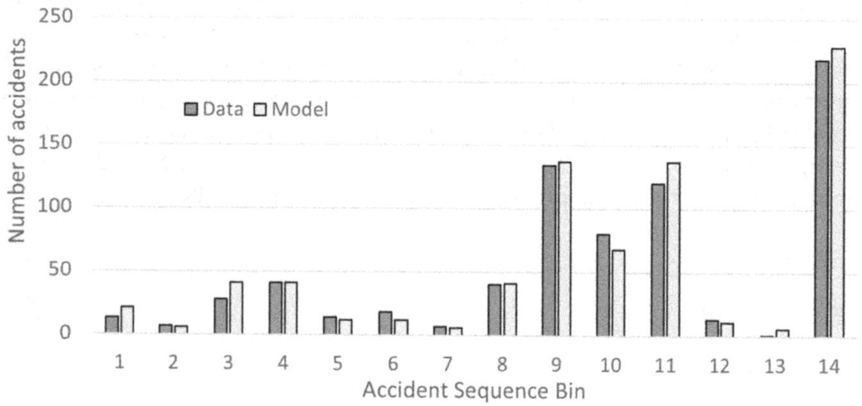

Figure 11.7 Number of accidents for data and model (Burns & Bonaceto, 2020).

promising results compared to the traditional approach of modifying data from the nuclear industry.

REFERENCES

Akyuz, E., & Celik, M. (2015). Application of CREAM human reliability model to cargo loading process of LPG tankers. *Journal of Loss Prevention in the Process Industries, 34*, 39–48.

Boring, R., & Gertman, D. (2016). *P-203: Human reliability analysis (HRA) training course.* US Nuclear Regulatory Commission. https://inldigitallibrary.inl. gov/sites/sti/sti/Sort_1048.pdf

Burns, K., & Bonaceto, C. (2020). An empirically benchmarked human reliability analysis of general aviation. *Reliability Engineering and Systems Safety, 194.*106227. https://scholar.google.com/citations?view_op=view_citation&hl =en&user=DyhtQV0AAAAJ&citation_for_view=DyhtQV0AAAAJ:9yKSN- GCB0IC

Chandler, F. T., Chang, Y. H., Mosleh, A., Marble, J. L., Boring, R. L., & Gertman, D. I. (2006). *Human reliability methods – Selection guidance for NASA.* National Aeronautics and Space Administration. https://www.sintef. no/globalassets/project/hfc/documents/nasa_hra_report.pdf

Embrey, D. E. (1986). *SHERPA: A systematic human error reduction and prediction approach.* International Meeting on Advances in Nuclear Power Systems, Knoxville, TN.

Embrey, D. E. (1993). Quantitative and qualitative prediction of human error in safety assessments. In (Anonymous) *Institution of chemical engineers symposium series* (Vol. 130, pp. 329–329).

Gertman, D., Blackman, H., Marble, J., Byers, J., & Smith, C. (2005). The SPAR-H human reliability analysis method. *US Nuclear Regulatory Commission, 230*(4), 35.

Hollnagel, E. (1998). *Cognitive Reliability and Error Analysis Method (CREAM).* Amsterdam: Elsevier.

Martins, M. R., & Maturana, M. C. (2010). Human error contribution in collision and grounding of oil tankers. *Risk Analysis: An International Journal, 30*(4), 674–698.

Øien, K., & R. Rosness, (1998), Methods for safety analysis in railway systems. Report no. STF38 A98426. Sintef Industrial Management – Safety and Reliability. Trondheim, Norway.

O'Hare, D., Wiggins, M., Batt, R., & Morrison, D. (1994). Cognitive failure analysis for aircraft accident investigation. *Ergonomics, 37*(11), 1855–1869.

Rasmussen, J., Pedersen, O. M., Mancini, G., Carnino, A., Griffon, M., & Gagnolet, P. (1981). Classification system for reporting events involving human malfunctions. *Report EUR.*

Rausand, M., & Haugen, S. (2020). *Risk assessment: Theory, methods and applications* (2nd ed.). Wiley.

Reason, J. T. (1990). *Human error.* Cambridge, UK: Cambridge University Press.

Sgobba, T. (2017). *Space safety and human performance.* Amsterdam: Butterworth-Heinemann.

Swain, A. D., & Guttman, H. E. (1983). *Handbook of human reliability analysis with emphasis on nuclear power plant applications. Final report (No. NUREG/CR--1278).* Sandia National Labs.

Ung, S. T. (2018). Human error assessment of oil tanker grounding. *Safety Science, 104*, 16–28.

Wickens, C. D., & Flach, J. M. (1988). Information processing. In E. L. Wiener & D. C. Nagel (Eds.), *Human factors in aviation* (pp. 111–155). Cambridge, Massachusetts: Academic Press.

Williams, J. C. (1986). HEART—A proposed method for assessing and reducing human error in ninth advances in reliability technology symposium. NEC, Birmingham, June, AEA, Technology, Culcheth, Warrington.

Wu, B., Yan, X., Wang, Y., & Soares, C. G. (2017). An evidential reasoning-based CREAM to human reliability analysis in maritime accident process. *Risk Analysis, 37*(10), 1936–1957.

Yang, K., Tao, L., & Bang, J. (2013). Assessment of flight crew errors based on THERP. In *3rd International Symposium on aircraft airworthiness*. (ISAA2013). Toulouse, France. https://trid.trb.org/view/1419288

Zhou, Q., Wong, Y. D., Loh, H. S., & Yuen, K. F. (2018). A fuzzy and Bayesian network CREAM model for human reliability analysis–The case of tanker shipping. *Safety Science, 105*, 149–157.

Zhou, Q., Wong, Y. D., Xu, H., Van Thai, V., Loh, H. S., & Yuen, K. F. (2017). An enhanced CREAM with stakeholder-graded protocols for tanker shipping safety application. *Safety Science, 95*, 140–147.

Chapter 12

Formal safety assessment

12.1 INTRODUCTION

The use of qualitative methods for risk assessment has a long history within the maritime industry. Quantitative studies were used only for very specific purposes, often as a result of requirements from other industries using ships for transportation and other purposes, e.g., the offshore oil and gas industry. However, since 1997, when IMO introduced Formal Safety Assessment, the use of quantitative risk assessment has increased.

The use of quantitative methods has opened for an opportunity to compare different concepts and safety measures on a common scale. In Chapter 8, several techniques, both quantitative and qualitative, have been described for analyses of event probabilities and consequences. In addition to these techniques, the cost-benefit analysis (CBA) described in Chapter 15 makes it possible to compare and implement the results of the risk analysis in terms of costs. Hence, by applying a sequence of all these methods, decisions can be made about which concepts to choose and which safety measures to implement based on a simple assessment of the costs involved. The validity of comparing different concepts or safety measures on a common scale is, however, not only related to the scale itself but also to the analysis process. Different approaches to the problem and different assessments of details may contribute to different results. As a result of this situation, a need for standardization and generalization of the analysis approach/process was brought to the surface. One proposed standard analysis approach that has gained wide recognition is the so-called Formal Safety Assessment (FSA) approach.

Formal Safety Assessment (FSA) is a designation used in a number of different contexts and industries (e.g., the nuclear industry) in order to describe a rational and systematic risk-based approach for safety assessment. In the maritime world the expression Formal Safety Assessment (FSA) is being used by the International Maritime Organization (IMO) and its members to describe an important part of the rule-making process for international shipping. It is this maritime type of FSA that will be described in this chapter.

DOI: 10.4324/9781003055464-12

FSA is sometimes also referred to as the safety case concept. Safety cases were first introduced in the UK and are among others used in the offshore oil and gas industry. A safety case in that context will however have a broader scope than an FSA, covering not only risk assessment but also the safety management system in place for the facility and descriptions of all safety-related features of the facility. See, e.g., Wang (2002) for a comparison of the two approaches.

The process described in the IMO guidelines can at times be slightly confusing, in particular due to the use of terminology. It has also been observed that FSAs that have been performed often deviate from the process described (Psaraftis, 2012). In this chapter, an attempt has been made at describing the process in a way that should give good results and that attempt to describe the intention behind FSA, even if it is not necessarily true to the process as described by IMO in all respects. For those who want to go deeper into this, it may be useful to read the FSA guidance (IMO, 2018) in parallel with reading this chapter.

12.1.1 Background to FSA

Large-scale maritime accidents, in particular the accidents with Herald of Free Enterprise (1987) and Exxon Valdez (1989), prompted a re-evaluation of the current prescriptive (i.e., rule-based) regime for marine safety. The regime was regarded as unfavorable compared to the safety regimes used in other industries, which were based on more scientific methods such as risk analyses and cost-benefit analyses. Especially within the UK, work was carried out to establish a more rational approach to rule development. In 1993, the UK Marine Safety Agency (currently called the Maritime and Coastguard Agency) proposed a five-step procedure for safety analysis named Formal Safety Assessment (FSA) to IMO. The main purpose of the FSA methodology was to provide a more systematic and proactive basis for the IMO rule-making process.

In 1997, IMO adopted "Interim Guidelines for the Application of Formal Safety Assessment (FSA) to the IMO Rule-Making Process". The use of the FSA methodology on helicopter landing areas on cruise/passenger ships in 1997 was influential in IMO's decision to abandon this risk reduction measure because the Implied Cost of Averting a Fatality (ICAF) was found to be far from cost-effective. The FSA guidelines were formally implemented in 2001. They have been revised several times and the most recent version was approved by IMO in 2018 (IMO, 2018).

12.1.2 Intended use of FSA

FSA is a tool designated to assist maritime regulators in the process of improving and deriving new rules and regulations. The main intention behind the development of the FSA methodology for maritime activities was that it should be used in a generic way for shipping in general. The methodology was initially derived with two potential users in mind:

- IMO committees: The application of the FSA methodology can provide helpful inputs into the review process of existing regulations and into the evaluation process of proposed new regulations.
- Individual maritime administrations: Application of the FSA methodology can be used in the process of evaluating/assessing proposed amendments to IMO regulations. It can also be used in order to evaluate whether additional regulations, that exceed the IMO requirements, should be introduced.

It is anticipated that FSA should be relied upon where proposals may have far-reaching implications in terms of safety, cost, and legislative burden. The application of FSA will enable the benefits of proposed changes to be properly established and to therefore give the decision-makers a clearer perception of the scope of the proposals and an improved basis on which to take decisions.

The classification societies also recognize that the FSA methodology can be used in the process of improving and developing classification rules. DNV GL has, e.g., included a statement in their rules for classification of ship (DNV GL, 2018) where they confirm that they may consider risk-based assessments as a means of documenting compliance with the rules (Section 2.5.8 in the rules) and that the FSA process may be used for that purpose. Applying FSA for developing classification rules should in principle not be fundamentally different from using the methodology for development of the IMO rules.

It is important to stress that the intention with the Formal Safety Assessment process from IMO was not to serve as a method for risk assessment of individual ships. As we shall see, the process is however largely a generic risk assessment process, but with certain specifics that are relevant only for the rule-making process. The most important example is the use of what is called a "generic ship" (Section 7.1.4). A generic ship is supposed to be representative for the whole group of ships that will be influenced by the proposed regulation being considered in the FSA. In a risk assessment for a specific ship, this is of course not relevant. Instead, the specific particulars of that ship will form the basis for the risk assessment.

12.2 FSA APPROACH

FSA is a standardized, holistic method for risk assessment. The method involves several of the elements that are found in any risk assessment process. The process is divided into five steps, as illustrated in Figure 12.1.

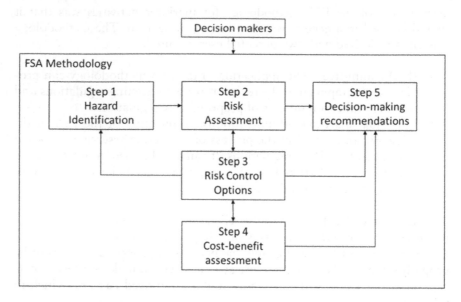

Figure 12.1 FSA method (IMO, 2018).

The figure illustrates that FSA is not a linear process, where each step is completed before being followed by the next step. Typically, there will be a number of iterations, e.g., as illustrated between steps 2 and 3. When the risk assessment has been completed, risk control options for reducing risk are identified and subsequently, the risk assessment must be updated to etermine the effect of the risk control options.

A couple of comments on the terminology used in the figure are relevant to make:

– Step 1 Hazard Identification in practice not just covers identification of hazards, but also ranking and screening, using a method similar to Preliminary Hazard Analysis (Chapter 8).
– From Chapter 3, we remember that Risk Assessment is the whole process of planning, preparing, and performing risk analysis, as well as comparing with risk acceptance criteria. Step 2 as described in the FSA guideline is primarily risk analysis but including comparison with risk acceptance criteria. Strictly speaking, it would therefore be more precise to call it Risk Analysis rather than Risk Assessment.

In Figure 12.2, a more detailed description of the five steps is provided. In this figure, the feedback loops have been removed, to simplify the illustration.

Each of the five steps is described in more detail in the following sections in this chapter. The FSA methodology is quite complex because it involves the use of a wide range of different techniques, some of which are described in

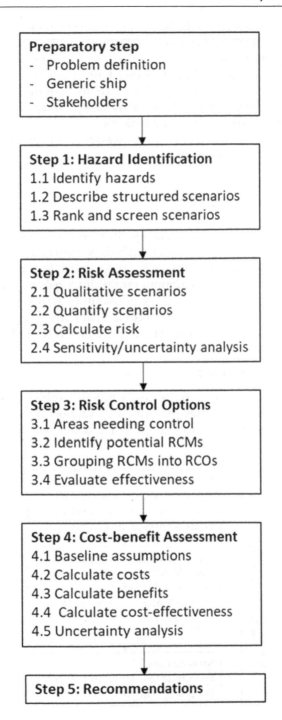

Figure 12.2 Detailed breakdown of FSA process.

earlier chapters. As a result, only the most important aspects are described and discussed here, and where appropriate, references to other chapters are made.

The description in this chapter is based on the approach suggested by IMO, for use in the rule-making process. This means that some steps may be performed differently and that other methods than those described here may be just as relevant to use if the methodology is applied for a specific ship or for other purposes than rule-making.

Before the five steps in the process are started, some preparatory work is also required, and this is described first.

12.3 PREPARATORY STEP

As for any risk assessment (see Chapter 8), there are some initial activities that need to be done before the main analysis is started. This includes defining the problem at hand, determining the scope, planning and organizing the work, providing a system description, establishing risk acceptance criteria, etc. In the FSA process, there are three preparatory activities that are worth mentioning specifically: Defining the problem, describing a generic ship, and identifying the stakeholders. The activities are illustrated in Figure 12.3.

12.3.1 Problem definition

Of particular importance is to prepare a precise problem definition. Risk assessments are in most cases performed to support decision-making. In the case of FSA as described by IMO, the decision is whether to implement

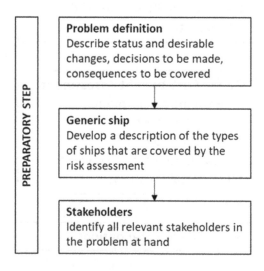

Figure 12.3 Sub-steps in the preparatory step of FSAs.

changes to the rules or not. To decide on this, we need to describe precisely what the status is and what the proposed changes are. The same will apply for any other type of decision that we may want to use FSA for. If the objective is to determine whether the risk level for a specific ship is acceptable or not, acceptance criteria need to be defined in advance and the decision that we want to make is whether risk is acceptable or not. In other situations, it may be that two alternative designs are available, and we want to decide which one has the lowest risk.

The decision that we want to make will determine what answers we need to get from the risk assessment and this in turn will influence how the risk assessment should be planned and performed. If the objective is to determine whether the risk is acceptable or not, all aspects of risk for the ship need to be covered. If the objective is to compare alternatives, it may be sufficient to consider only those aspects which are different between the alternatives and determine the differences in risk.

Part of the problem definition will also be to determine what types of consequences should be considered in the risk assessment. In some cases, the focus may be only on risk to people. An example may be if a proposal for strengthening the regulations regarding evacuation from ships is put forward. This is a proposal that only will impact on the risk to people, and it is natural to limit the risk assessment to only this. On the other hand, a proposal to reduce the probability that explosions will occur on oil tankers may have impact on both personnel and the environment. It may then be relevant to cover both risk to personnel and the environment in the risk assessment.

12.3.2 The generic ship

One of the first steps that we have to go through as part of the preparations for performing any risk assessment is to establish a description of the system that will be covered by the risk assessment. In the case of an FSA process, this is what is called the generic ship.

With the premise that FSA is a tool to support development of regulations, it also follows that an FSA normally will have to consider a large group of vessels. Regulations will normally be made for a type of vessel, e.g., cruise ships, or even for ships in general, e.g., COLREG. The generic ship that is defined should be representative for all the ships in the affected group.

In the FSA guidelines, it is stated that "a generic model should therefore be defined to describe the functions, features, characteristics and attributes which are common to all ships or areas relevant to the problem in question". It is also underlined that the description should not just cover technical and engineering features but should recognize that the ship is a part of a larger, integrated system comprising the environmental context, organizational/management infrastructure, and personnel in addition to the technical/engineering system. This is a strong indication of an expectation that human, organizational, and environmental factors are taken into account in the analysis.

To summarize, the generic ship description should address:

- Environmental factors
- Organizational context
- Human factors
- Main particulars of ship
- Types of cargo
- Key functions
- Operations
- Technical systems

Table 12.1 provides some examples of key systems and functions that may be relevant to include.

A transport operation can be characterized as a sequence of distinct phases or cycles where each phase requires the use of different systems and functions. Hence, the need for the various functions given in Table 12.1 will vary according to a specific ship's operational phase. The operational phases of a generic ship can be defined as follows:

- Entering and leaving port
- Berthing and unberthing
- Loading and unloading of cargo
- Passage/transit
- In repair yard/dry dock

Table 12.1 Systems and functions of a generic ship

Systems and functions	
Accommodation and hotel service	Anchoring
Communications	Carriage of payload
Control	Communications
Electrical	Emergency response and control
Crew	Habitable environment
Lifting	Maneuverability
Machinery	Mooring
Management systems	Navigation
Navigation	Pollution prevention
Piping and pumping	Power and propulsion
Pressure plant	Bunkering as storing
Ballast	Stability
Safety systems	Structure

Each element of the generic ship has more detailed descriptions and sub-categories which must be applied when utilizing the FSA methodology, but such descriptions go beyond the scope of this book.

Clearly, the larger the group of vessels is, the more difficult it is to describe a generic ship that is representative for all these. By necessity, the generic ships are therefore often not very specific with regard to detailed design features, operational patterns, cargo, etc. This is however a fine balance, because with few details about the ship, it is also more difficult to do a detailed risk assessment. By nature, FSA will therefore often not go into a lot of detail on, e.g., the specific systems on board. However, it is important that the description is sufficiently detailed in those aspects which are affected by the proposed regulations. If this is not adequate, the comparison of risk with and without the regulations implemented will have limited meaning.

The first efforts to establish a generic ship model were aimed at creating one generic ship model that should account for nearly all (merchant) ships. Later, the development did however move more in the direction of different generic models for different ship types, for example, different generic ship models for oil tankers, chemical tankers, gas tankers, etc. No matter what method is used, it is of crucial importance that the generic ship model applied in the FSA methodology is applicable to the problem examined.

Generic ship

In the SAFEDOR project (see, e.g., Breinholt et al., 2012), a number of FSA example studies were conducted, and the following is a brief summary of some elements of the description of a generic ship from one of these studies (IMO, 2007).

The risk assessment addresses container ships (UCC) of 100 GT and above that carry only containers (ships carrying combined cargo are not considered). A container ship is defined as a sea-going vessel specifically designed, constructed, and equipped with the appropriate facilities for carriage of cargo containers. These containers are stowed in cargo spaces, i.e., in cargo holds below or above deck.

A basic assumption is that a container ship is built according to technical regulations and rules of a recognized classification society.

For container vessels, there are two main operational patterns. Liner operation typically involves ships with large transportation capacities. They sail on a fixed route with a limited number of ports according to a schedule with fixed arrival and departure times. These schedules enable long-term planning for transport of large quantities. Their operating profile includes fewer stays in port and more open sea voyage. Loading and unloading require significant time. Major line trades are Europe–North America and Europe–East Asia.

Feeder operation typically involves much smaller ships on short distances, e.g., along coastlines. They are characterized by frequent

Table 12.2 Generic ship descriptions

Vessel type	Feeder	Liner
Capacity (TEU)	1,706	4,444
Length (m)	173	271
DWT	21,750	58,255
Speed (km)	20	25
Crew	20	20
Market price (2005)	USD 36 million	USD 67 million

port calls. Their routes, cargo, and departure times are dominated by short-term demands. Additionally, they are required for areas with limitations in draught or breadth.

Based on statistics, the two segments are equally important. Large liner vessels represent 60% of the total transport capacity, while nearly 55% of the ships are small feeder vessels.

Two vessels (one feeder and one liner) were selected as generic ships. Mostly, average values are used in the calculations, with some exceptions. The two ships are described in Table 12.2.

It may be noted that this description contains very little information about environmental, human, and organizational factors. There is also limited information on technical systems, although the description in the report provides some more information than is included here.

12.3.3 Stakeholders

In the FSA guideline, the terms "interested entities" and "interested parties" are being used to describe stakeholders.

One problem when applying cost-benefit analysis (CBA, step 4 in the FSA methodology) in risk assessment is the number of stakeholders (or parties) involved with regard to the vessel and their different roles. A stakeholder may be defined as (ISO 31000) a "person or organization that can affect, be affected by, or perceive themselves to be affected by a decision or activity".

In relation to the FSA process, it is primarily risks and costs that are relevant to consider. In many cases the stakeholder that imposes a risk is not necessarily the same stakeholder that carries these risks. An example is risk related to oil spills. One possible measure is to introduce a double hull. The cost associated with this will primarily be carried by the shipowner, due to increased construction costs. The risk reduction will also benefit the shipowner, although it is more likely that stakeholders like insurance companies, fisheries in the areas where a release may occur, and other users of the environmental resources will benefit more from the reduced risk.

This means that each group of stakeholders will have different benefits and costs associated with a risk reduction measure and this will obviously

Table 12.3 Examples of stakeholders and their costs, benefits, and risks

Stakeholder	Incurs costs	Receives benefits	Imposes risks	Carries risks
Owner/ charterer	Cost of vessel	Income	Choice of vessel specifications	Loss of vessel
Cargo owner	Pays for passage	Profit from trade	Dangerous cargoes	Loss of cargo
Operator	Running costs	Income	Operating practice	Loss of income
Crew	–	Employment	Lack of due diligence	Loss of life/property
Passenger	Fares	Transport, leisure	–	Loss of life/property
Flag state	Administration costs, employment	Fees	Inadequate local legislation	–
Port of call	Cost of infrastructure, operating costs	Fees	Navigational control, dredging levels	Damage to infrastructure, loss of trade
Coast state	Local navigation	–	Inadequate navigation aids	Pollution and clean-up
Insurer	–	Premiums	–	Claims
Other vessels	–	–	Impact, loss of life	Impact, loss of life
Classification societies	Operating costs	Fees	Lack of due diligence	Negligence claims
Designer/ constructor	Materials/labor	Fees	Lack of due diligence	–

also influence their view on whether a measure should be introduced or not. This is clearly a challenge for decision-making, although the cost-benefit analysis normally will consider the total costs and benefits without splitting these on different stakeholders. It may also be noted that the so-called "risk imposer pays principle" (or policy) implies that those stakeholders who, voluntarily or not, impose risk on others should pay for that privilege.

Table 12.3 indicate which stakeholders incur which costs, receives which benefits, and imposes and carries which risks.

12.4 STEP I: HAZARD IDENTIFICATION

Hazard identification is the first step of the FSA methodology. The main objective of this step as described by IMO (2018) "is to identify a list of hazards and associated scenarios prioritized by risk level". Further, hazard is defined as "a potential to threaten human life, health, property or the environment" and accident scenario is defined as "a sequence of events

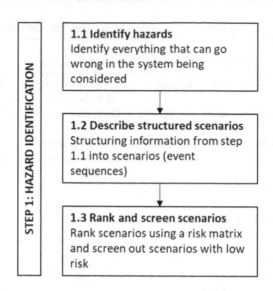

STEP 1: HAZARD IDENTIFICATION

1.1 Identify hazards
Identify everything that can go wrong in the system being considered

1.2 Describe structured scenarios
Structuring information from step 1.1 into scenarios (event sequences)

1.3 Rank and screen scenarios
Rank scenarios using a risk matrix and screen out scenarios with low risk

Figure 12.4 Step 1 of the FSA methodology.

from the initiating event to one of the final stages". Step 1 may be divided into three sub-steps as shown in Figure 12.4.

12.4.1 Step 1.1: Identify hazards

The first step is aimed at identifying as much as possible of "what can go wrong" (refer the definition of risk in Chapter 3). The process will typically be a combination of an analytical approach and a creative approach.

The analytical approach is based on identifying and collecting relevant sources of information about accidents, incidents, failures, and errors that have occurred earlier, either in the system that is being considered (e.g., the same type of ships) or in similar systems that may be of relevance (e.g., other types of ships, other systems operating in marine environment, or other types of transport systems). Examples of sources of information that can be consulted include industry statistics and other industry reports, accident and investigation reports, internal accident and incident statistics and reports, maintenance systems, generic sources about system failures, component failures, human errors, etc.

Often, this information is summarized into various forms of lists that contain generic descriptions of the information that can be found from the sources. For example, rather than including "failure of fuel pumps", "failure of ballast pumps", and "failure of cargo pumps", a generic failure type called "failure of pump" is included. This can in turn be applied to the different systems, to determine if this is relevant for each of the systems being considered.

Table 12.4 Example breakdown of bulk carrier and operation

Functional elements	Operations
Bridge and communication equipment	Transit
Accommodation – sleeping quarters	Approaching coast/port
Utility rooms in accommodation	Taking pilot onboard
Engine room	Berthing
Fuel tanks	Loading
Various utility tanks	Unloading
Various utility rooms in hull	Unberthing
Cargo holds	Leaving port
Deck area	Pilot departing
Lifeboats/rafts	Docking

Various generic lists that can support the process exist and Annex 2 in the FSA guidelines provide an example (IMO, 2018). What these lists often have in common is that they contain a mix of hazards, initiating events, accidents, consequences, etc.

In most cases, it will be beneficial to divide the ship into functional elements before starting the identification process. This can be, e.g., bridge, accommodation, engine room, and cargo holds. It may also be beneficial to consider different operations separately, e.g., approaching port, berthing, unloading, and transit. The advantage of this is that the hazard identification becomes more focused, and it is easier to identify all relevant hazards. The disadvantage may be that it takes longer to complete the process.

A bulk carrier may be subdivided into elements and operations as illustrated in Table 12.4.

The creative element in the process is usually conducted as a structured workshop involving various experts within the field. This is similar to what has been described for several of the hazard identification methods covered earlier in the book (Chapter 8). Typical examples of participants are as follows:

- An expert on hazard identification will lead the workshop.
- Usually, the workshop leader is accompanied by a secretary that records the discussions and the findings in a suitable format.
- Various experts on the technical systems, operations, human and organizational factors, etc. will provide the domain knowledge required in the process.

12.4.2 Step 1.2: Describe structured scenarios

As mentioned, the generic lists that are used to support the hazard identification process are often not structured in a way that makes the outcomes of the process easy to use directly in the risk assessment process that follows step 2. One reason for this is that what is identified often is a mix of causes, events, consequences, and hazards. It is therefore usually necessary to structure the information and establish accident scenarios that can be analyzed more easily.

An example of this can be that the hazard identification may identify the following as "hazards" that need to be considered in the risk assessment:

- Fuel oil
- Hot surfaces that can ignite flammable material
- Fire
- Explosion
- Damage to engines due to explosion
- Etc.

If these are considered systematically, we see that these "hazards" can be put together into an accident sequence, starting with release of fuel oil, followed by ignition, followed by a fire and/or explosion and that the explosion (if it takes place in the engine room) can damage the engines.

Step 1.2 is therefore about structuring the information from the hazard identification workshop into a set of event sequences that can be further analyzed. The sequences are also grouped into a set of generic accident types that form the basis for the quantitative analysis.

12.4.3 Step 1.3: Rank and screen scenarios

The last stage of the hazard identification comprises ranking and screening of the scenarios that have been identified. This has two objectives:

- The ranking implies that the scenarios that contribute most to risk are identified.
- The screening means that those scenarios that are considered to contribute very little to the total risk are not taken further into Step 2 of the process. This is done by excluding scenarios that are ranked lowest in risk.

The screening saves time in performing Step 2 (which is normally far more time-consuming than Step 1), without running the risk of not analyzing important risk contributors in sufficient detail.

The ranking can be done qualitatively, but a more systematic approach is to use a risk matrix (Chapter 9), meaning that we have to determine a rank for probability/frequency and consequence of each of the identified scenarios. Of course, what we are doing then is in principle a risk analysis

Table 12.5 Proposed severity classifications

SI	Severity	Description	Equivalent fatalities	Effects on ship and equipment
		Effects on crew (Injury/fatality)		
1	Minor	Single or minor injuries	0.01	Local equipment damage
2	Significant	Multiple or severe injuries	0.1	Non-severe ship damage
3	Severe	Single fatality or multiple severe injuries	1	Severe damage
4	Catastrophic	Multiple fatalities	10	Total loss

IMO (2018).

Table 12.6 Proposed frequency classifications

FI	Frequency	Definition	F (per ship year)
7	Frequent	Likely to occur one per month on one ship	10
5	Reasonably probable	Likely to occur once per year in a fleet of 10 ships, i.e., likely to occur a few times during the ship's life	0.1
3	Remote	Likely to occur once per year in a fleet of 1,000 ships, i.e., likely to occur in the total life of 50 similar ships	10^{-3}
1	Extremely remote	Likely to occur in the lifetime (20 years) of a world fleet of 5,000 ships	10^{-5}

IMO (2018).

since we are answering the three questions used to define risk (Chapter 3). Principally, it may therefore be argued that this should be part of Step 2 and not included in Step 1 Hazard Identification. In practice, this is however not important since Step 1 and Step 2 in any case are just (somewhat random) stages in a continuous process.

In the FSA guidelines, IMO has also suggested categories that can be used in the risk matrix. These are shown in Tables 12.5 and 12.6.

The guidelines also note that if other types of consequences are considered, e.g., environmental consequences, other definitions of severity index will have to be developed and applied. They also describe an index for oil spills, where SI=1 corresponds to a spill less than 1 ton, SI=2 corresponds to a spill between 1 and 10 tons, and so on. The maximum category is SI=6, with spill size above 10,000 tons.

The severity and frequency indexes can be regarded as logarithms since the definitions imply that the risk will increase by a factor of 10 when we increase the index value by 1. For the consequence, we see, e.g., that SI 4 corresponds to a consequence that is ten times higher than SI 3 (1 equivalent fatalities vs 10 fatalities). For frequency, FI 7 is 100 times higher than FI

5. It follows from the definitions of FI that we also could have defined FI 6 equal to a frequency of 1 per ship year, FI 4 equal to a frequency of 0.01 (or 10^{-2}), and so on.

In the tables, four index "levels" are defined for both frequency and severity, but if we want to, it is easy to define seven levels for frequency instead (since the index runs from 1 to 7).

We recall that risk was defined as frequency (or probability) times consequence: $R = f \cdot C$. This also means that $\log(R) = \log(f) + \log(C)$. As we have shown, FI is the same as $\log(f)$ and SI is the same as $\log(C)$. Therefore, we can calculate a risk index as follows:

$$RI = \log(R) = \log(f) + \log(C) = FI + SI \qquad (12.1)$$

The ranking of the scenarios can then be done based on RI.

It has been stated that this sub-step includes ranking and screening. This also means that we have to make a decision on what to leave out of the further analysis and what to include. With the ranking as a basis, this is in practice a question of what cut-off limit for RI we use for screening out scenarios. There is no rule or guidance for how to do this and it will depend on the discretion of the analyst. Some considerations to make:

- An initial screening can be done purely based on qualitative considerations; do we consider that a scenario should be analyzed more in detail? Of those that we consider should be included, the lowest RI can act as a cut-off limit and all scenarios with RI on this level or above should be included.
- The RIs for all scenarios that have been selected for further analysis can be added together and the RIs for scenarios that have been screened out are also added together. The "80-20 rule" can be applied also in this case: If the total RI for the included scenarios is more than 80% of the total RI for all scenarios, this is an indication that the majority of the risk contributors are included.
- This can be supplemented with evaluation of the total number of scenarios that are proposed and analyzed in more detail. If the total number becomes too high, this is an indication that the amount of work in Step 2 will be high.
- Finally, discretion can of course also be used to include scenarios that we are concerned about, even if the ranking shows that the risk is likely to be low.

The outcome of this step is eventually a list of scenarios with associated descriptions, ranked according to risk index.

A final comment is that Step 1 in total follows the method that was described as a Preliminary Hazard Analysis in Chapter 8.

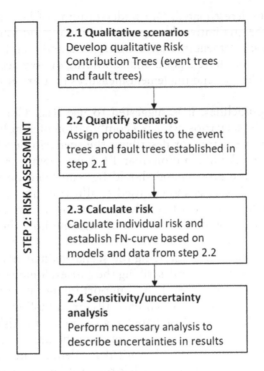

Figure 12.5 Sub-steps in Step 2 risk assessment.

12.5 STEP 2: RISK ASSESSMENT

According to the IMO guidelines (IMO, 2018), the purpose of Step 2 is to do "a detailed investigation of the causes and initiating events and consequences of the more important accident scenarios identified in step 1". Step 2 may be subdivided as shown in Figure 12.5.

The qualitative descriptions will typically use some form of tree structures (fault trees, Bayesian networks, event trees, influence diagrams, etc.) to show sequences and logical relations between causes, events, and consequences. The quantification is done by assigning probabilities, frequencies, and consequence descriptions to the tree structures, and this in turn is the basis for calculating risk. Finally, sensitivity and/or uncertainty analysis should also be performed, to provide the decision-makers with a better understanding of the risk.

12.5.1 Step 2.1 Qualitative scenario descriptions

Focus (and criticism) of FSA is often directed at the quantitative aspects of the analysis. However, the importance of the qualitative descriptions should not be underestimated. Our understanding of how accidents occur is reflected in the qualitative descriptions. Doing a good job at this stage is therefore a crucial element in understanding risk and thereby also in reducing risk. If we do not have a good understanding of how accidents occur, it is also hard to identify and implement measures that are effective in reducing risk.

In the FSA guidelines, it is suggested to use Risk Contribution Trees (RCT) to describe scenarios. RCT is a combination of a fault tree for causal analysis and an event tree for describing the scenario development to the end consequences. An illustration of an RCT is shown in Figure 12.6. The figure illustrates the process and should be read from bottom to top. At the bottom, a fault tree is shown, used to illustrate the causes of event i and in the middle is an event tree, illustrating how the event may develop. At the top is also shown an FN curve, illustrating how the results can be presented.

Alternative methods for describing the scenarios may also be used, e.g., using a Bayesian Network for describing the causes. Simpler methods are in practice not suitable since they do not support quantitative analysis.

IMO prefers risk profiles, which basically are simplified fault trees, for the qualitative risk analysis. Compared to a fault tree, the risk profile is simpler because there are no logical gates between the underlying causes. Risk profiles are deduced mainly from historical accidental outcomes rather than from underlying causes/failures, which is the case for fault trees. Figure 12.7 shows the risk profile for the accidental outcome of a collision. It is recommended to apply the common fault tree construction technique instead of risk profiles if a detailed analysis is performed/required. The risk profiles may also act as input to a fault tree analysis.

12.5.2 Step 2.2 Quantify scenarios

In this step, frequencies, probabilities, and quantitative expressions of consequences are provided as input to the RCTs. A variety of data may be required but the most important are:

- Frequencies of various events that are part of the RCTs. This can be various external events that are relevant to include occurrence of certain conditions, etc. An example can be occurrence of severe storms per year.
- Probabilities of failure/errors or the occurrence of certain conditions in a given situation. Examples are probability of failure of various equipment, human error probabilities (HEP), probability of low visibility when passing a narrow fairway, etc.
- Consequences of events, e.g., the number of fatalities given a specific event occurring or the size of a spill of oil.

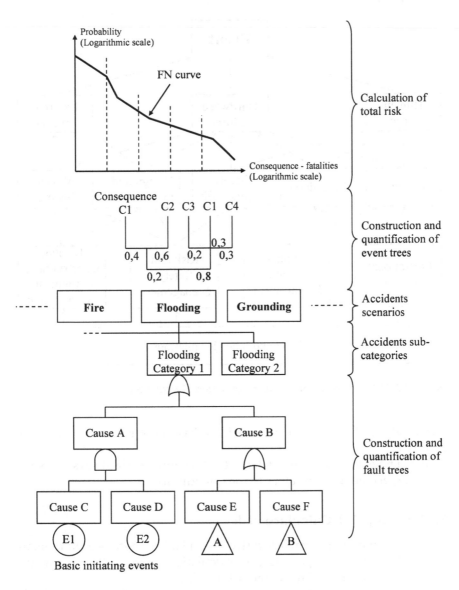

Figure 12.6 Illustration of a Risk Contribution Tree (RCT).

Finding the information to quantify can be quite hard, and in general the availability of data from maritime applications is poor. Databases of events and losses (fatalities, spills) exist, but they have several weaknesses, including unclear/different definitions of what different events are and sometimes serious underreporting. Data on causes of events are typically even harder to find, although using fault tree analysis may make it easier because we

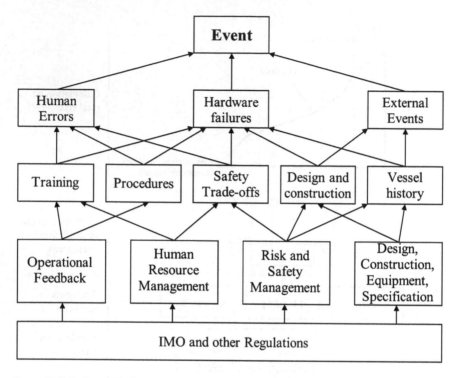

Figure 12.7 Risk profile for the accidental outcome of a collision.

may be able to find data for the individual basic events that are part of the fault tree.

As a supplement to empirical data, it is therefore nearly always necessary to use expert judgment as input to the quantification.

12.5.3 Step 2.3 Calculate risk

When risk models are established (RCTs) and all the data needed to quantify the RCTs has been provided, the calculation is essentially just a matter of "pressing the button" to get the risk results.

Risk can be expressed in many ways as described in Chapter 9, but in the FSA method, two ways of expressing risk are proposed used:

– FN curve
– Individual risk

Both have been described in Chapter 9. The FN curve can be constructed based on the end events in the event trees that form part of the RCT. For each end event, a consequence (expressed in number of fatalities) is given, and a frequency is calculated. By sorting all the end events according to the number of fatalities and adding together the frequencies of all end events with N number of fatalities or more, we arrive at a set of pairs of numbers (f_i, n_i) that can be plotted in the FN curve.

Calculation of individual risk is usually done by first calculating a total PLL (expected number of fatalities) value and then dividing by the number of people exposed. PLL is first calculated for each end event in the event tree by multiplying the number of fatalities for the end event with the frequency of the same event. Total PLL is arrived at by adding together the PLL contributions from all the end events. When determining the number of people exposed, it is necessary to take into account the time that the crew spends on the ship. If there are two crews that man the ship, each crew spending half the year on the ship, the total number of people exposed is the crew size multiplied by two.

If different personnel groups are exposed to risk (crew, passenger, third party), it is necessary to distinguish between these also in the event tree and the PLL calculation. Rather than giving the consequence as a total number of fatalities for each end event, we then have to specify number of fatalities for each personnel group.

> Consider a cruise ship that collides with another cruise ship. In a situation like this, the crew on the cruise ship that hits the other cruise ship will be first-party victims, the passengers on the same cruise ship will be second-party victims (passengers), while both crew and passengers on the other ship (the one being hit) are third-party victims.
>
> In Table 12.7, possible output from an event tree with five end events is shown. For each end event, the total number of fatalities is shown, plus a split of these on first-, second-, and third-party victims. Based on this, PLL is calculated for the three groups separately and in total.
>
> From this, we can establish an FN curve using the information in Table 12.8.
>
> To calculate the individual risk for average individuals in the three groups, information about the size of the groups is necessary. Let us assume that the first cruise ship (that hits the other ship) has a crew size of 500 and 2,000 passengers. The crew spends half the year on

Table 12.7 Example output from an event tree

End event no.	Freq.	Fat 1. party	Fat 2. party	Fat 3. party	Total fat	PLL 1. party	PLL 2. party	PLL 3. party	Total PLL
1	$3 \cdot 10^{-2}$	1	0	0	1	$3 \cdot 10^{-2}$	0	0	$3 \cdot 10^{-2}$
2	$6 \cdot 10^{-3}$	3	2	0	5	$1.8 \cdot 10^{-2}$	$1.2 \cdot 10^{-2}$	0	$3 \cdot 10^{-2}$
3	$3 \cdot 10^{-4}$	4	6	0	10	$1.2 \cdot 10^{-3}$	$1.8 \cdot 10^{-3}$	0	$3 \cdot 10^{-3}$
4	$1 \cdot 10^{-5}$	5	15	5	25	$5 \cdot 10^{-5}$	$1.5 \cdot 10^{-4}$	$5 \cdot 10^{-5}$	$2.5 \cdot 10^{-4}$
5	$2 \cdot 10^{-6}$	10	100	90	200	$2 \cdot 10^{-5}$	$2 \cdot 10^{-4}$	$1.8 \cdot 10^{-4}$	$4 \cdot 10^{-4}$
						$4.9 \cdot 10^{-2}$	$1.4 \cdot 10^{-2}$	$2.3 \cdot 10^{-4}$	$6.4 \cdot 10^{-2}$

Table 12.8 Input data to FN curve

N	Frequency of N fatalities	Frequency of N or more fatalities
1	$3 \cdot 10^{-2}$	$3.6 \cdot 10^{-2}$
5	$6 \cdot 10^{-3}$	$6.3 \cdot 10^{-3}$
10	$3 \cdot 10^{-4}$	$3.1 \cdot 10^{-4}$
25	$1 \cdot 10^{-5}$	$1.2 \cdot 10^{-5}$
200	$2 \cdot 10^{-6}$	$2 \cdot 10^{-6}$

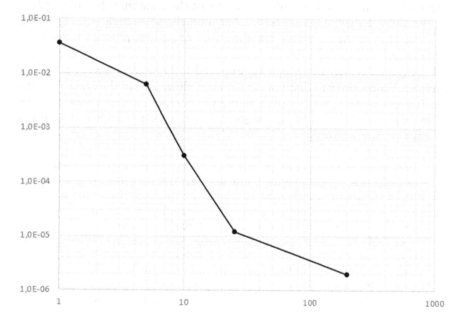

Figure 12.8 FN curve based on example.

the ship, implying that the total number of crew exposed is 1,000 persons. The passengers are assumed to spend one week on board, i.e., 52 cruises per year. The total number of passengers exposed therefore becomes

$$2,000 \frac{\text{persons}}{\text{cruise}} \cdot 52 \text{ cruises} = 104,000 \text{ persons}$$

For the other cruise ship, we assume 400 crew and 1,200 passengers. The crew spend half the year on the ship, and total number of persons exposed is 800. For this ship, the average length of cruises is two weeks, and the passengers thus spend twice as long on the ship. The total number of passengers exposed thus becomes

$$1,200 \frac{\text{persons}}{\text{cruise}} \cdot 26 \text{ cruises} = 31,200 \text{ persons}$$

Since both crew and passengers on this ship are third-party victims, the total group becomes 32,000 persons.

An observation that can be made in the calculation of the third-party victims is that some persons in this group are exposed to risk for 2 weeks (the passengers) while the crew are exposed for 26 weeks, i.e., the individual risk to the crew is 13 times higher than the passengers. This is not reflected in the calculation that has been done in this case and illustrates that one should be aware of potential differences within the group when considering a large group of people like this.

From the information provided, it is now possible to calculate several individual risk values:

$$IR_{\text{Total}} = \frac{6.4 \cdot 10^{-2}}{137,000} = 4.7 \cdot 10^{-7}$$

$$IR_{1.\text{ party}} = \frac{4.9 \cdot 10^{-2}}{1,000} = 4.9 \cdot 10^{-5}$$

$$IR_{2.\text{ party}} = \frac{1.4 \cdot 10^{-2}}{104,000} = 1.3 \cdot 10^{-7}$$

$$IR_{3.\text{ party}} = \frac{2.3 \cdot 10^{-4}}{32,000} = 7.2 \cdot 10^{-9}$$

This illustrates clearly how large the differences between different groups may be. The highest risk is nearly four orders of magnitude higher than the lowest risk in this particular case.

12.5.4 Step 2.4 Sensitivity and uncertainty analyses

Sensitivity and uncertainty analyses are different methods, but for both, the main purpose is to provide a better basis for understanding risk and thus for making decisions. A combination of the two methods may also be applied. The guidelines do not provide any information on how this should be done other than that the method chosen will depend on the risk analysis methods that have been applied.

Briefly, the methods entail:

- Sensitivity analysis is performed by changing the input to the risk assessment and determining the effect of this on the results. Often, a fixed change is applied, e.g., by increasing the input values by 10% compared to the original value (or another increase or decrease). The % change in the results is then determined and based on this, the "importance" of the parameters can be ranked based on how large change they give in the results.

- Formal uncertainty analysis implies that we establish probability distributions for all input parameters that are used in the risk analysis. This is subsequently used to calculate a probability distribution for the calculated results (e.g., individual risk or PLL). In practice, analytical solution is not feasible and Monte Carlo simulation is used to generate a probability distribution for the results. Formal uncertainty analysis is time-consuming and is not very much used, except in the nuclear industry.
- Simplified uncertainty analysis can be done in many ways, but a simple approach would be to determine a high and low value for each parameter going into the risk calculation. "High" and "low" can be defined qualitatively ("the highest/lowest value that the parameter is likely to have") or more formally (the 5%/95% confidence limits for the parameter). Next, sensitivity analysis can be performed using the high/low values to see the effect on the results and the parameters can be ranked based on the effect their high/low values have on the results. This gives a different ranking compared to a "pure" sensitivity analysis since the degree of change of the input values will depend on the uncertainty.

Simplified uncertainty analysis

In Figure 12.9, a simple event tree is shown.

In Table 12.9, the input values and the results from the analysis are shown. Notice that both frequency and conditional probabilities are subject to sensitivity analysis. The total PLL with the initial values given in the event tree is $2.8 \cdot 10^{-3}$.

In the leftmost column, the input parameters that are to be changed are listed. The next three columns contain the simplified uncertainty analysis, showing first the parameter values used in the original risk calculation and next the "Low" and "High" values. The final four

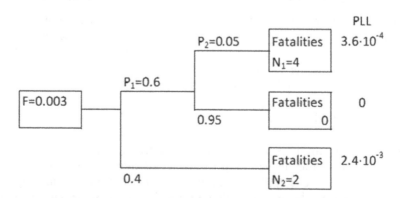

Figure 12.9 Example event tree for simplified uncertainty analysis.

Table 12.9 Uncertainty and sensitivity analysis results

	Uncertainty analysis			Sensitivity analysis			
Parameter	Value	Low	High	Low risk	% change	High risk	% change
Frequency	0.003	0.001	0.01	$9.2 \cdot 10^{-4}$	−67%	$9.2 \cdot 10^{-3}$	+233%
P_1	0.6	0.3	0.8	$4.4 \cdot 10^{-3}$	+59%	$1.7 \cdot 10^{-3}$	−39%
P_2	0.05	0.01	0.2	$2.5 \cdot 10^{-3}$	−10%	$3.8 \cdot 10^{-3}$	+39%
N_1	4	2	15	$2.6 \cdot 10^{-3}$	−7%	$3.8 \cdot 10^{-3}$	+36%
N_2	2	0	5	$3.6 \cdot 10^{-3}$	−87%	$6.4 \cdot 10^{-3}$	+130%

columns show the results of the sensitivity analysis, giving the updated risk values for the "Low" and "High" values respectively.

From the results, we can see that the largest change in risk is found when we apply the "High" value for frequency. In most cases, it would be the "High" values that are of interest, because we would be more interested in knowing how high the risk may be in the worst case rather than how low the risk may be. However, from the table, we can see that the "High" value for P_1 actually gives a decrease in risk. For this parameter, it is therefore the "Low" value which is relevant to consider.

Another thing worth noting is that a simple sensitivity analysis, increasing/decreasing all values by, e.g., 10% would give a different ranking than shown in this case example.

In the description and also in the example, changes in the parameter values going into the calculation are discussed. However, there will be uncertainty associated with other aspects of the risk calculation as well:

- Assumptions – In many cases, assumptions will be reflected directly in the risk model (e.g., assumptions about number of passengers on a cruise ship), but this is not always the case. Assumptions may still have considerable uncertainty associated with them, but this is not necessarily easy to reflect in an uncertainty analysis.
- Model uncertainty – The risk models will always be simplifications of the real world, and this also introduces uncertainty. This is also often difficult to take into account in uncertainty analysis.

Uncertainty analysis of the parameters will therefore normally not cover all aspects of uncertainty in a risk assessment. Quantifying this may be difficult, but it is important to be aware also of these uncertainties.

12.6 STEP 3: ESTABLISH SAFETY MEASURES

The purpose of step 3 is to first identify Risk Control Measures (RCMs) and then to group them into a limited number of Risk Control Options

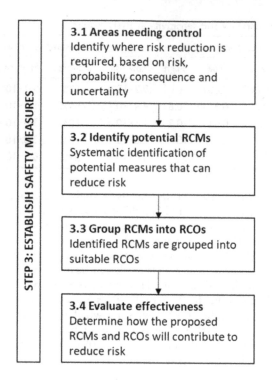

Figure 12.10 Sub-steps of Step 3 establish safety measures.

(RCOs) for use as practical regulatory options. Step 3 comprises the four sub-steps shown in Figure 12.10.

Before describing these steps in detail, some comments are given on the use of the terms RCM and RCO.

12.6.1 RCM and RCO

The terms Risk Control Measures (RCM) and Risk Control Options (RCO) are specific to the FSA methodology. In most other contexts, terms like barriers, risk reduction measures, safety functions, etc. are used instead of the term RCM. This is discussed in Chapter 15. RCO is a term that is tied very closely to the use of FSA in the regulatory processes.

The FSA guidelines (IMO, 2018) describe that RCMs should be grouped to form "practical regulatory options". However, it is not very clear from the IMO guidelines how this should be done. The purpose is stated to be to establish a set of "well thought out Risk Control Options (RCOs)". The underlying idea is presumably that regulatory changes often will imply more than changing just one specific regulation and one specific measure. RCOs can then be a means of addressing larger "packages" of changes as one proposal.

It is also stated that this can be done in many different ways, but that two approaches may be considered.

- A "general approach" – The idea behind this is that RCMs that contribute to reduce the probability of initiating evens are grouped together. In other contexts, this would also be called a "probability reducing approach". If this is related to the bow-tie model (see Chapter 3), this would cover RCMs that influence the left-hand side of the bow-tie.
- A "distributed approach" – This provides reduction of risk by reducing the negative development (escalation) of initiating events and reducing the consequence. This may be called "consequence reducing approach" and is aimed at the right-hand side of the bow-tie.

It is noted that the terms RCM and RCO are not often used in accordance with this terminology in applications. In the SAFEDOR project reports (SAFEDOR, 2021), the term RCO is in practice the only one used and the RCOs proposed in the reports are in most cases very specific and detailed and should probably rather be called RCMs in accordance with the FSA guidelines. Certainly, for other uses than regulatory decision-making, it does not seem useful to always group proposals for reducing risk together.

12.6.2 Step 3.1 Areas needing control

The purpose of this step is to focus effort in reducing risk on those areas where this is most needed. On an overall level, the ALARP principle will guide the need for risk control. If the risk is in the unacceptable (or intolerable) region, risk must be reduced. Further, if risk is in the ALARP region, risk should be reduced as far as reasonably practicable.

When searching for risk control measures, the FSA guidelines list four areas where one should focus effort:

1. Risk level: It is natural to first focus attention on the scenarios that have the highest risk level and contribute most to risk. This will normally also be the scenarios where the largest risk reduction can be achieved.
2. High probability: Some scenarios may have only a medium risk level, with low severity but a high probability. From an operational point of view, many "disruptions" may be unwanted even if the consequences are limited. It may therefore be beneficial to look for measures to reduce the probability of these scenarios.
3. High severity: There may also be scenarios that have a very low probability but very high consequences, potentially threatening the whole company (e.g., a major accident with many fatalities on a cruise ship). Even if these may have only a medium risk, it may still be necessary to consider measures to reduce the consequences.

4. <u>Considerably uncertainty:</u> Finally, large uncertainty in probability, severity, or both could be a reason to implement extra or redundant risk control options/measures. This may be regarded as being in accordance with a precautionary approach, although we do not have sufficient knowledge about a potential problem.

12.6.3 Step 3.2 Identify risk control measures

RCMs are essentially the same as are called barriers in most other applications. This is discussed in Chapter 9 and may give useful background for performing this step.

In the discussion of barriers, various ways of classifying barriers are given. In the FSA guidelines, a separate appendix (Appendix 6, IMO, 2018) describes a set of "attributes" of RCMs. The attributes are partly classifications (Category A and B) and partly characteristics (Category C) of RCMs. The attributes are shown in Table 12.10.

Table 12.10 Attributes of RCMs

Category	Attributes
A	**Preventive** – reducing the probability
	Mitigating – reducing the consequences
B	– **Inherent** control follows the principles of inherent safety, where we aim to remove and avoid hazards as much as possible, thus reducing the need for measures to control risk.
	– **Engineering** controls are technical solutions to reduce risk.
	– **Procedural** controls rely on personnel to follow defined procedures
C	– **Diverse** (vs concentrated) – where risk control is distributed in different ways
	– **Redundant** (vs single) – failure of one risk control will not lead to system failure or an accident
	– **Passive** (vs active) – no action is required for the risk control to function
	– **Independent** (vs dependent) – the risk control is not dependent on functioning (or failure) of another risk control
	– **Auditable** (vs not auditable) – the functioning of the risk control can be audited or verified
	– **Quantitative** (vs qualitative) – risk control based on quantitative risk analysis
	– **Established** (vs novel) – risk control that is established as a marine technology
	– **Developed** (vs non-developed) – The technology underlying the risk control is developed, both technically and from a cost point of view
	– **Critical** and **involved** human factors – Critical human factors are actions that directly can cause an accident or allow a scenario to progress. Involved human factors are actions that are required to control the risk, but failure to do so will not cause an accident or allow the scenario to progress. In simple terms, we may say that critical actions make the situation worse if performed incorrectly, and involved actions improve the situation if performed correctly

IMO (2018).

Table 12.11 Possible risk control measures

RCM	Attributes[a]
Audible alarm on the bridge when the bow door is open or not closed properly at departure.	A: Preventive B: Engineering C: Active, Independent, Auditable, Established
Procedure requiring control of whether the bow door is closed before departure.	A: Preventive B: Procedural C: Active, Auditable, Established, Involved
Strengthen bow door hinges to tolerate severe weather/sea conditions.	A: Preventive B: Engineering C: Passive, Independent, Established
Drainage of water on the vehicle deck to tanks equipped with pumps situated in the lower parts of the hull.	A: Mitigating B: Engineering C: Passive, Independent
Moveable transverse bulkheads on the vehicle decks.	A: Mitigating B: Engineering C: Passive, Established
Procedure to ensure that transverse bulkheads are closed before departure.	A: Mitigating B: Procedural C: Active, Auditable, Developed
Audible alarm on bridge if transverse bulkheads are open at departure.	A: Mitigating B: Engineering C: Active, Auditable, Developed

[a] Only a selection of "C attributes" are described.

These attributes may be used as a help in identifying RCMs and also in qualitatively evaluating the RCMs. The classifications presented in Chapter 9 may also be used.

> The flooding of the vehicle deck on Ro-Ro passenger vessels through an open or partly open bow door may result in rapid capsizing and the loss of many human lives. This was the accident scenario in the Herald of Free Enterprise disaster in 1987. The risk related to this scenario is found to be unacceptable for a given vessel and risk control measures must therefore be implemented in order to reduce the risk.
>
> Possible risk control measures may include one (or more) of the following.
>
> This list of risk control measures/options is not exhaustive.

12.6.4 Step 3.3: Grouping risk control measures

Grouping of the RCM into larger "packages" of Risk Control Options is a step that is included to support the regulatory process in IMO. However, in many cases, it may be relevant to implement combinations of RCMs to reduce risk. Alternative combinations may then be considered, and a convenient way of referring to these combinations is to group them into RCOs.

12.6.5 Step 3.4: Evaluating the effectiveness of RCMs/RCOs

In practice, this is a revision of the quantitative risk assessment performed in Step 2. Each of the RCMs/RCOs should be considered separately, going through the following steps:

- In what ways will the RCM/RCO influence risk? In practice, we have to identify where the risk model will be changed if the RCM/RCO is implemented. This can be changes to the fault tree structure, changes to the event tree structure, changes in the probability of the basic events, changes in the conditional probabilities applied in the event tree, or changes in the consequences (e.g., reduction in fatalities). An RCM/RCO may also remove a hazard completely, meaning that the risk contribution of course is eliminated.
- Next, we have to determine how efficient or reliable the RCM/RCO is. If a new technical system is introduced, e.g., an additional radar, we need to consider two aspects: What is the reliability of this new radar system and how much will it improve the probability of detecting obstructions or other ships? For other RCM/RCOs, other considerations may be relevant.
- Finally, when the risk model has been modified with any changes in structure or with updated input data, the risk calculation can be repeated. The effect of the proposed RCM/RCO can then be compared with the original risk calculation to find the effect.

In principle, the effect of RCOs could be calculated by adding together the effect of the individual RCMs that has been grouped together to form the RCO. In practice, it is however often not so easy, because there may be dependencies between the RCMs.

Assume that we have identified a set of RCMs, RCM_i, $i = 1,..,n$ and that the effects of these have been calculated to be Δr_{RCM_i}. Let us further assume that they are grouped together into one RCO. In general, the effect of the RCO then becomes

$$\Delta r_{RCO} \leq \sum_i \Delta r_{RCM_i}$$

If all RCMs are independent, the risk reduction effect of the RCO will be equal to the sum, otherwise it will be smaller.

12.7 STEP 4: COST-BENEFIT ASSESSMENT

The cost-benefit assessment (CBA) is important in any FSA because it provides input to the decision whether or not the suggested risk control

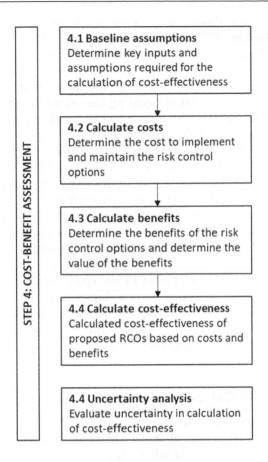

Figure 12.11 Sub-steps of cost-benefit assessment.

options/measures are feasible for implementation. A CBA determines if the benefits of implementing a given risk control option outweigh the cost of implementation. Cost-benefit assessment is described in Chapter 15, and in this chapter, we only describe CBA aspects and issues relevant in an FSA context. The CBA may be described in three sub-steps as shown in Figure 12.11.

12.7.1 Step 4.1: Baseline assumptions and conditions

There are a number of baseline assumptions and conditions that are needed as input to the CBA. This will include system description and geographical and operational boundaries. However, there are also some aspects that are not normally part of the problem definition and system description for a risk assessment. Some aspects that need to be established are as follows:

- What valuation should be used for personnel risk and environmental risk (benefits)? What values should we use for a reduction in loss of life of 1? And what value should we use for a reduction in oil spill of 1 ton?
- Which duration should we assume for the costs and benefits? This will often be a question of deciding the useful lifetime of the vessel. Some RCOs only imply an investment cost and no further costs, while other RCOs may require regular operating, inspection, and maintenance costs. Similarly, the risk reduction from the risk assessment is normally calculated for one year. This must be multiplied with the duration to get the total benefit or by adding together contributions from each year if the effect changes over time.
- What interest rate should be applied in the calculation of the net present value of the costs and benefits? This can have a large impact on the calculations, in particular if the initial investment is high.

In some cases, costs and benefits may be unevenly distributed, e.g., between different geographical areas or between ships within the group to be covered by the generic ship. If that is the case, it may be necessary to reconsider the generic ship and see if it is necessary to split it into several, more specific ship types. This is however not common to do.

12.7.2 Step 4.2: Calculate costs

The implementation of risk control options (or safety measures) may involve many different types of costs and benefits. Typical costs involved may include one or more of the following:

- Capital/investment
- Installation and commissioning
- Operating or recurrent
- Labor
- Maintenance
- Training
- Inspection, certification and auditing
- Downtime or delay

All the potential costs and benefits related to each of the relevant risk control options/measures should be identified in this stage of the CBA process. It is also important to identify potential negative effects that the implementation of risk control options can have on the system/activity in question. On a ship, such negative effects could for example include longer loading/unloading times, reduced speed, more downtime due to inspections and controls, etc.

One aspect of this calculation is that the costs (and the benefits) in almost all cases will be unevenly distributed between different stakeholders. From a societal point of view, this is not important because we will then consider all costs and benefits together. However, from the individual stakeholder's perspective, this can clearly be important. In particular, if there are direct economic benefits of the risk reduction, it may be relevant to investigate if costs and benefits are very unevenly distributed between stakeholders.

Costs and benefits for different stakeholders

Les us assume that stricter measures to reduce the risk of oil spills are introduced. The costs of this will initially be borne by the shipowners. Eventually, the cost will however have to be covered by those in need of oil transport capacity because the shipowners have to recover their costs and will in the long term not provide this service if their costs are not covered. On the other hand, the benefactors of the risk reduction may be fishermen that avoid damage to their livelihood, the fishing industry that processes the catch, the public using beaches and coast-line for recreation, etc.

In this case, it can be argued that the fishermen and the fishing indus-try get an economical benefit while the shipowners and transporters have to carry the cost. However, the benefit is achieved by a reduction in losses that is *caused* by the shipowners and transporters, and it can therefore be argued that this is a loss that the fishermen and fishing industry should not be exposed to in the first place.

12.7.3 Step 4.3: Calculate benefits

The benefit from implementing RCOs is of course reduction in potential losses (risk). The benefits of implementing an RCO on a ship usually include one or more of the following:

- Reduced number of injuries and fatalities
- Reduced casualties with vessel, including damage to and loss of cargo and damage to infrastructure (e.g., berths)
- Reduced environmental damage, including clean-up costs and impact on associated industries such as recreation and fisheries
- Increased availability of assets
- Reduction in costs related to search, rescue, and salvage
- Reduced cost of insurance

12.7.4 Step 4.4: Calculate cost-effectiveness

Different risk control measures result in different risk reductions, and each measure is associated with a set of distinct benefits and costs. To select the most cost-effective measures for implementation it is very advantageous to evaluate these against each other on a common scale, which normally implies monetary values. Approaches for cost-benefit analysis (CBA) of risk control/reduction measures are presented in Chapter 15, and these must be applied within the framework of the FSA. In particular, the Implied Cost of Averting a Fatality (ICAF) approach/methodology is very much used in FSAs. In essence, the ICAF methodology estimates the achieved risk reduction in terms of cost using the following equation:

$$ICAF = \frac{\text{Net total cost of measure}}{\text{Reduction in total fatality rate}} \qquad (12.2)$$

ICAF may also be calculated by dividing the annual cost of the measure and dividing by the annual reduction in fatalities (PLL), as long as the annual cost and risk reduction are constant. The ICAF value can be interpreted as the cost of averting a fatality. A decision criterion must be established for this value in order to evaluate whether a given risk control option/measure is cost-effective or not. A method for developing such a criterion is also presented in the previous chapter. Risk control measures with an ICAF value less than the target value of the criterion should be considered as cost-effective and therefore to be implemented.

12.7.5 Step 4.5: Evaluating uncertainty

Uncertainty and sensitivity were considered as part of the risk assessment in FSA. In the end, the outcome from the CBA is a very important basis for decision-making, and it is therefore necessary for the decision-makers to have a good understanding also of uncertainties related to the CBA results. To do that, we need to consider the uncertainty in the benefits (the risk assessment) and in the costs. In practice, sensitivity and uncertainty analyses can also be applied to evaluate the uncertainty in CBA.

One would perhaps expect that the results from the uncertainty analysis of risk can be used directly as input to the evaluation of uncertainty in the CBA results, but unfortunately it is not quite as simple as that. The reason is that the CBA uses changes in risk to express the benefits. This means that the uncertainty in the risk calculations to some extent is "canceled out" because we mainly use the same input values into both calculations and only change those (usually very few) parameters that are changed when an RCO is introduced. It is therefore primarily the uncertainty in how large the change is that is important in this case.

Uncertainty in risk reduction effect

Let us assume that we have calculated the risk of fire in the cabins on a cruise ship to be PLL = 0.05 per year. From statistics, it has been found that 20% of historical fires (and risk) have been caused by smoking. It is therefore proposed to prohibit smoking in the cabins and to enforce this with highly sensitive smoke detectors that will sound an alarm both in the cabin and to warn the crew. This is considered to reduce smoking by 90% and also to reduce the fires caused by smoking accordingly. The revised risk then becomes PLL = 0.041 and the risk reduction is $\Delta PLL = 0.009$.

The uncertainty in the risk reduction is mainly related to:

- The uncertainty in the proportion of fires that are caused by smoking. Since this is based on historical data, it is considered to be reasonably certain. The main change is that the proportion of the population that is smoking is declining, implying that this may reduce naturally over time.
- The uncertainty in how effective the smoking ban is (90%). It may be lower or higher but is not likely to be very different from the estimate. The uncertainty in this value is also limited.

There will be uncertainty also related to the baseline risk value (0.05) that has a certain impact on the uncertainty in the improvement, but this is much smaller than if we consider the uncertainty in the absolute risk value.

The cost estimates will also be uncertain, and the principles for evaluating the uncertainty in the cost will be similar to those we use to evaluate uncertainty in risk. Uncertainty in parameters, assumptions, and models need to be considered.

In the end, the most important answer that we want from performing uncertainty analysis on the CBA results is whether the uncertainty may impact on the conclusions we reach. If a proposed RCO has marginally higher cost than the benefits we achieve, it is important for the decision-maker to understand if this conclusion may be different when uncertainty is considered. It may then be argued that a precautionary approach should be adopted and that the RCO therefore should be implemented.

From this, we can also draw the conclusion that if the conclusions are very clear, with a wide margin to the point where a decision may be changed, uncertainty and sensitivity analysis is less important. This brings us back to the point that has been made several times that risk assessment is performed to help us make better decisions about risk. If the decision is clear, more detailed analysis is not necessary.

12.8 RECOMMENDATIONS FOR DECISION-MAKING

Step 5 of the FSA methodology involves proposing recommendations to the decision-makers on which RCO(s) that should be adopted to make the risks ALARP, i.e., as low as reasonably practicable. It is underlined that IMO has not recommended any risk acceptance criteria, although the guidelines contain examples of criteria. In practice, these also seem to be used as decision criteria in existing FSAs although not formally established as criteria.

The guidelines specify that there are three main outputs from Step 5:

- RCO(s) that are proposed implemented, described in SMART (Specific, Measurable, Achievable, Realistic, Time-bound) terms.
- A comparison and ranking of the proposed RCOs, based on their cost-effectiveness and potential for reducing risk.
- Sufficient information to enable detailed and critical review of the results generated in the previous steps and the basis for these results.

It is important that the information is presented in such a way that all stakeholders that may have an interest in the results are able to understand the information. This also includes those who are not experts in or have experience with risk assessment. The results should also be fully traceable and contain all information necessary to understand how the results have been arrived at.

The recommendations may be presented as a prioritized list of cost-effective risk control options/measures. Such a list should include a description of the options, including their cost-benefit ratios, and an evaluation of the uncertainties related to each of the options. The recommendations could also include suggestions for improvements to the analysis (or the FSA methodology) and advice on further work that should be carried out on the subject under consideration.

The guidelines (IMO, 2018) specify that the comparisons and ranking of RCOs *should take into account that, in ideal terms, all those [stakeholders] that are significantly influenced in the area of concern should be equitably affected by the introduction of the proposed new regulation.*

12.9 APPLICATION OF THE FSA METHODOLOGY

At the 70th session of IMO's Maritime Safety Committee (MSC), the topic of life-saving appliances (LSAs) for bulk carriers was discussed, and it was decided to include LSAs as part of the FSA process for these vessels. This case/example briefly summarizes some important aspects of a comprehensive FSA project on LSAs for bulk carriers that were performed by Norway (IMO, 2001).

The steps of the FSA methodology are described in succession below. Many details from the original report are obviously left out. This FSA study focuses solely on LSAs with the objective of identifying risk control options (RCOs) for bulk carriers that give improved life-saving capability in a cost-effective manner. The study is considered representative for all SOLAS bulk carriers over 85 meters of length.

The problem addressed in this study was related to the identification of effective RCOs that could bring down the fatality rates in evacuation associated with bulk carrier accidents, i.e., to improve the probability of evacuation success given different accident scenarios (e.g., a collision). Earlier individual and societal risk assessments for bulk carriers had shown that the risks are high in the ALARP region and that cost-effective risk control options therefore should be implemented. Several regulations in SOLAS 74 are affected by the recommendations of this particular FSA study. More detailed background information can be found in the reference source (IMO, 2001).

12.9.1 Step 1: Hazard identification

The process of hazard identification was carried out by multidisciplinary teams of relevant experts using the so-called "What if...?" technique. In this method a list of potential hazards is produced for a system or subsystem by asking "What if...?" something does not work as planned or something unexpected/undesirable happens. The hazards were identified and ranked separately for commonly used LSAs on bulk carriers, i.e., conventional lifeboats, free-fall lifeboats, davit launched liferafts, and throw overboard liferafts. All phases of an evacuation event were analyzed for hazards, from the occurrence of the initiating event, through mustering, abandoning, survival at sea, and the final rescue. Some, but not all, hazards were generic for all categories of survival crafts. Some hazards were also related to survival at sea and rescue.

The last task of the hazard identification step is to perform a hazard screening. In this analysis, the screening included a ranking of the hazards based on a qualitative assessment of their importance in terms of risk. The most important hazards were given particular consideration in the risk assessment step of the FSA.

12.9.2 Step 2: Risk assessment

Probabilities for various accident scenarios are fairly well established for bulk carriers through incident data sources such as Lloyd's Maritime Casualty Reports (LMIS) and Lloyd's Casualty Reports (LCR). Regarding evacuations, these data sources show that a total of 115 bulk carrier evacuations were recorded during the period from 1991 to 1998. The ship population exposed during this period was 44,732 ship years (i.e., an average fleet size of approximately 5,592 ships), identified through Lloyd's Register's World Fleet Statistics. This gives a total evacuation frequency of $2.6 \cdot 10^{-3}$ per ship year. This is approximately the same evacuation frequency as for merchant ships in general. Distributed with respect to type of accidental event, the resulting evacuation frequencies are listed in Table 12.12. The

Table 12.12 Evacuation frequencies and fatality probabilities

Type of accidental event	Number of events	Evacuation frequency [per ship year]	Fatalities	Number on board	Probability of fatality [%]
Collision	14	$3.1 \cdot 10^{-4}$	116	332	35
Contact	5	$1.1 \cdot 10^{-4}$	54	119	45
Fire/explosion	16	$3.6 \cdot 10^{-4}$	6	379	2
Foundered	51	$1.1 \cdot 10^{-3}$	618	1,209	51
Hull failure	5	$1.1 \cdot 10^{-4}$	0	119	0
Machinery failure	1	$2.2 \cdot 10^{-5}$	0	24	0
Wrecked/stranded	23	$5.1 \cdot 10^{-4}$	0	545	0
Total	115	$2.6 \cdot 10^{-3}$	794	2,727	29

number of crew members on board is obtained by multiplying the number of events with the average crew size per ship set to 23.7.

The objective of this study was to investigate improvements in evacuation from ships using Life-saving appliances (LSAs), i.e., what is happening after the accidental event has occurred. Because of this, it is not relevant to study the causes of the accidental events shown in Table 12.12. This is because the cause of, e.g., a collision is of limited interest when we have reached the stage where an evacuation is necessary. Because of this, fault trees describing the causes of the events were not developed in this study. This is a good illustration of how the method is adapted to the needs of the problem that is being studied.

Further, the event trees need to include all factors that can affect the evacuation, and in particular factors that are different for different types of LSAs. This is necessary to ensure that the risk assessment can distinguish between alternative options for evacuation. The event trees are therefore primarily an evacuation model, illustrating how the different accidental events will influence evacuation and the likelihood of successful evacuation.

The event trees for the different accidental events are structurally identical. This is because it is considered that the evacuation process and the factors influencing the evacuation process are the same in all cases. However, the probabilities applied in the event trees for the different events are not the same.

Improvements in LSAs will only influence risk to people and environmental and economic risk is therefore not covered. Potential loss of life (PLL) is therefore the only risk metric that is used in the calculations.

As mentioned above, an evacuation model is created in order to analyze how the different LSAs affect PLL. The model consists of the typical sequences of events that are associated with an evacuation using different LSAs. The sequence of events expected for evacuation using conventional lifeboats is shown in Figure 12.12. The sequence of events is slightly

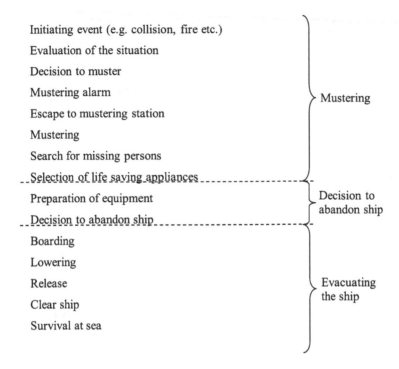

Initiating event (e.g. collision, fire etc.)

Evaluation of the situation

Decision to muster

Mustering alarm

Escape to mustering station

Mustering

Search for missing persons

Selection of life saving appliances

Preparation of equipment

Decision to abandon ship

Boarding

Lowering

Release

Clear ship

Survival at sea

Mustering

Decision to abandon ship

Evacuating the ship

Figure 12.12 Evacuation model for conventional lifeboat.

different for throw overboard liferafts and davit launched liferafts. For example, throw overboard liferafts are launched before boarding.

The evacuation sequence gives the underlying basis for the construction of an event tree, or several event trees if the sequence is divided into several sub-sequences as illustrated in Figure 12.12. The branch probabilities in these event trees are to some degree different for the different accident scenarios. For example, there is a slight probability of fatality as a result of the initiating event in the scenarios of ship collision or fire/explosion, while this probability is negligible for the accident scenario of hull/machinery failure. Statistics show that it is the evacuation sequence rather than the initiating event that has the highest contribution to fatalities. Another example is that there is a higher probability of not being able to escape to mustering stations in the event of foundering compared to, for example, hull/machinery failure.

Using statistics and expert judgment, Table 12.13 can be constructed for the undesirable events following an initiating event until preparation of the LSA. The scenario of jumping to sea, waiting, and being rescued may occur in the case that there is a faulty evaluation of the situation, an untimely decision to muster is taken, the crew is unable to reach mustering stations,

Table 12.13 Event tree branch probabilities for the undesirable events following a ship accident until preparation of the LSA

	Branch probabilities in event tree					
	Collision	Contact	Fire/ explosion	Foundered	Hull/ machinery	Wrecked/ stranded
Fatality as result of initiating event	0.0001	–	0.007	–	–	–
Wrong evaluation of situation	0.31	0.31	–	0.37	–	–
Fatality as a result of not jumping to sea – given faulty evaluation of the situation	I	I	–	I	–	–
Untimely decision to muster	0.03	0.03	0.015	0.03	0.03	0.03
Fatality as a result of not jumping to sea – given untimely decision to muster	0.90	0.90	0.90	0.95	0.95	0.95
Unable to reach mustering station	0.06	0.06	0.03	0.07	0.02	0.06
Fatality as a result of not jumping to sea – given being unable to reach mustering station	0.95	0.95	0.95	I	0.95	0.95
Not terminating search in time	0.04	0.04	0.04	0.04	0.04	0.04
Fatality as a result of not jumping to sea – given search for missing personnel not terminated in time	0.625	0.625	0.625	0.625	0.625	0.625
Fatality associated with jumping and awaiting rescue	0.358	0.323	0.420	0.970	0.420	0.323
Fatality as a result of not being successfully rescued from the sea	0.0016	0.0016	0.021	0.050	0.021	0.0016

and in the case that the search for missing personnel is not terminated in time.

Similar tables can be produced for the other parts of the event sequence (see Figure 12.12). Where different branch probabilities are expected for the different LSAs, separate tables and event trees must be constructed for each LSA. The event tree corresponding to Table 12.13 above is given in Figure 12.13.

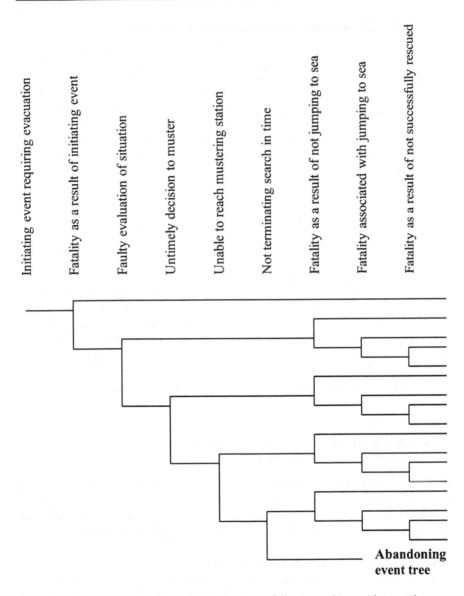

Initiating event requiring evacuation

Fatality as a result of initiating event

Faulty evaluation of situation

Untimely decision to muster

Unable to reach mustering station

Not terminating search in time

Fatality as a result of not jumping to sea

Fatality associated with jumping to sea

Fatality as a result of not successfully rescued

Abandoning event tree

Figure 12.13 Event tree for the undesirable events following a ship accident until preparation of the LSA.

The resulting probabilities of fatality and potential loss of life (PLL) for the different types of accidental events are summarized in Table 12.14, which shows that the established evacuation model reflects the real-world data fairly well. This is mainly due to the fact that actual data are used as the basic inputs and broken down into event tree branch probabilities. The

Table 12.14 Probability of fatality associated with evacuation

	Probability of fatality associated with evacuation			
	Based on evacuation model		Based on statistics	
Type of accidental event	Probability of fatality [%]	PLL [per ship year]	Probability of fatality [%]	PLL [per ship year]
Collision	45.7	$3.392 \cdot 10^{-3}$	35	$2.6 \cdot 10^{-3}$
Contact	44.1	$1.173 \cdot 10^{-3}$	45	$1.2 \cdot 10^{-3}$
Fire/explosion	27.7	$2.347 \cdot 10^{-3}$	2	$1.7 \cdot 10^{-4}$
Foundered	55.4	$1.497 \cdot 10^{-2}$	51	$1.3 \cdot 10^{-2}$
Hull/machinery failure	16.2	$5.180 \cdot 10^{-4}$	0	0
Wrecked/stranded	20.2	$2.461 \cdot 10^{-3}$	0	0

weaknesses of the model are that the statistical values are based on a very limited number of events, which result in uncertainties, and that the model does not take sufficient account for the time factor involved in evacuation (i.e., in some cases the crew have more time available than in other cases).

In Table 12.14, the PLL is calculated as the number of fatalities in the given time period (1991–1998) divided by the number of ship years in that period (44,732). The number of fatalities is found by multiplying the statistical number of fatalities (e.g., 116 for collisions) by the probability of fatality.

12.9.3 Step 3: Establish safety measures (risk control options)

Risk control options (RCOs), with the objective of improving the life-saving capability LSAs in a cost-effective manner, were identified and agreed upon by a multidisciplinary team of experts. The following RCOs were some of the measures identified for further assessment in this FSA study:

- Sheltered mustering and lifeboat area (SMA)
- Level alarms to monitor water ingress in all holds and forepeak (LA)
- Individual immersion suits to all personnel (IS)
- Free-fall lifeboats (FF)
- Marine evacuation system for throw overboard liferafts (MES)
- Redundant trained personnel (RTP)

The different RCOs affect the event trees modeled in Step 2 of the FSA (i.e., the risk assessment step), resulting in changes to the probability of fatalities associated with evacuation. The RCOs will affect each accident scenario differently, and consequently the relative changes in the potential loss of

Table 12.15 % reduction in the probability of fatality with RCO implemented

	Current	SMA	LA	IS	FF	MES	RTP
	Reduction in the probability of fatality [%] with RCO implementation						
Collision	45.7	−0.3	0.0	−1.2	−1.1	−0.3	−0.2
Contact	44.1	−0.3	0.0	−0.9	−1.3	−0.1	−0.2
Fire/explosion	27.7	−0.3	0.0	−4.5	+0.4	−1.0	−0.2
Foundering	55.4	−0.9	−14.8	−2.0	−4.9	−0.1	−0.4
Hull/machinery	16.2	−0.5	0.0	−1.6	−3.0	−0.2	−0.4
Wrecked/stranded	20.2	−0.5	0.0	−1.5	−1.8	−0.3	−0.3

life (PLL) value will vary both in terms of the RCOs implemented and the accidental event under consideration. Table 12.15 gives the relative change in percent for the probability of fatality in different accidental events for the RCOs listed above. This table is the underlying basis for calculating the cost-benefit relationship in the next step of the FSA methodology.

12.9.4 Step 4: Cost-benefit assessment

To establish a common and comparable cost-benefit ratio all the potential costs and benefits related to the different risk control options (RCOs) had to be identified. In this particular FSA study the cost estimation was done primarily by contacting suppliers of life-saving appliances, training centers, shipyards, and shipowners. Through these sources, estimates were established on relevant costs such as investment in equipment, installation on the ship, inspection, maintenance, training of personnel to operate the equipment installed, etc. A high- and a low-cost estimate were established to take account of, among other things, factual variability in cost in West Europe and the Far East, and the fact that the proposed technical solutions were not specified in detail. Depreciation of future costs was carried out at a rate of return of 5%.

The benefits obtained by implementing the RCOs were the reduced number of fatalities, which may be calculated as a reduction in the potential loss of life (PLL) parameter using data from Table 12.15. For each RCO, the total obtained reduction in the probability of fatality with RCO implementation was calculated as ΔPLL (in statistical terms the PLL is the expected loss of life). The ΔPLL was calculated as shown for the RCO of "sheltered mustering and lifeboat area" (SMA) in Table 12.16.

The Implied Cost of Averting a Fatality (ICAF) was calculated using the following equation:

$$ICAF = \frac{\text{Cost of RCO}}{\text{Reduction in PLL}} \qquad (12.3)$$

Table 12.16 Calculation of ΔPLL for the RCO "sheltered mustering and lifeboat area"

Type of accidental event	Reduction in probability of fatality [%]	Resulting probability of fatality [%]	Resulting PLL with RCO [per ship year]	Resulting PLL without RCO [per ship year]
Collision	−0.3	45.4	$3.37 \cdot 10^{-3}$	$3.39 \cdot 10^{-3}$
Contact	−0.3	43.8	$1.16 \cdot 10^{-3}$	$1.17 \cdot 10^{-3}$
Fire/explosion	−0.3	27.4	$2.32 \cdot 10^{-3}$	$2.35 \cdot 10^{-3}$
Foundered	−0.9	54.5	$1.47 \cdot 10^{-2}$	$1.50 \cdot 10^{-2}$
Hull/machinery failure	−0.5	15.7	$5.02 \cdot 10^{-4}$	$5.18 \cdot 10^{-4}$
Wrecked/stranded	−0.5	19.7	$2.40 \cdot 10^{-3}$	$2.46 \cdot 10^{-3}$
		Total PLL:	$2.45 \cdot 10^{-2}$	$2.49 \cdot 10^{-2}$
		ΔPLL =		$3.7 \cdot 10^{-4}$

The cost estimates and the reduction in PLL are in this example calculated for a lifetime expectancy of 25 years for all RCOs. This is a simplification of the approach used in the reference source (IMO, 2001). These simplifications are made to reduce the size and complexity of this example and results in slightly different ICAF values than those presented in original FSA report. The final recommendations are, however, the same.

Both a high and a low ICAF value were calculated for each RCO based on the high- and low-cost estimates, respectively. For this particular analysis, the decision criterion was based on an ICAF of £1 million. Other decision criteria may, however, have been selected, and this may have given different recommendations. Table 12.17 shows the calculation of the ICAF values. An RCO is recommended if its low (cost estimate) ICAF value is within the decision criteria of £1 million. A recommendation is considered robust if the high (cost estimate) ICAF value gives the same recommendation as the low (cost estimate) ICAF value.

12.9.5 Step 5: Recommendations for decision-making

Based on the results shown in Table 12.17, the following risk control options provide considerable improved life-saving capability in a cost-effective manner and are therefore recommended:

- Level alarms to monitor water ingress in all holds and forepeak (LA)
- Individual immersion suits to all personnel (IS)
- Free-fall lifeboat (FF)

The RCO of free-fall lifeboats (FF) is only relevant for implementation on new ships as an alternative to more traditional LSAs (e.g., conventional lifeboats). The costs of fitting this RCO to an existing ship would make it

Table 12.17 Calculation of ICAF values

RCO	ΔPLL [per ship year]	ΔPLL [per ship over 25 years]	Cost of RCO over 25 years [1,000£]		ICAF [£10^3]		Recommended?[a]	Robust recommendation?[b]
			Low	High	Low	High		
SMA	$3.70 \cdot 10^{-4}$	$9.25 \cdot 10^{-3}$	10	20	108.1	216.2	NO	YES
LA	$4.00 \cdot 10^{-3}$	$1.00 \cdot 10^{-1}$	14	21	140.0	210.0	YES	YES
IS	$1.27 \cdot 10^{-3}$	$3.18 \cdot 10^{-2}$	15	17.8	471.7	559.7	YES	YES
FF	$1.72 \cdot 10^{-3}$	$4.30 \cdot 10^{-2}$	-7.8[c]	18.2	-181.4	423.3	YES	YES
MES	$1.80 \cdot 10^{-3}$	$4.50 \cdot 10^{-3}$	4.50[d]		Criteria £1m		Not known	
RTP	$1.95 \cdot 10^{-4}$	$4.88 \cdot 10^{-3}$	8	10	1,639.3	2,049.2	NO	NO

[a] A RCO is recommended if its low (cost estimate) ICAF value is within the criterion of £1 million.
[b] A recommendation is robust if the high (cost estimate) ICAF value gives the same recommendation as the low (cost estimate) ICAF value.
[c] Free fall lifeboats have a lower cost than traditional LSAs (e.g., conventional lifeboats) when implemented instead of these on new ships.
[d] The maximum cost of a marine evacuation system to meet the ICAF criterion of £1 million (the costs are likely to be considerably higher).

unattractive in cost-benefit terms. The two other recommended RCOs are relevant for implementation on both new and existing ships.

If all the three recommended RCOs are implemented on a (new or existing) bulk carrier, the risk assessment procedure indicates that the evacuation success rate in the dominating foundering scenario increases from 44.6% (i.e., 1.0 – probability of fatality) to 66.2%, which is quite an improvement. Similar improvements are, however, not present for the other accident scenarios, and the general success rate in evacuation remains rather low also after implementing the recommended RCOs. This should call for additional measures, in particular measures with a focus on crew training and competence building.

12.10 FINAL COMMENTS

Before concluding this chapter, it is important to remind readers once more of the purpose of Formal Safety Assessment. The guidelines describe a process that is intended to support development of safety rules. This is a very different objective from what is the case in most risk assessments. The objective obviously also affects the way these studies are performed. Among others, it may well be that recommendations and conclusions may be very different when considering one specific ship compared to when looking at a large group of ships.

At the same time, the description of the process is also fairly generic and covers the steps that normally would be included in most risk assessments, even if the details are different. Wang (2002) has, e.g., compared it to the safety case approach widely applied in the UK, among others in the offshore industry.

One distinction that is worth noting is the considerable weight that FSA places on CBA. CBA is a tool that is used also in other risk assessments, but it is not common that it has such a central role as it does in the FSA process. It is more common to evaluate proposed risk reduction measures qualitatively and make a decision based on this rather than quantifying costs and benefits.

Sometime after introducing FSA, IMO established an FSA Expert Group to evaluate the FSAs that have been performed. Psaraftis (2012) has raised a number of criticisms regarding FSAs, based on the work in the expert group and some of his points are briefly discussed in the following.

- An industry problem that causes a problem for FSA is that the casualty databases that typically are used in the studies are not publicly available. This means that it is difficult to review and audit the results.
- Many studies seem to confuse cause and effect. In fairness, this is not specific for FSA but can be seen in many studies. The problem with

this may however be that inadequate focus is given in particular to the causal side. There seems to be a lack of understanding of the fact that a scenario in most cases is a whole sequence of events and conditions, leading up to the consequences (loss of life, spills, etc.).

- The use of expert opinion is often not performed in accordance with the recommendations in the FSA guidelines, and the qualifications of the experts applied are also sometimes questionable.
- The link between the hazard identification in Step 1 and the risk assessment in Step 2 is not always clear. An outcome of Step 1 is a ranking of hazards, and this should be used to determine where to focus the quantitative analysis. However, this is not always followed properly.
- The use of GCAF and NCAF is also criticized, and it is proposed that differences rather than ratios should be used to evaluate cost-effectiveness and rank RCOs.

In this chapter, one example of an FSA was included to illustrate the methodology. Many other examples can also be found, e.g., from the SAFEDOR project (SAFEDOR, 2021). Some other applications are:

- Yao-Tian and Jie (2008) – application to planning of VTS
- Zhang and Hu (2009) – traffic risk is coastal water and harbors
- Berle et al. (2011) – vulnerability assessment of a maritime transport system
- Zhang et al. (2013) – navigational risk on the Yangtze River
- Wang et al. (2020) – battery-powered high-speed ferry
- Trbojevic and Carr (2000) – safety improvements in ports

REFERENCES

Berle, Ø., Asbjørnslett, B. E., & Rice, J. B. (2011). Formal vulnerability assessment of a maritime transportation system. *Reliability Engineering & System Safety, 96*(6), 696–705.

Breinholt, C., Ehrke, K. C., Papanikolaou, A., Sames, P. C., Skjong, R., Strang, T., & Witolla, T. (2012). SAFEDOR – The implementation of risk-based ship design and approval. *Procedia - Social and Behavioral Sciences, 48*, 753–764.

DNV GL. (2018). *Rules for Classification of Ships (Part 1 in Chapter 1).* DNV GL AS. https://rules.dnv.com/docs/pdf/DNV/RU-SHIP/2018-01/DNVGL-RU-SHIP-Pt1Ch1.pdf

IMO. (2007). *Formal Safety Assessment-container vessels.* London: International Maritime Organization.

IMO. (2001). *Formal Safety Assessment of life saving appliances for bulk carriers.* MSC/74/5/5. London: International Maritime Organization.

IMO. (2018): *Revised guidelines for Formal Safety Assessment (FSA) for use in the IMO rule-making process.* MSC-MEPC.2/Circ.12/Rev.2, 9 April 2018.

ISO 31000. (2018). *Risk management - guidelines.* https://www.iso.org/standard/65694.html

Psaraftis, H. N. (2012). Formal Safety Assessment: An updated review. *Journal of Marine Science and Technology, 17*(3), 390–402.

SAFEDOR (2021). www.safedor.org

Trbojevic, V. M., & Carr, B. J. (2000). Risk based methodology for safety improvements in ports. *Journal of Hazardous Materials, 71*(1–3), 467–480.

Wang, J. (2002). Offshore safety case approach and formal safety assessment of ships. *Journal of Safety Research, 33*(2002), 81–115.

Wang, H., Boulougouris, E., Theotokatos, G., Priftis, A., Shi, G., Dahle, M., & Tolo, E. (2020). Risk assessment of a battery-powered high-speed ferry using formal safety assessment. *Safety, 6*(3), 39.

Yao-Tian, F., & Jie, W. (2008). Application of Formal Safety Assessment on planning VTS. In *2008 IEEE International Conference on Systems, Man and Cybernetics* (pp. 2207–2212). IEEE.

Zhang, D., Yan, X. P., Yang, Z. L., Wall, A., & Wang, J. (2013). Incorporation of formal safety assessment and Bayesian network in navigational risk estimation of the Yangtze River. *Reliability Engineering & System Safety, 118*, 93–105.

Zhang, J. P., & Hu, S. P. (2009). Application of formal safety assessment methodology on traffic risks in coastal waters & harbors. In *2009 IEEE International Conference on Industrial Engineering and Engineering Management* (pp. 2192–2196). IEEE.

Chapter 13

Security

13.1 INTRODUCTION

International seaborne transport today engages a large number of vessels operating globally and calling on numerous ports in Europe, Asia, and America. It involves feeder transport, off-loading, storage and loading in ports, and ocean transport. The transport process is further associated with complex logistical processes based on information technology. Finished products and non-bulk cargo are today almost exclusively transported in containers. The container concept raises a basic challenge with respect to the control or monitoring of the cargo transported in this manner.

There is a broad range of threats to the maritime trade that may affect ports, cargo, ships, and information systems. A brief list of the main threats is:

- Terrorism: Actions with the intent to cause death to civilians, intimidate populations or compel civil organizations to abstain from certain acts. Crew and passengers on vessels have also been targeted.
- Bombing: Placement of explosive devices or bombs in cargo in order to destroy vessels, port facilities, and even greater destruction.
- Piracy: Attack on merchant vessels, hijack, and demand of ransom for release of vessel, cargo, and crew.
- Stowaways: Persons found on board without authorization. Main categories are people seeking political asylum, immigrants with economic motives, and refugees.
- Drug traffic: Smuggling of drugs concealed in legal cargo or containers. Shipping by sea is attractive due to the long distances from production sites to the user markets.
- Theft may take place on board a vessel and in the port.
- Harmful cargo: Introduce biological or chemical agents that will harm personnel handling the cargo or ultimately the consumer.

DOI: 10.4324/9781003055464-13

13.1.1 Threats in the delivery phase

The security of seaborne transport cannot be seen isolated from the other elements in the cargo supply chain:

- Factory: Production and loading phase
  - Barriers, gates, and physical protection
  - Access control and monitoring systems
  - Personnel policies and procedures
- Export phase: Carriers and Seaports
  - Is the shipper known; any irregularities to type of cargo or consignee
  - Any changes to the transit phase
  - Physical inspection of containers upon arrival, control of documents
- Port
  - External and perimeter security ring (barriers and control)
  - Internal security procedures
  - Control and monitoring of internal assets and equipment
  - The security of the port–vessel interface
  - Personnel policies and procedures

13.1.2 Threats to seaborne transport

The security of tanker shipping was highlighted as a result of the Iran-Iraq war. Both port facilities and tankers were subject to more than 200 attacks in the Persian Gulf during the war. Although these events could not be characterized as piracy in the traditional way, it demonstrated that modern weapons have sufficient power to disable and sink modern tankers. On October 6, 2002, VLCC *Limburg* was carrying 397,000 barrels of crude oil from Iran to Malaysia and was in the Gulf of Aden off Yemen to pick up a cargo of oil. The vessel was hit by terrorist suicide bombers running a dinghy with explosives into the side of the tanker. The vessel caught fire and resulted in a spill of 90,000 barrels of crude oil. One crew member was killed and 12 were injured.

It is evident that oil tankers have large loss potential both physically, environmentally, and economically. Tankers are due to their large size, relatively moderate speed, and limited maneuvering ability vulnerable to attacks. LNG and LPG carriers have a similar or even greater loss potential. Jones (2006) indicates that a bomb like the one against Limburg would release up to half of an LNG load and ignite it within 3 minutes. Bulk Carriers seem to be less attractive for criminal attacks but may be attacked and sunk in vulnerable seaways. A sunk vessel in a port, canal, or narrow seaway may have serious and costly consequences for sea traffic and commercial activity. Although not a terrorist attack, the effect blocking of important waterways can have on trade was seen when the container ship Ever Given blocked the Suez Canal for six days in 2021.

Containerships are less subject to attack as they operate at higher speed and have greater freeboard. The risk related to container vessels is primarily related to the container cargo which is a potential vehicle for smuggling and delivery of explosives and harmful material targeted at the destination port or country.

13.1.3 Cyber threats

The majority of technical systems on a vessel today are computer-based and also to a large degree automated. Typical functions are propulsion, control, communication, cargo logistics, and navigation. Vessel systems are therefore susceptible to hostile penetration and manipulation or what may be termed as a cyberattack as a part of criminal action like theft, hijacking, or smuggling. The chance that cyberattacks will be detected, investigated, and even brought to justice is still today limited. A well-known and fascinating example of a cyberattack activity happened in the port of Antwerp in 2013 and made possible the smuggling and dispatch from the port of numerous containers of Colombian cocaine and heroin for a period of more than 2 years. In January 2014, a UK cybersecurity firm found flaws in one ECDIS system that would allow an attacker the ability to access and modify ship electronic navigation charts.

The vulnerability of marine IT systems is to a large degree related to the application of outdated technology and security philosophies and more specifically too simple firewall and antivirus software. This situation may also be explained by a general lack of competence and motivation to deal with the matter in the maritime sector. Currently the industry lacks serious standards, unlike other industries, in order to protect against cyberattacks.

13.2 IMPROVING THE SECURITY IN THE CARGO SUPPLY CHAIN

The focus on the security in airports and seaports too was raised after the attacks of September 11, 2001. The immediate effect was closure of seaports, which led to disruption and delay of cargo movement. In the aftermath of this tragedy, the US government took several initiatives to improve security. Important work has also taken place within the framework of IMO. Today, the primary focus is on the ports and cargo-related information processes rather than only the activities related to the seaborne transport phase.

13.2.1 US legislation

A fundamental element in the American security regulation is the *Maritime and Transportation Security Act* (MTSA) of 2002 (Jones, 2006) that addresses port security:

- Training of port officers and other personnel
- Security assessment and plans
- Authority of USCG (US Coast Guard) to issue additional programs and directives
- Flexibility: Acceptance to adjust the security level to the updated threat level
- Customs, trade, and security provisions

Other measures are:

- *Automated Manifest System* (AMS): Automated filing of manifests for cargo to and through US ports.
- *The Carrier Concept:* Requirement to take a formal role as carrier for a shipment and with specific requirements and registration as such.
- *Container Security Initiative* (CSI): A program for establishing agreement between the USA and trading partners for screening and inspecting containers.
- *Required Advance Electronic Presentation of Cargo Information* (RAECPI): US Customs and Border Protection (CBP) requires the manifest information 24 hours ahead of loading of cargo bound for US ports.
- *Customs-Trade Partnership Against Terrorism* (C-TPAT): A voluntary agreement that all parties in the cargo supply chain perform security self-assessments in order to reduce government security screenings.

Jones (2006) gives a more complete overview of US security measures.

13.2.2 ISPS Code

The International Ship and Port Facility Security Code (ISPS) Code was adopted by IMO and came into force in 2004. It is a set of measurements for international security by involving government authorities, ports, shipping companies, and seafarers. The primary objective of the ISPS Code is to provide a standardized, consistent global framework across the maritime world. The intention is to enable the countries to assess the security threats to the ships calling at their ports and take appropriate measures. The aims of ISPS are as follows:

- Define respective roles and responsibilities of all parties (governments, port administration, and the shipping and port agencies) concerned, at both a global and domestic level.
- Monitor the security relevant activities of people and cargo operation.
- Identify security threats on board vessel and in port and implement necessary measures.

- Define roles and responsibilities of the participants in the cargo supply chain: governments, agencies, local administrations, and the shipping and port industries.
- Define the roles and responsibilities of the Port State officer and onboard officers.
- Collect data concerning security threats and as a mean to eliminate them.
- Ensure that security-related information is exchanged by the parties.
- Provide methodologies for security assessment so that security plans are updated with changing threats.
- Shipping and ports are required to place appropriate security officers and personnel on each ship, in each shipping company, and in each port facility to prepare and to put into effect the security plans that will be implemented.
- Give guides for continuous improvement of ship and port security plans and measures.

The ISPS Code has two main sections: Part A which is mandatory and includes the maritime and port security-related requirements which should be followed by the governments, port authorities, and shipping companies. Part B provides guidelines on how to meet these requirements. Except for the non-mandatory Part B, the ISPS Code refers to a number of IMO resolutions, circulars, and letters, which are guides to be used by personnel with security duties.

The ISPS Code further defines three security levels:

- ISPS Security Level 1 – normal – when the ships and port facilities operate under normal conditions. Require minimum protective measures.
- Security Level 2 – heightened – will apply whenever there is a heightened risk of a security incident. Additional security measures will have to be implemented and maintained for that period.
- Security Level 3 – exceptional – when a security threat is imminent and SPECIFIC security measures will have to be implemented and maintained for that period. The security experts will work in close conjunction with government agencies and possibly follow specific protocols and instructions.

The responsibility for the implementation of ISPS requirements is the designated personnel established by the Code. Personnel assigned with these duties are in charge to assess, plan and implement the security measures:

- The Port Facility Security Officer (PFSO) is an authorized officer at the port who is responsible for the port facility security plan. PFSO is also responsible for liaison with the SSO and CSO.

- The Ship Security Officer (SSO) is the designated person on board the ship, responsible to implement and maintain the ship security plan and the overall security of the ship.
- The Company Security Officer (CSO) is the person overseeing that ship security assessment is carried out properly. Moreover, the CSO is responsible to develop the ship security plan, submit it for approval and ensure that it is implemented as appropriate.

A common issue related to ISPS Code has to do with its relationship to the ISM Code and about which is more important of the two – safety or security? In shipping, both safety and security are equally needed. Secondly, the implementation of both Codes means additional costs for operators and sometimes this leads to a dilemma on what is more important to be maintained. The general view seems to be that in cases of conflict between safety and security the former has first priority. This is also reflected in SOLAS, Chapter XI-2, Regulation 8 (IMO, 2009).

Another challenge related to security of the ship is the absence of an official monitoring procedure as there are no formal audits, either internal or external, in contrast to the ISM audits. This is also the case with most PSC inspections except for USCG ports where PSC officers also focus on security. But there is presently a positive development regarding PSC. Another issue is that because of confidential information included in Ship Security Plans, maintenance, and revisions are difficult to undertake. It has also been criticized that the ISPS Code is based solely on the 9/11 outcome and the early piracy activity in the area of Somalia, while no amendments or revisions have been made to the code based on experience from new types of security threats, including, e.g., cyberthreats.

13.2.3 SOLAS

SOLAS includes a separate chapter on security, Chapter XI-2 "Special measures to enhance maritime safety" (IMO, 2009). The chapter applies to ships engaged on international voyages and port facilities servicing such ships. A significant part of the SOLAS chapter place requirements on governments, but there are also specific requirements for ships. Some requirements worth mentioning are as follows:

- Administrations shall set security levels and ensure that this information is provided to all ships flying their flag.
- Ships shall comply with the security levels set by the Administration for that ship before entering port.
- All ships constructed after July 1, 2004, shall have a ship security alert system that can transmit a ship-to-shore security alert that identifies the ship, its location and that the security has been compromised.

- The Master can, at his own discretion, make and execute any decision deemed necessary to maintain the safety and security of the ship. If there is a conflict between safety and security, safety should have priority.

13.2.4 ISM Code

In 2017, IMO adopted an amendment to the ISM Code, including guidelines on maritime cyber risk management (IMO, 2018). The guidelines are brief and very general, and contain little detail about how cyber risk management should be approached. In principle, cyber risk management should be seen as an integrated part of the overall risk and safety management of the company and the ships.

The guidelines list potentially critical systems and underline the difference between what is called information systems and operational systems. Operational systems are systems that control a process that is part of the operation of the ship, e.g., the engine control system.

In risk assessment, we normally talk about hazards, while in security it is more common to use the terms threats and vulnerabilities. Threats (or threat actors) are more or less equivalent to hazards, while vulnerabilities are weaknesses in the systems that can be exploited by threats, potentially leading to negative consequences.

The guidelines refer to the following documents for more detailed information:

- The Guidelines on Cyber Security Onboard Ships produced and supported by BIMCO, CLIA, ICS, INTERCARGO, INTERTANKO, OCIMF, and IUMI (BIMCO, 2016).
- ISO/IEC 27001:2013: Information technology – Security techniques – Information security management systems – Requirements (ISO/IEC, 2013).
- Framework for Improving Critical Infrastructure Cybersecurity Version 1.1, April 2018, National Institute of Standards and Technology (Barrett, 2018).

13.2.5 Container security

Container transport is a central issue in the discussion of the security of seaborne transport. The integrity of the container is to a large degree dependent on the logistical management and control activities in the port. Scholliers et al. (2016) have proposed three methodological approaches:

1. Non-intrusive identification and detection technologies
2. Monitor the surroundings of the container
3. Adding intelligent equipment to the cargo or the container

Table 13.1 Container security issues and solutions

Gate process

Queues and parking	Booking of transports; assign time slots; safe parking areas outside of the port with CCTV surveillance.
Available documents	Prior to arrival in the port, all information about container, vehicle, driver, and seal should be available for risk assessment.
Data security	The information system in the port should have a high security level and be protected against access by unauthorized personnel.
Driver identification	Stopping of drivers and use of biometric identification.
Seal identification	Mechanical bolt seals may be replaced with electronic seals that allow identification without stopping the transport.
Physical damage inspection	Apply imaging techniques for the container and the undercarriage.
Container weight	Weighing with sensors in the port.

Customs process

Inspection for radiation	100% scanning is required for US ports. Scanning technology is present in some ports but has the drawback of giving many false positives due to the presence of Naturally Occurring Radioactive Materials (NORMs).
Inspection for gases	Putting a sensor through the container fittings. The method is labor and time intensive.

In the yard

Truck monitoring	Monitoring that driver does not deviate from the prescribed route by means of area surveillance system integrated with other systems. Camera images shall be linked to geographical location. C-ITS communication with driver.
Container in the yard	Door or intrusion sensors in the container. Container Security Devices (CSD). Standardization of these devices is needed.
CCTV and video analytics	Use of digital IP HD camera and Intelligent Video Analytics.
Link between information systems	Necessary to link video surveillance terminal operating system.
Drones for additional surveillance	Airborne micro-drones can supplement CCTV systems.

Scholliers et al. (2016).

Table 13.1 gives a summary of specific methods and techniques that are utilizing modern technology.

13.3 SECURITY ONBOARD

The CSO and SSO were mentioned earlier as having key roles in achieving security onboard, being responsible for respectively developing and implementing the security plan. The main tasks relating to ship security are as follows:

- Ship security assessment
- On-scene ship security survey (OSS)
- Establish and update ship security plan (SSP)
- Declaration of security (DoS): Facilitation cooperation between ship and port
- Records keeping: Log of all activities related to the security activities
- Continuous synopsis record (CSR): Provide transparency with respect to vessel ownership, security control organizations, and operating history
- Training and education
- Verification
- Certification

13.3.1 Compliance costs for the shipowner

The cost of the implementation of security regulations has been both analyzed and debated (Bichou, 2008). It is two opposing views on the joint relation between the commercial operation and the security processes: on the one hand, one may argue that the procedural requirements of the security regime act against the operational and logistical efficiency. The other view goes like this: The security procedures, despite from being mandatory, may be commercially rewarding by reducing sources of delay in cargo transport chain. It is also considerable disagreement about the cost level of the security regime. Studies have been undertaken both in the planning process and after its implementation. Even organizations like the World Bank and OECD disagree on both the initial and recurrent cost.

It has been proposed to apply a cost-benefit approach by the shipowner when establishing a security program for a specific ship or a fleet of similar vessels (Bichou, 2008). Based on expert opinion, surveys, and simulation studies, Diop et al. (2007) gave estimates on the efficiency of alternative security measures. The company may invest an amount of economical units, $S = S_1 + S_2 + \dots S_{15}$, in a program with 15 components. This investment will give the following security benefits in economical units: $Y = Y_1 + Y_2 + \dots + Y_{15}$. A summary of the benefits for a general case is shown in Table 13.2. The analysis indicates that the most efficient measure was access control to the ship (S11) by potentially deterring 45.8% of all relevant incidents. Patrol of cargo area (S10) is also efficient. It is more complicated to assess the finding of low efficiency of having a ship security officer (SSO). One may question the idea of incorporating the SSO function in the study as this mandatory anyway. Secondly, it is more complicated to model the effect of a management function than a technical system with limited or well-defined scope. The validity and complexity of the study is also reflected in the relatively high standard deviation for the estimates.

Table 13.2 Performance of security components

Component	Description	Performance:Y_i	Standard deviation
S1	Install security alarms	0.256	0.264
S2	Company security officer (CSO)	0.340	0.574
S3	Ship security officer (SSO)	0.254	0.535
S4	Surface radar	0.121	0.392
S5	Auto CCTV on ship	0.213	0.217
S6	Security patrol	0.283	0.237
S7	Auto CCTV cargo	0.153	0.134
S8	Security alarms – general	0.216	0.392
S9	Security alarm – stores	0.188	0.141
S10	Security patrol cargo areas	0.435	0.185
S11	Control of access to ship	0.458	0.315
S12	Control – embarkation of persons	0.354	0.371
S13	Monitoring restricted areas	0.138	0.123
S14	Monitoring of deck areas	0.175	0.154
S15	Security training drills	0.341	0.116

Bichou (2008).

13.4 PIRACY IN HISTORY

The history of piracy is almost as old as the first written historical sources (Beare, 2012). One of the first known attacks took place in the fourteenth century BC in Asia Minor where the Lukkans raided Cyprus. Later, when Athens raised in power, they were unable to fight and eliminate pirates in the Aegean Sea. The Romans were more successful, and around first century BC, a Pompey fleet eliminated the problem along the main trading routes. Even in these early periods piracy could flourish due to geography (narrow coastal waters), lack of naval power, and also support from rivaling states. The Mediterranean were after the collapse of the Western Roman Empire subject to piracy for centuries as long as up to early 1800. The Vikings were not classical pirates as they mainly attacked villages and towns near the coast. Their activity was eventually brought to an end by the Hanseatic League. In the sixteenth century, English pirates attacked Spanish vessels coming from the West Indies and the New World. They were later followed by French and Dutch pirates as these ships were attractive bounties laden with silver, gold, and attractive commodities. First it took place in the eastern Atlantic and later on in the Caribbean. Due to the economic potential of piracy, it was in fact legalized by rivaling states during periods. Piracy is also well known from early periods in the Persian Gulf, Cape Good Hope, the Malay Archipelago, and Hong Kong and Macao.

13.4.1 Definition of piracy today

During the 1980s, piracy became a major problem to international shipping. Piracy is defined in Article 101 of the 1982 United Nations Convention on the Law of the Sea (UNCLOS) as follows:

(a) *any illegal acts of violence or detention, or any act of depredation, committed for private ends by the crew or the passengers of a private ship or a private aircraft, and directed*
  (i) *on the high seas, against another ship or aircraft, or against persons or property on board such ship or aircraft;*
  (ii) *against a ship, aircraft, persons or property in a place outside the jurisdiction of any State;*
(b) *any act of voluntary participation in the operation of a ship or of an aircraft with knowledge of facts making it a pirate ship or aircraft;*
(c) *any act of inciting or of intentionally facilitating an act described in subparagraph (a) or (b).*

The IMO defines Armed Robbery in Resolution A.1025 (26) "Code of Practice for the Investigation of Crimes of Piracy and Armed Robbery against Ships" as:

1. *any illegal act of violence or detention or any act of depredation, or threat thereof, other than an act of piracy, committed for private ends and directed against a ship or against persons or property on board such a ship, within a State's internal waters, archipelagic waters and territorial sea;*
2. *any act of inciting or of intentionally facilitating an act described above.*

Piracy may be defined as follows in a less legal language (Tumbarska, 2018):

- Maritime robbery: Basically, the same as any type of armed robbery with focus on personal belongings or valuables and will primarily take place in a port. May also be the outcome when a greater operation went wrong.
- Hijacking: An attack with the intention is to steal a vessel or the cargo. The intention may be to sell both the ship and the cargo in an illegal market. The crew may be subject to violence during the operation.
- Kidnapping: Both crew and passengers may be taken hostage. Will require extensive planning, intelligence work, commitment to use force, and complicated negotiations.

13.5 PIRACY TODAY

In October 1985, four Palestinian men hijacked an Italian cruise ship, *Achille Lauro* (Figure 13.1), holding hundreds of hostages for two days.

Figure 13.1 Achille Lauro (<u>CC-BY-SA 3.0</u>, D.R.Walker).

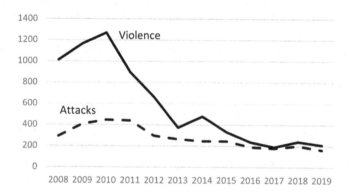

Figure 13.2 Annual number of piracy attacks and persons subject to violence (IMB, 2019).

The hijackers killed a disabled, 69-year-old Jewish American and threw his body into the sea.

In the coming years after these and other events, the security of merchant shipping became a major concern for both commercial organization and government organizations. The number of attacks increased to a peak in 2010. Statistics shows that the problem has decreased during the last decade. Figure 13.2 shows that piracy had its last peak in year 2010 with

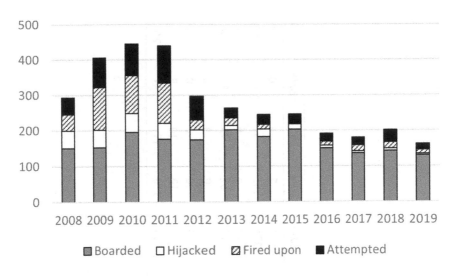

Figure 13.3 Annual number of terrorist attacks (IMB, 2019).

445 attacks that affected 1,270 persons (primarily crew). Since then, the number of attacks has decreased to one-third and the number of violent events to one-sixth.

The relative numbers of the type of attacks have changed during the last decade (Figure 13.3). The number of boardings has remained fairly constant whereas the number of hijackings and vessels fired upon has gone significantly down. The number of attempted attacks has also decreased but somewhat less. A breakdown of the type of violence is shown in Table 13.3. Around 2010 the major threat was to be taken as hostage. Recently, this has become less frequent but still represents more than 25% of the instances. Today kidnapping and ransom represent the greatest risk or almost 65% of the violence. The threat has moved from hostage taking to kidnap/ransom but have basically the same economic motive and together they represent 90–95% during the decade. The remaining forms like assaults, injured, killed, missing, or threatened are less probable outcomes.

13.5.1 Locations of piracy

The ICC International Maritime Bureau (IMB, 2019) divides their piracy statistics on the following regions and main countries:

- Southeast Asia: Indonesia, Malacca Straits, Malaysia, Philippines, Singapore Straits, and Thailand
- East Asia: China and Vietnam
- Indian sub-continent: Bangladesh and India

Table 13.3 Type of violence in relation to terrorist attacks

Type of violence	2008	2009	2010	2011	2012	2013	2014	2015	2016	2017	2018	2019
Assaulted	7	4	6	6	4	0	1	14	5	6	0	3
Hostage	889	1,050	1,174	802	585	304	442	271	151	91	141	59
Injured	32	69	37	42	28	21	13	14	8	6	8	7
Kidnap/ransom	42	12	27	10	26	36	9	19	62	75	83	134
Killed	11	10	8	8	6	1	4	1	0	3	0	1
Missing	21	8	0	0	0	1	1	0	0	0	0	0
Threatened	9	14	18	27	13	10	9	14	10	10	9	6
Total	1,011	1,167	1,270	895	662	373	479	333	236	191	241	210

Annual number (IMB, 2019).

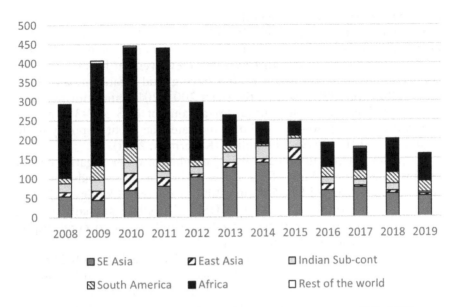

Figure 13.4 Regional distribution of piracy. Annual number of attacks (IMB, 2019).

- South America: Brazil, Colombia, Ecuador, Peru, and Venezuela
- Africa: Cameroon, Ghana, Mozambique, Nigeria, Congo, and Togo
- Rest of the world

During the period from 2008, the African coast has been the main location for piracy (Figure 13.4), in particular the period up to 2011. Around year 2010, the increase in attacks off the coast of Somalia raised international concern. The number of attacks in African waters has been reduced in recent years due to the success of controlling the activity in Somalia, and in recent years the highest number of attacks in Africa has been in the Gulf of Guinea and related to oil shipments from Nigeria and Angola. The second major region is SE Asia with Malaysia, the Philippines, and Singapore Straits as the critical areas. The number of attacks in this region has been stable, indicating that the anti-piracy measures have been less successful compared to Africa.

13.5.2 Somali piracy

During the early 1990s, piracy off the coast of Somalia became an increasing concern to international shipping. The number of attacks surged between 2005 and 2011 and got the attention of both the industry, governments, and international organizations. The pirates mainly attacked vessels and kept cargo and held crew hostage with the aim of negotiating release in exchange for ransom. During this period 3,741 crew members were held hostage for as long as 1,178 days (World Bank, 2013). In December 2009, the pirates held at least 14 vessels and hundreds of crew members hostage. At the time a Greek ship with Ukrainian crew were in hostage for nine months before being released and after having paid a ransom of US$ 2.6 million (Rotberg, 2009). During the years 2005–2011, an estimated number of 149 ships were held ransom for a value of US$ 315–385 million (World Bank, 2013). Another estimate indicates the profit of piracy in the single year of 2010 to US$ 300 million (Kellerman, 2011).

The piracy activity in Somalia had a number of other negative effects. Nearby economies in East Africa saw a decline in tourist industries and fishing catches dropped by 25% after 2006. Countries like Egypt, Kenya, Yemen, and Nigeria have lost around US$ 1.25 billion annually in trade revenue. The costs of insurance increased and similarly the cost of transport due to re-routing to avoid high-risk waters.

The pirates are primarily young men and have backgrounds like fishermen, ex-militiamen, and technical experts. They attack with relatively small skiffs that can reach speeds of 25 knots. They may operate in groups of three units and attack from the quarter or stern and usually in morning daylight. The target is mainly slow-speed vessels (less than 15 knots) and vessels with small freeboard, but even greater and more modern vessels have been targeted. Supported by motherships like captured fishing and merchant vessels, the pirates may operate up to 1,000 nm off the coast.

The piracy activity was supported by a well-developed land-based infrastructure supported by clan leaders and Islamist insurgent groups motivated by the profit from the activity. Apart from the armed offshore operations the onshore support provides basic goods and services in order to guard hijacked vessels and support lengthy ransom negotiating processes. The location of the pirate activity and anchorage of hijacked vessels are secured by local militia, clan members, and religious leaders. This activity has developed to a degree that led to the establishment of a 24-hour stock exchange in the town of Haradhere. It provided weapons, funding, and soldiers and got its share of the proceeds. It was even indicated that the piracy activity had economic ties to communities in neighboring countries like Saudi Arabia, Dubai, and Yemen, but also as far as Canada and the USA.

13.6 COMBATING PIRACY

13.6.1 Combating piracy in Somalia

It was gradually acknowledged that strong cooperation was required at the political, economic, legal, diplomatic, and military levels, as well as collaboration between diverse public and private sector stakeholders to fight piracy in the Gulf of Aden, Indian Ocean, and Arabian Sea. Cooperation efforts in the region have involved governments, regional organizations, intergovernmental organizations, and the shipping industry. It was realized that there is a number of obstacles to successful elimination of piracy:

- Lack of alternative employment: Decreased fishing resources and pollution
- Pirates have support and protection from local clans
- Insurance companies view protection measures too costly compared to the risk
- Political instability and widespread corruption
- Too much focus on combating piracy at sea rather than solving the political problems on land

13.6.2 Legal framework

Numerous attempts were made to codify law against piracy. The first relevant provisions were given in *Convention on the High Seas* that was adopted in 1958. Today the international legal framework for fighting piracy is laid down in United Nations Convention on the *Law of the Sea* (UNCLOS). However, the convention does not cover procedures for prosecution of pirates and liability issues relating to anti-piracy measures (UNCTAD, 2014). Some of the most relevant articles are shown in Table 13.4.

Other conventions that are relevant are as follows:

a) The Convention for the Suppression of Unlawful Acts Against the Safety of Maritime Navigation, 1988, and its Protocols;

Table 13.4 Relevant articles in UN Convention 1958 and UNCLOS 1982

Convention on the High Seas, 1958	United Nations Convention on the Law of the Sea (UNCLOS), 1982	Relevant articles
Article 14	Article 100	Duty to cooperate in the repression of piracy
Article 15	Article 101	Definition of piracy
Article 16	Article 102	Piracy by a warship, government ship, or government aircraft whose crew has mutinied
Article 17	Article 103	Definition of a pirate ship or aircraft
Article 18	Article 104	Retention or loss of the nationality of a pirate ship or aircraft
Article 19	Article 105	Seizure of a pirate ship or aircraft
Article 20	Article 106	Liability for seizure without adequate grounds
Article 21	Article 107	Ships and aircraft which are entitled to seize on account of piracy
Article 22	Article 110	Right of visit
Article 23	Article 111	Right of hot pursuit

UNCTAD (2014).

b) The International Convention Against the Taking of Hostages, 1979;
c) The United Nations Convention on Transnational Organized Crime, 2000.

The legal framework provides several key definitions (UNCTAD):

- Geographic scope: Piracy refers to an act "on the high seas" or in other words not in territorial waters.
- Private ends: Piracy is limited to acts for private ends, excluding politically motivated acts.
- Two-ship requirement: One private vessel is attacking another.
- Pirate ship: A ship is a pirate ship if the intention is to dominate another ship. A hijacked vessel will also be a pirate ship.
- Enforcement: Any state has universal jurisdiction to seize pirate ships and property on board and arrest persons.
- The right of visit: A government ship may board a ship on suspicion of being involved in pirate activity.
- The right of hot pursuit
- Liability and compensation provisions: Compensation for seizure without adequate grounds.

13.6.3 Military actions

Three international naval task forces have been deployed in the region:

- *Operation Enduring Freedom* by the Combined Task Force 150
- Counter-piracy operations by Combined Task Force 151
- *Operation Atalanta* by EU naval task force

The anti-piracy operations were coordinated through a monthly planning conference called *Shared Awareness and Deconfliction* (SHADE). It had representatives from NATO, the EU, and the *Combined Maritime Forces* (CMF) HQ in Bahrain and later on with representatives from more than 20 countries. From 2014 also Indian, Australian, and Chinese naval forces were participating. The broad participation was somewhat unique in the sense that countries in the UN Security Council which traditionally are opponents in foreign policy matters were able to cooperate and coordinate naval operations.

13.6.4 Economic reform

Policing of the waters with naval resources do not offer any long-term solution to piracy problem in Somalia. It is necessary to develop political stability and economic reform and thereby offer alternative occupations to piracy.

I

The Somali fishing industry is a central factor as the majority of people recruited as pirates are ex-fishermen (Glaser et al., 2015).

Before the outbreak of the civil war in 1991, the fishing industry had been in a period of development and growth. In order to compensate for a series of droughts in the 1970s the government actively stimulated the sector, and one measure was to establish village-based cooperatives.

Another reason for the collapse of the fishing industry was the over-fishing and depletion of stocks by illegal and unregulated activities by large foreign (European and Asian) longline vessels. This activity focused on tuna and local species like lobsters and squids. A second factor to the decline was a tsunami in 2004 due to an earthquake in the Indian Ocean. It resulted in destruction of fishing communities along 650 km of the coast. Looking to the future it was clear that the fishing industry still could have great potential for the Somali economy. A number of projects were therefore initiated to this end.

13.7 VESSEL SECURITY AGAINST PIRACY

Despite the fact that the frequency of pirate attacks has been reduced since 2011, it is still a major threat in West Africa (Gulf of Guinea) and Southeast Asia. It is evident that this is a challenge to local governments and inter-governmental organizations but also to the shipping companies themselves. *BIMCO, INTERTANKO,* and other commercial organizations are offering so-called best management practice (BMP5) for how to prepare to avoid and deal with piracy (ICS, 2018). The following approach is recommended:

- Understand the threat
  - Maritime threats are dynamic
  - Obtaining current threat information is critical for risk assessment and decision making
- Conduct risk assessments
  - Companies must conduct risk assessments
  - Identify ship protection measures
- Implement ship protection measures
  - Harden the ship
  - Brief and train the crew
  - Enhanced lookout
  - Follow Flag State and military guidance
- Report
  - Report to UKMTO (United Kingdom Maritime Trade Operations: www.ukmto.org) and register with MSCHOA (Maritime Security Centre – Horn of Africa: www.mschoa.org)
  - Report incidents and suspicious activity
  - Send distress signal when attacked

- Cooperate
  - Cooperate with other shipping and military forces
  - Cooperate with law enforcement to preserve evidence
  - Cooperate with welfare providers

13.7.1 Risk assessment of vessel

Assessment of the security risk ought to be addressed in the same manner as risk assessment in general and viewed as an integral part of the company safety management system. BMP5 refers also to *Global Counter-Piracy Guidance* (www.maritimeglobalsecurity.org). The risk assessment process should address:

- Requirements of the Flag State, company, charterers, and insurers.
- The threats related to the geographical areas of increased risk.
- Background factors like traffic patterns, local patterns of life, and fishing vessel activity.
- Cooperation with military. An understanding of presence should be obtained from UKMTO.
- The embarkation of Privately Contracted Armed Security Personnel (PCASP).
- The ship's characteristics, vulnerabilities, and inherent capabilities, including citadel and/or safe muster points to withstand the threat (freeboard, speed, general arrangement, etc.).
- The ship's and company's procedures (drills, watch rosters, chain of command, decision-making processes, etc.).

BMP5 stresses that the threats are dynamic and that it is essential that the risk assessment document is completed or updated ahead of each voyage or activity in the targeted waters. Following preventive measures are fundamental elements of the voyage plan:

- Review of the Ship Security Assessment (SSA), Ship Security Plan (SSP), and Vessel Hardening Plan (VHP). VHP is in practice a description of what measures are required to reduce the security risk to the ship. These may also be called security barriers (see Chapter 15).
- Guidance to the Master about the recommended route, updated plans, and requirements for group transits and national convoys.
- Company mandated Ship Protection Measures (SPM).
- Due diligence of Private Maritime Security Companies (PMSCs) for the possible use of PCASP.
- Companies should consider the placement of hidden position transmitting devices as one of the first actions of hijackers is to disable all visible communication and tracking devices and aerials.

- Review of company manning requirements. Consider disembarking of non-essential crew.
- Crew training plans.
- Information security: Communication related to the voyage plan should be kept at a minimum and restricted to as few parties as possible as required by contracts.

13.7.2 Operative measures

Before entering a Voluntary Reporting Area (VRA) BMP5 advocates the following actions:

- Update threat information and latest NAVAREA warnings. NAVAREA are the maritime geographic areas in which governments are responsible for navigation and weather warnings.
- Registration and reporting to VRA/MSCHOA.
- Confirm PCASP embarkation plan.
- Implement SSP before entering High-Risk Area (HRA).
- Perform drills of test access points, lockdown conditions, bridge teams' security knowledge, and roles of crew members.
- Testing of emergency communication plan.
- Define AIS policy.
- Reschedule planned maintenance when transiting an HRA.
- Avoid slow steaming and anchoring.
- Minimize VHF traffic and alternatively use satellite telephone or email.

13.7.3 Protection of vessel

The security protection measures (SPM) proposed by BMP5 are structured in three sections as outlined in Table 13.5 and involve both technical installations and procedures.

Table 13.5 Ship protection measures

Primary layer of defense	Secondary layer of defense	Last layer of defense
• Good lookout/vigilance • Arrange razor wire • Focus on maneuvering • Speed/freeboard • Armed security personnel	• Door hardening • Gate/grate • Motion sensor/CCTV	• Internal door hardening • Citadel/safe muster point • Communication

BMP5 ICS (2018).

13.7.4 Cyber security and risk management

Operation of ships is today heavily dependent on digital solutions for both operation of the ship as such and also the logistical management of cargo. International seaborne transport is therefore vulnerable to cyberattacks or failure of information (IT) and operational technology (OT). Shipping is based on the involvement of many stakeholders and thereby is subject to systems that may have security problems like outdated software or lack of focus on security.

BIMCO (not dated) has in cooperation with major shipping organizations written a guideline on cyber security within the framework of international regulations:

- IMO, 2017, Resolution MSC.428(98) on Maritime Cyber Risk Management in Safety Management Systems (SMS). International Maritime Organization (IMO).
- NIST, 2018, Cybersecurity Framework. Version 1.1. US National Institute of Standards and Technology.

The BIMCO guideline follows the general approach in risk management.

1. Cyber security and risk management: The starting point is to assess the relation between shipowner and manager, shipowner and agent, and shipowner and vendors and other external parties.
2. Identify threats: IT and OT systems may be subject to both unintended harm and malicious attacks with different motives like activism, criminal, or terrorism. The threats may be both untargeted (malware, scanning, etc.) or targeted (passwords, phishing, etc.).
3. Identify vulnerabilities: Common vulnerabilities are obsolete/unsupported operating systems, outdated antivirus software, and inadequate security configuration to mention a few forms. The vulnerability is both associated with the ship based systems but well as important is the interface with external systems. It is several onboard systems like cargo, bridge, propulsion/machinery, access control, and so on.
4. Assessing the likelihood: It is advocated to apply a risk assessment matrix where the probability of having a security breach is measured on an ordinal scale from 1 (unimaginable) to 5 (frequent). It is pointed out that relevant data sources are scarce and that data from similar sectors and technologies.
5. Impact assessment: The CIA model may be a guide for assessing the impact in terms of loss of confidentiality/integrity/availability. The risk matrix has similarly a five-level scale for the impact: 1 (no impact) to 5 (fatalities/significant damage). Factors to assess are fatality/injury and damage environment, assets, finance, and reputation.

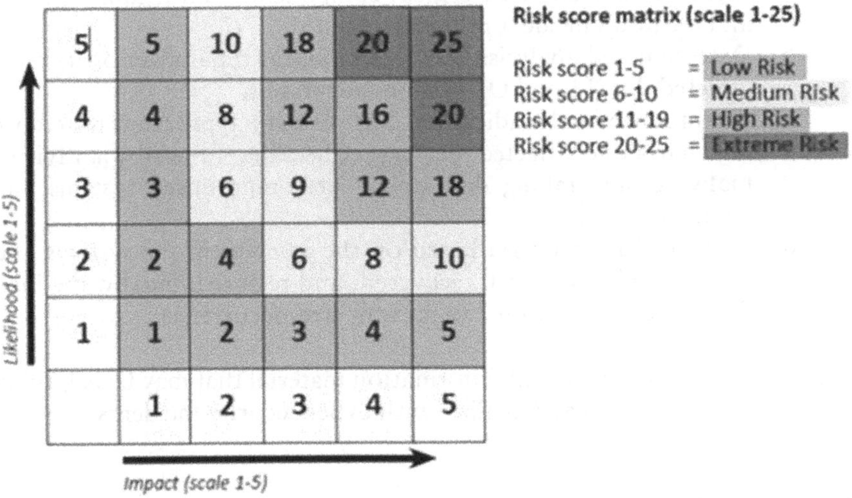

Figure 13.5 Risk Score Matrix.

6. Risk assessment: The vulnerability analysis and assessment of likelihood and impact is brought together in a risk matrix as illustrated in Figure 13.5. The risk score is expressed on an ordinal scale from Low Risk (1–5) to Extreme Risk (20–25). The cyber risk assessment process requires considerable technical competence, and it may be necessary to bring in third-party resources.

7. Develop protection measures: The guideline talks about defense both in depth and in breadth. Depth defenses may be physical security, protection of networks, firewall, testing of vulnerability, software whitelisting, to mention a few approaches. Defense in breadth involves analysis of the segments of the IT and OT network to assess the thrust of each segment. The defenses will both be of technical and procedural nature.

8. Develop detection measures: A key measure cyber risk management is detecting intrusions and infections. This can be supported with Intrusion Detection Systems (IDS) and Intrusion Prevention Systems (IPS). This should be supplemented with antivirus and anti-malware software.

9. Establish contingency plans: Loss of IT systems will most likely only affect business functions and not the safety of the vessel. The loss of OT systems will on the other side be more critical and effective emergency procedures must be in place. Under certain situations it will be necessary to disconnect the OT systems from shore network.

10. Respond to and recover from cyber security incidents: The Guide outlines four steps:

a. Preparation: Determine critical components, run regular backup, and create an incident response plan.
b. Detection and analysis: Locate and assess the impact on systems affected. List IT and OT systems separately.
c. Containment and eradication: The primary approach is to remove or quarantine affected devices. Checking firewalls, antivirus/malware, and taking disk image of compromised systems and memory dumps.
d. Post-Incident recovery: Based on the assessment the systems and data should be cleaned, recovered, and restored. Finally, the incident should be analyzed to prevent a re-occurrence.

The guide is supplemented with information material that may be helpful in the risk analysis process and dealing with cyber security incidents.

REFERENCES

Barrett, M. P. (2018). *Framework for improving critical infrastructure cybersecurity.* National Institute of Standards and Technology, Gaithersburg, MD, USA, Tech. Rep.

Beare, M. E. (Ed.). (2012). *Encyclopedia of transnational crime and justice.* Newbury Park, California: Sage.

Bichou, K. (2008). Security of ships and shipping operations. In W. K. Talley (Ed.), *Maritime safety, security and Piracy.* Boca Raton, Florida: CRC Press.

BIMCO. (2016). *The guidelines on cyber security onboard ships.* https://www. maritimeglobalsecurity.org/media/1014/c-users-jpl-onedrive-bimco-desktop-guidelines_on_cyber_security_onboard_ships_version_2-0_july2017.pdf

Diop, A. et al. (2007). Customs-trade partnership against terrorism: cost/benefit survey. *University of Virginia*, Report prepared for U.S. Customs and Border Protection.

Glaser, S. M., Roberts, P. M., Mazurek, R. H., Hurlburt, K. J., & Kane-Hartnett, L. (2015). *Securing Somali fisheries.* DOI: 10.18289/OEF.2015.001

ICS. (2018). *BMP5- Best management practices to deter piracy and enhance maritime security in the Red Sea, Gulf of Aden, Indian Ocean and Arabian Sea.* https:// www.ics-shipping.org/wp-content/uploads/2020/08/bmp5-hi-res-min.pdf

ISO/IEC 27001. (2013). *Information technology – Security techniques - Information security management systems – Requirements.* Geneva: ISO/IEC International Standards.

IMB. (2019). *Piracy and armed robbery against ships – Report for the period.* https:// www.icc-ccs.org/reports/2019_Annual_Piracy_Report.pdf

IMO. (2009). *SOLAS – Safety of Life at Sea – Consolidated Edition 2009.* London: International Maritime Organization.

IMO. (2017). *Resolution MSC.428(98) on Maritime Cyber Risk Management in Safety Management Systems (SMS).* https://wwwcdn.imo.org/localresources/en/OurWork/Security/Documents/Resolution%20MSC.428(98).pdf

IMO. (2018). *ISM Code - International Safety Management Code.* https://www.imo. org/en/OurWork/HumanElement/Pages/ISMCode.aspx

Jones, S. (2006). *Maritime security: A practical guide (1st ed.)*. London: The Nautical Institute.

Kellerman, M. G. (2011). Somali piracy: Causes and consequences. *Inquiries Journal, 3*(09), 1–2.

NIST. (2018). *Cybersecurity Framework. Version 1.1*. U.S. National Institute of Standards and Technology. https://www.nist.gov/news-events/news/2018/04/-nist-releases-version-11-its-popular-cybersecurity-framework

Rotberg, R. I. (2009). Fighting off the Somali pirates. *Boston Globe*. December 16. Boston.

Scholliers, J., Permala, A., Toivonen, S., & Salmela, H. (2016). Improving the security of containers in port related supply chains. *Transportation Research Procedia, 14*, 1374–1383.

Tumbarska, A. (2018). Maritime piracy and armed robbery evolution in 2008–2017. *Security & Future, 2*(1), 18–21.

UNCTAD. (2014). *Maritime piracy Part II: An overview of the international legal framework and multilateral cooperation to combat piracy*. https://digitallibrary.un.org/record/782373

World Bank. (2013). *The pirates of Somalia – Ending the threat, rebuilding a nation*. Washington, DC. http://hdl.handle.net/10986/16518

Chapter 14

Accident data

14.1 INTRODUCTION

Increased safety in shipping may be accomplished through different approaches such as improved ship design, better management both at shore and on board, and better training of human resources. Like in any decision problem, a key question is where to obtain the highest benefit-cost ratio with limited resources. This also applies for safety management, where the challenge is to allocate the time and money to obtain maximum reduction of losses.

Reduction of maritime risk is today to an increasing degree based on systematic methods like Formal Safety Assessment (FSA), Risk-Based Design (RBD), and various formal risk analysis methods. Historically, improved safety has been obtained through measures with a narrow scope in contrast to risk analysis where the ambition is to have a broader view where management, technology, and human resources are integrated.

Since risk analysis was developed as a tool in safety management, a key problem has been the lack of objective quantitative data. Too often, risk models are based on subjective expert estimates. Such data shall not be discarded but should rather be a supplement to "hard" data. Or as stated by Mazaheri et al. (2015): risk analysis should be evidence-based.

Even though commercial shipping is an international business involving most parts of the world, the knowledge base on risk is still fragmented and non-existent. A key problem is lack of an internationally accepted terminology for describing ship accidents, causal factors, and consequences. Another issue is that even among researchers, there is no agreement on an acceptable model for describing accidental processes.

During the past decades, we have witnessed a shift from sea-court to multidisciplinary investigation of ship accidents. This has also meant a shift in focus in the investigations, from trying to find out if someone is to blame to trying to find what we can learn from accidents to improve safety. The result is that there is today an increasing wealth of high-quality reports that may contribute to better understanding of how and why accidents happen. The remaining task is to agree on a causal model and calibrate this with systematic data from the investigation reports. A good overview of data sources for risk modeling is given by Ladan and Hänninen (2012).

This chapter addresses three main topics. Firstly, factual statistics related to the number of accidents and accident types. Secondly, how to model ship accidents or what is termed taxonomies. And thirdly, data sources to obtain evidence-based risk models.

14.2 TYPES OF DATA NEEDED IN RISK ANALYSIS

In a quantitative risk analysis, a number of different types of data are required and this section will give a brief overview of some of the most important types. Some comments are also provided and where/how data can be found.

14.2.1 Technical data

The technical data will mainly be descriptions of the ship and the systems onboard that are relevant for safety. This can be based on design information or as-built information, depending on the status of the ship being considered in the risk analysis. The degree of detail will depend on the level of detail in the analysis. For a detailed QRA, detailed information about the ship and systems is also required.

This type of information is usually readily available, and the uncertainty associated with the information is normally limited. The exception may be for ships that are still in early design phases, where detailed information not necessarily is available, and assumptions may have to be made about functionality and technical details.

14.2.2 Operational data

Supplementing the technical data is also operational data, covering aspects like types of cargo, trade, ports being visited, loading/offloading operations, crew size and composition, etc.

Most of this can also be determined with a high degree of certainty, although it is noted that the way that a ship is operated may change compared to the original intention. The information is normally available from the ship owner/operator.

14.2.3 Environmental data

Environmental data can be split into two groups: weather data and data about the waters, traffic, and other external influences.

Weather data is normally also easy to get hold of if the trading routes/ waters the ship will operate in are known. Various meteorological services around the world can provide this information, to whatever level of detail is required.

The uncertainty associated with this information is also limited. It may be argued that there is uncertainty associated with future climate changes, but this will typically be a limited uncertainty compared to the uncertainty in the frequencies and probabilities that are being used in the risk analysis.

Data about the waters the ship will operate in and about traffic is also relatively easy to get hold of. If the trade is known, information about the waters is easily available and traffic data are also easy to get hold with AIS data available. There will be some uncertainty related to future traffic, but this will in most cases also be relatively limited. More information about traffic data can be found in Section 14.7.

14.2.4 Event data

In some cases, one may choose to use event data directly (e.g., data on collisions and groundings) rather than modeling these events with fault trees illustrating their causes. Many organizations collect and publish event data, but there are several weaknesses that we often face when we are trying to use these data. This is discussed in Section 14.4.

14.2.5 Input to fault trees and event trees

This comprises a variety of data, including probability of failure of technical systems and components, human error probabilities, probabilities/ frequencies of various phenomena like ignition, and others. A whole range of sources therefore also need to be consulted, including use of expert judgment.

There are few reliable data sources for maritime equipment, but it may be possible to use generic databases for probabilities of technical systems/ components. It may be necessary to consider differences in operation and operating environment, but this may still be possible to use. This may be relevant for many types of mechanical, electromechanical, and electronic equipment. Human error probabilities are discussed in Chapter 11. Causal data in general are also discussed in Section 14.6.

The uncertainties associated with this type of data will typically be quite large, although it can vary significantly within this group. Failure and error probabilities will typically have large uncertainties associated with them.

14.2.6 Consequence data

The last group of data that may be mentioned is consequence data, e.g., number of people killed, extent of damage to the ship, or size of spill. Data on the number of fatalities per accident and spill sizes can be found, but the problem is that the circumstances in each accident will be very different. Therefore, the consequences can also vary significantly, and using statistical data can at best only give us an idea of the magnitude. In practice, we are therefore very often left to use expert judgment to determine this, based on scenario descriptions.

The uncertainty associated with this information may also be large, although normally not as large as uncertainty associated with the probabilistic data (event frequencies and failure probabilities).

14.3 EVALUATING DATA SOURCES

Even if we can find a potential data source that can be used as input to risk analysis, there are a number of aspects related to the quality of the data that should be considered before we make a decision to use them. The following may therefore be regarded as a "check list" of issues that should be evaluated when we are thinking about using a specific data source.

Age of data: The problem with any data source that is available is of course that the data represents history, while in a risk analysis we are attempting to predict the future. Both technology, operation, and management systems may change over time, meaning that historical data to some degree may not be relevant anymore. An example can be when autonomous ships are introduced. Statistics on navigational accidents from traditionally manned ships will not be relevant to use any more. This is of course a radical change, and more common is that there will be many small changes that occur over time, but that together can accumulate to make a big difference.

The time period that the data source covers is therefore important to know, and it may be necessary to consider whether large changes in technology or operation have occurred or can be expected to occur for the period that the risk analysis shall consider.

Relevance of data: An aspect that to some extent is similar is the relevance of the data source. This is mainly relevant for failure data and human error probabilities that may be based on completely different industries or applications. In Chapter 11, where human error probabilities are discussed, it was pointed out that most of the existing methods and data are from nuclear power plants. This is an environment that is very different from a ship. There may also be other data sources that suffer from similar weaknesses.

Completeness of data: This aspect is often difficult to judge but can be very important. Underreporting is often a major problem and the tradition for reporting incidents and accidents has not been very good in the maritime industry. Underreporting can obviously lead to underestimation of the risk and changes in reporting practice may also lead us to believe that there are trends in development of risk that in reality only is a result of changes in reporting.

Quality of the data: This is related to the reliability of the data that are reported: Are they classified correctly, are they sufficiently detailed and have they been recorded correctly in the database/publicly available records? Wrong classification and wrong recording are in practice impossible to detect for a user, and the only way of knowing the quality of the data is to find out what quality assurance the database operator has put in place.

Extent of data: Fortunately, relatively few accidents occur, and the extent of data can therefore in many cases be limited. This primarily affects the uncertainty in the resulting frequencies and probabilities and is therefore important input to uncertainty analysis.

Operator of the database: This may also be relevant information for judging a data source. Different operators will have different interests and objectives with collecting and publishing data, e.g., a national maritime authority vs a shipowner organization. This will influence how the data are collected, what data are collected, how they are classified, etc.

14.4 FREQUENCY AND CONSEQUENCE

14.4.1 Shipping statistics yearbook

One of the more readily available sources on accident frequency is the Shipping Statistics Yearbook which is published monthly and yearly (ISL, 2019). In the following tables, the format of the statistical information is indicated:

- Table 14.1 illustrates the size of the world fleet given in terms of number of vessels and gross tons since 1926.
- The fleet size is further broken down by the main ship types for largest national fleets as illustrated for Panama in Table 14.2. Note that both gross tons (gt) and deadweight (dwt) are used in the tables.
- Table 14.3 gives the ship losses by number and gt for the main fleets for the period 2012–2018. Data before this period can be found in earlier editions. Note that data on serious and less serious are not given in this publication. Total losses are broken down for age groups for the main ship types.
- Table 14.4 indicates the age distribution for Bulk Carriers and Combined Carriers by year. The losses for the world fleet are also broken down by age at loss and size in gross tons.

It should be noted that it is possible to estimate the loss frequency from these tables as both fleet size and loss numbers are given in this publication (Table 14.5).

Table 14.1 Table format – Total merchant fleet: 1926–2019. Ships of 100 gt and over

Year	No. of ships	1000 gt
1926	29,092	62,672
1927	28,967	63,267

2018	96,256	1,327,280
2019	97,318	1,371,570

ISL (2019).

14.4.2 Data from Sea-web

Sea-web™ is a major commercial online database covering most aspects of shipping like data on ships, companies, builders, ports, and casualties to mention a few areas. The service is a joint venture between the publication Fairplay and Lloyd's Register. Sea-web™ may be searched for individual data lookups or more complex analysis of aggregate data or time series.

Eliopoulou et al. (2016) have undertaken an analysis of basic risk figures for the world fleet based on numbers from the Sea-web™ database. The frequency of serious accidents for the main ship types through 2005–2012 is shown in Table 14.6.

The authors summarize the findings as follows:

- The lowest frequencies are observed for Fishing vessels, LNG, LPG, and Large tankers.
- Comparable higher frequencies are seen for Reefers, General cargo vessels, Cellular containerships, and Bulk carriers.
- Highest overall frequency is observed for RoRo cargo ships and Car carriers.

Based on data for the period of 2000–2012 the authors present the total loss frequency for vessel categories and main accident types (Table 14.7). The main findings may be summarized as follows:

- Collision: Highest frequencies are seen for General cargo ships and Car carriers. Bulk carriers, Fishing vessels, Reefer ships, and Passenger vessels exhibit lower frequencies.
- Contact events: General cargo, Fishing, and Passenger vessels are subject to this accident type.
- Fire and explosion: High frequency for RoRo cargo ships, Cruise vessels, Passenger RoRo vessels, and Fishing vessels. Lower frequencies are shown for Reefers and Passenger vessels. General cargo ships, Bulk carriers, and LPG ships are also subject to this accident.

Table 14.2 Table format – Merchant fleet of Panama by total and main ship types: 2010–2019. Ships of 300 gt and over

Year	Total fleet		Tankers		Bulk carriers		Container		General cargo		Passenger	
	No. of ships	1000 dwt	No. of ships	1000 dwt	No. of ships	1000 dwt	No. of ships	1000 dwt	No. of ships	1000 dwt	No. of ships	1000 dwt
2010	6,810	286,792	1,560	80,020	2,286	154,000	751	34,185	2,054	18,055	159	532
2011	6,713	300,800	1,461	74,717	2,475	174,435	749	35,877	1,899	18,297	129	474
-----	-----	-----	-----	-----	-----	-----	-----	-----	-----	-----	-----	-----
2018	6,395	326,118	1,499	81,622	2,401	186,324	589	36,009	1,800	21,726	106	437
2019	6398	323,031	1,489	80,510	2,387	182,764	610	37,658	1,805	21,622	107	477

ISL (2019).

Table 14.3 Table format – Total losses by flag and year. Ships of 500 gt and over

Year	2012		2013						2017		2018	
Flag	No.	1000 gt	No.	1000 gt	No.	1000 gt	No.	1000 gt	No.	1000 gt	No.	1000 gt
Panama	7	34.8	12	211.2	---	---	---	---	5	23.1	8	186.3
Marshall Islands	1	2.5	2	5.9	---	---	---	---	2	154.7	0	0

Russia	4	5.3	0	0	---	---	---	---	0	0	0	0
Norway	0	0	0	0	---	---	---	---	0	0	1	0.8
Others	9	34.2	13	63.2	---	---	---	---	18	65.7	10	51.8
Total	59	490.0	60	558.5	---	---	---	---	48	371.6	38	278.7

ISL (2019).

Table 14.4 Table format – Total losses by year and age of vessel for bulk and combined carriers. Ships of 500 gt and over

Age	0–2 years		5–9 years		10–14 years		15–19 years		20–34 years		25 years & over		Total	
Year	No.	1000 gt	No.	1000 gt	No.	1000 gt	No.	1000 gt	No.	1000 gt	No.	1000 gt	No.	1000 gt
2010	2	36	1	28	1	90	0	0	3	34	9	243	16	431
2011	1	40	2	82	0	0	0	0	5	143	12	187	20	452
2012	0	0	1	30	0	0	0	0	0	0	3	53	4	83

ISL (2019).

Table 14.5 Table format – World total losses by age and size class in 2018. Ships of 500 gt and over

Age	0–4 years		5–9 years		10–14 years		15–19 years		20–24 years		25 years -		Total	
Size class	No.	1000 gt	No.	1000 gt	No.	1000 gt	No.	1000 gt	No.	1000 gt	No.	1000 gt	No.	1000 gt
----														
1000<2000	0	0	2	3	0	0	2	3	0	0	4	6	8	12
2000<4000	0	0	2	6	0	0	1	3	1	3	6	17	10	29

Total	1	1	5	94	4	13	4	24	3	80	21	66	38	279

ISL (2019).

Table 14.6 Annual frequency of serious accidents per ship-year

	2005	2006	2007	2008	2009	2010	2011	2012
General cargo	2.50E–02	2.43E–02	3.78E–02	3.93E–02	3.54E–02	1.82E–02	1.71E–02	1.55E–02
Bulk carrier	2.78E–02	2.84E–02	3.70E–02	2.62E–02	3.20E–02	1.73E–02	1.22E–02	1.62E–02
Car carrier	2.86E–02	3.08E–02	5.04E–02	6.76E–02	2.11E–02	2.31E–02	1.75E–02	2.50E–02
Fishing vessel	5.65E–03	4.19E–03	6.79E–03	7.47E–02	7.53E–03	3.19E–03	3.38E–03	3.51E–03
LNG ship	1.56E–02	2.03E–02	1.15E–02	9.76E–02	7.78E–03	0.00E–00	0.00E–00	0.00E–00
LPG ship	1.54E–02	2.21E–02	2.09E–02	2.54E–03	6.61E–03	6.97E–03	1.32E–02	1.57E–03
Reefer vessel	1.81E–02	2.88E–02	4.31E–02	3.92E–02	2.59E–02	1.40E–02	1.36E–02	3.66E–03
RoRo cargo	2.66E–02	2.61E–02	5.25E–02	5.54E–02	3.90E–02	2.73E–02	4.51E–02	8.17E–03
Cell. containership	2.13E–02	2.97E–02	3.14E–02	3.65E–02	2.81E–02	2.51E–02	2.60E–02	1.68E–02
Large tanker	1.77E–02	1.19E–02	2.73E–02	1.32E–02	1.51E–02	1.22E–02	–	–
Cruise vessel	5.11E–02	5.25E–02	7.10E–02	8.17E–02	7.05E–02	9.47E–02	2.02E–02	2.70E–02
Passenger vessel	9.52E–03	6.38E–03	1.01E–02	1.56E–02	3.00E–02	2.57E–02	1.90E–02	1.92E–02
Passenger RoRo	2.91E–02	2.30E–02	5.78E–02	6.53E–02	7.23E–02	7.71E–02	4.80E–02	5.03E–02

Eliopoulou et al. (2016).

Table 14.7 Annual frequency of total losses by accident category

	Collision	Contact	Fire/ explosion	Foundered	Hull/ mach.	Wrecked/ stranded
General cargo	6.51E–04	4.23E–05	1.69E–04	1.33E–03	1.10E–04	6.08E–04
Bulk carrier	2.36E–04	0.00E+00	1.18E–04	2.95E–04	8.85E–05	4.72E–04
Car carrier	6.13E–04	0.00E+00	0.00E+00	0.00E+00	0.00E+00	4.60E–04
Fishing vessel	1.30E–04	3.55E–05	3.79E–04	7.70E–04	3.55E–05	2.49E–04
LNG ship	0.00E+00	0.00E+00	0.00E+00	0.00E+00	0.00E+00	0.00E+00
LPG ship	0.00E+00	0.00E+00	1.96E–04	1.96E–04	0.00E+00	1.96E–04
Reefer vessel	9.24E–05	0.00E+00	2.77E–04	3.70E–04	0.00E+00	5.55E–04
RoRo cargo	0.00E+00	0.00E+00	7.26E–04	5.45E–04	1.82E–04	1.82E–04
Cell. Container	0.00E+00	0.00E+00	0.00E+00	8.87E–05	4.43E–05	6.65E–05
Large tanker	0.00E+00	0.00E+00	0.00E+00	0.00E+00	0.00E+00	0.00E+00
Cruise vessel	0.00E+00	2.38E–04	4.77E–04	0.00E+00	0.00E+00	2.38E–04
Passenger vessel	1.30E–04	3.83E–05	2.30E–04	4.21E–04	7.65E–05	1.15E–04
Passenger RoRo	0.00E+00	0.00E+00	3.96E–04	4.45E–04	0.00E+00	2.47E–04

Eliopoulou et al. (2016).

- Foundered: General cargo ships have the highest frequency. Somewhat lower frequency is shown for Fishing vessels, RoRo cargo, Passenger RoRo, Passenger vessels, Reefer vessels, and Bulk carriers.
- Wrecked/stranded: General cargo, Reefer vessels, Bulk carriers, and Car carriers show the highest frequency.

14.4.3 Annual overview from EMSA

The European Maritime Safety Agency (EMSA) is responsible for the maintenance and enhancement of EMCIP, the *European Marine Casualty Information Platform*, a tool to store and analyze casualty data and investigation reports provided by the Member States (MS). Each MS is responsible for investigating accidents in their territorial waters and for their own vessels as Flag State. EMSA is reporting maritime accidents in European waters on an annual basis in a systematic format based on data from EMCIP (EMSA, 2019).

The electronic version of EMCIP was initiated operationally in 2011 and has later undergone some modifications and additions. An overview of the data items or the taxonomy of the EMCIP database is shown in Figure 14.1. The database serves as the basis for producing reports with statistical material for various EU and MS policymaking on maritime safety (EMSA, 2019). EMSA has also published special reports like "Safety Analysis of

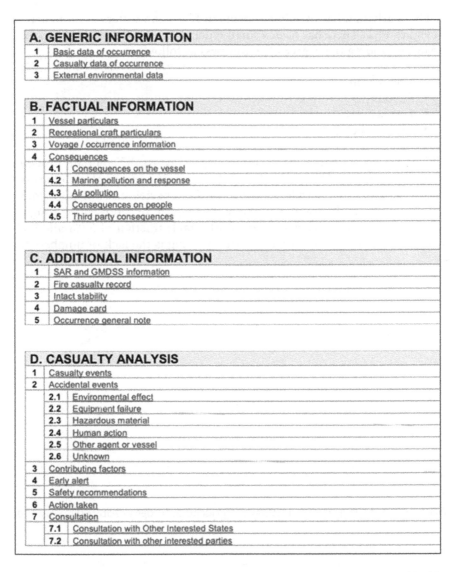

A. GENERIC INFORMATION	
1	Basic data of occurrence
2	Casualty data of occurrence
3	External environmental data

B. FACTUAL INFORMATION		
1	Vessel particulars	
2	Recreational craft particulars	
3	Voyage / occurrence information	
4	Consequences	
	4.1	Consequences on the vessel
	4.2	Marine pollution and response
	4.3	Air pollution
	4.4	Consequences on people
	4.5	Third party consequences

C. ADDITIONAL INFORMATION	
1	SAR and GMDSS information
2	Fire casualty record
3	Intact stability
4	Damage card
5	Occurrence general note

D. CASUALTY ANALYSIS		
1	Casualty events	
2	Accidental events	
	2.1	Environmental effect
	2.2	Equipment failure
	2.3	Hazardous material
	2.4	Human action
	2.5	Other agent or vessel
	2.6	Unknown
3	Contributing factors	
4	Early alert	
5	Safety recommendations	
6	Action taken	
7	Consultation	
	7.1	Consultation with Other Interested States
	7.2	Consultation with other interested parties

Figure 14.1 EMCIP taxonomy. Source: EMCIP taxonomy – List of attributes (EMSA, March 2017).

Data Reported in EMCIP – Analysis of Marine Casualties and Incidents involving Container vessels" (EMSA, 2020). Similar reports have also been prepared for Ro-Ro vessels and fishing vessels.

The database itself is solely accessible to authorized agencies within EU and Member States and Norway. Research institutions may get access upon request to EMSA.

The yearly number of ship accidents in European waters is summarized in Figure 14.2. There has been a small reduction (5.6%) from 2014 to 2019. EMSA is grading the degree of seriousness as follows:

- Very serious
- Serious
- Less serious
- Marine incident

This contrasts with the more common practice to differentiate between casualty categories. It is assumed that very serious is identical with the normal term *losses* and stands for 2.5% of the reported cases (Figure 14.3). The serious cases are roughly ten times as frequent or representing roughly 25%. Less serious accidents show a fairly stable fraction of 57% and incidents 15%. A weakness in the EMSA reporting is the lack of numbers for

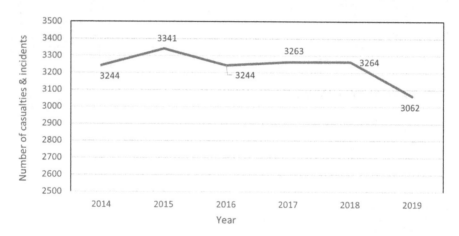

Figure 14.2 Annual number of casualties and incidents in European waters (EMSA, 2019).

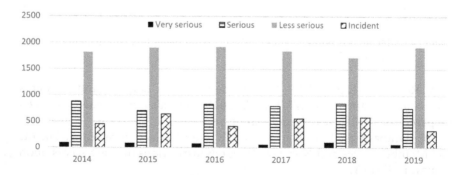

Figure 14.3 Degree of seriousness of ship accidents (EMSA, 2019).

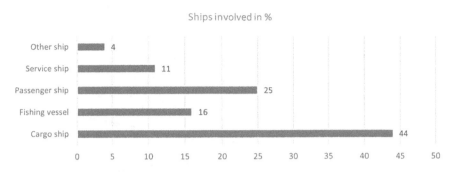

Figure 14.4 Casualties by main ship type category (EMSA, 2019).

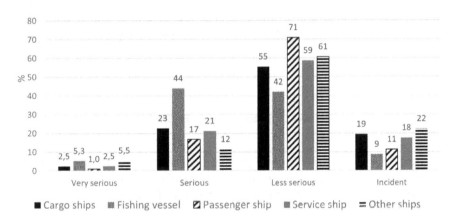

Figure 14.5 Accidents by degree of seriousness and ship type (in %) (EMSA, 2019).

vessels at risk (exposure), which means that the accident probability is not revealed. The data therefore primarily gives the relative importance or weight of the categories used.

The most frequent ship type involved in accidents is cargo ships (44%), but passenger ships are also quite frequent (25%) as shown in Figure 14.4. Fishing vessels are more exposed to both very serious (5.3%) and serious accidents (44%) than other ship types (Figure 14.5). The opposite picture is shown for passenger ships with fractions of 1% and 17%.

Looking at each accident type, it can be concluded that all types show a small reduction in number of annual cases, with two exceptions. The number of collisions seems to be fairly stable whereas the number loss of propulsion events has shown a steep increase (see Figure 14.6). The latter category is strictly speaking not an accident but rather a matter of loss of availability or reduced reliability. One may also suspect that the increase may be explained by better reporting made possible by the introduction of AIS.

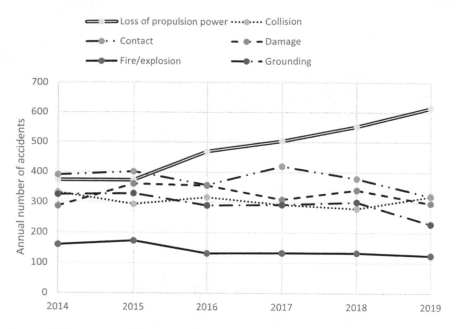

Figure 14.6 Distribution of events by accident type (EMSA, 2019).

Figure 14.7 gives an overview of the main consequences. It is difficult to see any distinct pattern in development over time. Some of the categories seem quite stable whereas others show large fluctuations from one year to the other.

14.4.4 Data from national maritime administrations

Because the EMCIP database has been set up by EMSA, the majority of European countries have discontinued their national statistical services. The national administrations currently limit their own services to investigation of accidents and publishing of special reports on specific safety issues.

A complication of using different national databases is that different accident definitions are used and often structured differently from one source to another. We see different ways of classifying even basic entities like ship type, accident category, and consequence classification.

We will in the following sections comment on a few examples of national databases but not pretend that these are in any way more unique or outstanding than those of other national administrations.

14.4.4.1 The United Kingdom and Canada

The Marine Accident Investigation Branch (MAIB) is investigating and reporting marine accidents in UK waters and UK-flagged vessels. Apart

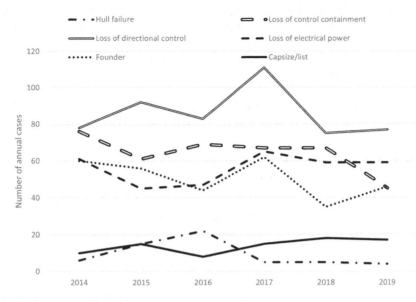

Figure 14.7 Consequence of ship accident (EMSA, 2019).

from reporting individual accidents MAIB produces an annual statistical summary with the main facts (MAIB, 2019). The information focused on is ship types, accident types, and consequences. Analysis or statistics on causal factors are not covered in this report.

MAIB also publishes so-called Safety Digests on any topic highlighted in recent accident investigations. The topics presented may be seen as lessons learned from individual accident cases. These reports are published on a regular basis (MAIB, 2021).

MAIB gives definitions of the degrees of seriousness of marine accidents that seem to have broad acceptance:

- **Very Serious Marine Casualty (VSMC):** Marine Casualty which involves total loss of the ship, loss of life, or severe pollution.
- **Serious Marine Casualty (SMC):** Marine Casualty where an event results in one of: immobilization of main engines, extensive accommodation damage, severe structural damage, such as penetration of the hull underwater, etc., rendering the ship unfit to proceed: pollution or a breakdown necessitating towage or shore assistance.
- **Less Serious Marine Casualty (LSMC):** This term is used by MAIB to describe any Marine Casualty that does not qualify as a VSMC or an SMC.
- **Marine Incident (MI):** A Marine Incident is an event or sequence of events other than those listed above which has occurred directly in connection with the operation of a ship that endangered, or if not corrected

would endanger, the safety of a ship, its occupants or any other person or the environment (e.g., close quarters situations are Marine Incidents).

- **Accident:** Any Marine Casualty or Marine Incident. In historic data, accident had a specific meaning, broadly equivalent to (but not identical to) Marine Casualty.

The Transportation Safety Board of Canada (TSB) is issuing a similar annual report with the same scope as MAIB (TSB, 2019).

14.4.5 Marine Accident Inquiry Agency (MAIA) of Japan

Japan Transport Safety Board (JTSB) publishes high-level statistics on ship accidents with respect to accident categories and ship types in a similar way like other national administrations. This has limited value as data for risk analysis. However, the Japanese publish studies of specific safety issues called MAIA Digest. One investigation focused on marine incidents in fog (MAIA, 2007). The background is that from spring until summer the Japanese archipelago has a high frequency of dense fog. The distribution of main causes in fog-related collisions is shown in Figure 14.8.

14.5 ACCIDENT TAXONOMIES

Systematic coding and use of accident data in risk analysis should preferably be based on some kind of data classification or taxonomy.

14.5.1 Skill-rule-knowledge model

One of the more influential approaches to the modeling of accident causes is the Skill-Rule-Knowledge model proposed by Rasmussen (1982) making a distinction between the following behavior types:

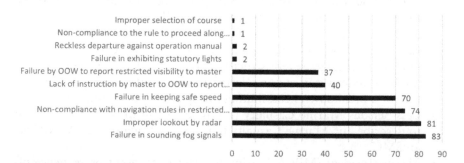

Figure 14.8 Causes of collision in fog in Japanese waters: 2001–2005. Number of cases (MAIA, 2007).

- Skill based: Well-learned behaviors almost performed automatically. Error sources are inattention or over-attention.
- Rule based: Behavior is controlled by rules. Situation A is tackled with response B. Error sources are misapplication of a good rule or application of a bad rule.
- Knowledge based: Require problem solving and are subject to a broad range of error sources.

The strength of the concept was demonstrated by Burns and Bonaceto (2020) in an analysis of general aviation accidents (see Section 11.8).

14.5.2 SHEL model

The SHEL model (Hawkins & Orlady, 1993) focuses on the sources of error (Figure 14.9). The operator or person on the "sharp end" is influenced by:

- Organization or software (S)
- Equipment or hardware (H)
- Environment (E)
- Humans or liveware (L)

14.5.3 Swiss Cheese Model (SCM)

Reason (1990) introduced a model that focused on the fact that the majority of accidents happen as a result of the failure of a number of safety barriers (Figure 14.10). The concept has later been presented in different versions and is also known as the Swiss Cheese Model.

The Swiss Cheese Model has had strong influence on later accident modeling. The model sees error mechanisms in an organizational perspective where top-level decisions influence middle management and subsequently preconditions for unsafe acts. Latent failures in these defenses will lead to active failures or unsafe acts.

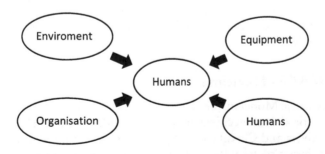

Figure 14.9 SHEL model (Hawkins & Orlady, 1993).

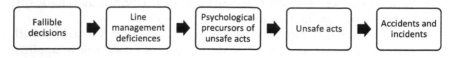

Figure 14.10 Safety barriers (Reason, 1990).

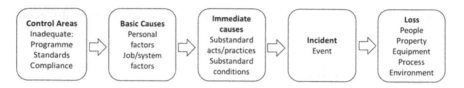

Figure 14.11 Marine Systematic Cause Analysis Technique (MSCAT) (ISRS, 2015).

14.5.4 MSCAT

One of the early methods for systematic accident analysis and documentation was the Systematic Causal Analysis Technique (SCAT) by the International Loss Control Institute – ILCI (Bird & Germain, 1985). The approach covered a broad approach to risk by discussing causal factors, management of safety, accident investigation, and safe work practices to mention a few. The Institute was later taken over by DNV and modified in order to accommodate for the marine sector and renamed MSCAT. The MSCAT model (ISRS, 2015) has much in common with the Swiss Cheese Model and primarily differs in the naming of the barriers shown in Figure 14.11. The coding of causal factors is based on a predefined set or taxonomy as illustrated in Table 14.8.

14.5.5 MaRCAT

Like MSCAT, the ABS Marine Root Cause Analysis Map (MaRCAT) applies a map of predefined factors to code causal factors as shown in Figure 14.12. On the top level of the analytical tree, there are three main branches for technical, human, and external factors that are broken down to the coding level.

14.5.6 HFACS- Maritime

The Swiss Cheese Model inspired Shappel and Wiegmann (2000) to develop a taxonomy for aircraft accident analysis and coding called HFACS (Human Factors Analysis and Classification System). It has found widespread application in different sectors. Reinach and Viale (2006) proposed a version for railroad and added a fifth level termed "External factors".

Table 14.8 Systematic Causal Analysis Technique (SCAT). Section of checklist

Personal factors

Inadequate physical/physiological capability

- Inappropriate height, weight, strength, etc.
- Restricted range of body movement
- Limited ability to sustain body positions
- Sensitivity to sensory extremes
- Vision deficiency
- Etc.

Mental or Psychological stress

- Emotional overload
- Fatigue due to mental task load or speed
- Extreme judgment/decision demands
- Routine, monotony, demand for vigilance
- Extreme concentration/perception demands
- Etc.

Lack of knowledge

- Lack of experience
- Inadequate orientation
- Inadequate initial training
- Inadequate update training
- Etc.

Job factors

Inadequate leadership and/or supervision

- Unclear/conflicting reporting lines
- Unclear/conflicting responsibility
- Improper/insufficient delegation
- Inadequate policy, procedures, practices, etc.
- Etc.

Inadequate tools and equipment

- Inadequate assessment of needs and risks
- Inadequate human factors considerations
- Inadequate standards and specifications
- Inadequate availability of tools
- Etc.

Inadequate engineering

- Inadequate assessment of loss exposure
- Inadequate consideration of human factors
- Inadequate standards, specifications, and/or design criteria
- Inadequate monitoring of construction
- Etc.

Inadequate work standards

- Inadequate development of standards
- Inadequate communication of standards
- Etc.

Bird and Germain (1985).

A modified version of HFACS for application in the marine sector was proposed by Chen et al. (2013) (Table 14.9). Unlike the original HFACS model it also incorporates external factors like legislation gaps, administrative oversights, and design flaws. The middle management level is termed "Unsafe Supervision" which is more relevant from the ship accident perspective, and covers the role of the senior officers onboard. The Preconditions level is structured in line with the SHEL model. Finally, the "Unsafe Acts" level is structured in accordance with the *Generic Error Modeling System*

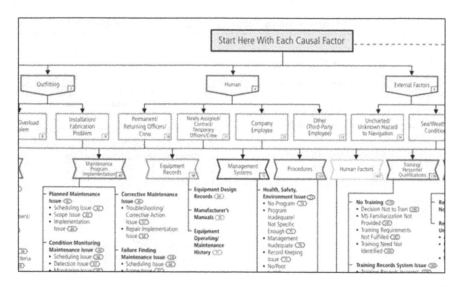

Figure 14.12 Section of MaRCAT coding scheme for causal factors (ABS, 2005).

Table 14.9 Overview of the HFACS-MA framework

Level	1st layer	2nd layer	3rd layer
5	External factors	Legislation gaps Administration oversights Design flaws	
4	Organizational influence	Resource management Organizational climate Organizational process	
3	Unsafe supervision	Inadequate supervision Planned inappropriate operations Failed to correct know problems Supervisory violation	
2	Preconditions (SHEL)	Software	
		Hardware	
		Condition of operators	
		Environment	Physical Technological
		LIveware	
1	Unsafe acts	Errors	Skill-based errors Rule-based mistakes Knowledge-based mistakes
		Violations	Routine Exceptional

Chen et al. (2013).

(GEMS) proposed by Reason (1990) and thereby also slightly different from the original HFACS version.

14.6 CAUSAL FACTOR DATA

14.6.1 Port state control findings

Port State Control (PSC) has become an important element in regulation of commercial shipping and generates a large pool of data on non-conformities relative to safety rules and regulations. The role of PSC is to inspect the safety conditions of a foreign vessel visiting a port state. The basis for PSC in Europe is the Paris Memorandum of Understanding (Paris MOU) which secure harmonized procedures for control in European coast states and some non-European states. Similar MOUs have been instituted in other parts of the world. More information can be found in Chapter 4.9.

The port state inspection findings are gathered and stored in the THETIS database (EMSA, 2010). An overview of findings is shown in Figure 14.13. This kind of data has limited application in risk analysis as they only focus indirectly on causal factors. PSC findings are best characterized as aggregate data and are primarily relevant in a discussion of safety measures.

14.6.2 Data from accident investigations

During the last decades, accident investigations are conducted by multidisciplinary teams organized as independent units relative to maritime

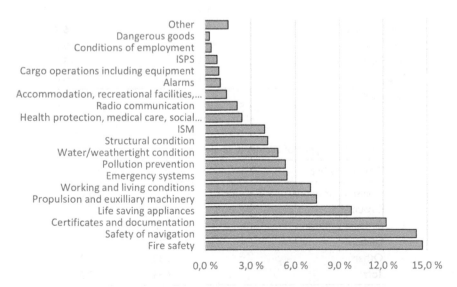

Figure 14.13 Deficiency categories in PSC (Kretschmann, 2020).

administrations and the shipping business. The pioneering example is the Marine Accident Investigation Branch (MAIB) in the UK. This means that today we have an increasing number of high-quality reports covering facts, events, and causes of maritime accidents. This knowledge source may be the basis for statistical analysis. This kind of data is already applied in different studies of the causation of maritime accidents.

14.6.3 SIRC study

Seafarers International Research Centre (SRC) has analyzed 693 accidents from different countries with the majority from the UK (Acejo et al., 2018). A breakdown of the countries is presented is shown in Table 14.10. The investigation covered different accident types, but the largest group was traffic-related accidents with collision and grounding constituting more than 50% of the cases (Table 14.11).

The reports were analyzed by two independent teams and the objective was to identify and classify the cause of the accident. The findings were discussed together with a moderator to find an agreement. A distinction was made between the following causal categories:

Table 14.10 Number of investigated accidents by origin of the report

Country of origin	Frequency	In %
UK	203	29.3
Australia	145	20.9
USA	57	8.2
New Zealand	43	6.2
Germany	137	19.8
Denmark	108	15.6
Total	693	100.0

Acejo et al. (2018).

Table 14.11 Types of accidents analyzed

Type of accident	Frequency	In %
Collision, close quarters & contact	248	35.8
Grounding	118	17.0
Fire and explosion	66	9.8
Lifeboat	23	3.3
Other	238	34.2
Total	693	100.0

Acejo et al. (2018).

- *Immediate cause*: A cause that directly leads to the accident at the end of the error chain.
- *Contributory cause*: One or more causes leading to the immediate cause or creating the conditions for the immediate or other contributory causes to happen.

An overview of the identified immediate causes of collisions is shown in Figure 14.14. It appears that the four most frequent causal factors represent almost two-thirds of the material:

- Inadequate lookout: 24.6%
- Failure in communication: 15.3%
- Poor judgment: 14.1%
- Pilot error/mishandling: 12.9%

The contributory causes in collisions show greater variation but even here the four most dominating factors represent almost 79% of a total observation material 194.8%:

- Ineffective use of technology: 24.2%
- Failure in communication: 20.6%
- Weather/environmental factors: 19.8%
- Third-party deficiency: 14.1%

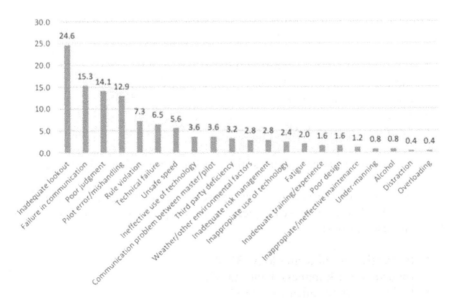

Figure 14.14 Immediate cause in collisions (Acejo et al., 2018).

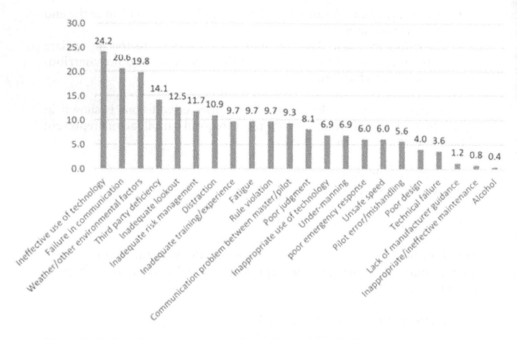

Figure 14.15 Contributory causes in collision (Acejo et al., 2018).

Notice also that failure in communication is both a frequent immediate and contributory cause. On average, 1.9 contributory causes were identified per case (Figure 14.15).

The five most frequent immediate causes in grounding represent 57% of the observations (Figure 14.16):

- Failure in communication: 16.1%
- Inadequate lookout: 11.9%
- Poor judgment: 11.0%
- Fatigue: 9.3%
- Technical failure: 8.5%

The three first immediate causes also had a dominating role in collision accidents.

An overview of the contributory causes in groundings is presented in Figure 14.17. The four most frequent factors represent 10.6% of the total 215.7% observations:

- Ineffective use of technology: 31.4%
- Inadequate risk management: 28.8%
- Failure of communication: 24.6%
- Third-party deficiency: 17.8%

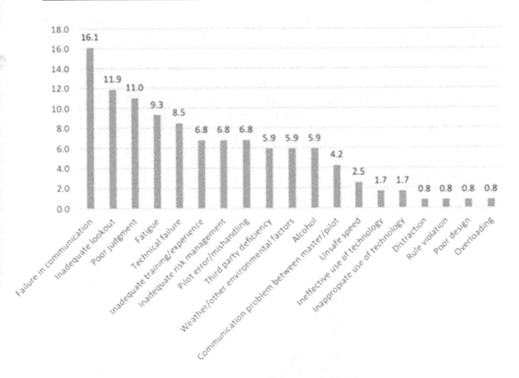

Figure 14.16 Immediate causes in grounding (Acejo et al., 2018).

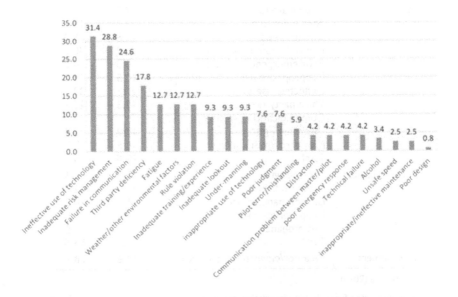

Figure 14.17 Contributory causes in groundings (Acejo et al., 2018).

As for immediate causes it has also strong similarities with groundings with respect to contributory causes. The average number of causes per grounding case was 2.2.

In order to get a better picture of how the identified factors group under the main functional headings, the data were structured as follows: Personal state, job factor, disturbance, technical factor, and environment. See Table 14.12. The following crude frequency observations can be made for collision and grounding taken together:

- Personal factors: 186.7%
- Job factors: 376.7%

Table 14.12 Functional grouping of causes for collision and grounding

Group	Causal factor	Frequency (%)	
		Collision	Grounding
Personal state	Distraction	11.3	5.0
	Overloading	0.4	0.8
	Fatigue	11.7	22.0
	Rule violation	17.0	13.5
	Inadequate training/experience	11.3	16.1
	Under-manning	7.7	9.3
	Inadequate risk management	14.5	35.6
	Alcohol	1.2	9.3
	Total	**75.1**	**111.6**
Job factor	Inadequate lookout	37.5	21.2
	Poor judgment	22.2	18.6
	Unsafe speed	11.6	5.0
	Failure in communication	35.9	40.7
	Poor emergency response	6.0	4.2
	Inappropriate use of technology	10.5	9.3
	Ineffective use of technology	27.8	33.1
	Third-party deficiency	17.3	23.7
	Communication problem between Master/pilot	12.9	8.4
	Pilot error/mishandling	18.5	12.7
	Total	**199.8**	**176.9**
Technical factor	Poor design	5.6	1.6
	Technical failure	10.1	12.7
	Lack of manufacturer guidance	1.2	0.0
	Inappropriate/ineffective maintenance	2.0	2.5
	Total	**18.9**	**16.8**
Environment	Weather/environmental factors	**22.6**	**18.6**

Acejo et al. (2018)

The percentage figure is seen in relation to the total number of observations from both accident types and both immediate and contributing causes.

- Technical factors: 35.7%
- Environment: 41.2%

This investigation indicates that Job factors are roughly twice as frequent as Personal factors and that Technical and environmental factors are marginal relatively speaking.

14.6.4 Finnish study

Mazaheri et al. (2015) have performed a more comprehensive study of the potential of both accident and incident reports as a knowledge source for accident modeling. Their objective was to introduce evidence-based risk modeling where subjective expert opinion is replaced by statistical data from reports. The authors chose grounding accidents as a case, primarily because the category is one of the more frequent ones. Acknowledging the need for a taxonomy they proposed their own version of the HFACS model called HFACS-Ground.

The model is slightly different from the maritime version proposed by Chen et al. (2013). Rather than applying the SHEL model on the Preconditions level, they use a structure based on Environmental factors, Condition of the operator, and Personal factors.

Table 14.13 presents the findings from the accident. An immediate observation is the fact that most findings or causal factors are related to *Preconditions* and *Unsafe acts*. This can partly be explained by the nature of existing investigations which focus more on the events onboard and less on the role of the managing company. The authors identify *Judgment/decision* as the most frequent active failures and explain it mainly by the lack of proper Coordination/communication/planning.

A comparison of findings from accident and incident reports is given in Table 14.14. There is a lack of consistency between the two information sources on several factors. This may be explained in different ways. The accident investigation team has an independent role whereas the incident report to a large degree is written by the personnel involved in the event. This is illustrated by the fact that accident investigations place more emphasis on *Inappropriate communication and cooperation*, whereas incident reports often focus on *Organizational factors and support* and thereby the importance of management support.

Another aspect is the fact that the alternative approaches have different resources available. Accident investigations are undertaken by multidisciplinary teams with little time pressure whereas the analysis of incidents

Table 14.13 Analysis results of accidents with HFACS-Ground

Level	1st Layer	%	2nd layer	%	3rd layer	%
Active failures	1 Unsafe acts	26.6	Error	83.3	Skill-based	34
					Judgment/decision	56
					Perceptional	10
			Violation	16.7	Routine	57.8
					Exceptional	42.2
Latent failure/ condition	2 Preconditions	49.1	Environmental factors	40.7	Physical environment	37.9
					Technological environment	41
					Infrastructures	21.1
			Condition of operator	16.9	Cognitive factors	34.3
					Psycho-behavioral factors	30.3
					Adverse physiological states	22.9
					Physical/mental limitations	9.8
					Perceptual factors	2.7
			Personal factors	42.4	Coordination/ comm./planning	93.7
					Personal readiness	6.3
	3 Unsafe supervision	8.5	Inadequate supervision	52.7		
			Planned inappropriate operations	20.3		
			Failed to correct known problems	15		
			Supervisory violations	12		
	4 Organizational influence	12	Resource management	41.2		
			Organizational climate	13.1		
			Organizational process	45.7		
	5 External factors	2.8	Regulation gaps	15.9		
			Other factors	84.1		

Mazaheri et al. (2015).

is performed by one or few persons who already are busy operating the vessel.

14.6.5 Powered grounding accidents

An ongoing study of causal mechanisms in powered grounding accidents (PGA) is based on 96 cases investigated in UK, Canadian, and Australian waters (Kristiansen, 2021). Accident characteristics and causal factors were

Table 14.14 Comparison of frequency of causes from incident and accident reports

General category	Incident reports (%)	Accident reports (%)
Accidental loss of control	4.4	0.8
Alarm missing or not clear	0.3	4.0
Bad visibility	0.3	3.3
Darkness	0.3	1.7
Errors (skill-based/judgement/decision)	6.8	8.9
Fairway	3.6	4.1
Hazardous natural environment	5.1	5.2
Inappropriate communication and cooperation	2.4	10.3
Inappropriate maintenance	1.2	0.6
Inappropriate regulations and practices	20.5	7.9
Inappropriate route planning	0.9	6.5
Inappropriate ship/bridge system design or equipment	3.3	3.8
Inappropriate training	2.9	5.1
Lack of redundancy	4.1	1.0
Lack of situational awareness	0.3	3.0
Mechanical failure or unexpected behavior	20.1	3.5
Organizational factors and support	19.4	7.0
Other personal factors	1.2	6.2
Ship moving off course	0.3	3.7
Traffic	1.2	0.6
Under-manning of necessary stations	0.0	5.6
Violation of good seamanship practices	1.5	7.0

Mazaheri et al. (2015).

coded from findings from the investigation reports. The objective is to establish a BN model through inspection of the statistical material and a series of correlation analyses.

14.6.5.1 Causal factors

One way of characterizing the data material is to make a distinction between cases where the main problem was inadequate navigation or failure of vessel control (Table 14.15). It is evident that PGAs may stem from quite different situations. Navigation-related accidents were slightly more frequent than so-called control accidents. Safe operation requires that both route planning, navigation, and control are performed adequately. The inspection of the data showed either failure of a single or a multitude of functions (Table 14.16). Inadequate navigation was a causal factor in 71% of the cases, 64% of the cases showed control failure, and planning failure was present in 37% of the cases. The analysis also found distinction

Table 14.15 Subgroups of investigated PGA accidents

Scenario	Cases	Scenario	Cases
Navigation related	55	Did not turn at waypoint	17
		Drift-off from intended course	25
		Wrong course set	13
Control related	41	Loss of speed or directional control	19
		In turning maneuver	20
		Disturbed/influenced by traffic	2

Kristiansen (2021).

Table 14.16 Failure of main functional tasks in PGA

Inadequate task performance	Freq. (%)
Navigation	23
Navigation – Control	17
Control	23
Planning – Navigation	13
Planning – Control	6
Planning – Navigation Control	18

Kristiansen (2021).

between cases where the function was performed erroneously or not at all (omission):

- Route planning: Wrong plan (23) – No planning (13)
- Navigation: Navigation error (37) – Navigation omission (32)
- Control: Control error (53) – Control omission (21)

Wrong execution of tasks seems slightly more frequent than omissions.

The investigation reports primarily focused on factors related to the events onboard and only marginally on the role of the company. The causal factors were structured in the following manner:

- Human and organizational factors (HOF) – 71%
- Environmental factors (weather and sea) – 23%
- Technical failure or non-available system – 6%

The main grouping of the HOFs are *Personal factors* and *Job demand* as shown in Table 14.17. They were found to be equally frequent in this investigation. The coded factors (layer 3) revealed the following dominating causes: Bridge Resource management (BRM), Situational understanding, Competence, Motivation, and Experience.

Table 14.17 Human and organizational factors in PGA

Layer 1	No.	Layer 2	Freq.	Layer 3	Freq.
Personal factors/ readiness	187	Personal state	74	Motivation	26
				Mental state	4
				Situational understanding	44
		Cognitive factors	72	Competence	30
				Experience	25
				Wrong focus	17
		Fitness	41	Fatigue/exhausted	20
				Asleep	13
				Alcohol/drug	8
Job demand	181	Mental load	54	Distraction	12
				Alone on bridge	24
				Task load/stress	18
		Cooperation	87	Language/culture	4
				Bridge resource management (BRM)	57
				Pilot monitoring	26
		Non-routine event	40	Sudden event	22
				Wrong instruction	4
				Unfamiliar situation	14

Kristiansen (2021).

In addition to the HOFs, navigation and control are also influenced by environmental and technical factors, although navigation is influenced by relatively few factors. It is interesting to note that technical failure of navigation equipment is very small (Table 14.18).

Vessel control on the other hand is influenced by environmental factors, seaway conditions, and technical factors as shown in Table 14.19. The dominating factors were: Narrow seaway, Current, Control system failure, Shallow water, and Wind.

14.6.5.2 Track history of vessel

In addition to the identification of causal factors from the investigation report, it is also possible to analyze the track of the vessel. Most reports on traffic-related accidents will have a presentation of course lines and way points in a nautical chart as illustrated in Figure 14.18. This gives the opportunity to code key parameters that characterize the accident scenario. In the present analysis of PGA, the following factors were seen as relevant:

– Vessel speed at time of grounding
– Deviation from intended heading (degrees)
– Time frame from loss of control until grounding
– Characteristic width of the seaway

Table 14.18 Technical and environmental factors influencing navigation

Factor Group	Freq.	Factor	Freq.
Information	32	Visibility	8
		Unavailable systems	12
		Missing seamarks	12
Technical	3	Navigation equipment failure	3

Kristiansen (2021).

Table 14.19 Technical and environmental factors influencing vessel control

Factor Group	Freq.	Factor	Freq.
Physical influence	51	Current	20
		Wind	14
		Shallow water	17
Seaway conditions	36	Narrow seaway	26
		Traffic disturbance	6
		Frequent course change	4
Vessel Control	27	Course stability	3
		Turning ability	6
		Control system failure	18

Kristiansen (2021).

This kind of data makes it possible to compare characteristics of the different accident scenarios. As an indication of the potential of this kind of analysis, two PGA scenarios have been compared (Figures 14.19 and 14.20).

- Mean value for Drift-off: Critical time: 13 minutes. Deviation: 12 degrees
- Mean value for Did-not-turn: Critical time: 23 minutes. Deviation: 34 degrees

The data indicate that Did-not-turn scenarios are slightly more serious in the sense that the situation may develop longer both in time and heading error before the vessel grounds.

14.7 TRAFFIC DATA

Risk analysis models for grounding and collision accidents need input about traffic pattern and traffic volume as a key element (Chapter 10). Rather than just observing the number of passages in a fairway, the introduction of Automatic Identification System (AIS) offers the possibility to log data about

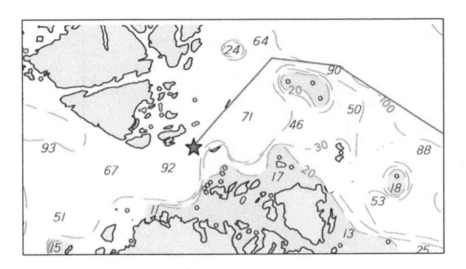

Figure 14.18 Illustration of track line of a grounded vessel.

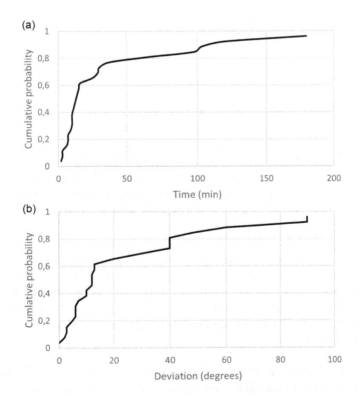

Figure 14.19 Drift-off scenario of PGA (Kristiansen, 2021).

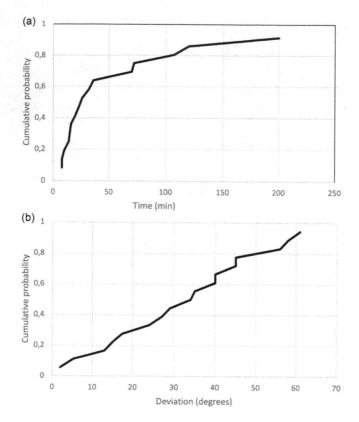

Figure 14.20 Did-not-turn scenario for PGA (Kristiansen, 2021).

the traffic for a selected fairway for a specified period. AIS was at the outset introduced as an aid to navigation and better identification of approaching vessels. Today it is also a key facility for traffic monitoring by Vessel Traffic Services (VTS). The AIS message is updated with a certain rate (Ladan & Hänninen, 2012):

- Static information: Identification of vessel is updated every 6 minutes.
- Dynamic information: Position, speed, heading, rate-of-turn, etc. is updated every 2–180 seconds (depending on speed and course alteration).
- Voyage-related information: Vessel loading, cargo, destination, way-points, and ETA is not updated automatically, but depends on input by the crew.

Historical AIS data are saved and can be used to analyze traffic patterns and density with much higher precision and detail than earlier methods. Figure 14.21 shows the traffic off the coast of Portugal for one month in

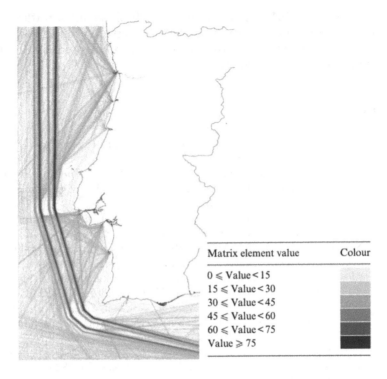

Figure 14.21 AIS traffic off the coast of Portugal from July 9 to August 9, 2008 (Silveira et al., 2013).

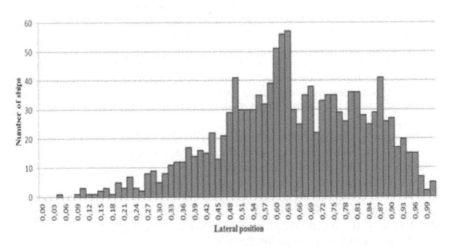

Figure 14.22 Traffic distribution for non-dangerous traffic off Cape Roca TSS traffic lanes (Silveira et al., 2013).

2008 (Silveira et al., 2013). The map indicates both the main traffic routes and a visual presentation of the traffic density. Based on the AIS logs, it is also possible to compute the traffic frequency for a cross section of a fairway as indicated in Figure 14.22.

REFERENCES

ABS. (2005). *Guidance notes on the investigation of marine incidents.* Houston, Texas: American Bureau of Shipping.

Acejo, I., Sampson, H., Turgo, N., Ellis, N., & Tang, L. (2018). *The causes of maritime accidents in the period 2002–2016.* Seafarers International Research Centre (SIRC), Cardiff University. Cardiff.

Bird, F. E. & Germain, G. L. (1985). *Practical loss control leadership.* Loganville, Georgia: International Loss Control Institute Inc.

Burns, K., & Bonaceto, C. (2020). An empirically benchmarked human reliability analysis of general aviation. R*eliability Engineering and Systems Safety,* 194.106227. https://scholar.google.com/citations?view_op=view_citation&hl=en&user=Dy htQV0AAAAJ&citation_for_view=DyhtQV0AAAAJ:9yKSN-GCB0IC

Chen, S. T., Wall, A., Davies, P., Yang, Z., Wang, J., & Chou, Y. H. (2013). A human and organizational factors (HOFs) analysis method for marine casualties using HFACS-Maritime Accidents (HFACS-MA). *Safety Science,* 60, 105–114.

Eliopoulou, E., Papanikolaou, A., & Voulgarellis, M. (2016). Statistical analysis of ship accidents and review of safety level. *Safety Science,* 85, 282–292.

EMSA. (2010). *THETIS* (online). https://portal.emsa.europa.eu/web/thetis-eu

EMSA. (2019). *Annual overview of marine casualties and incidents.* https://www.emsa. europa.eu/publications/item/3734-annual-overview-of-marine-casualties-and-incidents-2019.html

EMSA. (2020). EMCIP–Analysis of Marine Casualties and Incidents involving Container vessels. https://www.emsa.europa.eu/we-do/safety/accident-investigation/item/ 4266-annual-overview-of-marine-casualties-and-incidents-2020.html

Hawkins, F. H., & Orlady, H. W. (1993). *Human factors in flight.* Milton Park, Oxfordshire: Routledge.

ISL. (2019). *Shipping statistics yearbook.* Institute of Shipping Economics and Logistics. https://shop.isl.org/media/products/TOC_SSMR_63-5-6.pdf

ISRS. (2015). *Marine systematic cause analysis technique – An introduction.* DNV. http://www.isrs.net/files/SCAT/Introduction%20to%20M-SCAT%20rev4. 2.pdf

Kretschmann, L. (2020). Leading indicators and maritime safety: Predicting future risk with a machine learning approach. *Journal of Shipping and Trade,* 5(1), 1–22.

Kristiansen, S., 2021, Unpublished material.

Ladan, M., & Hänninen, M. (2012). Data sources for quantitative marine traffic accident modeling. Department of Applied Mechanics. Aalto University publication 11/2012. Alto.

MAIB. (2007). *Message for the prevention of marine incidents in fog.* Southampton: Marine Accident Investigation Branch.

MAIB. (2019). *Marine accident statistics – 2019.* Southampton: Marine Accident Investigation Branch.

MAIB. (2021). *Safety Digest – lessons from marine accident reports 1/2021.* Southampton: Marine Accident Investigation Branch.

Mazaheri, A., Montewka, J., Nisula, J., & Kujala, P. (2015). Usability of accident and incident reports for evidence-based risk modelling – A case study on ship grounding reports. *Safety Science, 76,* 202–214.

Rasmussen, J. (1982). Human errors – a taxonomy for describing human malfunction in industrial installations. *Journal of Occupational Accidents,* 4(2–4), 311–333.

Reason, J. (1990). Human error. Cambridge: Cambridge University Press.

Reinach, S. & Viale, A. (2006). Application of a human error framework to conduct train accident/incident investigations. *Accident Analysis & Prevention,* 38(2), 396–406.

Shappel, S. A. & Wiegmann, D. A. (2000). The Human factors analysis and classification system - HFACS. Federal aviation administration, Civil Aeromedical Institute. Oklahoma City, OK, USA.

Silveira, P. A. M., Teixeira, A. P., & Soares, C. G. (2013). Use of AIS data to characterize marine traffic patterns and ship collision risk off the coast of Portugal. *The Journal of Navigation, 66*(6), 879–898.

TSB. (2019). *Marine transportation occurrences in 2019.* Gatineau QC, Canada: Transportation Safety Board of Canada.

Chapter 15

Risk reduction measures

15.1 INTRODUCTION

Identifying and selecting risk reduction measures is an important task in the safety management process since this is one of the key activities in improving the risk level in the industry.

Improvements in the maritime industry have traditionally been driven by accidents, where risk reduction measures have been introduced to avoid or mitigate accident types that already have occurred and that have had serious consequences. Risk-informed safety management requires a more systematic approach to the identification of risk reduction measures, where risk analysis should form the basis for the process. Compliance with the intention behind the ALARP principle also requires a systematic process for identification of risk reduction measures.

In this section, the term barrier will be used as a general term to describe risk reduction measures. We will therefore start by explaining this term and different ways of classifying barriers. IMO has chosen to use the terms Risk Control Measures (RCM) and Risk Control Options (RCO) in their FSA guidance document (IMO, 2018), and these terms will also be explained and elaborated in more detail. We will then move on to describe systematic processes for identification and evaluation of risk reduction measures in more detail. Cost-benefit analysis plays an important role in deciding on what risk reduction measures to implement, and this will be presented in some detail.

15.2 BARRIERS AND BARRIER CLASSIFICATION

15.2.1 What is a barrier?

The term barrier is becoming increasingly common in many industries, and this term will therefore be used in this chapter. Although the maritime industry applies different terms, it is useful to know and understand the terms that are being used in other industries also, making it easier to find information that can be applied also in the maritime industry.

DOI: 10.4324/9781003055464-15

The term itself can be linked to the energy-barrier principle (see Chapter 19) that is perhaps the most commonly used accident model. The principle is based on hazards being related to some sort of energy (kinetic, potential, heat, electrical, etc.) and that there are assets (people, environment, economic values, etc.) that we want to protect. To provide protection, we use barriers that we place "between" the hazard and the asset.

Initially, barrier had a very physical meaning, covering objects that could be used to physically protect assets. This could be walls, fences, protective clothing, and many other types of protection. However, as safety management developed, it has taken on a much wider meaning. In particular, James Reason (1997) was instrumental in this development, when he introduced his Swiss Cheese model and active failures and latent conditions that could lead to accidents.

Today, barrier is therefore often used with a much wider meaning, covering more or less anything that contributes to reduce risk one way or another. This means that it is far more than just physical measures that are included in the term, but also human actions and organizational measures. Some examples of what may be labeled as barriers are human intervention to prevent accidents, maintenance to prevent equipment from failing, training to make people better equipped for doing their job safely, signs warning of danger, and procedures describing correct and safe performance of work.

A possible definition of barrier is as follows:

> A physical and/or nonphysical means planned to prevent, control, or mitigate undesired events or accidents.
>
> *(Sklet, 2006)*

This definition extends the barrier concept to include both physical and nonphysical means and it also specifies that the barriers can act anywhere in the event sequence.

15.2.2 RCM and RCO

IMO (2018) uses the terms risk control measures and risk control options more or less synonymously with the term barriers. However, the definitions are quite different (IMO, 2018):

- Risk control measure (RCM): A means of controlling a single element of risk.
- Risk control option (RCO): A combination of risk control measures.

It is not very clear what is meant by a "single element of risk", but it is assumed that this is a measure that is aimed at a specific undesired event and a specific part of the accident process (i.e., to prevent, control or mitigate). The idea behind the term RCO seems to be that this is a way of combining several RCMs into a robust combination of measures that can reduce the risk associated with a specific hazard to an acceptable level. An example can be, e.g., stability of RO-RO-vessels. There are many possible RCMs that can be proposed, either to prevent, control, or mitigate accidents due to loss of stability. Based on a large number of RCMs, several RCOs can be proposed, combining RCMs in different ways. By evaluating different combinations, it may be possible to find the optimal RCO both from a safety and cost point of view.

It is noted that in the literature and in FSAs that have been performed, the distinction between RCM and RCO is not very clear, and often just one of the terms is being used. This may be because this originated from IMOs need for methods and tools to support their rule-making process. Often when rules are modified, it may be a combination of different measures, and finding the best solution is then important before agreeing on what to implement. It may therefore be that this primarily is a useful distinction for IMO and not in many other contexts.

In the following, the term barrier will be used.

15.2.3 Classification of barriers

A system for classification of barriers can be a help in identification of proposals for risk-reducing measures since it helps to systematize and structure proposals, and at the same time acting as a trigger for coming up with ideas.

A number of proposals for how barriers can be classified have been put forward and we will only go through three examples. When we look at the process of identifying risk reduction measures, we will see how we can use these in practice.

15.2.3.1 Classifications based on accident sequence

Several classifications are based on the accident sequence and where in the sequence a barrier will have an effect. The simplest classification of this type is to distinguish between proactive and reactive barriers. In simple terms, we can say that proactive barriers are those that prevent undesired events from occurring while reactive barriers are those that act after the event and reduce or prevent negative consequences if the event has occurred. This can also be tied to the bowtie (Chapter 3), where proactive barriers work on the left-hand side of the bowtie and reactive on the right-hand side.

This is a simple way of distinguishing, although a weakness is that defining what should be the center of the bowtie in a continuous event sequence is not clear. Depending on the preferences of individuals and the purpose of the analysis, the center may be chosen differently and that also means that what is defined as a proactive barrier in one case may end up as a reactive barrier in another case. The classification is therefore far from being unambiguous. However, it may still be useful in many situations.

A more detailed classification based on the accident sequence can be found from the ARAMIS project (Salvi & Debray, 2006), where they use avoid, prevent, control, and mitigate. This can be interpreted as follows:

- Avoid: Measures that remove the possibility that an event can occur completely, by removing the hazard. This can be, e.g., to stop using a certain flammable material and replacing it with a non-flammable material instead.
- Prevent: Measures that reduce the probability of an event from occurring. On pressurized containers, overpressure protection is normally implemented to avoid too high pressure. This reduces the probability of collapse occurring but has not removed the hazard.
- Control: Measures that control the development of the accident sequence. If flammable liquid has been released, e.g., diesel or petrol, measures to prevent ignition may be introduced.
- Mitigate: Measures that limit the consequences of an event. An example at the very end of the accident sequence can be first aid to injured people.

One of the advantages of using a classification based on accident sequence is that it is normally considered to be better to avoid or prevent events rather than try to control or mitigate. Stopping the accident development as early as possible is an advantage and this classification helps us to identify how early in the accident sequence a barrier has effect.

15.2.3.2 Classifications based on type of barrier

A different approach is to classify barriers based on their type, according to the MTO principle (Man Technology Organization). Sklet (2006) has, e.g., proposed to distinguish between technical, human, and organizational barriers. Technical barriers are all engineered solutions, human barriers are where human operators perform some action that stops or intervenes in the accident development, and organizational barriers are various measures such as procedures and training.

Technical barriers are also split on active and passive barriers. Active barriers are those barriers that have to be activated, by a signal, a human intervention, or by some sort of energy, while passive barriers do not need to be activated to work. An example of an active barrier is a fire extinguishing system that must be activated before it has any effect. A passive barrier is, e.g., a fire-resistant wall that is always functioning and requires no activation.

This classification can be particularly useful for following up the barriers in operation. Technical barriers require inspection, testing, and maintenance while human barriers need training, competence checking, etc. Organizational barriers will on the other typically require regular revisions and audits.

15.2.3.3 Classification based on function

Reason (1997) has proposed a classification that is partly based on the function of the barrier and partly on the accident sequence. The classification also has a wider scope than the other classifications that have been described.

- **Understanding and awareness:** This is aimed at all who make decisions or execute work related to hazardous systems and operations. Ensuring a good understanding of what the hazards are is necessary to make the right decisions in all situations.
- **Guidance on safe operation:** Even if we understand the hazards, we all need help in performing the work in a safe manner. We are seldom able to think about everything that may go wrong even if we understand the hazards. Guidance, usually in the form of procedures or job descriptions act as a condensed "knowledge base" that provides best practice descriptions to avoid accidents.
- **Alarms and warnings:** We also need adequate warnings if a situation is starting to come out of control, e.g., if an accident sequence has "started".
- **System restoration:** Barriers that can contribute to recover from situations that are about to come out of control are covered by this. This can be many types of systems and solutions, e.g., the brakes on a car.
- **Barriers between hazards and losses:** This is in line with the physical understanding of barriers in the original energy-barrier model. An example can be a fire-resistant wall that separates inhabited areas from hazardous areas.
- **Containment and elimination of hazard:** This can be seen as barriers that intervene in the next step of the accident sequence. Some gas systems have double wall piping, containing the leak even if the inner piping should fail.
- **Means of escape and rescue:** These are barriers related to moving people away from a hazardous situation if it comes out of control.

15.2.4 Barrier properties

Different sources and authors have provided descriptions of properties of barriers that are important. The properties highlighted are not the same in all sources and the number of properties also varies. In the following, we have tried to condense the information and summarize this into four key properties that we should look for when considering barriers. This is based on a variety of sources, most notably CCPS (1993), Hollnagel (2016), HSE (2008), NORSOK S-001 (2020), and Rausand and Haugen (2020).

15.2.4.1 Specific

Barriers should be implemented for a specific purpose, linked to the risk analysis, and we should know what that purpose is. General barriers that are implemented to improve safety "in general" are seldom effective and it is hard to know what effect they have. The specific purpose is important to know as part of the safety management process – to understand what the risk is and if there are weaknesses in our protection against accidents, we also need to know what barriers are in place to protect us. If we do not have this understanding, we will not see what effect weakening of barriers has on the risk level. This is a common mistake that has been shown to have an impact in accidents – barriers have been modified or removed because operators or designers have not had a full understanding of why they have been put there in the first place.

15.2.4.2 Functional

This follows on from the previous requirement. When we know the purpose of the barrier, we need to design it and maintain it so that it is able to perform the intended task. The risk analysis may have identified that in a certain compartment on a ship, there may be a fire of a certain magnitude and of a certain type. If we have decided that we need a barrier to extinguish this fire within, e.g., 5 minutes after detection, we need to design our system to act quickly enough, to have sufficient capacity to extinguish the largest fire that can occur in the compartment, and to use a technology that is efficient against the type of fires that will occur. It may also be that we conclude that some fires may be very large or of long duration but have so low probability that they represent a very small risk. The choice may then also be to say that we do not need to design the barriers to extinguish this fire (due to low risk). These evaluations are all part of the process of determining what a "functional" barrier is.

15.2.4.3 Reliable

The next requirement is that the barrier must work when we need it. This is different from functionality. A system may be able to do what we want it to do if it is activated, but that does not help us very much if it fails to activate. On the other hand, a barrier may be highly reliable, but again this is of no use if it does not do the job when activated. A reliable system will usually have several properties that help us achieve the goal:

- Reliable components – the individual components that the barrier is composed of need to be reliable. A sprinkler system will consist of many components, including valves, pumps, piping, nozzles, etc., and all of these need to be reliable to ensure that the water reaches the target.
- Redundancy – redundant systems will normally be more reliable than single systems.
- Independence – this may be seen as a property of the whole system and not just an individual barrier but is still important. An independent barrier is not dependent on the functioning of any other barriers.
- Robustness – this has to do with the ability of the barrier to work under all relevant conditions, and in particular that it works in the conditions it is exposed to when the event that the barrier should protect against occurs. A fire extinguishing system that is damaged by the fire it is supposed to protect against is of limited value.

15.2.4.4 Verifiable

Finally, it must be possible to verify the functioning of the system also in situations where it is not needed. A system that can only be tested properly in a real emergency will always have considerable uncertainty associated with it. How can we know that it will function when needed? Barriers should therefore be possible to test, preferably in situations that are as close as possible to the real situations. Clearly, this can be a challenge to do since activation of barriers also can have negative effects. An example is a full-scale testing of a sprinkler system. This will inevitably lead to serious water damage to equipment and systems and is not something we want to do unless an emergency occurs. As far as possible, testing of parts of the functionality or parts of the system should still be possible to do. For sprinkler systems, testing the pumps is, e.g., one activity that should be done regularly, without releasing water. Similarly, it may be possible to shut off part of the system and test opening of valves and nozzles with small amounts of water.

15.2.5 Attributes of RCMs from IMO

IMO (2018) has defined a set of attributes of RCMs. The attributes are partly a classification scheme and partly a set of properties of RCMs. There is therefore considerable overlap with what has been presented in this chapter already, but it is useful to understand how IMO looks at barriers. The attributes are divided into three categories, A, B, and C.

Category A attributes divide RCMs into two groups: for preventing and mitigating risk control. Preventive RCMs reduce the probability of events, while mitigating RCMs reduce the consequences. This is for all practical purposes the same as proactive and reactive barriers.

Category B attributes have some similarity with the earlier classification into technical, human, and organizational, but is divided in a somewhat different manner. IMO also defines three types:

- Engineering risk controls which are part of the design. This broadly corresponds to technical barriers.
- Inherent risk controls. These are described as "where at the highest conceptual level in the design process, choices are made that restrict the level of potential risk" (IMO, 2018). This can be difficult to grasp, but the term inherent safety is commonly used in design and is primarily meant to imply solutions whereby we avoid hazards completely, e.g., by replacing a poisonous chemical with a non-poisonous chemical.
- Procedural risk control is where operator intervention is required, i.e., corresponding to human barriers as defined earlier.

The third group of attributes, category C, are more comparable to properties, although not only for single measures but also for the complete set of RCMs that are put in place. In the following, a slightly modified version of the category C attributes is shown. The modifications are mainly to give a somewhat better structure of the definitions of the attributes:

- **Diverse vs concentrated risk control:** The control may be distributed in different ways across aspects of the system, or it may be similar across aspects of the system.
- **Redundant vs single risk control:** Redundant risk control means that the risk control is duplicated (either by similar or different risk controls), while single risk control means that if this control fails, the accident sequence will continue to progress.
- **Passive vs active risk control:** As defined earlier, where passive control requires no action while an active control must be activated.
- **Independent vs dependent risk control:** In the first case, the risk control is not influenced by or has no influence on other risk controls. If risk controls are dependent, they may influence each other (one way or both ways), thus potentially increasing the probability that they may fail simultaneously. This is relevant in relation to redundant controls. If redundant controls are dependent, they will not be as reliable as independent redundant controls.
- **Involved vs critical human factors:** Human actions may be involved in preventing and controlling accidents but can also cause accidents. The first group is called involved human factors. This is where an action can stop an accident sequence from progressing, but failure to

do so will not cause an accident in itself or cause the accident sequence to progress further. Raising an alarm can contribute to reduce the consequences of an event but failing to do so will not in itself cause an accident or make the situation worse. Critical human factors are on the other hand actions where an accident may result if the action is performed incorrectly. An example can be opening a wrong valve, releasing flammable gas into the atmosphere instead of transferring the gas to another closed system.

- **Auditable vs not auditable** reflects whether the risk control measure can be audited or not. This is similar to what earlier was called verifiable.
- **Quantitative vs qualitative** reflects whether the risk control measure has been based on a quantitative or qualitative assessment of risk. It may be noted that this is not a characterization of a property of the RCM, only how it has been identified.
- **Established vs novel:** This touches upon the uncertainty related to whether a risk control can be expected to be implemented and work as intended. There will always be uncertainty about this when implementing something new compared to using well-known, tried, and tested technology. It may also be that a technology is established for industrial, shore-based applications but may be novel for use on ships.
- **Developed vs non-developed:** To a certain degree, this is the same attribute as the previous one. IMO focuses mainly on technical effectiveness and basic cost in the description of this attribute.

15.3 IDENTIFYING RISK REDUCTION MEASURES

A requirement for a systematic approach to safety management is also that we systematically identify proposals for risk reduction measures. This can be done in different ways, but it is recommended that the identification is based on scenario descriptions from risk analysis.

To make this feasible and efficient, it is obvious that the risk analysis must contain good scenario descriptions. This is unfortunately not always the case, at least not as systematic and good as we would like and may need for identifying risk reduction measures. It may be noted that this has little to do with whether the analysis is qualitative or quantitative. For this purpose, it is the descriptions that are important, not the quantification, and a qualitative analysis may be just as good as a quantitative analysis in this respect.

A common problem in many quantitative risk analyses is that causes of events are not sufficiently well described. Experience is that the frequency of events (e.g., collision or fire) often is based on statistics and not going

Figure 15.1 Process for identification of risk reduction measures.

into detail on the causes. This means that it is harder to identify risk reduction measures that can contribute to reduce the probability of events occurring.

In quantitative analyses, methods like event tree analysis are often used and sequences are then readily identified. However, in qualitative analyses, sequences are not often described, instead events and causes are described as isolated events.

In Figure 15.1, a process for identification of risk reduction measures is illustrated.

Input to the process are scenario descriptions with event sequences described. The process is started by selecting the scenario with the highest risk. It is not necessary to do so, the process can also be started with a randomly selected event from those that require that further risk reduction is considered (i.e., events in the unacceptable region or the ALARP zone if the ALARP principle is applied). In practice, this can however be a good starting point.

The next step (2) is then to select the first event in the event sequence. The identification of risk reduction measures is based on asking what measures can be introduced (or improved) to reduce the probability that this event will occur. To trigger the identification process, classification according to type (Technical, Human, Organizational) may be used as a help. We then ask questions like "Are there any technical measures that can contribute to reduce the probability that this event will happen?". Reason's (1997) classification may also be used in the same way. At this stage, we do not evaluate the feasibility, efficiency, cost, or other aspects of the proposals. The initial objective is to come up with as many ideas as possible and then the evaluation will be performed later. This process is best performed in a workshop with personnel with relevant competence about the risk, technical systems, and operations present.

When we have completed the identification for the first event, we go back to step 2 to select the next event in the sequence and repeat step 3 for this event. This process is repeated until all events in the scenario have been covered. We then move to step 5 and the whole process from step 1 is repeated until all scenarios that need to be covered have been analyzed.

The final step (6) entails a systematization and structuring of the proposed risk reduction measures. Often, we will find that a measure can influence several scenarios and thus are identified several times. Further, it may also be that some proposals that are slightly different can be joined to a common

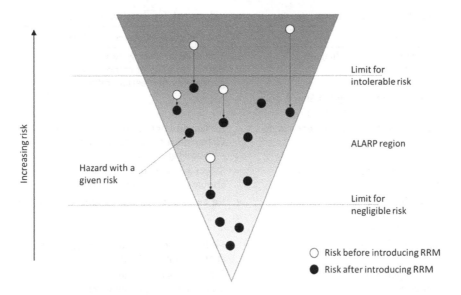

Figure 15.2 Illustration of the risk reduction process when using the ALARP principle.

risk reduction measure that has a wider effect than initially thought. This structuring will act as a preparation for performing the evaluation of the proposed risk reduction measures.

Figure 15.2 illustrates the process of reducing risk when applying the ALARP principle. The black dots in the figure represent the risk picture after risk reduction measures have been introduced. The figure shows that only some of the hazards have a reduced risk. However, all the hazards that originally were in the upper region have been reduced and moved below the limit for intolerable risk. In the ALARP region, only some of the hazards have been moved, and it is not necessarily those that have the highest risk since this will depend on where reasonably practicable measures are identified. In the lower region, below the limit of negligible risk, no change has occurred because we will not normally look for ways to reduce risk in this region.

Collision scenario

To illustrate steps 2 and 3, consider a specific collision scenario described in Table 15.1. The first column describes the event sequence, while the second column describes possible measures to reduce risk. A tabular format like this can be useful to present all possible measures in a systematic manner. It is noted that this is just for illustration – there may be other measures and some of the proposed measures in this table are not necessarily very practical to implement. The feasibility should however be considered in the next step.

Table 15.1 Example of possible measures to reduce risk for a collision scenario

Event sequence	Possible measures to reduce risk
Two ships are approaching on a collision course	– Traffic separation system – Better planning of route to reduce probability of meeting traffic
Ship A has right of way, but Ship B does not change course or speed	– None
Ship A contacts Ship B but receives no response	– Earlier attempt at contact by radio – Other means of attracting attention – sound/light signals
Ship A changes course to avoid collision	– Select action that will not lead to collision should other ship react in time
Ship B changes course at the same time, leading to A and B still being on a collision course	– Be extra observant of actions of other ship
Ship A and B collides	– Reduce speed or optimize impact angle to reduce consequences
Ship A is penetrated, and two compartments are open to sea	– Stronger plating in sides – Improved compartmentalization
Ship A quickly develops a 15 degree list	– Smaller compartments – Filling compartments on opposite side if sufficient reserve buoyancy and if possible

15.4 EVALUATING AND PRIORITIZING RISK REDUCTION MEASURES

The output from the identification process is a structured list of proposals for risk reduction, linked to the scenarios where they will have an effect. The next step is then to evaluate the risk reduction measures, as a basis for making decisions about what to implement or not.

An important input to the decision making is of course the cost of the proposed measure versus the effect that we can expect. Cost-benefit analysis will be covered in Section 15.5. In this section, a set of criteria or factors to consider when evaluating risk reduction measures are presented and discussed. These can be used to do a more systematic evaluation than just considering costs and benefits (risk reduction). These criteria are partly related to the barrier properties that were discussed in Section 15.2.4.

15.4.1 Effects on risk

The first and most obvious factor is what effect the proposed risk reduction measure will have on risk. This may seem straightforward, in particular if we have a quantitative risk model where we take into account the effect and recalculate the risk. However, there are several effects on risk that should be considered.

Effect on specific event: This will normally be the first thing we think about and try to evaluate. An example can be an improved fire extinguishing system to reduce the risk associated with fire. This will reduce the consequence should a fire occur and if we know the types and sizes of fires that can occur, and we know the technical details of the system we can say something about how much risk is reduced. In an event tree, this can, e.g., be modeled as an increased probability of extinguishing the fire.

Effect on other events: The way that we have described the process of identifying risk reduction measures, each measure will be associated with a specific event. However, some measures will influence more than one type of events. An example of this can be, e.g., installation of an improved radar system that can give warning about potentially critical situations. This will influence both collision and grounding and when the total effect of a measure is determined, all effects need to be considered.

Negative effects: Introduction of risk reduction measures may also have negative effects, usually not on the event that they are mainly aimed at but on other events. A well-known example from the oil and gas industry is the introduction of fire-resistant walls to separate hazardous areas from other areas. Clearly, this is positive because it reduces the probability that a fire will spread from one area to another. However, a negative effect is that natural ventilation due to wind is reduced and gas will therefore be more likely to accumulate if a leak occurs. This can lead to more severe explosions should it ignite. Another issue is whether extensive automation of navigation functions reduces the ability of the mate on watch to deal with emergencies. A navigator not fully involved in the control of the vessel may be inattentive and less able to detect the first indications of an emerging threat.

Risk associated with implementing the measure: A final effect that also may be relevant in some cases is that implementing a measure may imply a risk in itself, in particular when extensive construction/modification work is required. Modifications that require high-risk activities, e.g., work at height, can sometimes represent a significant risk increase in the period when they are taking place. This is also an effect that should be considered.

When considering the total effects on risk of a risk reduction measure, all four of these aspects should be added together, to get the total, net risk reduction. With a comprehensive quantitative risk model available, the first three can usually be quantified by modifying the risk model. The last contribution is often more difficult since this risk is normally not included in the risk analysis.

15.4.2 Reliability

The reliability of a barrier, or how likely it is that it will function when needed, is obviously a relevant parameter to consider. When determining the risk reduction effect, the reliability should normally be considered. Determining the reliability of a risk reduction measure is however not necessarily easy, and some simple rules of the thumb may be useful to do a coarse ranking:

1. The most reliable measures are normally passive, technical measures. These require no activation and as long as they have not deteriorated with time, the probability that they will function is normally very high.
2. Active technical measures are ranked second. The reliability of an active system can vary a lot, but with careful design and proper maintenance the reliability will normally be high.
3. Human actions are next. The probability that humans make mistakes can often be quite high, in particular in emergency situations where we have to make decisions and react quickly.
4. Organizational/procedural measures are considered to be the least reliable.

This ranking is not absolute, and examples that violate this can be found, but this can still be helpful as a good indication.

15.4.3 Verifiability

This was also one of the properties discussed earlier. In practice, this influences the reliability of the risk reduction measure. For this purpose, we can distinguish between reliability "as designed" and reliability "in operation". In operation, the important thing is that the risk reduction measure works when needed and the key activity we use to verify that it is working is testing. Inspection and maintenance will normally increase the probability that a barrier is working, but only through testing can we actually verify that it is working. Some barriers cannot be tested at all, without physically destroying them (e.g., heat detectors that are activated by melting a fusible material). Other tests can cause damage to equipment (e.g., releasing sprinkler systems). On the other hand, there are also systems that can continually test themselves, e.g., control systems (although this not necessarily will test all functionality in a system).

15.4.4 Independence

Independence is an important property of a risk reduction measure and should be considered as part of the evaluation. This is a property that will affect the reliability of all the barriers seen together. If two barriers are independent, the combined probability of failure of those two barriers can be found by multiplying the probabilities of failure. However, if they are not independent, the combined probability will be higher. Risk management is commonly based on the principle of "defense-in-depth", i.e., that there are several barriers in place to control risk, in particular when the consequences potentially can be severe. If the barriers are not independent, the protection will be weakened since we may experience simultaneous failure of several barriers.

It should be underlined that what is important in this case is that a risk reduction measure is independent of the barriers that are intended to reduce risk associated with the same event. Even if barriers that work against different events fail simultaneously, this is not critical since it is extremely unlikely to have two events at the same time. An example can be that two barriers, one protecting against loss of stability and one protecting against fire are dependent on the same utility system. They are thus not independent. However, this is not a problem, since the events are different and not likely to occur at the same time.

15.4.5 Where in event chain

We can use the event chains to systematically identify risk reduction measures, and it is also quite common to use the event chains to prioritize the risk reduction measures. The philosophy is that barriers that act early in the event chain (e.g., to avoid hazards) should be preferred over barriers that act late (e.g., to mitigate consequences). In most cases it is fairly obvious that this is a better strategy. If we can avoid oil spills, that is clearly a better approach than to invest in equipment to clean up oil spills on beaches. We can then, e.g., use Avoid – Prevent – Control – Mitigate (see Section 15.2.3.1) to distinguish different barriers.

15.4.6 Duration

It is often easy to assume that a proposed risk reduction measure that is implemented will continue working as long as the system that it is protecting is in operation. If we implement a barrier on a ship, we may end up assuming that this will be operational for the lifetime of the ship.

However, this is not always the case. For technical barriers, there is the obvious need for maintenance, to ensure that it remains operational. For human and organizational barriers, maintenance is often also necessary, although in a different form. A typical example is training in handling emergencies. This is knowledge and skills that we fortunately need to use very seldom. However, this also means that we will forget what we have learned and lose the skills with time. This type of training therefore needs to be repeated quite often, to continue having an effect.

15.4.7 Cost

Last, but not least important, is the cost of the risk reduction measure. Clearly, the more expensive a measure is, the less likely are we to accept it. Cost-benefit analysis will be described in Section 15.5 and is the most common method used to evaluate whether a proposed risk reduction measure is cost-effective or not.

15.4.8 Summary

As can be seen from the above, a number of properties need to be considered when evaluating proposals for reducing risk. In many cases, it will be possible to determine whether a proposal should be implemented purely based on a qualitative or semi-quantitative scoring of these criteria.

However, in all cases, this is not possible. One proposal may score high on some properties and low on others, while another proposal may score in a very different way, but perhaps with the same number of high and low scores. Comparison may thus become difficult, and also deciding whether to implement it or not.

In practice, most of the properties discussed above will eventually influence how much effect a proposed measure will have on risk or what the cost of implementing a measure will be (or both). Comparing costs with effects on risk is therefore the most systematic (and common) method for deciding on whether a proposal should be implemented or not.

Illustration of a simple method for evaluation of risk reduction measures

To illustrate a semi-quantitative method for evaluating a proposed risk reduction measure, consider the following example. In a risk analysis of a RO-RO ferry, it has been found that the risk associated with collision is high and that improved evacuation means are necessary. It is not specified what this is, but it should be possible to deploy quickly and be functional up to storm conditions.

Table 15.2 illustrates how a proposal can be systematically evaluated on the properties discussed in this section.

In this table, a score and a description are provided for each criterion. The score is used mainly to simplify comparison but is not necessary to be included. If scores are being used, we need to define what we mean by the scores, e.g., when we say that "Reliability" is "High", we should define categories, e.g., "High" corresponds to more than 95% probability of functioning, "Medium" is 85–95% probability, and "Low" is less than 85%. Similar categories need to be defined for all criteria. Alternatively, if we are doing a quantitative analysis, we can also put the numbers in the table instead of using a score.

Table 15.2 Evaluation of improved evacuation means

Criterion	Score	Comments
Effect on specific event	High	Consequences are expected to be reduced significantly.
Effect on other events	High	Will influence all events where evacuation is required, e.g., grounding, stability problems, and fires.
Negative effects	None	No negative effects are foreseen but need to be considered in detail for specific a proposal.
Risk in implementation	Small	Requires modifications to ferry, but this is performed in yard and the risk is comparable to ordinary construction work.
Reliability	High	Highly reliable system is specified.
Verification	Medium	Testing all functionalities of evacuation systems under realistic conditions is hard. Only partial verification is likely to be possible.
Independence	Complete	No dependence on other barriers.
Where in event chain	Mitigate	This reduces the consequences when evacuation is required, so late in the event sequence.
Duration	Permanent	If a proper maintenance scheme is in place, the effect is permanent.
Cost	High	Modification of the ferry and installation of new equipment will be expensive.

15.5 COST-BENEFIT ANALYSIS

Cost-benefit analysis (CBA) is a technique for comparing the costs and benefits of a project. The technique was originally developed to help appraise public sector projects to ensure that one achieved the greatest possible value for the money invested, but the concept of cost-benefit analysis has applications far beyond the public sector domain. In this chapter we are concerned with CBA in the context of safety management.

The concept of cost-benefit analysis is quite simple and need not necessarily involve complex mathematical methods. We all perform cost-benefit analyses daily, for example, when we are shopping for groceries. Most people decide on which items to buy based on a trade-off between their perceived benefits and costs, such as quality, price, personal preferences, etc. One should recognize that costs and benefits can be understood in general terms and not just in monetary terms. The costs of buying a certain product involve of course its monetary value but may also involve, e.g., poor quality, health hazards, and effect on the environment. It is important to bear this in mind when performing all kinds of CBAs, as it may be difficult to examine all costs and benefits on a common scale.

First, some basic theory will be presented that is necessary to understand before performing CBA. This involves basic economic and cost-optimization theory that enables us to calculate monetary costs and benefits. This is followed by a closer look at CBA in a safety management context. The main principles and a general approach are presented together with some useful CBA methodologies.

15.5.1 Economic theory

To do cost-benefit analyses we need to express costs and benefits on a common scale. For now, let us assume that we can express both costs and benefits in monetary terms. Both costs and benefits may arise at different points in time, and because the value of money changes over time due to inflation and "time value of money" factors, one must be able to calculate the value of all monetary costs and benefits at the same point in time. To calculate the value of a present amount P at some future point in time, the following simple equation can be used:

$$F = P \cdot (1+i)^n \tag{15.1}$$

where

P = Present monetary amount [currency]
F = The future value of P after n years/periods [currency]
i = Rate of interest per year/period (corrected for inflation) given as a decimal fraction, e.g., 0.05 (5%).
n = Number of years/periods

We can also calculate the present value P of a future amount F by using the following equation (deducted from Eq. 15.1):

$$P = \frac{F}{(1+i)^n} = F \cdot (1+i)^{-n} \tag{15.2}$$

If a chain of n equal future amounts F are incurred at regular intervals (e.g., inspection costs, maintenance costs, loan payments), and the rate of interest i for those intervals can be assumed as constant, it can be shown that the present value P of this chain of future amounts F is as follows (present value of annuity):

$$P = F \cdot \frac{(1+i)^n - 1}{i \cdot (1+i)^n} = \sum_{i=1}^{n} F \cdot (1+i)^{-n} \tag{15.3}$$

These equations are simple but very important tools in relation to CBA.

15.5.2 Cost optimization

Safety measures are implemented in a system to reduce the risk. The costs involved in implementing safety measures may as such be understood as preventive costs, i.e., costs related to preventing danger to people, property, and the environment. Safety can be improved by a wide range of measures such as physical safety equipment/systems, organizational safety programs, improved operating procedures, etc., and preventive costs could therefore include:

- Costs related to the design and development of safety measures/programs
- Cost of equipment and installation
- Costs related to the inspection and maintenance of safety equipment in all its life
- Staff operating costs
- Training costs
- Enforcement costs
- Inspection and auditing costs
- Administration costs
- etc.

On the other hand, the implementation of safety measures also results in benefits of reduced cost of losses related to the economic consequences that are more likely to be avoided because of reduced risks. In a maritime context typical cost of losses include:

- Total loss of ship, additional costs of getting a new vessel into operation
- Degraded operability/operation resulting in unscheduled delays
- Loss of future income (due to total loss or ineffective operation)
- Repair costs
- Fines and penalties
- Compensation to third parties
- Negative publicity (may be difficult to quantify)
- etc.

Based on the above discussion a pure economical exercise can be performed to establish the optimal level of preventive safety measures that should be implemented. This can be done by studying the total safety cost, being the sum of preventive costs and cost of losses, and finding where this total cost is at its lowest. Such a cost optimization is illustrated in Figure 15.3. For this to be valid, it must be possible to directly value the costs and benefits of safety measures in economic terms. In practice the economic consequences of accidents have a dominant position, but as commented earlier other factors may be taken into consideration as well.

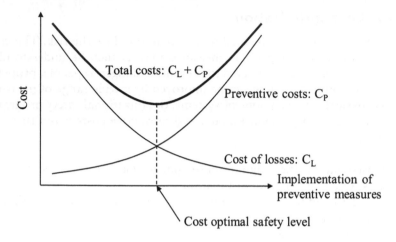

Figure 15.3 Optimal implementation of safety measures.

The cost curves for preventive costs (C_P) and cost of losses (C_L) in Figure 15.3 are symmetrical. However, this will not be the case in most practical applications, and the curves will generally vary greatly depending on the type of system under consideration.

In this section we have focused on establishing the cost-optimal safety level, but the type of cost optimization illustrated by Figure 15.3 has many applications, e.g., in terms of establishing an economically optimal level of preventive maintenance. Cost optimization can be a very useful tool in relation to cost-benefit analysis. In the following example, some results from a risk assessment study of bridges with respect to ship collisions are presented (Sexsmith, 1983). This case/example shows how basic mathematical theory can be used to perform simple cost-benefit analyses.

Design of bridge against ship collision

A bridge designer is to establish the optimal energy capacity of a bridge design that is exposed to ship traffic. The energy capacity denotes the energy that the bridge can absorb without collapsing. If the impact energy from a ship colliding with the bridge is larger than the bridge's energy capacity, the bridge will experience catastrophic collapse. Increasing the energy capacity of the bridge requires the implementation of protective measures that come at a considerable extra cost, but increased capacity also means reduced probability of collapse (increased return period). Table 15.3 provides information about three different bridge design alternatives.

A loss of USD $2 \cdot 10^8$ in the event of catastrophic collapse is assumed, including both the economic losses associated with the bridge structure and losses that occur because of loss of use of

Table 15.3 Key bridge design parameters and cost of protective measures (C_c)

	Bridge energy capacity [MJ]	Return period for exceedance T [years]	Annual rate of exceedance $\lambda = 1/T$ [-]	C_c [USD]
Concept 1	800	500	0.002	$3.0 \cdot 10^6$
Concept 2	1,000	1,000	0.001	$6.0 \cdot 10^6$
Concept 3	5,000	5,000	0.0002	$12.0 \cdot 10^6$

the bridge. The real rate of interest (i) is assumed to be 3% = 0.03 [per year].

Assuming a loss C_f will occur at a specific time in the future, the present value of this loss can be expressed as follows:

$$C_0 = C_f \cdot e^{-it} \tag{15.4}$$

where
  C_0 = The present value of the future loss
  C_f = The future loss in present monetary units (not inflated)
  i = The real interest rate per year (excluding inflation)
  t = The time to the future loss (years)

When the time until occurrence of the loss is a random variable (i.e., loss is stochastically distributed), the present expected value of the future loss (i.e., C_0) can be expressed as follows:

$$C_0 = E\left(C_f \cdot e^{-it}\right) = C_f \int_0^T e^{-it} f(t)\, dt \tag{15.5}$$

where
  $E()$ = The expected value function
  $f(t)$ = The probability density function for the time t (i.e., the time to occurrence of catastrophic collapse)
  T = The period being considered (lifetime of the bridge)

Ship collisions with bridges are rare and independent random events in time. The events can therefore be considered as Poisson events, and the time to first occurrence is therefore exponentially distributed:

$$f(t) = \lambda e^{-\lambda t} \tag{15.6}$$

where
  λ = Annual rate of exceedance of bridge energy capacity
  t = The time to the future loss

Table 15.4 Calculation of total cost for the three alternatives

	Bridge energy capacity [MJ]	Return period for exceedance T [years]	Annual rate of exceedance $\lambda = 1/T$ [-]	C_c [USD]	C_0 [USD]	$C_T = C_C + C_0$ [USD]
Concept 1	800	500	0.002	$3.0 \cdot 10^6$	$12.5 \cdot 10^6$	$15.5 \cdot 10^6$
Concept 2	1,000	1,000	0.001	$6.0 \cdot 10^6$	$6.5 \cdot 10^6$	$12.5 \cdot 10^6$
Concept 3	5,000	5,000	0.0002	$12.0 \cdot 10^6$	$1.3 \cdot 10^6$	$13.3 \cdot 10^6$

It then follows that the present value of the loss can be expressed as follows:

$$C_0 = C_f \int_o^T e^{-it} \lambda e^{-\lambda t} dt = \frac{C_f \cdot \lambda}{i + \lambda} \qquad (15.7)$$

Equation 15.7 can be implemented into a CBA to establish the optimum total cost consisting of the cost of loss (C_0) and the cost of protective measures (C_C). Both costs will depend on the concept selected:

$$C_T = C_0 + C_C \qquad (15.8)$$

Increased capacity of the bridge means decreasing cost of losses (because losses are rarer), but the cost of the preventive measures increases. The total cost (C_T) for the three design concepts is calculated in Table 15.4. Here C_0, i.e., the present value of future loss of the bridge, is calculated using Eq. 15.7 with $C_f = $ USD $2 \cdot 10^8$ and $i = 3\% = 0.03$.

Based on model and the assumptions, Concept 2 has the lowest total cost of the three concepts, resulting in an energy capacity of 1,000 [MJ] for the bridge. However, it must be recognized that such a model is simplified and sensitive to the assumed parameters. In this example, Concept 3 has the lowest cost if the rate of interest is reduced to 2%, while concept 1 has the lowest cost if the rate is increased to 7%.

Seeing how sensitive the results are to the input values and at the same time recognizing that many of the input values are highly uncertain, sensitivity analysis is very important when doing cost-optimization studies and cost-benefit analysis.

15.5.3 CBA in safety management

As briefly mentioned in the introduction to this chapter, the implementation of safety measures directed at reducing the risks of severe accidents related to a system unavoidably incurs costs. However, there are also benefits related to improved safety, predominantly reduced cost of losses. In this context there are two views on safety management and the use of cost-benefit analysis (CBA) that we should understand.

The first view considers the system as a whole and recognizes that for all systems there will be a limit to how much that can be spent on safety measures before the system (e.g., an oil tanker) becomes uneconomic or uncompetitive. With this starting point, the challenge for safety management will be to achieve maximum risk reduction (i.e., best possible safety level) within given economic limits. This is primarily an optimization problem.

In this case, it is not the risk level but the total costs that determine what should be implemented. This may mean that we can end up with systems with very high risk, because the economy is marginal. The argument against this is that a system that only can be profitable if the risk is very high should not be accepted by society. In most cases, prescriptive regulations will therefore also ensure that the risk is not excessively high. An example can be lifeboats – even if the operation of a ship is economically marginal and it is not making money, the ship still cannot be operated without lifeboats because there are prescriptive regulations specifying that lifeboats have to be installed on the ship.

In the second view, decisions about reducing risk are based on comparing the costs of implementing a safety measure with the benefits gained through this implementation. If this comparison is favorable, the measure should be implemented, regardless of the total cost of all measures that are implemented.

In principle, this view will not place any limitation on how much money should be spent on reducing risk. In practice, the prescriptive regulations will however normally bring the risk down to a level where relatively few additional measures are cost-effective. In a risk assessment context this favors the second view on CBA, in which the main principle is to find safety measures that cost-effectively reduces the risk. Regarding risk reduction, one often has several different and potential safety measures to choose from, and CBA enables us to identify for implementation those that are cost-effective.

A cost-benefit analysis is simple to perform if the costs and benefits for the suggested safety measures are known. The challenge is how to establish these costs and benefits based on the risk analysis model developed.

The costs of safety measures are mainly associated with the costs of implementing, operating (including inspections, audits, and maintenance), and managing the safety measure.

Estimating the benefits of safety measures is, in general, more complicated. First, we need to agree on how to calculate the benefits on a monetary scale. This implies among others to place a value on reduced risk of loss of life. This is a complex problem. Next, we calculate the reduction in risk using the quantitative risk analysis model that has been developed. In practice, this is done by estimating what effect the risk reduction measure has on the fault trees and event trees, mainly by updating the probabilities going into the calculations or also by modifying the structure of the risk model.

In cost-benefit analysis, we normally use changes in PLL (Expected number of fatalities, see Chapter 9) to express the reduction in risk. This is because we often use various approaches to expressing an economic value for a saved

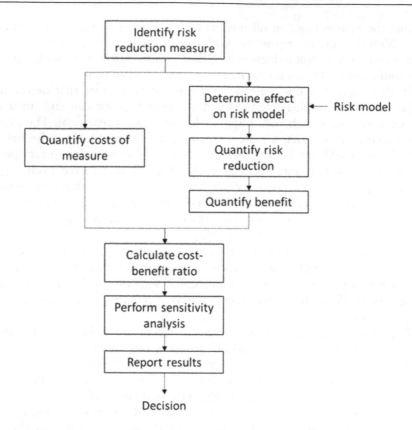

Figure 15.4 CBA in a risk assessment context.

life. Reduction in PLL is a direct expression of statistically expected saved lives and can therefore be used directly in the cost-benefit calculation.

When the costs are known and the benefits have been expressed in monetary terms, a cost-benefit ratio (i.e., the costs divided by the benefits) or other measures can be calculated. By comparing different measures, they can be ranked and cut-off limits for what should be implemented or not can also be established.

As already pointed out, there is considerable uncertainty in these calculations, both for the risk reduction and other assumptions applied. Sensitivity analyses should therefore be performed to test the robustness of the initial recommendation. This CBA approach can be illustrated by Figure 15.4.

CBAs regarding safety-related matters are normally based on marginal considerations, which means that the preventive measures are implemented as long as the estimated benefits of reduced risk at least equals the expected costs (i.e., costs≤benefits). In guidance to the implementation of the ALARP principle (HSE, 2001), it is suggested that measures should be implemented as long as the costs are not "grossly disproportionate" to the benefits. This means that even when the costs are higher than the benefits, it may be argued that the measures should be implemented. HSE (2001) also gives some indication of what they mean by "grossly disproportionate" and states that when the risk is close to the unacceptable region (i.e., near the upper limit of the ALARP region), the factor could be as high as 10, i.e., even if the costs are ten times higher than the calculated benefit, one should consider implementing the measure. On the other hand, if the risk is low, close to the lower limit of the ALARP region, a disproportionality factor of 1 (i.e., cost equal to benefit) could be used.

A problem related to assessing the benefits of averted and/or reduced consequences, because of introduced safety measures, is often the large number of consequence types that may be affected and the fact that the safety improving effects of a particular safety measure may vary strongly among the consequences. One particular safety measure may, therefore, affect many types of damage extents for several accident types as illustrated by the general accident model given in Figure 15.5. The total effect of the measure can as a result be difficult to establish and quantify for CBA applications.

Oil spill during tanker offloading at a refinery

A shore-based oil refinery receiving crude oil from shuttle tankers has recently carried out risk analyses for the parts of its operation that may result in oil spills. One area where such spills occur is during tanker offloading. The emergency response unit at the refinery is well prepared to initiate fast and effective clean-ups of smaller spills during offloading, but spills larger than 10 tons are a major concern as they will often be difficult to contain and the harm to the surrounding environment can be serious. The probability of such large spills was in the risk analysis estimated to $2.0 \cdot 10^{-4}$ per offloading using fault tree analysis of the offloading. The refinery has an average of 120 tanker offloadings per year. The average size of large oil spills was estimated to 50 tons using event tree development from the initiating event of "offloading equipment failure", and the cost of such spills is estimated

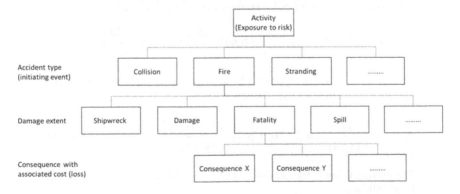

Figure 15.5 General accident model.

to USD 40,000 per ton. This cost includes clean-up costs, fines, compensation to the local fishing community and affected landowners, extra public relation costs, etc.

The management at the refinery has concluded that the risk involved in offloading is acceptable, but practicing the ALARP principle, they decide to perform a cost-benefit analysis for available safety measures. One possible safety measure is a newly developed offloading installation. This installation is more reliable against spills and reduces the average size of large spills. It is estimated that implementation of this new equipment will reduce the probability of large spills to $1.2 \cdot 10^{-4}$ per offloading and the average size of large spill down to 15 tons. The cost of the installation is USD 750,000 and it has an expected lifetime of 12 years. It is more efficient and requires less maintenance than the existing offloading installation, resulting in an estimated cost reduction of USD 20,000 per year, and maintenance of the new installation included. We assume a 5% rate of interest per year.

The existing risk for large spills (in economic terms) can be calculated as follows:

$$R_0 = C \cdot P = \frac{50 \text{ t}}{\text{spill}} \cdot 40,000 \text{USD} \cdot \frac{2.0 \cdot 10^{-4} \text{ spills}}{\text{offloading}} \cdot 120 \frac{\text{offloadings}}{\text{year}}$$

$$= 48,000 \frac{\text{USD}}{\text{year}}$$

The risk after implementation of the safety measure, i.e., the offloading installation, will be as follows:

$$R_1 = C \cdot P = \frac{15 \text{t}}{\text{spill}} \cdot 40,000 \text{USD} \cdot \frac{1.2 \cdot 10^{-4} \text{ spills}}{\text{offloading}} \cdot 120 \frac{\text{offloadings}}{\text{year}}$$

$$= 8,640 \frac{\text{USD}}{\text{year}}$$

The benefits of reduced risk can then easily be calculated:

$$\Delta R = R_0 - R_1 = 48,000 \frac{\text{USD}}{\text{year}} - 8,640 \frac{\text{USD}}{\text{year}} = 39,360 \frac{\text{USD}}{\text{year}}$$

The net present value P_b of this benefit for the 12-year lifetime of the equipment can then be calculated:

$$P_b = F \cdot \frac{(1+i)^n - 1}{i \cdot (1+i)^n} = 39,360 \text{USD} \cdot \frac{(1+0.05)^{12} - 1}{0.05 \cdot (1+0.05)^{12}} = 348,858 \text{USD}$$

The present value P_c for the costs of the new offloading installation will be as follows:

$$P_c = 750{,}000 - 20{,}000 \cdot \frac{(1+0.05)^{12} - 1}{0.05 \cdot (1+0.05)^{12}} = 572{,}735 \text{USD}$$

The cost-benefit ratio will then be:

$$\frac{C}{B} = \frac{P_c}{P_b} = \frac{572{,}735 \text{USD}}{348{,}858 \text{USD}} = 1.64$$

The cost-benefit ratio is larger than 1.0, i.e., the cost of implementing the measure is higher than the calculated benefits. Hence, the proposed new offloading installation is not found to be cost-efficient. The conclusion will then be that it should not be implemented unless there are other arguments for implementing it.

15.5.4 Cost-benefit analysis methodologies

All suggested risk control measures result in different risk reductions, different benefits and negative effects, and different implementation costs. Without a method to evaluate these risk control measures against each other on a similar basis, it would be very difficult to select the most cost-effective measures for implementation, i.e., the measures that result in the greatest benefits compared to the costs. As long as most of the costs and benefits can be quantified in monetary terms one should attempt to evaluate the measures on a similar basis. In the previous section, the cost-benefit ratio was used as a measure, but many other methods are also being used. The main principles of two popular approaches will be described here: the Cost per Unit Risk Reduction (CURR) and the Implied Cost of Averting a Fatality (ICAF).

CURR was initially developed for use by IMO in the international context where one may expect very differing views on how reduced fatalities and injuries should be valued. The approach adopted is to value the cost and benefit items, except from the economic benefits of reduced fatalities, in monetary terms and to separately establish the number of equivalent lives lost over the lifetime of the measure assuming an equivalence between minor injuries, major injuries, and death (e.g., 100 minor injuries accounts for 10 major injuries, which again accounts for one death, see Chapter 9). The net present value (NPV) of implementing a risk control measure is calculated using the following equation:

$$\text{NPV} = \sum_{t=0}^{n} (B_t - C_t)(1+r)^{-t} \tag{15.9}$$

C_t = The sum of costs in period t

B_t = The sum of benefits in period t (excluding economic benefits of reduced fatalities)

r = The interest rate per period
t = The time period being considered, starting in period (e.g., year) 0 and finishing in period n

In principle, this calculation is the same as the one shown in Eq. 15.3.

The resulting NPV is then used to calculate a Cost per Unit Risk Reduction (CURR) by dividing NPV by the estimated number of reduced equivalent fatalities (ΔPLL):

$$CURR = \frac{NPV}{\Delta PLL}$$

CURR values for different risk reduction measures can then be compared and provides a ranking of cost-effectiveness in improving human safety.

The Implied Cost of Averting a Fatality (ICAF) methodology is a much-used methodology for studying risk control measures on a common scale. Principally, it is similar to CURR, but the calculation is done per year instead of for the whole period. The methodology calculates/estimates the achieved risk reduction in terms of cost using the following equation:

$$ICAF = \frac{\text{Net annual cost of measure}}{\text{Reduction in annual fatality rate}} \tag{15.10}$$

The net annual cost of a measure is calculated by distributing all the costs related to the implementation and operation of a measure over the measure's lifetime. This is achieved by calculating the yearly annuity. ICAF may also be calculated by dividing the net present value for the whole lifetime of the safety measure by the total reduction in fatalities for that particular period. The ICAF value can be interpreted as the economic benefits of averting a fatality. A decision criterion must be established for this value to evaluate whether a given risk control option/measure is cost-effective or not, and this criterion would in a way involve pricing a human life. A method for determining criteria for ICAF is presented below.

IMO (2018) distinguishes between two alternative ways of calculating ICAF, calling them GCAF and NCAF, respectively. GCAF (Gross Cost of Averting a Fatality) is calculated based on the total cost of implementing a risk reduction measure, while NCAF (Net Cost of Averting a Fatality) subtracts any economic benefits from the cost before dividing by the reduction in fatality rate. An example can be that implementing a risk reduction measure also may lead to more efficient operation. NCAF will obviously always be the same or smaller than GCAF (NCAF $\leq$ GCAF).

IMO recommends that GCAF is calculated first and that decisions primarily are based on this. If a risk reduction measure is marginally above the criterion set, NCAF may be calculated to see if this brings the measure within the limit, leading to the conclusion that it should be implemented.

A third and less comprehensive method for adaptation onto a common scale is to only examine the net present value (NPV) of the different safety measures. Safety measures (or risk control options) with a positive NPV, which do not have other negative effects on the system under consideration, should always be implemented. However, only a few safety measures will normally have a positive economic NPV, and the method has a major weakness in not addressing the relative differences in risk reduction effect between different safety measures.

15.5.5 Establishing criteria for ICAF

The benefits of averting a fatality are difficult to quantify. Some even mean that such quantification is impossible because it involves associating economic value with human lives. It is however important to underline that this is a criterion that enables us to make decisions about whether to reduce risk or not, and not an expression of how much a life is worth. An analogy to this may be that accepting risk is not the same as accepting that people are killed – what we accept is a possibility that this may occur in the future.

The advantage with applying criteria like this is that it helps us to rank and prioritize measures, which in turn give better utilization of limited resources. For decision making, it is therefore clearly useful to have a criterion like this.

There are different ways of calculating ICAF criteria, but Skjong and Ronold (1998) propose that the ICAF value can be calculated through analysis of a so-called Life Quality Index, which is a compound social indicator. Skjong and Ronold define the Life Quality Index as follows:

$$L = \gamma^w \cdot \varepsilon^{1-w} \tag{15.11}$$

where
L = Life Quality Index
γ = Gross domestic product per person per year
ε = Life expectancy [years]
w = Proportion of life spent in economic activity (in developed countries $w \approx 1/8$)

The principle of ICAF as a criterion for risk reduction is to implement safety measures as long as the change in L is positive. By partially differentiating L and requiring that $L > 0$, the following equation is established:

$$\frac{\Delta\varepsilon}{\varepsilon} > -\frac{\Delta\gamma}{\gamma} \cdot \frac{w}{1-w} \tag{15.12}$$

It can be assumed that the prevention of a fatality in average will save $\Delta\varepsilon = \frac{\varepsilon}{2}$, which equals half of the life expectancy. The largest change in gross domestic product, $|\Delta\gamma|_{max}$, is gained by implementing this expression for

$\Delta\varepsilon$ in Eq. 15.11. This can be interpreted as the optimum acceptable cost per life year saved. The optimum acceptable implied cost of averting a fatality, $ICAF_0$, can then be calculated by Eq. 15.13.

$$ICAF_0 = \left|\Delta\gamma\right|_{max} \cdot \Delta\varepsilon = \frac{\gamma \cdot \varepsilon}{4} \cdot \frac{1-w}{w} \tag{15.13}$$

where

$$\left|\Delta\gamma\right|_{max} = -\frac{\gamma(1-w)}{2w} \text{ (based on Eq. 15.12)}$$

$\Delta\varepsilon$ = Years saved by averting a fatality = $\varepsilon/2$

Based on this criterion proposed/suggested safety (or risk control) measures should be implemented as long as the calculated ICAF value does not exceed $ICAF_0$ (i.e., the criterion).

$ICAF_0$ will vary significantly between different countries based on this approach. If we assume that $w = 1/8$ applies to all countries we can calculate some example values, based on data from the International Monetary Fund for GDP (estimates for 2021) and World Health Organization for life expectancy (data from 2020):

Norway: $ICAF_0$ = 10 million USD
Greece: $ICAF_0$ = 4.3 million USD
Philippines: $ICAF_0$ = 1.1 million USD

These calculations raise some principal problems. There may be arguments for taking different national economic standards into consideration. One argument is that we can regard protection against accidents as a "luxury item", that will be important for us only when our basic needs for food and drink, housing, clothes, etc. have been satisfied. The richer we are (higher GDP), the more we will value safety also.

However, it is more questionable from an ethical point of view. We can also imagine situations where we have, e.g., a ship with Norwegian officers on the bridge and a Philippine crew. Using the above calculation means that we are more willing to spend money to protect the officers than the crew, something which immediately appears unethical. Using the same value for everyone is therefore also an option.

IMO (2018) presents common values for NCAF and GCAF, not tied to nationalities, although it is underlined that these are presented for illustrative purposes only. The values are shown in Table 15.5.

It is also suggested that the values should be increased by 5% annually. The values above are based on documents discussed at MSC Session 72, in 2000. Increasing these values by 5% annually up until the time of writing this book (2021) means that the highest values have increased from 3 million to 8.4 million USD.

Table 15.5 Illustrative criteria for NCAF and GCAF

	NCAF [Million USD]	GCAF [Million USD]
Covering risk of fatality, injury, and ill health	3	3
Covering risk of fatality only	1.5	1.5
Covering risk of injury and ill health only	1.5	1.5

Although there is no agreement on precisely what values to use, we can also see that the variation after all is fairly limited and falls within an order of magnitude. If we consider the uncertainty in calculation from risk analysis, these are in most cases at least as large as this, probably larger. Based on this, choosing a value somewhere in the middle of the range indicated may be a reasonable starting point.

A final comment may be that the valuation of both spills and fatalities may change significantly over time. The costs associated with accidents may increase significantly due to more focus on safety in society in general and in media. These may be regarded as intangible costs (related mainly to reputation), but there may also be increased expectations with regard to paying compensation and covering the full costs of an accident.

15.6 CASE STUDY: OIL SPILL PREVENTION MEASURES FOR TANKERS

Preventing pollution from maritime activities has been a major priority in later decades. On an international basis concern about the environmental impact of shipping has resulted in MARPOL, i.e., the International Convention for the Prevention of Pollution from Ships. MARPOL, which is one of the more important agreements achieved within the context of IMO, comprises design and operational regulations and requirements geared toward reducing pollution to both air and sea from shipping.

In addition to MARPOL, several regional agreements and regulations exist on the prevention of pollution from shipping. One such set of regulations is the U.S. Oil Pollution Act of 1990 (OPA 90) that was established as a direct consequence of the Exxon Valdez grounding accident in 1989 that resulted in a spill of 33,000 tons of crude oil in Prince William Sound on the coast of Alaska. OPA 90 gives the ship owners full economical liability of spills in U.S. coastal waters (strict liability).

There are several safety measures that may reduce potential oil spills because of ship collisions and groundings. The National Research Council (1991) therefore performed a cost-benefit analysis on some of these possible safety measures for tankers, and the following is a short summary of this analysis.

The objective of the analysis is to calculate the cost-effectiveness of alternative designs for oil spill prevention. These are compared to a standard MARPOL tanker with Protectively Located Segregated Ballast Tanks.

Segregated ballast tanks (SBT) mean that there shall be designated tanks for ballast and that ballast is not to be carried in cargo tanks (except in very severe weather conditions in which case the water must be processed and discharged in accordance with specific regulations). Protective location (PL) of SBT means that the required SBT must be arranged to cover a specified percentage of the side and the bottom shell of the cargo section to provide protection against oil outflow in the case of groundings and collisions. The alternative designs (i.e., safety measures) studied in this analysis are explained below:

1) **Double Bottom (DB)**

The double bottom (DB) constitutes the void space between the cargo tank plating, often referred to as the tank top, and the bottom hull plating. MARPOL requires a DB of 2 [m] or B/15, whichever is less (B = Breadth of the vessel). The DB space gives protection against low-energy grounding. Another benefit is the very smooth inner cargo tank surface that facilitates discharge suction and tank cleaning. Drawbacks to a double bottom include increased risks associated with poor workmanship, corrosion, and obstacles to personnel access (to the DB). Other drawbacks are related to reduced side protection relative to the PL/SBT configuration, and increased explosion risk related to cargo leak into the DB. The DB configuration is shown in Figure 15.6.

2) **Double Sides (DS)**

The double sides (DS) constitute the void space between cargo side/wing tanks and hull side plating. The minimum width of the DS is

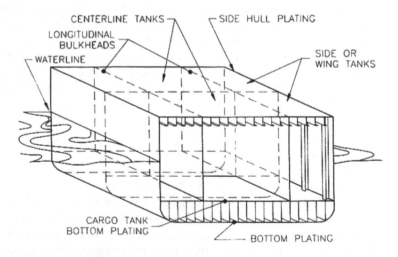

Figure 15.6 Double bottom (DB) configuration.

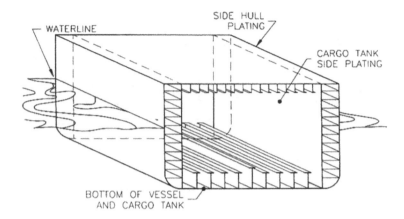

Figure 15.7 Double side (DS) configuration.

equal to that of the DB (i.e., 2 [m] or B/15). However, to meet ballast requirements, the width is likely to be larger (normally B/7–B/9). The design offers good protection against collisions, and also some of the advantages from double bottoms as the side tanks protect the outboard region of the bottom. Drawbacks of the DS configuration are related to bottom damages that result in direct spills, as well as higher susceptibility to asymmetric flooding. The double-side configuration is shown in Figure 15.7.

3) **Double Hull (DH)**

The double hull (DH) constitutes the void space between tank and hull plating in both the sides and bottom. Compared to DS (and PL/SBT) the side protection is reduced, as the width of the tanks may be less than in the double sides design because ballast can be divided among the side and bottom spaces. From a cargo operations point of view the design is excellent. Drawbacks are similar to those of DS and DB. In addition, the configuration/construction is more exposed to crack damages due to more plating. Corrosion may also be a larger problem. This places high demands on access for inspections and maintenance. The double hull configuration is shown in Figure 15.8.

4) **Hydrostatic-Driven Passive Vacuum (HDPV)**

The hydrostatic-driven passive vacuum (HDPV) construction consists of making openings to cargo tanks airtight. This results in a progressive drop in pressure with cargo outflow, and thereby reducing pollution as the vacuum "holds the oil back" in the tank. Drawbacks are related to air tightening, and the ability to quickly locate the damaged tank for closure of all necessary openings (e.g., vent pipes). It may also introduce additional hazards, e.g., if the vent pipes do not

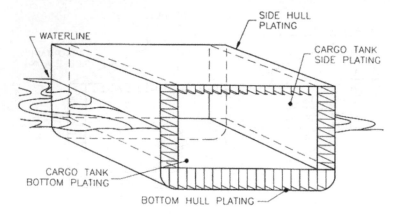

Figure 15.8 Double hull (DH) configuration.

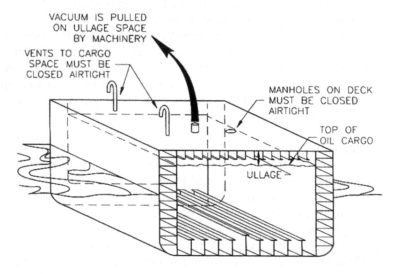

Figure 15.9 The hydrostatic-driven passive vacuum (HDPV) configuration.

open properly when needed (during loading of the tanks). Structural strengthening of the deck may also be necessary to avoid structural collapse because of tank vacuum. The hydrostatic-driven passive vacuum (HDPV) configuration can be illustrated by Figure 15.9.

5) Smaller Tanks (ST)

This design alternative is based on reducing the volume of the individual cargo tanks, which will reduce the potential oil outflow volume. The main drawback is related to more plating and thus an increased risk of plate cracking.

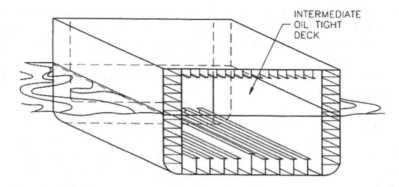

Figure 15.10 The interior oil-tight deck (IOTD) configuration.

6) **Interior Oil-Tight Deck (IOTD)**

An interior (horizontal) oil-tight deck (IOTD) essentially has the same effect as smaller tanks, in that it reduces the volume of oil that is available for release. The main difference is that with smaller tanks, there will be vertical separations but, in this case, horizontal separation is used. This is also favorable from a hydrostatic point of view. The need for ballast tanks makes double sides (DS) necessary, resulting in extra protection of sides. The major drawbacks to this configuration are complex operation as well as corrosion damage. The interior oil-tight deck (IOTD) configuration may be illustrated by Figure 15.10.

7) **Double Sides and Hydrostatic-Driven Passive Vacuum (DS/HDPV)**

This design configuration is a combination of double sides (DS) and hydrostatic-driven passive vacuum (HDPV).

8) **Double Hull and Hydrostatic-Driven Passive Vacuum (DH/HDPV)**

This design configuration is a combination of double hull (DH) and hydrostatic-driven passive vacuum (HDPV).

The eight design alternatives presented above are different with respect to the following costs:

- Capital cost: The deadweight of the alternative designs is equal. However, the cost of design and construction will vary because of different complexity.
- Maintenance and repair costs: Some designs require more maintenance and repair because of higher exposure to salt water, resulting in more corrosion, increasing need for inspections, and higher steel replacement costs. In addition, some designs are more exposed to cracking damages, resulting in the same types of costs.

- Insurance costs: Insurance will vary slightly between the design alternatives in that hull and machinery insurance is proportional to capital cost. In addition, less risk for serious spill accidents may reduce insurance costs.
- Fuel consumption: Higher tanker lightweight will increase fuel consumption.

The alternative designs were analyzed relative to a MARPOL tanker with protective location of segregated ballast tanks (PL/SBT). The volume of oil spills averted by implementing the different design configurations was estimated through an analysis of 38 large spill accidents, and the spill volume averted is considered as the benefit of the implementation. The tons of spill averted are presented in Table 15.6. It is distinguished between a typical and a major spill year as well as between small and large tankers. The analyzed accidents showed that the economic claims clustered around USD 28,000 per ton of oil spilled (1990 USD). However, some claims could reach up to USD 90,000 per ton (Exxon Valdez). Because of this variation the tons of oil spill averted is not converted into economic figures in Table 15.6.

The increased transport costs because of the design alternatives were calculated through a realistic weighting of three typical (and realistic) transport scenarios. All the alternative designs had higher transport costs than the base transport cost of a MARPOL tanker with PL/SBT. Based on 600 million tons of oil carried per year, which approximated the total U.S. seaborne oil transport, the increased transport cost associated with each design alternative was established. The results of this analysis are presented in Table 15.7.

The cost-effectiveness of the different tanker design alternatives can be found by dividing the incremental transport costs for each design alternative

Table 15.6 Tons of oil spill averted for the different design alternatives

	Typical spill year performance		Major spill year performance	
Design alternatives	Small tanker [tons]	Large tanker [tons]	Small tanker [tons]	Large tanker [tons]
Double Bottom (DB)	2,600	4,500	13,600	24,000
Double Sides (DS)	None	None	None	None
Double Hull (DH)	3,300	5,300	17,600	28,400
Hydrostatic Control-Passive (HDPV)	4,300	3,700	22,800	19,600
Smaller Tanks	2,100	2,700	11,200	14,400
Interior Oil-Tight Deck with DS	4,000	5,400	21,200	28,800
DS with HDPV	3,800	5,500	17,200	29,200
DH with HDPV	3,800	5,600	20,000	29,600

Table 15.7 Incremental transport costs for the design alternatives

Design alternatives	Incremental cost [Million USD per year]
Double Bottom	462
Double Sides	339
Double Hull	712
HDPV	1,080
Smaller Tanks	430
IOTD	872
DS/HDPV	1,102
DH/HDPV	2,047

Table 15.8 Added transport cost per ton of oil saved

Design alternatives	Typical spill year performance		Major spill year performance	
	Small tanker [10^3 USD/ton]	Large tanker [10^3 USD/ton]	Small tanker [10^3 USD/ton]	Large tanker [10^3 USD/ton]
Double Bottom	178	103	34	19
Double Sides	No oil saved	No oil saved	No oil saved	No oil saved
Double Hull	216	134	40	25
HDPV	251	292	55	47
Smaller Tanks	205	159	38	30
IOTD	218	161	41	30
DS/HDPV	344	200	64	38
DH/HDPV	539	366	102	69

with the amount of oil each design prevents from being spilled, presented in Table 15.6. The results are presented in Table 15.8.

The most expensive ways to prevent oil spill are double sides (DS), which does not prevent any additional oil spill, and double hull with hydrostatic-driven passive vacuum (DH/HDPV). Two of the design alternatives could be described as medium cost, namely the double sides with hydrostatic-driven passive vacuum (DS/HDPV) alternative and MARPOL ships with HDPV. The most cost-effective alternatives are double bottom, double hulls, smaller tanks, and interior oil-tight deck (IOTD).

Assuming that the costs of oil spill vary from USD 28,000 to 90,000 per ton of oil spilled, none of the design alternatives are cost-effective in a typical year. However, in major spill years, all the design alternatives can be cost-effective. Other cost-effectiveness studies carried out on the implementation of double hull tankers have, however, shown that this measure is not cost-effective. A study performed by the Transportation Centre of Northwestern University (Brown & Savage, 1996) shows that the expected (i.e., most likely) benefits of double hull on tankers were only about 18% of the costs expected.

15.7 ALTERNATIVE APPROACHES TO SELECTION

In this section some alternative approaches for choosing among risk-reducing measures will be presented. Which method to apply will largely depend upon the amount of information available regarding the activity or system under consideration.

15.7.1 Ranking of concepts

For risk-based CBA, as illustrated by Figure 15.4, there are several tasks that must be performed. Firstly, the effect of a specific safety measure on occurrence probability and the consequences must be assessed. This can be done by analyzing the effects of the safety measure on the probabilities in the fault trees and event trees. However, utilizing these techniques requires lots of detailed information about the activity or system under consideration, as well as a comprehensive risk analysis model for that activity/system. For an existing system, such information can often be obtained, and risk analysis models developed, but this may be costly and the uncertainties involved in the information and models are often relatively large. In addition, for an activity not yet carried out or for a system being designed, neither detailed information nor risk analysis models may be easily available. In this situation, ranking of different alternatives may be used.

With this approach, no attempt to calculate costs and benefits explicitly is made. Instead, the alternatives are ranked based on a set of carefully selected parameters. The parameters reflect the costs and benefits that are to be taken into consideration and compared as part of the CBA, and it is important that the number of cost and benefit parameters is balanced. The ranking is performed by giving each parameter for each alternative a grade that reflects how good the concept is relative to the other concepts for the various parameters.

Consider an offshore development project in which there are three possible design concepts for the oil production. In the initial phase of the design a preliminary CBA is to be performed to identify the best concept and show how the different design concepts relate to each other in terms of costs and benefits. The project management has selected a set of parameters or criteria on which they want the concepts to be assessed, and these parameters include central economical-, technical-, and safety-related factors that should give a reasonably good total picture of how the different concepts (or alternatives) differ. The assessment parameters, and the information estimated/gathered on each of these for the three design concepts, are shown in Table 15.9.

The different concepts must be weighed against each other for each of the assessment criteria/parameters. This can, for example, be done by implementing an ordinal grade scale for each parameter ranging from 1 to 3, where 1 denotes the best concept, 2 the second-best concept, and 3 the

Table 15.9 Concept alternatives for offshore oil production

Assessment criteria	Design concepts		
	Fixed platform	Floating platform	Underwater production
Prod. cost per barrel	USD 14.00	USD 12.00	USD 9.00
Investment cost	NOK 5 billion	NOK 6 billion	NOK 4 billion
Development time	3 years	4 years	5 years
Technological status	Established	Quite known	Problematic
Availability (up-time)	0.99	0.94	0.90
Est. accident frequency	10^{-3}	10^{-2}	10^{-4}
Est. risk of fatality	10^{-5}	10^{-4}	10^{-10}
Fatalities, large accident	30	50	0
Annual spill volume	10 tons	22 tons	30 tons
Accidental spill volume	5,000 tons	100 tons	500 tons

Table 15.10 Concept alternatives for offshore oil production

Assessment criteria	Design concepts		
	Fixed platform	Floating platform	Underwater production
Prod. cost per barrel	3	2	1
Investment cost	2	3	1
Development time	1	2	3
Technological status	1	2	3
Availability	1	2	3
Est. accident frequency	2	3	1
Est. risk of fatality	2	3	1
Fatalities, large accident	2	3	1
Annual spill volume	1	2	3
Accidental spill volume	3	1	2
Sum	18	23	19

worst concept. This is done for all the assessment criteria and then the sum of all the grades is found for each concept. Consequently, the concept with the lowest total grade is considered the best one. The ranking of the three offshore development concepts is performed in Table 15.10.

By assuming that all of the assessment criteria/parameters have equal importance the fixed platform concept is considered the best concept, slightly better than the underwater production concept. This does not mean that the fixed platform is "approved". The project management and its associated team of engineers should now make improvements on all the concepts based on the information provided by Tables 15.9 and 15.10. Later in the design process, when the concepts have been improved and more information is available about them, a more detailed ranking process should be performed, maybe involving more detailed assessment criteria/parameters.

The main advantage of the ranking technique described above is its simplicity. The technique is, however, very sensible to the assessment parameters that are included. In addition, all the parameters have been given equal weighting, which may not be a correct reflection of how we view their importance in reality. For example, the estimated risk of fatality may be considered more important than the annual spill volume. Finally, the comparison between the concepts becomes quite rough because the absolute differences in each of the assessment parameters have little impact. For example, if the accidental spill volume for the fixed platform concept was ten times as high, this fact would not have changed the total result of the ranking. The latter drawbacks may be diminished if a grading scale that opens for an assessment of absolute differences within a certain parameter is implemented. Such a grading scale can, for instance, be defined as follows:

1. Very good performance
2. Good performance
3. Acceptable performance
4. Poor performance
5. Very poor performance
6. Unacceptable performance

The use of such a scale could possibly have resulted in a different end result of the ranking performed in Table 15.10.

15.7.2 Relative importance ranking

In previous chapters it has been shown that it often is practical to express safety by a set of consequence parameters. Such parameters may be fatalities per 10^8 working hours, economical and material loss, spill volume, etc. The relative importance of these consequence parameters is, however, not intuitive and introduces difficulties for the analyst. One possible method that may be used to estimate the relative weights of importance is presented here. This method is not ideal but deal with the problem in a concise manner. The lack of consideration of the relative importance of different assessment parameters was one of the main drawbacks of the ranking technique presented in the previous section.

Let us consider a case in which the safety of oil transport from an offshore installation to a shore-based refining facility is to be analyzed. It is considered that the following safety parameters are relevant:

- Spill volume (i.e., oil pollution of the environment)
- Economical/material loss (i.e., damage to and loss of vessel)
- Number of fatalities aboard
- Population exposed by an explosion

Table 15.11 Questionnaire for ranking of safety parameters

	5	4	3	2	1	0	1	2	3	4	5	
Spill volume				×								Economical loss
Spill volume								×				Number of fatalities
Spill volume							×					Exposed population
Economical loss									×			Number of fatalities
Economical loss										×		Exposed population
Number of fatalities								×				Exposed population

Two different oil transportation concepts are to be assessed based on their risk characteristics. Other parameters than safety related may have been included in the analysis, such as the investment cost and development time, but these are not considered of importance in this particular analysis.

The first task of the cost-benefit assessment process is to establish the relative weights of importance that the assessment criteria/parameters are to be given. One possible approach in estimating these relative weights of importance is to gather/organize a group of experts holding excellent system knowledge about the systems and operations under consideration, as well as substantial familiarity with the preferences of governments and the society at large (i.e., the other stakeholders). How these experts perceive the relative importance of the different assessment parameters, which here are exclusively related to safety, can then be measured using, for example, questionnaires. Based on this information the relative weights of importance for the group of experts can be established.

Table 15.11 presents a questionnaire that is applied when estimating the relative weights of importance for the different safety parameters related to the case of oil transportation. So-called paired comparison is applied in this technique. A value of 0 implies equal importance, while the value of 1 to 5 favors the relative importance of one of the parameters concerned. The values given in Table 15.11 are examples of answers that might have been collected from the experts, and these answers indicate, for example, that both the parameters of exposed population and number of fatalities are far more important than economical loss.

By allowing each member of the group of experts to answer such a questionnaire a quantitative estimate of the resulting relative weights of importance for the safety parameters can be calculated. In practice this calculation is certain matrix operations on the questionnaire data (i.e., Table 15.11). The method is described in Section 11.7. The relative weights of importance are then calculated on a common scale and may be presented as shown in Figure 15.11.

The next step is to establish a utility function for each of the four safety parameters. In this context utility denotes the decision-maker's scale of preference, and a utility function describes graphically how the perceived utility changes with changing consequences. It is assumed here that the utility can

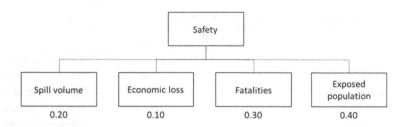

Figure 15.11 Relative weights of importance for the set of safety assessment parameters.

be expressed as a continuous function with values between 0 and 1. The higher the utility (i.e., close to 1) the more acceptable a specific consequence is for a given accident. Similarly, as the consequence becomes more and more unacceptable and undesirable, the lower the utility will be (i.e., close to 0). Thus, the utility functions express the perceived risks, with the risk increasing as the utility number decreases. The utility functions can also be established using a group of experts. The utility functions for the four safety parameters in the oil transportation case are shown in Figure 15.12. The graphs presented require the following comments:

- Spill volume: No spill volume gives the highest utility (i.e., 1.0) as no environmental damage can be regarded as acceptable. However, small oil spills do not cause substantial damage to the environment, and the utility remains relatively high. Medium oil spill in the order of 100–1,000 [tons] will, on the other hand, reduce the utility significantly as the consequences to the environment increases dramatically. Even larger spills (i.e., >1,000 [tons]) will have very detrimental effects on the environment and cannot be accepted, hence the low utility value for such spills.
- Economical loss: Economical losses less than NOK 10 million have relatively little effect on the utility, which remains relatively high. This shows that such material losses in this particular case are regarded as quite acceptable. One possible explanation to this is that such costs may be involved even in smaller accidents (i.e., a threshold cost). Losses larger than NOK 10 million have a significant effect on the utility function, and costs in the order of NOK 100 million are considered totally unacceptable – hence the utility value of 0.
- Fatalities aboard: Accident statistics reveal difficulties in avoiding 1–3 fatalities, and although very undesirable the utility therefore stays relatively high. Accidents of size 3–10 fatalities are considered much more serious, hence a dramatic reduction in the utility value. Accidents with more than 10 fatalities are catastrophes that must be avoided. As can be seen in Figure 15.12, the utility value is set at 0 for accidents involving 20 or more fatalities.

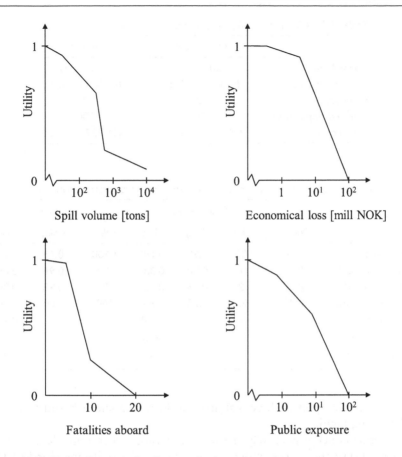

Figure 15.12 Utility functions for the set of safety parameters.

- Public exposure: With regard to public exposure the utility function is close to being a linear function of the logarithm of the exposed population. An exposed person may in this context be defined as a person within a given distance from the center of a fire or explosion, resulting in that person being subjected to serious danger. As would be expected the utility decreases considerably as the number of exposed individuals increases.

The relative weights of importance and the utility functions for the selected set of safety assessment parameters give us the necessary basis for estimating the cost-benefit ratio for the different system concepts (e.g., designs of the oil transportation vessel/system). Let us assume that two concepts are to be compared against each other based on the four safety parameters and the defined utility functions. The two oil transportation concepts are described in Table 15.12. The oil is loaded from a buoy

Table 15.12 Alternative oil transportation concepts

	Basis concept	Alternative
Spill volume	500 tons	1,500 tons
Economical loss	NOK 16 million	NOK 4 million
Fatalities aboard	8	2
Exposed population	50	100
Total costs	NOK 800 million	NOK 770 million

Table 15.13 Cost-benefit analysis (CBA) of the two oil transportation concepts

Criterion	Priority	Basis concept			Alternative		
		Value	Utility	Weight	Value	Utility	Weight
Spill volume	0.20	500	0.95	0.190	1,500	0.65	0.130
Economical loss	0.10	16.0	0.80	0.080	4.0	0.96	0.096
Fatalities aboard	0.30	8	0.55	0.165	2	0.98	0.294
Exposed population	0.40	50	0.75	0.300	100	0.60	0.240
Total utility				0.735			0.760
Costs [million]				800			770
C-B ratio				1,088			1,013

offshore and shipped with two shuttle tankers to a shore-based terminal and refining facility.

The basic concept is clearly the most favorable when considering environmental damage as the expected spill volume is only one-third of the spill volume for the alternative concept. In addition, the expected size of the exposed population is only half for the basic concept. The alternative is, on the other hand, more oriented to reducing ship damage in terms of expected economic loss, and the expected number of fatalities aboard is also considerably lower. In Table 15.13, a cost-benefit analysis of the two concepts is performed using the weights of importance and the utility graphs. The total utility, which is considered as a measure of the benefit/ quality of a specific concept, is calculated and compared to the total costs, i.e., the net present value of the necessary investments and future operational costs. A cost-benefit ratio is calculated, enabling a comparison to be made on a common scale. In Table 15.13, the "priority" is the relative weights of importance for the safety parameters/criteria, and the utilities are read off the graphs in Figure 15.12 depending on the characteristics of the two concepts. The weight is the product of the priority figure and the utility. Table 15.13 shows that the alternative concept has a lower cost-benefit ratio than the basic concept and is hence considered better in this analysis.

15.7.3 Valuation of consequence parameters

In earlier sections of this chapter, it is assumed that the potential consequences or losses for a particular concept can be expressed as concise values such as the number of fatalities. However, this is sometimes problematic and not always desirable, especially in relation to accidents that involve both injuries and fatalities. In the context of CBA, it is often necessary to establish an economical cost of both injuries and fatalities in order to compare different costs and benefits on a common scale. In Chapter 9, a simple method for converting injuries to fatalities was presented. This section will look closer at the valuation of people/personnel-related consequences parameters. In analyses of the safety of people/personnel, the following consequence parameters may be considered:

- Fatality
- Permanent disability
- Temporary disability

The costs associated with these people-related consequences may be valued according to the following factors:

- Insurance payments
- Estimated remaining life income
- Claim for compensation
- Implicit social costs

The valuation of people-related consequences can be controversial, both politically and ethically. In Table 15.14, a possible valuation method for such consequences is applied to two different concepts having different potential consequences in terms of fatalities as well as permanent and temporary disability. Unitary economic costs are applied to express the cost of one fatality, one permanent disability, and one temporary disability. These values can be calculated based on the cost factors stated above. The total

Table 15.14 Total people-related loss costs of two different concepts

Safety parameters	Unitary cost [1,000 USD]	Concept 1		Concept 2	
		Number	Cost	Number	Cost
Fatality	1,000	2	2,000	1	1,000
Permanent disability	400	15	6,000	10	4,000
Temporary disability	20	60	1,200	100	2,000
Calculated people-related loss cost			9,200		7,000
Costs of implementation			5,500		6,100
Total Costs			14,700		13,100

costs for the two concepts are calculated as the sum of the preventive costs (i.e., costs of implementation) and the consequences (i.e., average losses in an accident). Concept 2 gives the lowest total cost in this study.

McCormick (1981) introduced the term social costs, which is to express the society's costs of an injury or fatality. The following equation was suggested:

$$
\text{Social cost} = \begin{cases} NC(1+i)^t, & \text{when } t < 6{,}000 \\ NC(i+1)^{6{,}000}, & \text{when } t \geq 6{,}000 \end{cases} \tag{15.14}
$$

where
   N = Number of injuries or fatalities
   C = Cost of damage per day
   i = Daily rate of interest
   t = Duration of damage or sick leave in days (6,000 days is equivalent with a fatality)

This method of calculating the social cost of injuries and fatalities is only one of many different models that may be used. A problem not solved with Eq. 15.14 is that of establishing C, the cost of damage per day, which may among other things vary considerably depending on the type of injuries suffered, the country in question, etc.

In an investigation made by O'Rathaille and Wiedemann (1980) it was attempted to establish the average social cost for ship collisions and groundings based on statistical consequences. It was focused on oil spill and loss of lives, and the statistical data basis is presented in Tables 15.15 and 15.16.

Table 15.15 Ship accidents and oil spills (1976)

	Number of accidents		Pollution rate, % of accidents leading to pollution
Primary cause	Total	Of which led to pollution	
Collisions	44	1	2.27
Groundings	121	14	11.57

Table 15.16 Fatality risk in ship collisions and groundings (1976)

	Number of accidents		Number of lives lost per accident
Primary cause	Total	Number of lives lost	
Collisions	44	41	0.93
Groundings	121	4	0.03

Table 15.17 Average total costs of collisions and groundings (1977)

Primary cause	Cost of spills per accident [£]	Cost of fatalities per accident [£]	Average total cost per accident [£]
Collisions	5,100–6,100	79,363–91,416	84,463–97,516
Groundings	50,000–280,000	2,816–3,243	52,816–283,243

These tables show that the likelihood of oil spill is largest in groundings, while collisions much more frequently result in fatalities.

Based on the experience of known accidents the cost of oil spill per accident was estimated to £5,100–£6,100 for collisions and £50,000–£280,000 for groundings, reflecting that groundings are more likely to result in oil spill and that the spills on the average are larger. However, as these figures show, the costs related to groundings tend to vary greatly.

In an assessment combining both economical and non-economic factors related to fatalities, the cost of a fatality was estimated to £85,170–£98,105 in 1977 prices. Based on this, the average total costs of collisions and groundings, respectively, are estimated in Table 15.17.

It can be concluded from Table 15.17 that the average total social cost of a grounding accident seems to be higher than that of a collision.

Insurance payments to people as well as the involved company/ organization in the aftermath of accidents must also be considered a cost related to accidents. Insurance companies give compensation to the bereaved and tend to vary considerably from case to case depending on the circumstances and the insurance schemes. There also tend to be quite different insurance practices in different countries. All these aspects make it difficult to generalize about insurance payments. The same accounts for claims of compensation that often surface in the wake of accidents. Such claims are often based on the lost (entirely or partly) remaining life income by reaching nominal age. Methods used in calculating such figures are often referred to as human capital methods.

The willingness to pay for preventive safety measures focusing on reducing fatalities differs between industries and the types of activities. Table 15.18 shows an American overview of estimated preventive measures costs per saved human life for different activities. The table shows that the nuclear industry, for example, is willing to pay more to save a human life than most other activities.

When studying the cost-benefit values in Table 15.18, it must be recognized that such values usually have a limited period of validity because of factors such as regulatory changes, new technology, and changed public risk perception. Cost-benefit values must therefore be used or referred to with great care.

Table 15.18 Cost-benefit ratios (C/B) for different safety measures, million USD per life saved

Industry/activity	Safety measure	C/B
Nuclear industry	Radwaste effluent treatment systems	10
	Containment	4
	Hydrogen re-combiners	>3,000
Occupational health and safety	OSHA coke fume regulations	4.5
	OSHA benzene regulations	300
Environmental protection	EPA vinyl chloride regulations	4
	Proposed EPA drinking water regulations	2.5
Fire protection	Proposed CPSC upholstered furniture flammability standards	0.5
	Smoke detectors	0.05–0.08
Automotive and highway safety	Highway safety programs	0.14
	Auto safety improvements, 1966–1970	0.13
	Airbags	0.32
	Seat belts	0.08
Medical and health care programs	Kidney dialysis treatment units	0.2
	Mobile cardiac emergency treatment units	0.03
	Cancer screening programs	0.01–0.08

15.8 BARRIER MANAGEMENT IN OPERATION

In the previous sections in this chapter, identification and selection of barriers has been discussed. This has to do primarily with how we design the technical systems and also how we design operations and activities to be performed. This ensures that we have a system and operations that are as safe as reasonably practicable.

However, it is not sufficient that we establish this and then expect everything to continue working perfectly, without any problems or deterioration of performance. An important activity in safety management is also to follow up the functioning of these barriers on a day-to-day basis. This is in some contexts also called barrier management. A good illustration of this is the Exxon Valdez grounding and spill. Before Exxon Shipping got the permission to start the oil transport from Valdez an elaborate risk analysis was undertaken. The arctic region was deemed very sensitive to oil pollution and one of the measures taken was to set up a VTMS (Vessel Traffic Management System) manned with nautical competent personnel. However, years later and due to reduced public funding the VTMS was manned with non-nautical personnel at the time of the accident. The accident investigation concluded that the service had become non-functional.

Barrier management is a widely used concept in the offshore oil and gas industry, and ABS has published guidelines for barrier management (ABS, 2020). Several objectives may be identified of barrier management:

- Ensuring that all barriers remain operational and effective whenever needed. This implies that we need a system for checking that barriers are working and for correcting errors if they are not working. For technical systems, this will be provided by testing, inspection, and maintenance of barriers. For human and organizational barriers, it is however not quite as straightforward. For humans, training and testing can be used while for organizations, audits is perhaps the primary means.
- Ensuring that all personnel making decisions that influence risk are aware of what barriers exist and how they are functioning. This is firstly about training people, so they understand what barriers are in place. Secondly, we also need to have a system that keeps track of the status of the barriers and let relevant people know when they are not functioning.
- Ensuring that adequate compensating actions are taken whenever a barrier is impaired or not functioning. To do this, we need to understand the effect on risk, when a barrier is not functioning. Next, we need to understand what other measures can be put in place (usually temporarily) to compensate.

Barrier management is an integrated, and key, part of any safety management system. It is also noted that many activities that normally already are part of a safety management system may be labeled as barrier management. It may therefore be questioned whether this really adds anything new or is just a new name for something that is already being done.

The advantage is that barrier management thinking gives a much more comprehensive overview of the barriers and ensures that nothing "falls between two chairs" or is not followed up properly.

In the oil and gas industry, barrier management is often computerized and can be linked to various other systems, such as the maintenance system. For each barrier a description typically containing the following elements is provided:

- What is the barrier, i.e., the system, procedure, training, etc. that describes the barrier.
- What is the function of the barrier, i.e., what is it intended to do? A barrier may have only one function, e.g., a fire detector only is put in place to detect fire. Other barriers may have several functions. An example is the lifeboats of a ship that both serves to evacuate people safely off the ship and also provides protection until those onboard can be rescued. These are two very different functions, although we are not always aware of this fact.
- When is the barrier needed? This ties in with risk analysis and emergency response plans. Is the barrier needed for specific hazards, specific situations, or more generally? Lifeboats are not aimed at a specific

hazard but are relevant for all situations when abandoning the ship is required.

- What is the effect if the barrier is not functioning? This is important to understand so that we know if certain activities have increased risk and whether compensating measures are required.
- What are the performance requirements for the barrier? What is that we expect it to do? This can be expressed in many ways and many dimensions. For lifeboats we may, e.g., specify the reliability of the launching mechanism, the reliability of the engine/propulsion system, the duration that the engine should continue running, the number of people it can hold, the weather conditions under which it can be safely launched, the weather conditions under which it can protect people, etc.
- How do we verify that the performance requirements are met and how often? With the wide range of possible performance criteria, the ways of verifying that they are met will also vary accordingly. Functioning of the engine and the launching mechanism of a lifeboat can be tested, e.g., SOLAS specifies requirements for testing of lifeboats. Many of the other performance requirements must be verified in design and will typically not change during the lifetime of the lifeboat. Regular testing is therefore not required, but we should maintain documentation showing that they are designed to meet the requirements.

REFERENCES

ABS. (2020). *Advisory on barrier management, American Bureau of Shipping.* https://ww2.eagle.org/en/Products-and-Services/offshore-energy/exploration/Barrier-Management-Advisory.html

Brown, S., & Savage, I. (1996). The economics of double-hull tankers. *Maritime Policy and Management, 23*(2), 167–175.

CCPS. (1993). *Guidelines for Safety Automation of Chemical Processes.* American Institute of Chemical Engineers.

Hollnagel, E. (2016). *Barriers and Accident Prevention.* Milton Park, Oxfordshire: Routledge.

HSE. (2001). *Reducing risks, protecting people.* https://www.hse.gov.uk/risk/theory/r2p2.pdf

HSE. (2008). *Optimizing hazard management by workforce engagement and supervision.* https://www.hse.gov.uk/research/rrpdf/rr637.pdf

IMO. (2018). *Revised Guidelines for Formal Safety Assessment (FSA) for use in the IMO Rule-Making Process.* https://wwwcdn.imo.org/localresources/en/OurWork/HumanElement/Documents/MSC-MEPC.2-Circ.12-Rev.2%20-%20Revised%20Guidelines%20For%20Formal%20Safety%20Assessment%20(Fsa)For%20Use%20In%20The%20Imo%20Rule-Making%20Proces...%20(Secretariat).pdf

McCormick, N. J. (1981). *Reliability and risk analysis: methods and nuclear power applications*. New York: Academic Press.

National Research Council. (1991). *Tanker spills: Prevention by design*. Washington, DC: The National Academies Press.

NORSOK S-001. (2020). *Technical safety, NORSOK standard*. https://www.standard.no/en/sectors/energi-og-klima/petroleum/norsok-standard-categories/s-safety-she/s-0013/

O'Rathaille, M., & Wiedemann, P. (1980). The social cost of marine accidents and marine traffic management systems. *The Journal of Navigation, 33*(1), 30–39.

Rausand, M., & Haugen, S. (2020). *Risk assessment: Theory, methods and applications (2nd ed.)*. Hoboken, New Jersey: Wiley.

Reason, J. (1997). *Managing the risks of organizational accidents*. Aldershot, Hampshire: Ashgate.

Salvi, O., & Debray, B. (2006). A global view on ARAMIS, a risk assessment methodology for industries in the framework of the SEVESO II directive. *Journal of Hazardous Materials, 130*(3), 187–199.

Sexsmith, R. G. (1983). Bridge risk assessment and protective design for ship collision. In *IABSE Colloquium* (pp. 425–433). https://www.e-periodica.ch/cntmng?pid=bse-re-003%3A1983%3A42%3A%3A59

Skjong, R., & Ronold, K. O. (1998). Social indicators and risk acceptance. In *Offshore mechanics and arctic engineering conference, 1998*. OMAE.

Sklet, S. (2006). Safety barriers: Definition, classification, and performance. *Journal of Loss Prevention in the Process Industries, 19*(5), 494–506.

Chapter 16

Emergency preparedness and response

16.1 INTRODUCTION

A substantial part of this book has so far focused on how to improve the safety of maritime activities, primarily through the implementation of risk-reducing measures. If the safety work is effective, the number of accidents will be reduced, but the chance of an accident occurring will always exist as no activity or system is 100% reliable and safe. We therefore also have to prepare for the possibility that accidents will occur, and we should be able to handle accidents and other emergencies in the best possible way.

A number of terms are being used to describe the activities that we undertake to reduce the negative effects of accidents. Emergency preparedness, emergency response, contingency planning, etc. are all being used to describe the same things. In this chapter, we will mainly use three terms:

- Emergency planning: This covers the activities that we do to determine what we need to be prepared for handling, what resources we need, how to act, etc.
- Emergency preparedness: This covers the resources that we have in place, based on the emergency planning. This can include emergency procedures, emergency organization with roles and responsibilities, fire-fighters, first-aid personnel, etc. and also the training that all involved personnel go through to be prepared.
- Emergency response: These are the concrete actions that we take in an emergency.

It is often easy to focus on the actions that are taken as part of the immediate response to an accident as emergency response. However, in many situations, there will be far more resources required and it will be far more time-consuming to get back to a normal situation again. To the extent that this is critical, the emergency planning and preparedness should also consider this.

DOI: 10.4324/9781003055464-16

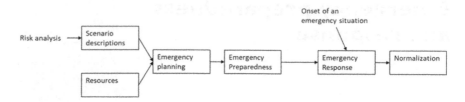

Figure 16.1 The emergency planning, preparedness, and response process.

Figure 16.1 schematically illustrates the process. From the risk analysis, scenario descriptions are already available. This information, combined with information about available resources (internal and external), provides input to the emergency planning which in turn determines what emergency preparedness we establish, in terms of procedures, personnel, training, equipment, etc. If an emergency situation occurs, we need to respond to this emergency, in accordance with plans and with available resources. After the emergency has been handled, the normalization phase follows. In this chapter, we will not go into details of the normalization phase.

Inadequate preparations for handling emergencies can have serious consequences. The lack of emergency preparedness was, for instance, devastatingly evident in the grounding of the oil tanker Exxon Valdez in Prince William Sound (Alaska) in 1989. The accident took place in protected water and under favorable weather conditions, and the catastrophic environmental and economic consequences of the accident were largely the result of inadequate handling of the situation. The mobilization of necessary resources for the clean-up operation was seriously delayed, and in addition the coordination of the containment and clean-up operation was poor. The result is now a tragic chapter in maritime history: What was initially only a moderate-sized spill resulted in one of the worst man-made environmental disasters of modern times.

This chapter will examine some key aspects related to emergency preparedness. After a brief presentation of a few maritime accidents in which improved and appropriate emergency preparedness could have reduced the consequences, the focus will be on the following main topics:

- Emergency and life-saving regulations (i.e., SOLAS, the ISM Code, and STCW)
- Emergency preparedness activities and functions
- Human behavior in catastrophes
- Evacuation risk and evacuation simulation
- Pollution emergency planning

16.2 EXAMPLES OF ACCIDENTS

16.2.1 Amoco Cadiz

The VLCC (Very Large Crude Carrier) Amoco Cadiz was on a laden voyage when it lost rudder control 10 miles off the coast of Brittany (France) on the 16th of March 1978. The loss of control happened at 09:46 and was due to failure of the steering engine (Hooke, 1989). The weather was harsh, and the vessel immediately started drifting toward the shore. The Master of Amoco Cadiz was not prepared for this very undesirable event, and over the course of the next couple of hours he made a number of decisions that in the end contributed to the grounding off the village of Portsall almost 12 hours after the initial failure. Subsequently the vessel lost its integrity and broke up, resulting in the entire cargo of 223,000 tons of crude oil being spilt. Figure 16.2 shows Amoco Cadiz after sinking.

After the failure of the steering engine, an unsuccessful attempt to repair the engine was done. Given the harsh weather conditions, it was obvious that a salvage operation would be complicated. By pure chance the radio traffic from the Amoco Cadiz was intercepted by the ocean tug "Pacific", which immediately started to steam toward the disabled vessel. After some delay, for reasons described below, the tanker was taken under tow at 14:25, but due to the hard weather the tow broke at 17:19. A second tow started

Figure 16.2 Amoco Cadiz after sinking (Wikipedia, public domain picture).

at 20:35, but it was not able to take control of the drifting vessel and the vessel went aground.

After the steering engine failure, the captain made several poor decisions that contributed to the loss of the vessel. These included the following:

- It took 1 hour and 45 minutes before a call for tug assistance was sent out.
- The main engine was stopped.
- It took 1 hour and 30 minutes to negotiate a towing contract. The Master of Amoco Cadiz wanted to avoid Lloyd's Open Form (a standard contract for marine salvage operations), and the negotiations were complicated by language problems on both sides.
- The initiation of the second tow was inadequately coordinated.

Research initiated because of the accident later found that the vessel would have been easier to control had the propulsion and forward speed been kept. Having the superstructure at the aft, Amoco Cadiz could have sailed into the wind, which was blowing toward land, and thereby kept the vessel offshore. Valuable time was also lost as the Master was reluctant to accept the Lloyd's Open Form, primarily for economic reasons.

Having a single propeller and a single rudder Amoco Cadiz was obviously at risk for the hazard of steering engine failure becoming reality. The tragic fact was that the preparedness for the emergency situation of steering engine failure was inadequate and that such preparedness could have resulted in a much less unfortunate outcome.

16.2.2 Capitaine Tasman

One of the reasons why incidents sometimes lead to serious accidents is that a seemingly non-serious initiating event is not handled with necessary determination. This happened in the engine room fire onboard the cargo ship Capitaine Tasman (Cowley, 1994). The key events of this accident are outlined in Figure 16.3. It was the motorman that first detected smoke from the fuel oil heater. An attempt was made to extinguish the fire with the use of a powder unit, but when this failed, the chief engineer was called upon. The chief sounded the general alarm and various measures were then taken to isolate the heater electrically, but these mainly failed. Forty-five minutes after the smoke was first detected the fire was put out by a party of four fire-fighters in SCBA (Self-Contained Breathing Apparatus) by means of powder and foam. However, the fire reignited and the fire-fighting team had to return to fight the fire using water. The fire hose was left with spraying water to prevent further re-ignition. It was then observed that the fire had spread to the workshop above the heater, and only then was it decided to activate the CO_2 flooding system. Being without power and with empty SCBA bottles, it was finally decided to

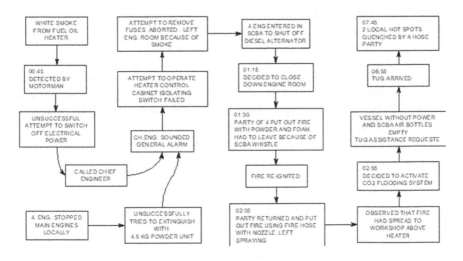

Figure 16.3 Sequence of events in the engine room fire onboard Capitaine Tasman.

request tug assistance. Seven hours after the initial smoke detection, a hose party quenched the local hot spots.

The response to the fire was inadequate in a number of ways:

- The general alarm was not sounded immediately
- Oil supply to the heater was not shut off
- Failure to isolate the fuel oil heater electrically
- Delayed start of fire pumps
- Persistent use of portable equipment
- Hot spots and secondary fires were not detected

It can be concluded that the crew was never in control of the situation during the fire-fighting operation.

16.2.3 HSC Sleipner

The HSC Sleipner (High-Speed Craft) stranded on a small rock/shoal on the west coast of Norway on a dark autumn evening in 1999 (NOU, 2000). Forty-five minutes after the impact the seriously damaged vessel floated/slid off the rock, disintegrated, and sunk to about 150 meters depth. The evacuation of the 85 passengers and crew on board was totally out of control, and 69 people had to jump into the water when the vessel sank. The majority of these were picked up by nearby vessels, but a total of 16 persons perished. This accident demonstrated that inadequate certification of life-saving appliances and lack of emergency training can result in fatal consequences when an unexpected accident occurs. Although the accident was caused by navigational failure, the dramatic nature of the consequences was, in addition to poor emergency preparedness and improper life-saving equipment, to a large degree a result of the heavy damage to the hull. Considering the potential impact forces and the extent of damage in accidents involving high-speed crafts, the present design requirements for high-speed crafts should be questioned. IMO revised the High-Speed Craft code (IMO, 2000) in 2000, to reflect experience since the introduction of the code and considering that craft were becoming bigger and faster.

The main events in the grounding/stranding of HSC Sleipner were as follows:

1. Stranding:
   - Damage to the bottom on both hulls
   - Water ingress, also in engine room
   - Progressive list
   - Starboard generator stopped
   - Port generator started but stopped almost immediately
   - Loss of internal communication
   - Transitional emergency power did not function

2. Attempt to release starboard life rafts:
   - Raft containers under the waterline
   - Did not release due to lack of hydrostatic release units
   - Manual release system did not function
   - The release system was fairly complicated
   - Lack of training in use of the system

3. Attempt to release port life rafts:
   - Fore unit did not initially release upon activation. After it released, it overturned in the sea
   - Aft unit released, but the container was filled with water
   - The release line was tangled and did not function

The failure to complete a safe evacuation was also related to the following factors:

   - The lifejackets were stowed in enclosed recesses
   - The life raft release system was brand new and inadequately tested
   - The organization of the evacuation was chaotic due to shock and lack of training

In the accident investigation report, critical remarks were made regarding the lifejackets. It was found that the lifejackets were difficult to put on, did not have a good fit, and tended to slip off over the head. They also had limited buoyancy and thermal protection. In the aftermath of the Sleipner accident there were discussions within the Norwegian Maritime Authority of whether the lifejackets should have been approved, despite the fact that the approval process was formally in order. HSC Sleipner was equipped with immersion suits for the crew, but the crew lacked knowledge of their existence, as well as training in the use of these immersion suits, and only a few succeeded in putting them on.

Given the fact that HSC Sleipner was certified to carry 380 passengers, it is not difficult to envision the potential for a major catastrophe under these circumstances. After the accident the ship owner and operator of HSC Sleipner has been heavily criticized for inadequate safety management and lack of emergency preparedness. There is no doubt that proper execution of these activities would have reduced the consequences of the accident.

16.2.4 Costa Concordia

Costa Concordia was an Italian registered cruise ship that operated in the Mediterranean Sea. The ship went on its maiden voyage in 2006 and had a capacity of 3,780 passengers and a crew of 1,100. The description of the accident is largely based on Wikipedia (https://en.wikipedia. org/wiki/Costa_Concordia_disaster).

Figure 16.4 Costa Concordia in Palma, Mallorca (Photo: Jean-Phillipe Boulet, CC BY 3.0).

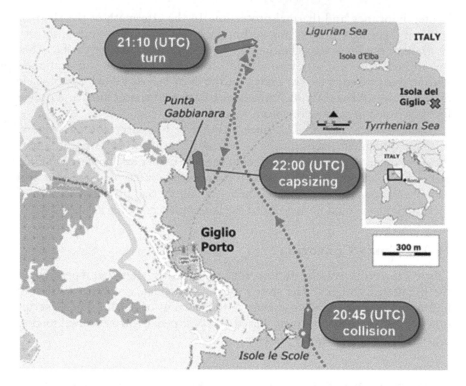

Figure 16.5 Route of Costa Concordia from impact until capsizing (CC BY-SA 3.0).

On 13 January 2012, Coast Concordia was cruising northbound along the western coast of Italy when it struck a rock off the island of Giglio. The impact created a 50 m long hole in the ship side, stretching over three compartments. As a result, the engine room was flooded, and power was lost. The ship also started listing as a result of the flooding and was drifting without proper steering or power. After about one hour, the ship drifted onto the Punta Gabbianara and came to rest with an angle of heel of about 70 degrees. At the time of the impact, there were 3,206 passengers and 1,023 crew members on board. In total, there were 32 fatalities (although numbers vary somewhat) – 27 passengers and five crew. In addition, there were 64 non-fatal injuries.

Several aspects of the response seem to have failed:

- Initially, the passengers were told that everything was under control. However, when the ship lost power, passengers started panicking.
- The order to abandon ship was not given until more than 1 hour after the initial impact. Despite this, several crew members started preparing before the order was given.

- No mustering/lifeboat evacuation drill had taken place for the passengers.
- The evacuation was noted as "complete" in the log of the Harbor Master in Livorno about 6 hours after the order to abandon ship was given. SOLAS specifies that evacuation should be possible to complete within 30 minutes. Lifeboat launching was very difficult even with the initial heel of around 20 degrees and became impossible as the ship came to rest at 70 degrees.

The accident with Costa Concordia also serves to illustrate how different phases of the accident and the aftermath have completely different timescales:

- The evacuation and rescue operation were completed in 6 hours.
- The search for missing people was first conducted in the period 14–30 January, over a period of 16 days. During this period, 17 bodies were found. The search continued intermittently also after this, with the final body not being found until after the wreck had been salvaged.
- The salvage operation was completed when the ship arrived in Genova in July 2014, approximately 2.5 years after the accident. The remediation work at the salvage site continued until May 2018, more than 6 years after the accident.

This illustrates that the timescale typically moves from minutes/hours to days/weeks to years. This also means that although it may appear that very large resources are required for the first phase, it will in fact be the last phase that is the most expensive and by far the one that requires most resources in total.

16.3 PRINCIPLES OF EMERGENCY RESPONSE

In Norway, emergency response generally builds on four key principles, and these are worth elaborating on.

Responsibility – The first principle is that the authority, organization, company, unit, etc. that has day-to-day responsibility for an operation, a system, or an activity also should be responsible for all aspects of preparing and handling of emergencies, including emergency planning, preparedness, and response. This also includes responsibility of the normalization phase. The advantage of this is that those with responsibility on a day-to-day basis will know the systems and activities best, they will know the risks and scenarios best, and they will have the best overview of what resources are available and their capabilities.

Similarity – This can be seen as an extension of the first principle and states the emergency response organization should be as close to the normal, operating organization as practicable. The reasons are much the same as for the previous principle, that individuals in the emergency organization should know the parts that they are responsible for as closely as possible and that they are used to reporting lines and who they interact with normally.

Proximity – The third principle deals with what levels of the organization should be involved in emergency response. Emergencies should be dealt with at the lowest possible level in the organization that is affected. For a ship, this would imply that emergencies primarily need to be handled by the crew. Obviously, there are usually wider implications (e.g., damage to cargo or delays in delivery) that will involve the shipowner, but it is primarily the crew that has to deal with the immediate response.

Collaboration – In major emergencies, there will often be many different organizations involved. For an accident in port, the crew itself, police, fire-fighters, medical personnel, port authorities, etc. will be involved. Close collaboration is a necessity to make the most out of available resources and ensure that everything runs smoothly. The principle states that collaboration is not the responsibility of one specific actor, but that all individuals involved have a responsibility to ensure good collaboration. This means that no one can stand back and claim that "this was not our responsibility".

These principles have implications for all phases of preparation, from planning, through preparedness to response.

16.4 EMERGENCY AND LIFE-SAVING REGULATIONS

16.4.1 SOLAS

SOLAS has no section dedicated to emergency preparedness or response as such but has an extensive Chapter III that covers Life-saving appliances and arrangements. This chapter is organized into three parts (IMO, 2001a):

A. General
B. Requirements for ships and life-saving appliances
C. Alternative designs and arrangements

The content of Chapter III is outlined in Table 16.1. The regulation has special requirements for passenger ships on top of the general provisions for cargo ships. The regulation focuses on two main aspects, namely design requirements and guidelines for operation.

SOLAS is very specific in many respects, setting out very detailed requirements for ships. Some examples are as follows:

– Regulation 6 Communications: "At least 3 two-way VHF radiotelephone apparatus shall be provided on every passenger ship and on every cargo ship of 500 gross tonnage and upwards."

Table 16.1 SOLAS Chapter III: Life-saving appliances and arrangements

Part	Content
A – General	Contains Regulation 1–5, with general regulations covering Application of the regulation, Exemptions, Definitions, Evaluation, testing and approval, Production tests.
B – Requirements for ships and life-saving appliances	Consists of five sections: Section I: Passenger ships and cargo ships (Regulation 6–20) – contains specific requirements to a wide range of systems, such as communications, personal life-saving appliances, muster lists, arrangements and launching stations, survival crafts, rescue craft, evacuation means, etc. Section II: Passenger ships (additional requirements) (Regulation 21–30) – specific requirements relevant for passenger ships, covering many of the same aspects as Section I Section III: Cargo ships (additional requirements) (Regulation 31–33) – specific requirements for cargo ships relating to survival craft, rescue boats, personal life-saving appliances and survival craft embarkation and launching arrangements Section IV: Life-saving appliances and arrangements requirements (Regulation 34) Section V: Miscellaneous (Regulation 35–37) – Covers various aspects of emergency preparedness and response, covering training manuals, maintenance of life-saving appliances and muster lists/emergency instructions.
C – Alternative designs and arrangements	Provides a methodology for getting approval for alternative solutions that do not meet the requirements of Part B. Specifies that deviations are possible, "provided that the alternative design and arrangements meet the intent of the requirements concerned and provide an equivalent level of safety to this chapter."

- Regulation 7 Personal life-saving appliances: "for passenger ships on voyages of less than 24 hours, a number of infant lifejackets equal to at least 2.5% of the number of passengers on board shall be provided;"
- Regulation 17 Rescue boat embarkation, launching and recovery arrangements: "Recovery time of the rescue boat shall be not more than 5 min in moderate sea conditions..."

However, there are also more functional requirements, such as:

- Regulation 11: "Lifeboats and liferafts for which approved launching appliances are required shall be stowed as close to accommodation and service spaces as possible."
- Regulation 11: "Muster stations shall be provided close to the embarkation stations."
- Regulation 17: "The rescue boat embarkation and launching arrangements shall be such that the rescue boat can be boarded and launched in the shortest possible time."

These regulations specify an intention rather than a specific requirement: "As close as possible", "close to", and "shortest possible time".

It is not our intent to describe all the regulations, but some elaboration may be provided on some of the aspects that would normally fall under the heading emergency preparedness and response.

Regulation 8: Muster list and emergency instructions – specifies that instructions of what to do in an emergency should be available for all persons on board. Passengers should have information in their cabins regarding muster stations, what to do in emergencies and how they put on lifejackets.

Regulation 19: Emergency training and drills – Specifies requirements for mustering, instructions to passengers about what actions to take and how to put on lifejackets. All crew members should take part in at least one abandon ship drill and one fire drill every month. All drills shall, as far as practicable, be conducted as if it was a real situation. There are also specific requirements to what an abandon ship and fire drill should include. The crew members should further be given training in the use of the life-saving appliances. The regulation also contains a requirement for record-keeping.

Regulation 20: Operational readiness, maintenance, and inspections – This regulation is aimed at ensuring that all life-saving appliances will function when needed. Among others, weekly inspections shall be conducted (and recorded) of survival craft, rescue boats, and launching appliances, engines in lifeboats and rescue boats shall be started, lifeboats shall be moved, and the general emergency alarm shall be tested.

Like for IMO's regulations in general, criticism of SOLAS has focused on the following aspects:

- Too much concerned about the technical details of life-saving appliances and systems
- Too little focus on the overall performance of life-saving appliances, i.e., the ability to save people
- Unrealistic testing conditions – primarily in calm weather in protected waters

These aspects have led to some ambivalence among seafarers: They know that the risk of evacuation is high, but on the other hand they see no point in training with inadequate systems under unrealistic conditions. In addition, the average passenger seems to have an unrealistic perception of the effectiveness of evacuation and life-saving appliances and systems. This is shown by the shock and anger found among the general public in the aftermath of maritime catastrophes such as the loss of Herald of Free Enterprise, Scandinavian Star, and Estonia.

The increasing number of high-speed ferries and high-capacity cruise vessels has brought to the surface the problem of inadequate approaches for the verification of evacuation systems. There are obvious ethical problems related to realistic full-scale testing of such systems, most importantly the significant risk of injury when testing the systems with people involved. In this context the computer simulation approach to testing has obtained considerable interest. The simulation approach is based on models of the vessel arrangement and the flow of people toward mustering and lifeboat stations. The approach is very much dependent on the ability to model and simulate both individuals and crowd behavior in emergency situations. As a response to this situation IMO has introduced regulations that address the use of simulation approaches in the assessment of life-saving effectiveness. We will return to this later in this chapter.

16.4.2 ISM Code: emergency preparedness

The International Safety Management (ISM) Code focuses on the implementation of systematic safety management, and as part of this, there are also requirements to emergency preparedness. The ISM Code is incorporated as chapter IX in the SOLAS Convention (IMO, 2001a). The requirements of the ISM Code were adopted by IMO in 1993 through Resolution A.741(18). Guidelines on the implementation of ISM are found in Resolution A.788(19) (IMO, 1995). A general description of the ISM Code can be found in Chapter 4 of this book.

Chapter 8 Emergency Preparedness in the ISM Code states the following:

8.1 The company should identify potential emergency shipboard situations and establish procedures to respond to them.
8.2 The company should establish programs for drills and exercises to prepare for emergency actions.
8.3 The safety management system should provide for measures ensuring that the company's organization can respond at any time to hazards, accidents and emergency situations involving its ships.

These requirements are very different from the requirements in SOLAS. Rather than specifying in detail what is required, it is in practice a simplified form of the emergency planning process described in Section 16.1.8. Emergencies should be identified, procedures to respond to emergencies should be established and relevant training programs should be implemented. Further, necessary resources required for the company to respond should also be in place.

It is clear that these requirements go much further than the SOLAS regulations in the sense that the company has to identify potential emergency situations and respond to those, and not only equip its vessels in accordance with certain standardized (prescriptive) requirements. These regulations also indicate that a shipping company or manager should undertake emergency planning in terms of the following aspects (ICS, 1994):

- Duties of personnel
- Procedures and checklists
- Lists of contacts, reporting methods
- Actions to be taken in different situations
- Emergency drills

16.4.3 STCW requirements

STCW is short for the International Convention on Standards of Training, Certification and Watch-keeping for Seafarers. Chapter VI of the STCW Code specifies "standards regarding emergency, occupational safety, medical care and survival functions" for crewmembers (IMO, 2002a). Key elements in securing minimum emergency preparedness are:

- Familiarization training:
  - Communicate with other persons on board on elementary safety matters
  - Ensure understanding of safety information symbols, signs, and alarm signals
  - Know what to do if:
    - a person falls overboard
    - fire or smoke is detected
    - the fire or abandon ship alarm is sounded
  - Identify muster and embarkation stations and emergency escape routes
  - Locate and learn how to use lifejackets
  - Learn how to raise the alarm
  - Have basic knowledge of the use of portable fire extinguishers
  - Learn to take immediate action upon encountering an accident or other medical emergencies before seeking further medical assistance on board
  - Identify the location of fire- and watertight doors fitted in the particular ship
- Basic training for crew with designated safety or pollution prevention duties:
  - Personal survival techniques
  - Fire prevention and fire-fighting
  - Elementary first aid
  - Personal safety and social responsibilities

- Crew competence requirements:
  - Competence to undertake defined tasks, duties, and responsibilities
  - Competence evaluation in accordance with accepted methods and criteria
  - Examination or continuous assessment as part of an approved training program

16.5 EMERGENCY PREPAREDNESS ACTIVITIES AND FUNCTIONS

16.5.1 Planning

The ISM Code requires that emergency preparedness should be based on an identification of hazards, estimation of risks, and the introduction of safety (or risk reduction) measures. This requirement has obvious implications for how a company plans and prepares for emergency situations. Key activities in the planning process include the following:

- Risk assessment:
  - Identify/locate hazards
  - Outline accident scenarios
  - Estimate probabilities and consequences
- Establish resources:
  - Ship arrangement
  - Safety-related equipment and systems
  - Manning and safety functions
- Outline emergency plan objectives:
  - Evacuation
  - Safeguard people
  - Mobilization of rescue operations
  - Control and mitigation of incidents
  - Salvage of vessel
  - Rehabilitation of conditions onboard
- Maintain plan:
  - Train
  - Arrange drills/exercises
  - Audit/review plan

Risk assessment may be undertaken using well-established risk analysis techniques and accumulated experience. The description of the likely and relevant accident scenarios should emphasize both the development of events and the role of equipment and human resources. Figure 16.6 shows how the key elements of a fire accident scenario may be outlined as a basis for the planning process.

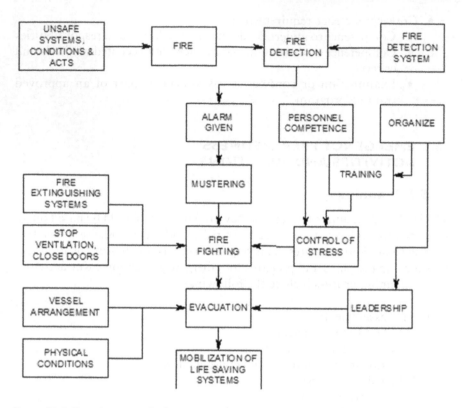

Figure 16.6 Key elements of a fire accident scenario.

The emergency plan should cover all the main accident scenarios, which may include the following:

- Fire
- Explosion
- Collision
- Grounding/stranding
- Engine breakdown
- Disabled vessel, loss of power and control
- Cargo related accidents
- Person overboard
- Emergency assistance to other ships
- First aid
- Unlawful acts threatening safety and security

It is underlined that for specific ships carrying special cargo or performing special operations, other scenarios may also be relevant. Further, the scenarios need to be specific also. "Fire" can cover a lot of different scenarios that need to be responded to in very different ways, e.g., fire in the accommodation, electrical fire, fire in cargo, and fire in engine room. Specific procedures will be required for these, and they can also vary a lot from one ship to another. For a Ro-Ro ship, fire in cargo will be very different from an LNG tanker which again will be very different from a bulk carrier transporting ore.

At the same time, it is not possible to develop procedures for absolutely all scenarios that can occur. The main thing to consider is how the response should be – if the response is the same or at least very similar for two scenarios it is sufficient to have one procedure that covers both.

Emergency preparedness plans must be based on a realistic time frame and should take into consideration factors such as the likely speed of escalation, how damage may propagate, and possible energy releases. The key elements in an emergency plan may be as follows:

1. Preface
2. Safety systems, life-saving equipment
3. Information systems, decision support systems
4. Organization of emergency teams, job descriptions
5. Distress signals
6. Information to crew and passengers
7. For each accident scenario:
   - Situation assessment
     - Category 1: Minor accident
     - Category 2: Alert situation
     - Category 3: Distress situation
   - Decision criteria
   - Defense and containment measures
   - External resources
8. Whom to contact depending on situation assessment
9. Evacuation plans:
   - Muster plan, boat stations
   - Evacuation routes
   - Information systems, control
   - Lifeboat/liferaft manning
10. Training
    - Familiarization, basic training
    - Specialist training: Fire-fighting, lifeboat coxswains, first aid
    - Drills
11. Revision of plan

In Section 16.1.18, the requirements for a plan according to MARPOL are shown, and it may be noted that these are simpler than shown here.

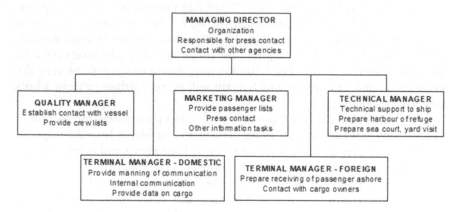

Figure 16.7 Land support team.

16.5.2 Land support

Experience has shown that the engagement of shipowner and manager is vital in the case of a serious accident. They shall serve as support and coordinator for the crew onboard and supply information on a continuous basis to families, official agencies, and media. The psychological effect of prompt and truthful information should not be underestimated. In an otherwise difficult or even tragic situation this may have a considerable positive impact on the company's goodwill. An example of how the managing company sets up the organization ashore is given in Figure 16.7.

16.5.3 Decision support

Regulation 29 in Chapter III of SOLAS states that a decision support system for emergency management shall be provided on the navigation bridge of passenger vessels. The system shall, as a minimum, consist of printed emergency plans. All foreseeable emergency situations shall be identified in the emergency plans. In addition to the printed emergency plans, one may also accept the use of a computer-based decision support system (DSS) on the navigation bridge. A DSS provides all the information contained in the emergency plans, procedures, and checklists. The DSS should also be able to present a list of recommended actions to be carried out in foreseeable emergencies. The main objectives of a DSS include:

- Issue warnings of dangerous situations and damage
- Detect critical trends
- Give a quick presentation of critical information
- Presentation of contingencies
- Enhance the overall understanding of the emergency situation

The emergency DSS shall have a uniform structure and be easy to use, and the following data from sensors and alarm systems might be presented in time series:

- Draught, heel, trim, freeboard
- Water level in tanks and compartments
- Status of watertight doors and fire doors
- Temperature and smoke concentration
- Status of all emergency systems

An emergency DSS may also integrate input from operators, external sources, and static information such as hydrostatic calculations (curve sheet and stability). Figure 16.8 outlines the data structure for an integrated fire-fighting system (IFFS), which may be one module within a more comprehensive decision support system (Rensvik & Kristiansen, 1994). The IFFS may, for example, support the crew, provide information to passengers and external parties, and be used to control remotely operated fire-fighting systems.

A very important requirement for all decision support systems is fast and user-friendly input and output of information. The use of graphical interfaces in the DSS is therefore highly recommended. Two examples of how such graphical interfaces might be configured are given in Figures 16.9 and 16.10. In addition to presenting the instantaneous fire situation, the integrated fire-fighting system (IFFS) shown in the figures below may also be used for preview (prognosis), maintenance of emergency information systems, as well as logging of events and developments.

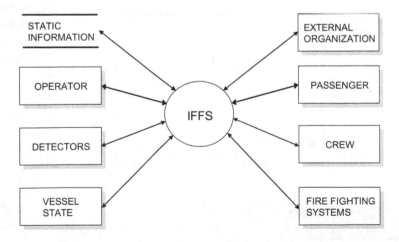

Figure 16.8 The data structure for an integrated fire-fighting system (Rensvik & Kristiansen, 1994).

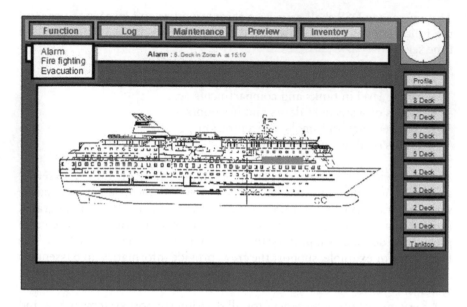

Figure 16.9 IFFS – Localization of fire on deck plan (Rensvik & Kristiansen, 1994).

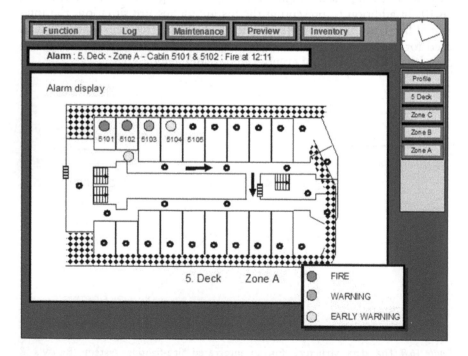

Figure 16.10 IFFS – Localization of fire in section (Rensvik & Kristiansen, 1994).

16.6 HUMAN BEHAVIOR IN EMERGENCY SITUATIONS

As a result of investigation of and research on catastrophic accidents, we have improved our understanding of how humans react in emergency situations. This enables us to make realistic assessments of what can be expected in terms of evacuation effectiveness, which again gives us a better basis for design of life-saving appliances. Particularly, investigations and research have confirmed that evacuation training is of great importance in terms of reducing the potential consequences of serious accidents.

16.6.1 General characterization

Maritime accidents often have a very dramatic nature. Some important characteristics of emergency situations are presented in Table 16.2. The emergency situation is a function of the physical nature of the accident, the dramatic and uncontrolled development (or escalation) of events, and the perceived threat to people's own safety. The degree of drama in maritime

Table 16.2 Characteristics of emergency situations

Parameter	Characteristics
Accident nature	– Degree of immediate threat to own life – How fast the events develop, or the situations change
Lack of warning	– People are unprepared for the next event – High degree of uncertainty – Influenced by rumors and "hear-say"
Time pressure	– Quickly changing situations
Drama	– Degree of injury and number of fatalities – Despair, fear, and other stress reactions
Physical chaos	– Trapped in enclosed areas, moving objects – Blocking of escape routes – Darkness, smoke
Vessel state	– Fire, explosion, water ingress, sea motion – Heeling, sinking
Threat to own life	– Heat, lack of oxygen, drowning – Impact from explosion, structural failure – Extreme weather
Lack of control	– Strong feeling of anxiety – Impaired by own stress reactions and trauma – Lack of information – Lack of leadership, team spirit, and solidarity – Influenced by reactions of other people
Isolation	– The vessel is an "island" in the ocean – No or limited assistance from other vessels or land-based resources – Critical delay of rescue and salvage – No safe haven: Forced to evacuate into the sea

accidents is further compounded by the degree of isolation that is experienced on a ship and the limited availability of assistance from external resources (e.g., other ships or salvage units). The degree of rescue and salvage help is often limited or delayed, and there is nowhere to flee other than evacuating into the sea.

The degree of drama that can be experienced onboard a ship may be illustrated by the loss of the Ro-Ro passenger ferry Estonia (Laur et al., 1997). The vessel sailed with a speed of about 14 knots in head seas, with a significant wave height of 4.3 meters, when the bow visor's three locking devices failed. When the bow door fell off the locks on the inner ramp failed, allowing water to flow into the vehicle deck. It has been estimated that the water inflow might have been in the order of 300–600 tons per minute. Because of free-surface effect the ship heeled to 30° within minutes, and simulations indicate that the vessel reached a heel of 60° in only 16 minutes. At 40° heel the water reached the windows on deck 4, probably resulting in progressive flooding. Due to the rapid development of this event, there were neither given any alarm nor organized any evacuation from the bridge. Later studies indicated some local attempts to assist passengers, and that some passengers managed to reach the boat deck, although without being able to launch any lifeboats. Many of these saved their lives when they managed to get into the liferafts that were released as the ship sank. Almost immediately after the vessel started to heel, people had problems with leaving their cabins and movement in the narrow corridors (1.2 m wide) was difficult. Many passengers were trapped inside their cabins, and many of those that were able to get out of their cabins got stuck in staircases. Loose furniture and large objects also hampered movement in public spaces. It has been confirmed by many of those that survived that the people onboard were struck by well-known emergency reactions, ranging from panic to apathy, despite early attempts to take responsibility and assist each other. Of the 989 people onboard, it is judged that only 300 reached the outer decks, and only 160 of these succeeded in boarding a liferaft or lifeboat when the vessel sank. In the end, helicopters or vessels picked up 138 people, giving a survival rate of only 14%.

Studies of stress reactions under emergency situations show that it is feasible to make a distinction between four phases of an emergency or catastrophic event/situation (Sund, 1985). These phases and corresponding stress reactions are presented in Table 16.3. The early phases of "pre-accident" and "warning" are characterized by denial and/or a feeling of being invulnerable. This may lead to a critical delay of necessary response or fighting of the accident. In the "acute" phase people are typically subjected to more dramatic reactions like shock, panic, or becoming paralyzed. If these reactions strike the majority of the crew and passengers onboard a ship, the consequences may be severely worsened. It is important to note that there is a risk of developing so-called posttraumatic reactions. This knowledge

Table 16.3 Stress reactions in different phases of an emergency event/situation

Phase	Stress reactions
Pre-accident	– Denial: "This will not happen to me"
Warning	– Denial, illusion of being invulnerable
Acute	– Shock and stress reactions: Alarm, psychosomatic, passiveness, uncontrolled behavior
Intermediate	– Development of syndromes: Emotionally unstable, depression, guilt, isolation, overreaction
Post	– Posttraumatic disturbance such as stress and neurosis. Regaining emotional stability, control, and good health. Continued need of treatment

Based on Sund (1985).

Table 16.4 Negative stress reactions

Behavior	Characteristics
Sensing	– Narrow-minded, selective focus – "Everything or nothing behavior"
Cognition	– Stereotypic, "frozen attitude" – Short-term oriented, loss of perspective/overview – Unable to solve complicated problems
Reaction	– Limited search for information – Stereotypic behavior – Perseverance/persistence – Impulsive or lamed

has led to a greater focus on treatment and counseling in the aftermath of accidents and catastrophes.

Research has shown that persons involved in emergencies have a limited ability to deal with challenges related to evacuation and salvage operations. As indicated in Table 16.4, people in emergencies tend to become narrow-minded and stereotypic and become unable to deal adequately with complicated problems. An immediate lesson of this fact should be to design simple evacuation systems and other life-saving appliances. For example, in a number of emergency situations there have been accidents related to the release of lifeboats, such as inability to activate the system and premature release leading to uncontrolled fall.

Sund (1985) has also given indicative numbers on the relative distribution of how people manage emergencies. These are presented in Table 16.5. The group that behaves optimally and takes leadership may be from 10% to 30% of the total. A larger group, of about 50% to 75% of the total, will be slightly reduced but will function reasonably well with adequate leadership. Strong psychic reactions can be seen for as much 25% of a group, while

Table 16.5 Stress reactions in emergencies – distribution within a group

Part of population	Characteristic behavior
10–30%	– Behaves balanced – Realistic perception and assessment of situation – Helps others in the group, able to cooperate – Takes leadership, demonstrates initiative
50–75%	– Light psychic lameness or apathy – Slightly puzzled or confused – Becomes active under leadership – No need for medical help
10–25%	– Strong psychic reactions needing medical treatment
1–3%	– Loss of mental control – Symptoms of serious nervous breakdown – Acute mental disorder or panic

Based on Sund (1985).

between 1% and 3% will lose mental control and/or experience nervous breakdown. In addition to the characteristics of the emergency situation under consideration, the following background factors may determine the degree of adequate/balanced behavior:

- Earlier experience with emergencies
- Personality type and psychic health
- Duration of employment and age
- Leadership experience
- Intelligence

16.6.2 Emergency behavior

Over time, considerable knowledge has been accumulated about concrete behavior in emergency situations. Reisser-Weston (1996) proposes that one should see the total evacuation time as a function of the following phases:

- Detection
- Decision
- Non-evacuation behavior
- Physical evacuation

The author has further proposed a task structure in emergency situations that is outlined in Figure 16.11. In the event of an alarm, one has three basic options: act, investigate, or wait. Reisser-Weston further points out that the emergency behavior is influenced or determined by a set of Performance Shaping Factors (PSF), which are presented in Table 16.6. Performance shaping factors are factors assumed to have an effect on human behavior (see also Chapter 11).

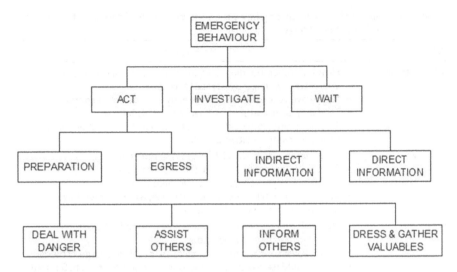

Figure 16.11 Hierarchical task structure of an emergency (Reisser-Weston, 1996).

Table 16.6 Performance Shaping Factors (PSF) in an emergency situation

PSF	Description
Structural	– Organization of the workplace – Physical characteristics, rules
Effective	– Emotional, cultural, and social factors – Behavior is affected by stress, perceived risk, trust, and cultural norms
Informational	– Direct information – Warning cues and information about escape routes – Communication and advice
Task and resource characteristics	– Possible conflict between current task or job function and the need for evacuation

Reisser-Weston (1996).

Reisser-Weston (1996) has also summarized the results of a number of studies of human behavior in emergency situations, primarily relating to fires in office buildings. Some of the findings of this study are briefly presented in Table 16.7.

16.7 EVACUATION RISK

One of the first investigations of the risks associated with evacuation from ships was undertaken by Pyman and Lyon (1985). The main findings of their investigation are summarized in Table 16.8. It was found that the

Table 16.7 Human behavior in emergency situations, summary of research findings for fires in office buildings

PSF	Findings
Informational	• High rise office building: 14% interpreted the alarm as genuine • Informative warning system: 81% responded • Many have to observe a fire directly in order to respond • Tendency to investigate ambiguous signals further • 45% were unable to differentiate fire alarm from other signals • False alarms desensitize people
Effective	• Investigated alarm signal: Men: 15%, Women: 6% • Called fire department: Men: 6%, Women: 11% • Got family together: Men: 3%, Women: 11% • Women will warn others and evacuate, whereas men have a tendency to deal with the danger
Structural	• Hospitals are hierarchical: Individuals respond adequately in accordance with their position • Persons with authority are critical for mobilizing large crowds • Time to initiate evacuation without direction from staff: 8 minutes 15 seconds • Time to initiate evacuation when directed by staff: 2 minutes 15 seconds • In public places people are slower to break out of the normal routine

Reisser-Weston (1996).

Table 16.8 Evacuation risk – worldwide 1970–1980

	Merchant vessels	Fishing vessels
Accidents with fatality during evacuation	37%	32%
Fatalities among those attempted evacuation	13%	15%
Hard weather accidents with fatality	78%	73%
Fatalities in hard weather evacuation	35%	36%
Calm/moderate weather accidents with fatality	16%	26%
Fatalities in calm/moderate weather evacuation	5%	8%
Fatalities among those in a fast-sinking accident	86%	–

Pyman and Lyon (1985).

average probability of one or more fatalities occurring during an evacuation was in the order of 35%. The effect of weather conditions was, not surprisingly, found to be considerable. According to this study the probability of fatalities occurring during evacuation is more than three times higher during hard weather conditions compared to calm or moderate weather. The average proportion of fatalities during evacuation is 14%, i.e., one out of seven trying to evacuate does not survive. The effect of hard weather conditions on the number of fatalities is even more dramatic: Approximately five times higher in hard weather compared to calm/moderate weather.

Table 16.9 Evacuation risk for bulk carriers

Type of accidental event	Number of events	Evacuation frequency [per ship year]	Fatalities	Number on board	Probability of fatality [%]
Collision	14	$3.1 \cdot 10^{-4}$	116	332	35
Contact	5	$1.1 \cdot 10^{-4}$	54	119	45
Fire/explosion	16	$3.6 \cdot 10^{-4}$	6	379	2
Foundered	51	$1.1 \cdot 10^{-3}$	618	1,209	51
Hull failure	5	$1.1 \cdot 10^{-4}$	0	119	0
Machinery failure	1	$2.2 \cdot 10^{-5}$	0	24	0
Wrecked/stranded	23	$5.1 \cdot 10^{-4}$	0	545	0
Total	115	$2.6 \cdot 10^{-3}$	794	2,727	29

DNV (2001).

During the 1980s and 1990s, there was an increasing concern about the safety of bulk carriers. At the 70th session of IMO's Maritime Safety Committee (MSC) the topic of life-saving appliances (LSAs) for bulk carriers was discussed. As a response to this a comprehensive FSA project on LSAs for bulk carriers was performed (DNV, 2001), and evacuation risk was estimated based on reported evacuations for the period of 1991–1998. The results of this study are presented in Table 16.9.

The evacuation frequency for bulk carriers is $2.6 \cdot 10^{-3}$ per ship year. The average probability of fatality in evacuation is 29%, which is defined as the ratio of fatalities to the number of crew at risk. This figure is much higher than the earlier cited estimate for merchant ships in calm/moderate weather evacuation, but more comparable with the hard weather figure (Pyman & Lyon, 1985). This supports the assessment that bulk carrier losses often happen under dramatic conditions like hard weather and fast sinking.

The data material from which Table 16.9 was established and has also been analyzed with respect to the effectiveness of different evacuation methods, and the results of this analysis are shown in Table 16.10 in terms of probability of fatality. The data indicates that direct transfer to another vessel is one of the safest evacuation means with a fatality rate less than 1%. It is further clear that liferafts are safer than lifeboats. It should not be a big surprise that the least preferred method of evacuation is directly into the sea, i.e., so-called "wet" evacuation.

One important piece of knowledge to be drawn from the evacuation risk data presented above is that evacuation is a very risky activity with a high probability of fatality. Given the high fatality rate in evacuation, it should be clear that appropriate emergency/evacuation training and preparedness is of essential importance for seafarers. Learned responses to critical events and situations, as well as a degree of familiarity with simulated situations (i.e., training), can be of significant importance in terms of saving lives in

Table 16.10 Evacuation of bulk carriers – probability of fatality for different evacuation methods

Evacuation method	Number of events	Fatalities	Number to be evacuated	Probability of fatality
Transferred by helicopter	8	17	219	0.078
Transferred to vessel	8	1	201	0.005
Lifeboat	4	57	112	0.509
– then picked up by helicopter	0	–	–	–
– then picked up by vessel	3	24	79	0.304
– unknown further salvage	1	33	33	1.000
Liferaft	3	10	68	0.147
– then picked up by helicopter	0	–	–	–
– then picked up by vessel	3	10	68	0.147
Both lifeboat and liferaft	13	81	310	0.261
– then picked up by helicopter	1	21	25	0.840
– then picked up by vessel	9	4	209	0.019
– then by helicopter and vessel	1	5	25	0200
– unknown further salvage	2	51	51	1.000
Direct into sea (wet evacuation)	6	99	127	0.780
– then picked up by helicopter	0	–	–	–
– then picked up by vessel	4	63	90	0.700
– unknown further salvage	2	36	37	0.973
Transferred to helicopter and evacuation to survival craft	3	1	78	0.013
Transferred to helicopter and picked up by vessel	1	0	25	0.000

DNV (2001).

evacuation. This accounts for all phases of an evacuation, from calm and controlled behavior at muster stations, correct use of personal life-saving appliances such as lifejackets and immersion suits, proper use of lifeboats and life rafts, behavior in these evacuation crafts, use of first aid, etc.

In the 1990s, the so-called free-fall lifeboats were introduced as an alternative to conventional lifeboats lowered by davits on the sides of the vessel. See Figure 16.12. Free-fall lifeboats are launched on skids on the aft end of the vessel and are therefore less influenced by rolling motion. However, like the conventional type, it is subject to failure of critical steps before the actual release of the unit: evacuation to the boat, decision to board,

Figure 16.12 Free-fall lifeboat (Wikimedia Commons).

Table 16.11 Probability of fatality during evacuation

Type of event	Conventional lifeboats		Free-fall lifeboat	
	Probability of fatality (%)	PLL (per ship year)	Probability of fatality (%)	PLL (per ship year)
Collision	45.7	$3.36 \cdot 10^{-3}$	44.6	$3.28 \cdot 10^{-3}$
Contact	44.1	$1.15 \cdot 10^{-3}$	42.8	$1.12 \cdot 10^{-3}$
Fire/explosion	27.7	$2.36 \cdot 10^{-3}$	28.1	$2.40 \cdot 10^{-3}$
Foundered	55.4	$1.44 \cdot 10^{-2}$	51.3	$1.34 \cdot 10^{-2}$
Hull/machinery	16.2	$5.12 \cdot 10^{-4}$	13.2	$4.17 \cdot 10^{-4}$
Wrecked/stranded	20.2	$2.45 \cdot 10^{-3}$	18.4	$2.22 \cdot 10^{-3}$

DNV (2001).

untimely launching, and unsuccessful clearing the ship. This is illustrated in one of the first of the risk-based comparisons of the two concepts as shown in Table 16.11. The study showed that the probability of fatality was mainly governed by the type of accident and only marginally influenced by the type of lifeboat. The study did not investigate the effect of weather conditions and may explain the last finding, namely that free-fall lifeboats are less influenced wind and sea state. This will be discussed further in the next section.

16.8 EVACUATION SIMULATION

SOLAS specifies the following maximum times for key evacuation phases on passenger ships:

- The maximum time from abandon ship signal is given to all survival crafts are ready for evacuation is 30 minutes (Chapter III, Regulation 11)
- The maximum time for abandonment of mustered people is 30 minutes (Chapter III, Regulation 21.1.4)

Simulation technology prompted IMO to develop standards for the adoption of such techniques in the assessment of evacuation effectiveness. In 2002, the Maritime Safety Committee (MSC) of IMO formally adopted the "Interim guidelines for evacuation analysis of new and existing passenger ships including Ro-Ro" (IMO, 2002b). These guidelines only address the assembly/mustering part of the evacuation process, and two scenarios are defined, namely day and night conditions. The SOLAS performance requirements are based on calm weather conditions, no list, and no effect of fire. It is evident that in harsh weather conditions, with list and/or the effects of fire, it will be much more difficult to achieve evacuation within the given performance requirements.

This section of this chapter will take a closer look on evacuation simulations. First, however, a brief introduction of crowd behavior is given. Not considering crowd behavior is considered to be a major shortcoming of many evacuation simulation models.

16.8.1 Crowd behavior

Jørgensen and May (2002) have discussed a number of important issues related to crowd behavior. The authors point to the fact that evacuation simulation models basically estimate individual behavior and more or less neglect crowd behavior, which according to them is a major shortcoming of these models. They have defined the concept of group-binding, which expresses the fact that people both rationally and emotionally have an interest in finding their relatives before being evacuated. The crew ideally manages the mustering of crowds in an emergency situation, but due to group-binding people will often be non-compliant to the instructions given by the crew and rather focus on finding their relatives. The degree of group-binding will be a function of the social composition of the passengers: Singles, couples, families, and groups of friends. The effect of group-binding will also be related to the size and arrangement of the vessel as well as the time of the day.

Jørgensen and May (2002) studied the social composition of the passengers on different Danish vessels and interviewed people about their willingness to disobey crew instructions. An average of 30% of the passengers would disobey crew instructions in order to find family members and other people they felt closely connected or related to. With an estimated probability of actually being separated of 0.2, the group-binding problem would affect 6% of the passengers in a given situation.

The authors also discuss panic in relation to emergency situations. As cited earlier (see Table 16.5) it has been estimated that 1–3% of group will panic and/or loose mental control. Jørgensen and May (2002) challenge this view based on psychiatric generalizations and the fact that crowd behavior is not taken into consideration. Panic behavior should also be seen as a sociological phenomenon, and they refer to work by Berlonghi (1996) who makes a distinction between different crowd phenomena:

- Passive crowd (e.g., spectators)
- Active crowds
  - Hostile crowd (e.g., mobs)
  - Escape crowds (often characterized by panic)
  - Acquisitive crowds (often characterized by craze)
  - Expressive crowds (often characterized by mass hysteria)

Other aspects of the realism of evacuation simulations are also discussed by Jørgensen and May (2002). Firstly, verification of a numerical simulation model requires that full-scale evacuation exercises with a large number of people in actual ships are performed. This is, however, impossible in practice, mainly for economic reasons. Secondly, in terms of arranging such evacuation exercises it is problematic to make the exercise fully realistic, as people will not be influenced by the perception of danger and feel of urgency that characterizes real emergency situations. It may also be dangerous to arrange such realistic evacuation exercises as real panic may arise.

16.8.2 Modeling the evacuation process

The evacuation of crew and passengers is a process involving a number of phases. The following phases can be used for modeling purposes:

1. Detection and acknowledgment of an emergency
2. Sound the alarm
3. Recognition (by people) of the alarm
4. Collection of life jacket, orientate oneself about the situation
5. Search for and unite with family and friends
6. Evacuate to safe place or mustering station
7. Prepare and deploy survival craft or escape system
8. Board survival craft
9. Launch craft or leave the vessel
10. Rescue by external resource

The simulation approach replicates the evacuation process described above in the time domain. In a time-stepping mode the behavior of each individual is estimated, taking the physical conditions and interaction with other people into consideration. The vessel is normally incorporated into the simulation model by a two-dimensional space grid. The result of the simulation is an estimation of the total time of evacuation. As pointed out by Galea et al. (2002) the simulation must address a number of aspects:

- Configurational: The physical layout and arrangement of the vessel with dimensions of rooms, corridors, and stairways.
- Environmental: Environmental factors that affect people under the evacuation, such as list, ship motion, presence of debris, heat, smoke, toxic substances, etc.
- Procedural: Basic rules for the phases in the evacuation process, for example, related to the guidance of passengers by crew, the organization at mustering stations, etc.
- Behavioral: Characteristics of how individuals behave and perform. The group of people onboard should reflect a realistic composition in terms of sex, age, walking speed, and ability to respond adequately. Some of these attributes may be dynamic and change value during the evacuation.

The EXODUS numerical simulation tool (Galea et al., 2002) consists of a number of interacting program modules as illustrated in Figure 16.13. The model considers the interaction of people relative to other people, physical arrangement, the state of the vessel, and fire threat. The models are rule-based, and the behavior of individuals is based on heuristic rules.

The "Behavior" module in EXODUS is critical for the realism of the simulation. It controls how people respond to the changing situation and

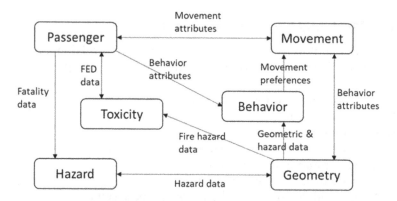

Figure 16.13 Module interactions in the EXODUS numerical simulation tool (Galea et al., 2002).

controls the "Movement" module. It functions on two levels, globally and locally, where the former addresses the decision on escape strategy whereas the latter determines behavioral responses and decisions made locally by individuals. For example, through the "Behavior" module EXODUS reflects reactions to such phenomena as congestion and group ties.

The EXODUS model has several output formats that visualize the development of the evacuation. A so-called footfall contour map indicates the most heavily used routes that passengers use during the evacuation. Another format presents the final assembly and density of people at the designated mustering stations. The simulation package also offers an option where parts of the evacuation can be viewed "live" in 3D, which makes it possible to study the effects of the ship's arrangement and potential "bottlenecks".

The effectiveness of an evacuation may be expressed by a number of variables:

- Times:
  - For individuals to muster
  - Total time used to muster and evacuate
  - Time wasted in congestions
  - Time to clear particular compartments or decks
- Distance traveled by individuals
- Flow rate through doors or openings

An experimental evacuation exercise has been conducted for a so-called Thames pleasure boat with 2 decks (Galea et al., 2002). A total of 111 volunteers, from 16 to 65 years of age, were engaged in the exercise. Forty-nine of these were located on the lower deck and 62 on the upper deck. For each deck there were four exits, two forward and two aft, and a twin set of staircases connected the two decks. During the experiment, the vessel was moored with starboard side to the jetty. Several evacuation tests were performed with different restrictions on access to the exits. The results from these evacuation exercises were then compared to the EXODUS simulation model where the predicted evacuation times were within 7% of the experiments.

16.8.3 A Simulation case

The evacuation time for a large passenger ship with a total of 650 passengers has been estimated using the EXODUS simulation tool (Galea et al., 2002), and below some interesting aspects of this simulation case are presented. The vessel in question had 10 decks divided into three vertical fire zones. The muster deck (i.e., deck 8) and the deck below are shown in Figure 16.14. The initial distribution of passengers within the ship before

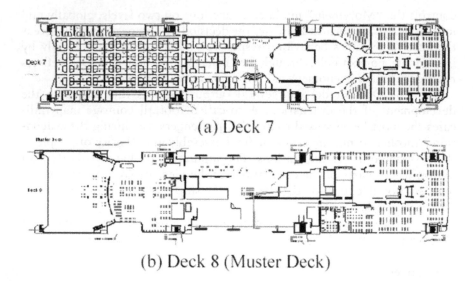

(a) Deck 7

(b) Deck 8 (Muster Deck)

Figure 16.14 Deck arrangement (Galea et al., 2002).

Table 16.12 Initial location of passengers

Deck	Fire zone 1	Fire zone 2	Fire zone 3
6	172	28	
7	176	24	
8			
9			150
10			100

Galea et al. (2002).

evacuation was as shown in Table 16.12. Passengers in fire zones 1 and 3 are assumed to be in their cabins.

Fire zones 1 and 3 have four staircases each, with each staircase located in the far corners and only allowing a single lane of passengers. Fire zone 2 has a single centrally located staircase allowing two lanes of people to move.

The simulation was done for nighttime conditions with a response time from 7 to 13 minutes, allowing for people sleeping in their cabins to wake up and get dressed (an IMO requirement). Travel speeds are also specified by IMO, and these are summarized in Table 16.13. Depending on gender, age, and degree of impairment, the travel speed in flat terrain varies with a factor of 3. The speed of walking down stairs is 30% lower than that on flat terrain, and around 50% lower walking up stairs. The IMO regulation specifies that the simulation should be run 50 times with random values. The MSC

Table 16.13 Passenger travel speed [m/s] as specified by MSC circ. 1033

Age [years]/impairment		Walking on flat terrain [m/s]	Walking down stairs [m/s]	Walking up stairs [m/s]
Female	<30	1.24	0.75	0.63
	30–50	0.95	0.65	0.59
	50 +	0.75	0.60	0.49
	Impaired 1	0.57	0.45	0.37
	Impaired 2	0.49	0.39	0.31
Male	<30	1.48	1.01	0.67
	30–50	1.30	0.86	0.63
	50+	1.12	0.67	0.51
	Impaired 1	0.85	0.51	0.39
	Impaired 2	0.73	0.44	0.33

IMO (2002).

Table 16.14 Range of mustering times with even keel vessel condition

Estimate	Fire zone 1	Fire zone 2	Fire zone 3
Minimum	14′ 59″	13′ 34″	13′ 42″
Average	15′ 32″	14′ 00″	14′ 32″
Maximum	15′ 58″	14′ 43″	15′ 24″

Galea et al. (2002).

circular 1033 (IMO, 2002) gives ranges of variation for the travel speeds cited in Table 16.13, and the range of variation is in the order of ±25%.

The estimated mustering times for an even keel (i.e., no heel) vessel condition are given in Table 16.14. The fact that fire zone 1 has the longest mustering time can be explained by the relatively high number of passengers, which may result in congestion, and that the evacuation includes walking up the stairs from decks 6 and 7 to the muster deck (i.e., deck 8).

IMO specifies that the dimensioning evacuation time should be taken as the highest value of four scenarios and added an extra 10 minutes to account for the assumptions and uncertainties in the model. The highest value in each scenario is to be taken as the 95%. For the given scenario (Table 16.14), the estimated maximum mustering time was 15 minutes and 58 seconds, and given an added safety margin of 10 minutes, the regulation says that the vessel needs 25 minutes and 58 seconds to evacuate and muster.

Congestion (of people) in specific areas of the ship can be studied in the EXODUS model, and for the case described above there was congestion in the range of 2.2–3.5 [persons/m²] at the base of the staircases for

19 seconds. IMO defines a congested area as an area where there is a passenger density of 4 [persons/m²] or more.

The effect of heel on the mustering time is not very significant in the present version of the EXODUS simulation model. Muster times for fire zone 1 for 0°, 10°, and 20° heel were found to be 15 minutes 32 seconds, 15 minutes 34 seconds, and 16 minutes, respectively. With 10° heel the mustering time only increases with 2 seconds, and for a heel of 20° the increase in time is still marginal with 26 seconds. 20° heel is quite dramatic and makes it considerably more difficult to move around the vessel. The heel itself may, in addition, result in increased stress and even panic. The increase in mustering time will therefore most certainly be much larger than 26 seconds. Evacuation from partly capsized passenger vessels is discussed in more detail below. The results confirm the inability of the EXODUS model to take factors like change in human behavior, physical chaos, and potential loss of electricity because of heel into consideration. When using such simulation tools, it is important to always have a clear understanding of the inherent limitations.

16.8.4 Evacuation from partly capsized vessels

Planning and training for evacuation of passenger vessels normally assumes that the vessel is in an upright or only moderately heeled condition. However, experience shows that this is not necessarily the case in real emergency situations:

- European Gateway (1982): Collided with another ship off Felixstowe (England) as a result of confusion at a bend in the channel. The collision resulted in puncturing of the main vehicle deck and the generator room below the waterline. Because of asymmetric flooding the vessel started to heel, reaching 40° in only three minutes, at which point the bilge grounded. During the next 10–20 minutes the ship rolled onto its side. Six of the 70 people on board drowned.
- Herald of Free Enterprise (1987): Uncontrolled flooding through the open bow door resulted in rapid heel to 30° only minutes after leaving port at Zeebrugge (Belgium). Within 90 seconds the vessel heeled/capsized to 90°, at which point the side of the vessel was resting on the seabed in the shallow water. At least 193 passengers and crew died.
- Estonia (1994): Failure of the bow door and ramp in a severe storm in the north Baltic Sea led to rapid water ingress onto the vehicle deck. Because of free-surface effect the ship heeled to 30° within minutes, increasing to 90° only 20 minutes after the bow ramp opened. About 10 minutes later the ship sank completely, resulting in 852 fatalities.

According to Spouge (1996), the difficulty of evacuation increases dramatically when a vessel heels beyond 45°. The main causes of death in such situations are as follows:

- Falling headlong with extreme heel
- Shock of water immersion results in heart diseases or other paralyzing illnesses
- Drowning due to rising water in compartments and inability to swim or escape (primarily to higher level)

After capsizing a vessel will usually come to rest in a stable position for a period of time. This will give some time for evacuation from inside the ship as well as rescue away from the ship. After a while, depending on the vessel's construction and the extent of damage, further water ingress will result in heel to 180° or sinking. Spouge (1996) has assessed the fatality risk for this accident scenario. The consequences of an evacuation of a passenger ferry, primarily consisting of large public spaces (i.e., type A), after capsizing to 90° are summarized in Table 16.15. Of the estimated 45% fatality ratio most people perished inside the vessel. The data also emphasize the importance of dry compared to wet evacuation. For a ship with cabins (i.e., type B) the fatality rate during night conditions will typically be 56%, considerably more than for type A vessels with mainly large open public spaces.

Spouge (1996) also proposed technical measures to improve the evacuation success rate. It was estimated that the survival rate could be improved with 3–7% for capsize scenarios beyond 45° if additional escape equipment and arrangement features were implemented. The following types of equipment/features were proposed: Ladders, bridges, ropes, escape windows, and

Table 16.15 Fatalities and survivors on a passenger ferry for short crossings in the case of 90° capsize

Outcome	Relative number of people [%]
Killed by fall to side of compartment	5
Killed by shock of immersion	4
Drowned in rising water	26
Drowned awaiting rescue	10
Total fatalities	**45**
Escaped on own	3
Rescued from dry by survivors	23
Rescued from water by survivors	1
Rescued from dry by rescuers	26
Rescued from water by rescuers	2
Total surviving	**55**

Spouge (1996).

elimination of full height partitions in public areas. In addition, limiting heel is considered very important in terms of saving lives.

16.8.5 Designing for safe evacuation

IMO has evidently been focusing on safe evacuation and especially for passenger and cruise vessels. However, the first attempt to regulate lacked the necessary degree of systematics (IMO, 2001c). Vanem and Skjong (2006) discussed a novel approach for a framework that would give a more balanced assessment of relevant evacuation scenarios.

An evacuation scenario is related to a given accident type and a number of factors like day/night conditions, limitation of escape routes, and whether a vessel intact is sinking upright or capsizes. One of the main findings of the authors was that the time for evacuation differs considerably with accident type as shown in Figure 16.15.

An important finding was that collisions and groundings result in shorter available time for evacuation than fires. In the latter situations, the crucial matter is to evacuate fire zones than abandoning the vessel. Fires on ships do not immediately result in water ingress and sinking. The authors illustrate this point with following estimates for fatality risks (PLL) as outlined in Table 16.16. Although the probability of evacuation for the accident

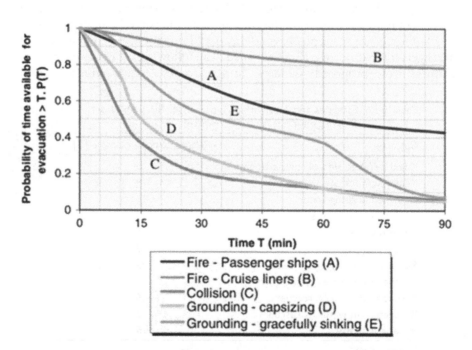

Figure 16.15 Expected available evacuation times (Vanem & Skjong, 2006).

Table 16.16 Probability of emergency evacuation

Type of scenario	Probability of emergency evacuation	PLL (N=3,000)
Fire Ro-Ro passenger ship	$4.4 \cdot 10^{-4}$	$1.4 \cdot 10^{-2}$
Fire cruiser liner	$2.6 \cdot 10^{-3}$	$1.4 \cdot 10^{-2}$
Grounding	$1.1 \cdot 10^{-4}$	$1.5 \cdot 10^{-1}$
Collision	$6.9 \cdot 10^{-4}$	1.3

Vanem and Skjong (2006).

Table 16.17 Evacuation scenarios

Scenario no.	Accident	Accident scenario
1	Fire	Fire in accommodation
2	Fire	Escalating fire in engine room
3	Collision & grounding	Sinking ship developing list
4	Collision & grounding	Ship sinking gracefully

Vanem and Skjong (2006).

types is comparable, the fatality risk is significantly greater for grounding and collision.

The authors propose four basic evacuation scenarios as shown in Table 16.17. It was found necessary to make a distinction between fires in the accommodation and the more time critical fires in the engine room leaving less time for evacuation. Collision and grounding accident lead to two scenarios namely sinking with increasing list or sinking upright.

These main scenarios were broken down into 14 resulting scenarios by differentiating on the basis of daytime/night conditions and alternatively Ro-Ro passenger vessels and cruise ships. The resulting scenarios with associated probabilities are shown in Table 16.18. Keep in mind that the probability means the probability that the scenario described may happen.

In a separate analysis the probability of safe evacuation from the vessel is analyzed. The number of people safely evacuated increases with the time to evacuate and this is illustrated for the scenario for evacuation at daytime for a vessel listing in Table 16.19. Successful evacuation will require at least 30 minutes for 540 people. The expected number of fatalities is 249. The probability of the scenario is $4.5 \cdot 10^{-4}$ and results in the following risk contribution:

$$\text{Risk}_{\text{scenario}} = 4.5 \cdot 10^{-4} \cdot 249 = 0.11 \text{ fatalities per year}$$

By summing the risk contributions for all scenarios, the evacuation risk for the vessel concept is found.

Table 16.18 Evacuation scenarios with associated probabilities

Main evacuation scenario	Case	Resulting scenario	Vessel	Probability
Escaping from fire zone	Day	Escaping at daytime	Ro-ro	$2.5 \cdot 10^{-4}$
			Cruise	$1.6 \cdot 10^{-3}$
	Night	Escaping at night	Ro-ro	$1.3 \cdot 10^{-4}$
			Cruise	$8.0 \cdot 10^{-4}$
Abandoning ship on fire	Day	Abandoning at daytime	Ro-ro	$1.9 \cdot 10^{-4}$
			Cruise	$1.2 \cdot 10^{-3}$
	Night	Abandoning at night	Ro-ro	$9.7 \cdot 10^{-5}$
			Cruise	$6.0 \cdot 10^{-4}$
Evacuation from sinking ship, listing	Day	Evacuation at daytime, listing		$4.5 \cdot 10^{-4}$
	Night	Evacuation at night, listing		$2.2 \cdot 10^{-4}$
Evacuation from sinking ship, upright	Day	Evacuation at daytime, upright		$3.9 \cdot 10^{-5}$
	Night	Evacuation at night, upright		$1.9 \cdot 10^{-4}$
Precautionary evacuation	Day	Precautionary evacuation at daytime		
	Night	Precautionary evacuation at night		

Vanem and Skjong (2006).

Table 16.19 Consequence associated with scenario: Evacuation at daytime, listing

Time (min)	<5	5–10	10–15	15–30	30–60	60–90	>90
Probability	0.18	0.19	0.20	0.18	0.13	0.07	0.05
Fatalities	530	420	290	90	0	0	0
Risk contribution	95	80	58	16	0	Total: 249	

Vanem and Skjong (2006).

16.9 POLLUTION EMERGENCY PLANNING

16.9.1 MARPOL

MARPOL is focusing on pollution and specifies requirements to ship emergency plans for various types of pollution from ships. The requirements are thus not general for any accident or incident, but specific for pollution only. MARPOL Annex 1, Chapter 5, Regulation 37 specifies requirements for a shipboard oil pollution emergency plan (IMO, 2017). This applies to all tankers above 150 gross tons and to other ships above 400 gross tons.

It is also stated in the regulation that the plan shall be developed in accordance with guidance given in the guidelines for the development of shipboard oil pollution emergency plans (IMO, 1992). The guidelines specify in greater detail what the plan should address.

The guidelines cover both operational and accidental spills. The primary purpose is to initiate actions to stop or minimize a spill and to mitigate the effects of the spill. The guidelines underline that emergency plans should be as simple as possible to make them easy to use, at the same time as being fit for the purpose.

Part 2 elaborates on the four mandatory requirements in MARPOL. These are the minimum that the emergency plan should contain:

- Procedure for reporting oil spills
- List of authorities/persons to contact
- Actions to be taken
- Procedures/contact point for coordination with authorities in combating pollution

Some countries (Coastal states) define additional measures to be taken against marine pollution (ICS, 2002). All ships operating in the territorial waters of these countries must adhere to these requirements. For instance, the USA requires the following additional measures:

- The ship must identify and ensure, through contract or other approved means, the availability of private fire-fighting, salvage, lightering and clean-up resources
- A qualified individual with full authority to implement the response plan, including the activation and funding of contracted clean-up resources, must be identified
- Training and drill procedures shall be described

Part 3 describes additional, non-mandatory provisions. It is suggested that plans and diagrams of the ship, types and quantities of response equipment, contact details for shoreside spill response coordinator, public affairs, and record-keeping may be included. There are also provisions for review of the plan and for exercises.

Appendix 2 of the guidelines contains an example of an oil pollution emergency plan. The suggested sections largely follow the mandatory requirements as specified in MARPOL:

1. Preamble
2. Reporting requirements
   2.1. When to report
   2.2. Information required
   2.3. Who to contact
3. Steps to control discharge
   3.1. Operational spills
   3.2. Spills resulting from casualties
4. National and local coordination
5. Additional information (non-mandatory)

APPENDICES

ITOPF also provides information about oil spill contingency planning in a technical information paper on Contingency planning for marine oil spills (ITOPF, 2016).

IMO (2001b) and the International Chamber of Shipping (ICS, 2002) have published guidelines that may assist companies in setting up a

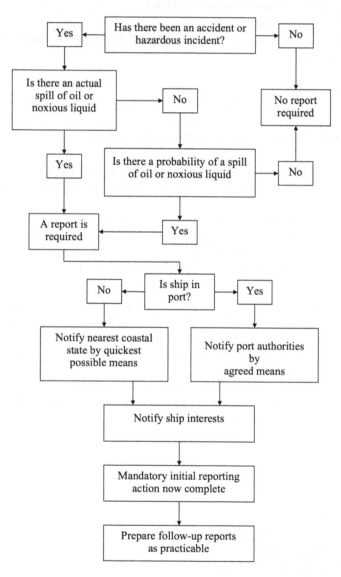

Figure 16.16 Decision-making process for the reporting of a polluting discharge (ICS, 2002).

SOPEP (Ship Oil Pollution Emergency Plan) or SMPEP (Shipboard Marine Pollution Emergency Plan). The ICS guideline gives flowcharts that support the decision-making process during an emergency. Figure 16.16 shows the guideline for the reporting of a polluting discharge.

Another important step in the pollution emergency plan is to mobilize the vessel's pollution prevention team, which involves all key officers onboard the vessel. The emergency plan shall have detailed job descriptions for each team member. For a specific spill scenario, the pollution emergency plan, which gives a description of measures to be taken, shall be given in both plain text and as a checklist. The plan is to be categorized according to the source of spill and the causes. Examples of an emergency plan from the ICS guideline are given in Tables 16.20 and 16.21 in plain text and as a checklist, respectively.

Maritime pollution accidents may directly involve a number of parties:

- Master and crew
- Shipping company
- Salvage vessel
- Port administration
- Fire-fighting brigade
- Pollution prevention agency
- Etc.

Table 16.20 Pollution emergency plan for the event of tank overflow during loading or bunkering – Measures to be taken

3.1.2 Tank Overflow During Loading or Bunkering

Measures to be implemented immediately:
- Stop all cargo and bunkering operations, and close manifold valves
- Sound the emergency alarm, and initiate emergency response procedures
- Inform terminal/loading master/bunkering personnel about the incident

Further measures:
- Consider whether to stop air intake into accommodation and non-essential air intake to engine room
- In the case of a noxious liquid substance, consider what protection from vapor or liquid contact is necessary for the response team and for other crew members
- Consider mitigating activities such as decontamination of personnel who have been exposed
- Reduce the tank level by dropping cargo or bunkers into an empty or slack tank
- Prepare pumps for transfer of cargo/bunkers to shore if necessary
- Begin clean-up procedures
- Prepare portable pumps if it is possible to transfer the spilled liquid into a slack or empty tank

If the spilled liquid is contained on board and can be handled by the pollution prevention team, then:
- Use absorbents and permissible solvents to clean up the liquid spilled on board.
- Ensure that any residues collected, and any contaminated absorbent materials used in the clean-up operation, are stored carefully prior to disposal.

ICS (2002).

Table 16.21 Checklist for response to operational spill of oil or a noxious liquid substance

This checklist is intended for response guidance when dealing with a spill of oil or a noxious liquid substance during cargo or bunkering operations. Responsibility for action to deal with other emergencies which result from the liquid spill will be as laid down in existing plans, such as the Emergency Muster List.

Actions to be considered (Person responsible)	Action taken?	
	YES	NO
Immediate Action		
• Sound emergency alarm (Person discovering incident)	☐	☐
• Initiate ship's emergency response procedure (Officer on duty)	☐	☐
Initial Response		
• Stop all cargo and bunkering operations (Officer on duty)	☐	☐
• Close manifold valves (Officer on duty)	☐	☐
• Stop air intake to accommodation (Officer on duty)	☐	☐
• Stop non-essential air intake to machinery spaces (Engineer on duty)	☐	☐
• Locate source of leakage (Officer on duty)	☐	☐
• Close all tank valves and pipeline master valves (Officer on duty)	☐	☐
• Commence clean-up procedures using absorbents and permitted solvents (Chief Officer)	☐	☐
• Comply with reporting procedures (Responsible: Master)	☐	☐
Secondary Response		
• Assess fire risk from release of flammable liquids or vapor (Chief Officer)	☐	☐
• Reduce liquid level in relevant tank by dropping into an empty or slack tank (Chief Officer)	☐	☐
• Reduce liquid levels in tanks in suspect area (Chief Officer)	☐	☐
• Drain affected pipeline to empty or slack tank (Chief Officer)	☐	☐
• Reduce inert gas pressure to zero (Chief Engineer)	☐	☐
• If leakage is at pump room sea-valve, relieve pipeline pressure (Chief Officer)	☐	☐
• Prepare pumps for transfer of liquid to other tanks or to shore or to lighter (Chief Engineer)	☐	☐
• Prepare portable pumps for transfer of spilt liquid to empty tank (Chief Engineer)	☐	☐
Further Response		
• Consider mitigating activities to reduce effect of spilt liquid (Master)	☐	☐
• Pump water into leaking tank to create water cushion under oil or light chemical to prevent further loss (Chief Officer)	☐	☐
• If leakage is below waterline, arrange divers to investigate (Master)	☐	☐
• Calculate stresses and stability, requesting shore assistance if necessary (Chief Officer)	☐	☐
• Transfer cargo or bunkers to alleviate high stresses (Chief Officer)	☐	☐
• Designate stowage for residues from clean-up prior to disposal (Officer on duty)	☐	☐

ICS (2002).

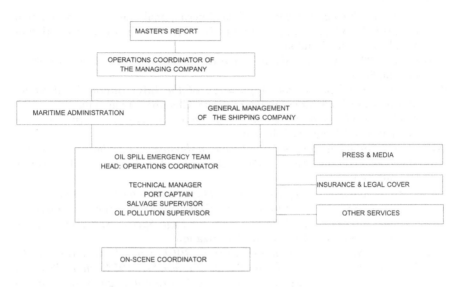

Figure 16.17 Oil spill response organization.

In addition, there are always several concerned parties in the case of maritime pollution accidents, such as the shipowner, cargo owner, local community (e.g., local fishermen), and government. Because of the interests involved, maritime pollution accidents often raise both political and legal issues with respect to overall management. Figure 16.17 indicates how an oil spill accident may be organized to be both effective and well-coordinated. The on-scene coordinator should have the necessary authority to direct the vessel and manage the available resources relating to salvage, oil spill containment, and clean-up. To coordinate the mobilization of various resources, an oil spill emergency team should support the on-scene coordinator with representatives from the involved parties.

REFERENCES

Berlonghi, A. E. (1996). Understanding and planning for different spectator crowds. *Journal of Safety Research, 2*(27), 134.

Cowley, J. (1994). Fire casualties and their regulatory repercussions, *Proceedings IMAS 94 Fire Safety on Ships*. Institute of Marine Engineers, London. ISBN 0-907206-57-3.

DNV. (2001). *Formal safety assessment of life saving appliances for bulk carriers (FSA/LSA/BC)*. http://research.dnv.com/skj/FsaLsaBc/FsaLsaBc.htm

Galea, E., Filippidis, L., Gwynne, S., Lawrence, P., Sharp, G., Blackshields, D., & Glen, I. (2002). *The development of an advanced ship evacuation simulation software product and associated large scale testing facility for the collection of human shipboard behaviour data*. International Conference on Human Factors in Ship Design and Operation. 2–3 October. The Royal Institution of Naval Architects, London.

Hooke, N. (1989). *Modern Shipping Disasters, 1963–87*. London: Lloyd's of London Press.

ICS. (1994). *Guidelines on the Application of the IMO International Safety Management Code*. London: International Chamber of Shipping.

ICS. (2002). *Model shipboard marine pollution emergency plan*. London: International Chamber of Shipping.

IMO. (1992). *Guidelines for the development of shipboard oil pollution emergency plans (amended by MEPC.86(44), 2000)*. London: International Maritime Organization.

IMO. (1995). *Guidelines on Implementation of the International Safety Management (ISM) Code by Administrations*. London: International Maritime Organization.

IMO. (2000). *International Code of Safety for High-Speed Craft (HSC Code)*. London: International Maritime Organization.

IMO. (2001a). *SOLAS: consolidated edition, 2001: Consolidated text of the International Convention for the Safety of Life at Sea, 1974, and its Protocol of 1988: articles, annexes and certificates: Incorporating all amendments in effect from 1 January 2001*. London: International Maritime Organization.

IMO. (2001b). *Guidelines for the Development of Shipboard Marine Pollution Emergency Plans*.

IMO (2001c). *SOLAS Amendments 2000*. London: International Maritime Organization, ISBN 92-801-5110-X.

IMO. (2002a). *STCW - International Convention on Standards of Training, Certification and Watchkeeping for Seafarers*. London: International Maritime Organization.

IMO. (2002b). *MSC circular 1033*. London: International Maritime Organization.

IMO. (2017). *International Convention for the Prevention of Pollution from Ships (MARPOL)* (6th edition). London: International Maritime Organization.

Jørgensen, H. D., & May, M. (2002). Human factors management of passenger ship evacuation. In *International conference on human factors in ship design and operation*. 2-3 October. The Royal Institution of Naval Architects, London.

Laur, U., Lehtola, K., & Eksborh, A. L. (1997). *Final Report on the capsizing on 28 september 1994 in the Baltic Sea of the ro-ro passenger vessel MV Estonia*. Final Report, ISBN: 951-53-1611-1.

NOU. (2000). *Hurtigbåten MS Sleipners forlis 26. november 1999* [The High-Speed Craft MS Sleipner Disaster–26 November 1999]. Statens forvaltningstjeneste Informasjonsforvaltning. https://www.regjeringen.no/contentassets/bbd5ba0 4f83a4d7c9c07793062a693d2/no/pdfa/nou200020000031000dddpdfa.pdf

Pyman, M. A. F., & Lyon, P. R. (1985). Casualty rates in abandoning ships at sea. *The Royal Institution of Naval Architects*, Joint Evening Meeting, London, January 17. ISSN: 0035-8967.

Rensvik, E., & Kristiansen, S. (1994). Development of an integrated fire-fighting system. In *IMAS 94, Fire Safety on Ships - Developments into the 21st Century*. The Institute of Marine Engineers.

Reisser-Weston, E. (1996). Simulating human behaviour in emergency evacuations. In *International Conference: Escape, evacuation & rescue – design for the future*. The Royal Institution of Naval Architects.

Spouge, J. (1996). Escape from partly capsized ferries. In *International Conference: Escape, evacuation & rescue – design for the future*. The Royal Institution of Naval Architects.

Sund, A. (1985). *Ulykker, katastrofer og stress In Norwegian*. [Accidents, catastrophes and stress]. Oslo: Gyldendal Norsk Forlag.

Vanem, E., & Skjong, R. (2006). Designing for safety in passenger ships utilizing advanced evacuation analysis – A risk based approach. *Safety Science, 44,* 111–135.

Chapter 17

Risk-based design

17.1 INTRODUCTION

Risk-based design is a relatively new term in the maritime community and did not really come to the forefront in the industry until the SAFEDOR project and the outcomes from this project were available (Papanikolaou, 2009).

Different terms are being used in different industries and applications to describe this concept. Initially when it was introduced, the term risk-based was commonly applied. However, in recent years, it has become more common to talk about risk-informed rather than risk-based. The reason for this is that risk is by no means the only parameter that is considered when we are designing something, first and foremost we will design a system to do the function we want it to do effectively and correctly. Risk will inform this process and can place restrictions on the design, but it is just one of many factors that must be considered. When searching for more literature on this topic, using the term risk-informed (with or without a hyphen) will therefore give more results than just searching for risk-based. However, in the maritime community, risk-based still seems to be the most commonly applied term and we will therefore use this term in this chapter.

Most of this book is in fact about how we make decisions based on risk. Among the applications that already have been discussed are:

- Safety management, as implemented in the ISM code. The basis for managing safety (in design and operation) should be good knowledge of risk, acquired through the use of risk assessment.
- Risk-based development of rules and regulations, as described by IMOs FSA regime. Development and improvement in rules and regulations is based on comprehensive risk analysis (Formal Safety Assessment).
- Risk-based inspection, e.g., as described by Paris MoU, where a risk model forms the basis for prioritizing which ships to inspect and how often. Many flag states have also implemented risk-based inspection when prioritizing their own inspections.

DOI: 10.4324/9781003055464-17

The principle of risk-based design is not any different from any of these applications. Risk assessment and thereby a good understanding of what may go wrong forms the basis, and the design is developed with information from this as additional input and constraints to the design process. An important difference from the traditional way of thinking is that risk is not seen as a constraint on the design, but rather that safety is an objective, along with other objectives like cost-efficiency and reliability. The implication of this is that safety still should meet minimum requirements, but that it also should be optimized beyond that.

Based on this, two motivations for using risk-based design can be identified:

- To obtain approval of designs that challenge rules (that may be outdated), by proving that they are equally safe as a design that complies with the rules.
- To optimize the level of safety on vessels that comply with the rules or to optimize the earning potential at the same time as maintaining the same level of safety.

Why is this important? Prescriptive regulations work well in situations where technology is stable and well tested and tried and where applications and operations vary to a limited degree. However, with more variation and faster technological development, it is harder for prescriptive regulations to keep up with the development. This has been seen with the development of larger and more innovative cruise ships and at the time of writing, the same problem has arisen with autonomous ships. Current regulations are not adapted to this new technology and both IMO and flag state authorities are struggling with the approval process for autonomous ships. Updating regulations is normally a slow process and is not always able to keep up with technological and operational development. Risk-based approaches can then supplement or replace prescriptive regulations.

Another aspect is that expectations from society in general and from various stakeholders are increasing. Serious accidents get more and more attention in media and causes more public reactions than ever before. To an increasing degree, society expects an "accident free" world and this also applies to maritime transport. This will often mean increased cost and the best way to optimize different objectives, where safety is one, is the introduction of risk-based design. In this way the use of human and economic resources are optimized and at the same time meeting a number of (competing) objectives.

On the other hand, rules are simple to apply and easy to check and it is unlikely to be practical to move to a situation where there are no prescriptive rules. The Norwegian oil and gas industry has for a long time had a regulatory framework that is largely functional and assumes extensive use of risk analysis. However, there are also prescriptive requirements that are part of the framework. Further, the regulations are supplemented with a number of (prescriptive) standards that can be used as a basis for design and operation. The greatest advantage of prescriptive rules is that the design and verification process is comparatively easy. Risk analysis as a tool is not nearly as well developed and requires different competence and expertise.

This chapter will first look at some relevant IMO documents and regulations that open up for use of risk assessment as a basis for approval of alternative designs. In particular, MSC 1455 (IMO, 2013) is important. Secondly, a general process for risk-based design will be described, based on the SAFEDOR project. Finally, a brief introduction to some specific applications where risk assessment is used to support design is provided, namely risk-based damage stability and the safe return to port rule.

17.2 IMO REGULATIONS

The first attempts at developing risk-based standards in the maritime industry were related to damage stability in the early sixties. Methods for probabilistic damage stability were developed, where the probability of what extent of damage a ship could experience determined what the ship should be designed to tolerate. However, this was not introduced in SOLAS until a decade later, in SOLAS 74, when probabilistic damage stability was given as an alternative to the traditional deterministic damage stability requirements. It was then stipulated that alternative designs could be accepted, as long as the risk was the same as, or lower than, the deterministic requirements would give.

Later, similar options have also been given for other design requirements. SOLAS II.2/17 Alternative Design and Arrangement for Fire Safety (IMO, 2002a) also allows alternative designs to be approved. This development was triggered by new cruise ship designs (specifically "Sovereign of the Seas") that had large atriums extending over several decks and that could not meet existing regulations for fire zones.

The regulations refer to MSC/Circ. 1002 (IMO, 2001). Some of the requirements from this guideline are as follows:

- All relevant fire hazards should be identified, a range of hazards should be selected that are representative and cover all relevant hazards and scenarios should be developed for all the selected hazards.
- This is followed by quantitative analysis, where the fire risk associated with the alternative design and a design complying with all prescriptive rules are compared.
- The results from the quantitative analysis should be compared with a set of performance criteria, that can be related to life, environment, or the ship/equipment.

MARPOL Regulation 19 (IMO, 2002b) describes requirements for double hull/bottom for oil tankers. However, the same paragraph also states that other designs may be approved, as long as they result in a risk level that is at least as low as the prescribed design.

> Other methods of design and construction of oil tankers may also be accepted as alternatives to the requirements prescribed in paragraph 3 of this regulation, provided that such methods ensure at least the same level of protection against oil pollution in the event of collision or stranding and are approved in principle by the Marine Environment Protection Committee based on guidelines developed by the Organization.

In 2013, IMO published guidelines for how the process of approving alternative designs should be performed (IMO, 2013). These guidelines are intended to serve as general guides for approval of all kinds of alternatives and equivalents, where prescriptive regulations are not met, but the ship-owner seeks approval for deviations. The process is described in more detail in the following section.

17.3 APPROACH TO RISK-BASED DESIGN

Ship design models were initially simple, synthesis models (Nowacki, 2019) where the optimization was based on a single parameter, usually the economic performance of the design. All other factors, including safety, were only requirements that had to be met. These developed into multi-objective design models, where several objectives, often measured in different ways were taken into account in the design. Next were holistic design models, where optimization with respect to all objectives is done simultaneously.

Risk-based design is in practice a multi-objective design method, where safety (or risk) is one of the objectives. Also in risk-based design, there will be a large number of constraints that have to be met. A multitude of specific requirements for safety equipment will typically form a large group of the constraints that have to be met, although developments in regulations increasingly allow for alternative solutions where these constraints are removed, as long as the risk level is equal to what would be achieved following the specific requirements.

Traditional ship design:

- Build a ship that is optimized with respect to functionality, operations, and cost at the same time as meeting the requirements of the regulations. The regulations act as constraints on the optimization process.

Risk-based ship design:

- Build a ship that is optimized with respect to functionality, operations, cost, and safety, at the same time as meeting any specific requirements and keeping within an upper acceptable limit for risk.

Rules are meant to maintain a minimum safety level, but this is for a given design, operation, and environment, and any changes or deviations will mean that the safety level deviates from this, in a positive or negative direction. Risk-based design will to a large degree ensure conformity in risk levels, also for novel concepts and "unusual" solutions.

Rule-based design preserves status quo and may prevent development of cheaper, safer, and more efficient solutions, simply because the rules did not consider the possibility when they were implemented. Further, it does not in itself contribute to the development of safer solutions. The link between risk and design may be lost in rule-based design as we do not necessarily appreciate what function a certain rule has in preserving present safety level. Without this knowledge, we run the risk of approving deviations without fully understanding the effects. Given rules are not necessarily the most cost-effective way of achieving a safe design.

Figure 17.1 illustrates a design process where safety is included as an objective. For clarity, only design objectives that relate to safety are shown. In practice, these should however be integrated with other objectives.

At the top of the figure, the context is shown. This includes a variety of aspects from a range of stakeholders and the figure is not complete in showing everything that may be relevant. Important however is the expectations from society with regard to performance of ships.

Without elaborating in detail, society will expect safe transport, but increasingly also focus on environmentally friendly transport. Further, society also look for cheap transport. We can already at this stage see very different objectives that may compete. Similarly, the company that is planning to build the ship also has its own values, preferences, and expectations. The commercial context within which the design takes place is also important to mention.

All of this leads into a specific set of performance expectations and safety goals. Safety goals can also be regarded as a performance expectation, although it is shown separately. Effectively, it will however feed into the complete set of expectations. This sets the framework for the design process, and this will now continue with establishment of requirements and constraints. For risk-based design to be feasible as a method for ship design, there are a set of principles or preconditions that have to be in place.

First of all, we must be able to measure "safety" in a consistent and complete manner. In practice, what we have ended up measuring is risk, i.e., we seek to minimize rather than maximize what we measure. In IMOs FSA guidelines (IMO, 2018), FN curves and individual risk are mentioned as possible risk measures. We have also looked at a number of other risk measures in this book that can be used (Chapter 9). A problem with all of these measures is that they do not necessarily measure precisely the same thing and they also have different properties and behave differently under changing conditions. The perspective that we have on risk (either risk to society as a whole or the risk to individuals working on the ship that we

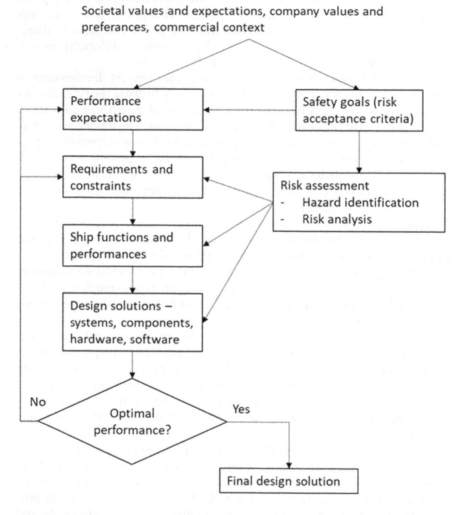

Societal values and expectations, company values and preferances, commercial context

Performance expectations

Safety goals (risk acceptance criteria)

Requirements and constraints

Risk assessment
- Hazard identification
- Risk analysis

Ship functions and performances

Design solutions – systems, components, hardware, software

No

Optimal performance?

Yes

Final design solution

Figure 17.1 A framework for risk-based design (adapted from Papanikolaou, 2009).

want to optimize) will also influence what parameter to choose and thereby influence the outcome of the design process.

Experience from other industries has also shown that the common measures of fatality risk (e.g., FN-curve, individual risk, and PLL) not necessarily are well suited in a design optimization process. In the offshore oil and gas industry in Norway, risk in design is among others measured in terms of loss of key safety functions. One such safety function may be "Evacuation means" and the risk is measured in terms of the frequency of loss of this function. Clearly, loss of the ability to evacuate has an impact on the safety of the people on board and this is therefore an indirect measure of fatality risk. At the same time, it can be regarded as a direct measure of the "quality" of the evacuation systems, and it is easy to measure improvements if we are able to reduce the frequency of loss.

Secondly, we need to have an agreed procedure for quantifying the measure, i.e., a common view on what risk assessment methods to use and how to be applied. An important part of this is also what data to use and how to use them. Again, the FSA guidelines (IMO, 2018) give some guidance on this, although it also leaves very large room for interpretation and differences in implementation.

Experience from other industries where QRA has been used for much longer and where data and methods have been established has shown that different analysts looking at the same system can arrive at very different results when doing a QRA. The offshore oil and gas industry in the North Sea has developed methods and common databases over several decades, but there may still be significant variation in the details of how studies are performed.

This is clearly a major challenge and in practice this is hard to overcome without using the methods over many years and developing a common, good practice. This may seem like a lot of work and perhaps with a questionable benefit, but experience from the oil and gas industry has also shown us that the practice of using risk-based design has significantly reduced the risk levels in the industry and has clearly led to the development of much safer design solutions. The principle of risk-based design therefore seems to work well in terms of the outcome of the design process, even if the methods and data may be questioned.

A third key principle is that the risk assessment process should be an integrated part of the design process. This is the same principle as was mentioned in connection with safety management in general, i.e., that the safety management process must be an integration of the general management process.

17.4 APPROVAL PROCESS ACCORDING TO MSC 1455

In 2013, the Maritime Safety Committee of IMO published "Guidelines for the approval of alternatives and equivalents as provided for in various IMO instruments" (IMO, 2013). The purpose of this document is to provide "a consistent process for the coordination, review and approval of alternatives and equivalents with regard to ship and system design".

The guidelines describe two alternative processes that can be used. One alternative is to compare with existing designs and "to demonstrate that the design has an equivalent level of safety". This is suggested by establishing functional requirements and performance criteria for essential ship functions for the existing design. Subsequently, it should be shown that the alternative design meets the same requirements and criteria. This does not require that an explicit risk analysis has to be performed.

The other alternative is to carry out a risk analysis for the alternative design and compare with risk evaluation criteria. This will of course require that such criteria are established and that they also reflect the risk level for existing ships.

Figure 17.2 illustrates the process to be followed to get approval of an alternative design, where the responsibilities of the submitter is shown on

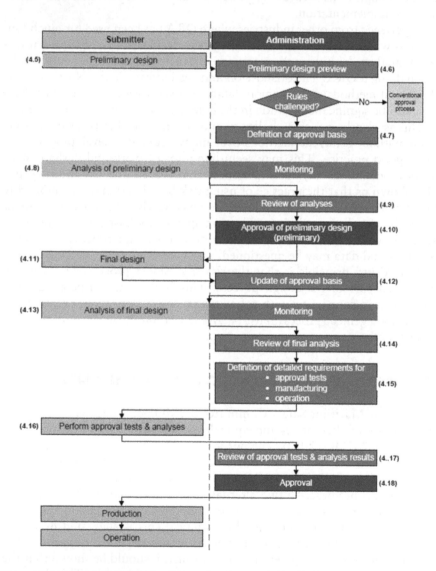

Figure 17.2 Approval process for alternative designs (IMO, 2013).

the left-hand side of the figure and the maritime administration on the right-hand side. The process can be divided into five main phases:

- Preliminary design
- Final design
- Perform approval tests & analyses
- Production
- Operation

Risk-based design is mainly related to the first two phases, although the results from these two phases obviously have impact on the final phases too.

There are detailed descriptions of the documentation to be submitted as part of the different phases and some of the key ones are mentioned here (documents that are part of a general approval process are not mentioned):

- Description of the alternative and/or equivalency design, including design basis. The purpose is to ensure a good understanding of how the regulations are being challenged by the proposed design.
- Risk assessment plans, describing which risk assessments are planned and performed, their scope, and timing.
- Risk analysis report(s) covering:
  - Identified hazards
  - Safeguards (barriers) included in the design
  - Risk evaluation criteria that have been applied (MSC 1455 contains examples of criteria that can be applied)
  - Description of analysis methods and applied and any workshops that have been conducted to identify/analyze risk
  - Frequencies and consequences of identified hazards
  - Risk models
  - References to data sources, use of expert judgment, description of key assumptions, uncertainty, and sensitivity analysis
  - Cost-benefit assessment
  - Risk reducing measures (barriers) that have been decided implemented
  - Design casualty scenarios
  - Information on the competence and experience of the experts that have participated in the analysis

It is noted that this largely is a description of the elements to be found in any (quantitative) risk assessment that is performed.

MSC 1455 (IMO, 2013) provides more specific details on individual documents and their content, but this is not repeated here. The information is largely collected in Chapter 6 of MSC 1455.

17.5 PROBABILISTIC DAMAGE STABILITY

The first regulations that allowed the use of probabilistic methods in design were related to probabilistic damage stability. The development started in the late 1960s but came first into force with Resolution A.265 (VIII) of SOLAS-74. The regulations were given as an alternative approach to the prescriptive regulations method.

The regulations have since been revised and the final proposal was adopted in 2005, which entered into force in 2009. IMO even stated that prescriptive regulations for damage stability had no future after this (Papanikolaou & Eliopoulou, 2008; Vassalos, 2014).

The principle of the regulations is that one has to show that the proposed design is at least equivalent to a design based on prescriptive regulations. This is not expressed explicitly through the risk level, but through a "Subdivision index", where the "Attained subdivision index" must be greater than the "Required subdivision index". The subdivision index can be interpreted as an expression of the probability of surviving a collision that causes damage to the hull. The attained index is the value that is calculated for the alternative design while the required index is the value achieved when using the prescriptive regulations.

The calculation is fairly elaborate and will not be described in detail here. However, the calculation is based on determining the probability of flooding of individual compartments or compartment groups and the probability of survival after flooding of the compartment(s). The consequences to crew/passengers, cargo, ship, or the environment are not taken into account explicitly.

REFERENCES

IMO (2001). *Guidelines on alternative design and arrangements for fire safety*, MSC/Circ. 1002, 26 June 2001. London: International Maritime Organization.

IMO (2002a). *SOLAS II-2/17: Construction – fire protection, fire detection and fire extinction: Alternative design and arrangements*, July 2002. London: International Maritime Organization.

IMO (2002b). *MARPOL 73/78 consolidated edition 2002*. London: International Maritime Organization.

IMO (2013). *Guidelines for the approval of alternatives and equivalents as provided for in various IMO instruments*, MSC.1/Circ. 1455, 24 June 2013.

IMO. (2018). *Revised guidelines for Formal Safety Assessment (FSA) for use in the IMO rule-making process*. MSC-MEPC.2/Circ.12/Rev.2, 9 April 2018.

Nowacki, H. (2019). On the history of ship design for the life cycle. In A. D. Papanikolaou (Ed.), *A holistic approach to ship design* (pp. 43–73). Switzerland: Springer Nature.

Papanikolaou, A., & Eliopoulou, E. (2008). On the development of the new harmonised damage stability regulations for dry cargo and passenger ships. *Reliability*

Engineering and System Safety, 93(9), 1305–1316. DOI: 10.1016/j.ress.2007. 07.009

Papanikolaou, A. D. (Ed.) (2009). *Risk-based ship design – Methods, tools and applications,* Berlin Heidelberg: Springer Science & Business Media.

Vassalos, D. (2014). Damage stability and survivability – 'nailing' passenger ship safety problems. *Ships and Offshore Structures, 9*(3), 237–256.

Chapter 18

Monitoring risk level

18.1 INTRODUCTION

One of the elements of safety management is to monitor the risk level, aimed at detecting trends and developments. Mainly, this is done by recording and monitoring failures, errors, incidents, accidents, and losses. A hazard identification process or a risk analysis project will in many instances be initiated due to uninspected non-conformities or serious accidents not anticipated on the background of the existing risk picture of the operation in question. The safety level of an activity may however deteriorate without showing dramatic change of accident frequency or seriousness of consequences. And despite such changes the operation may still be in the ALARP region. Prudent safety management should detect such changes as early as possible and before they result in serious accidents.

Systematic recording of event frequencies and consequence parameters offers a source of information for proactive monitoring of changes in the risk of an operation. This chapter describes various tools and techniques for statistical analysis of this type of data.

18.2 MONITORING LOSS NUMBERS

18.2.1 Time series of grounding accidents

A key parameter for monitoring the risk level of sea transport will be the number of accidents per year. Figure 18.1 shows the number of groundings of merchant and fishing vessels in the Norwegian fleet in the period from 2006 to 2015. It can be noticed that the extreme minimum and maximum are 76 and 104 accidents per year, respectively.

Counting the number of accidents can be misleading because the number of ships may vary in the period. This should therefore be taken into account, by normalizing the number of accidents with respect to the number of ships at risk. This has been done in Figure 18.1. However, in this case we see that the impact of this is very limited. The average annual number of groundings is 93 and the standard deviation is 8.1.

DOI: 10.4324/9781003055464-18

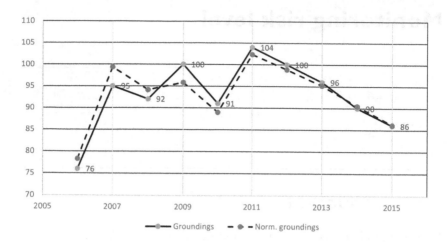

Figure 18.1 Annual number of groundings for Norwegian commercial vessels.

Source: The Norwegian Maritime Authority and Statistics Norway.

Assuming that the number of accidents follows a Poisson, distribution and setting the confidence interval (variation range) to ±2 times the standard deviation, we find that the number of accidents can vary between 77 and 109, purely due to random variation. With the exception of 2006, the number of accidents in all years fall within this interval. It may therefore be argued that the variation within the period is largely random and not indicating any change in safety level.

Another case can be to study the annual number of losses of the merchant fleet of Panama. This is reported regularly and Figure 18.2 shows a set of data for a 14-year period (2005–2018). The number of losses per year varied between 2 and 22 with a mean value of 13.5. The number of losses has in other words varied by a factor of 11 within this period. However, the figure also indicates a clear downward trend in annual losses, with a sharp drop from 2011 to 2012. The average loss number in the first seven years was 19.86 compared to 7.14 in the last period. This fact may be explained by improved safety management both within the maritime administration and the industry itself. With such a large difference, it seems misleading to characterize the period with the average number of 13.5 losses per year.

We can also use statistical methods in this case, to determine whether the reduction in number of losses is statistically significant or whether it may be due to random variation.

Given n observations drawn from a Normal distribution with unknown mean μ and unknown standard deviation σ, the range of uncertainty for the true mean is given by the Student-t distribution:

$$\mu = \pm t(n-1)\left(\frac{\hat{\sigma}}{\sqrt{n}}\right) + \bar{X} \tag{18.1}$$

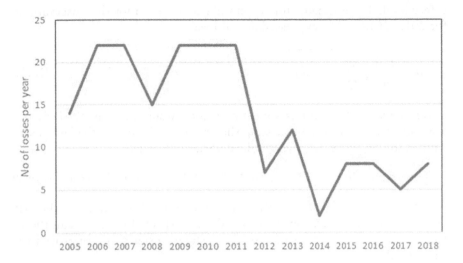

Figure 18.2 Number of losses per year (Panama).

or

$$\mu = \pm t(n-1)\left(\frac{s}{\sqrt{n-1}}\right) + \bar{X} \qquad (18.2)$$

where $t(n-1)$ is the *Student-t distribution* with $(n-1)$ degrees of freedom, s is the sample standard deviation, and $\hat{\sigma}$ is the unbiased single point estimate of the true standard deviation. The Student-t distribution is symmetric and unimodal about zero. The distribution is somewhat flatter than the Normal distribution. The first period had an average loss number of 19.86 and a sample standard deviation of 3.40 which gives the following estimate for the uncertainty of μ based on seven observations (7 years):

$$\mu = \pm t(7-1)\left(\frac{3.40}{\sqrt{7-1}}\right) + 19.86 = \pm 1.943 \cdot 1.39 + 19.86$$

by assuming a confidence interval of 90%, we get the minimum and maximum estimates for μ:

$$\mu_{min} = -2.70 + 19.86 = 17.16 \qquad \mu_{max} = +2.70 + 19.86 = 22.56$$

The 90% confidence interval for μ (0.05–0.95) is therefore:

$$\mu = 17.16 - 22.56 \text{(losses/year)}$$

Figure 18.2 shows that the loss number decreased dramatically from 2012 and one may question whether this represented a true improvement of the safety level. The estimator for the standard deviation for the observations for 2012–2018 can be expressed as follows:

$$\hat{\sigma} = \sqrt{\frac{1}{n-1} \sum_{i=1}^{n} (x_i - \bar{x})^2}. \tag{18.3}$$

Given the $n=7$ observations we get $\hat{\sigma} = 3.08$. Assuming conservatively the low estimate of μ (17.16) and setting the area of variation for the last observations to two times the true standard variation we get:

$$[\mu - 2\sigma; \mu + 2\sigma] = [17.16 - 6.16; 17.16 + 6.16] = 11.00 - 23.32 \text{(losses/year)}$$

Apart from the 12 losses in 2013, the remaining observations in the last period are all distinctly lower or outside the expected range. It can therefore be concluded that the safety level has shown a significant improvement since 2012.

An alternative approach is simply to compare the area of uncertainty for the mean loss number for the two periods and check the degree of overlap. For the last period the mean loss number was 7.14 and the sample standard deviation 3.08. The range of uncertainty for the true mean is given by:

$$\mu = \pm t \, (7-1) \left(\frac{3.08}{\sqrt{7-1}} \right) + 7.14 = \pm 1.943 \cdot 1.26 + 7.14 = [4.69 - 9.59]$$

It can be concluded that lowest estimate for the first period (17.16) is considerably greater than the highest estimate for last period (9.59). This is an even stronger indication that the safety level is significantly improved in the last 7 years compared to the first period.

This shows that the uncertainty related to the mean loss number is considerable and that one should be cautious about drawing any conclusion about changes in risk level from simple observations of loss numbers.

18.3 ANALYSIS OF TIME SERIES

In certain cases, it may be acceptable to identify a steady trend in the data material simply by inspection of the graphical presentation. This is the situation for annual total losses for the world merchant fleet through the period from 2005 to 2018 as shown in Figure 18.3. The diagram shows that the losses have gone down from around 100 to 40 per year, or a reduction of 60%. It is noted however that there was a steep rise from 2008 to 2010.

Given the fact that there is a clear reduction in the loss numbers, we may apply regression analysis to make a simple model for this trend. In the

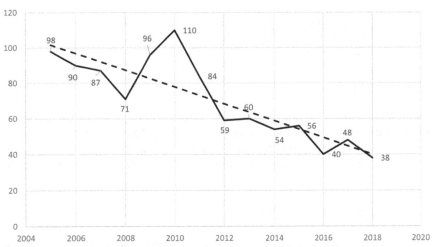

Figure 18.3 Number of total losses for the world fleet (500 gt and above).

Source: Shipping Statistics Yearbook, ISL.

current case, linear regression may be feasible. A linear regression model expresses the stochastic variable Y as a function of X (year):

$$Y = \beta_0 + \beta_1 \cdot X + \varepsilon \tag{18.4}$$

where

β_0=intercept parameter
β_1=slope parameter
ε=random error

The random error represents the difference between the true value of Y and the value given by the regression model. The basis for estimation of the model parameters is the following expression:

$$\hat{Y} = b_0 + b_1 \cdot X \tag{18.5}$$

The model assumes a linear relation between X and $\hat{Y}$. The parameters b_0 and b_1 are estimated by the least squares method which minimizes the sum of squares of difference (SSD) of the estimated value of Y and the measured value, or the residual SSD:

$$SSD = \Sigma \left(\hat{Y} - Y \right)^2 \tag{18.6}$$

It can be shown that the parameters are given by:

$$b_1 = \frac{\Sigma\left(X_m - \bar{X}\right)\left(Y_m - \bar{Y}\right)}{\Sigma\left(X_m - \bar{X}\right)^2} \tag{18.7}$$

$$b_0 = \bar{Y} - b_1 \cdot \bar{X} \tag{18.8}$$

A simpler approach for computing these parameters is to apply the *LINEST* function in MS Excel to find values for b_0 and b_1 that minimize the expression for SSD. By applying the least squares method, the following linear model is estimated for the annual number of losses (N_L) as a function of the year (Year):

$$N_L = 102 + 4.74 \cdot (2005 - \text{Year})$$

One question that is relevant to ask is how well the actual loss number is explained by this model (the trend line). One measure often used for linear models, is the Coefficient of determination, expressing the ratio between SSD for the regression to the total SSD:

$$r^2 = \frac{\Sigma\left(\hat{Y} - \bar{Y}\right)}{\Sigma\left(Y - \bar{Y}\right)} = \frac{\text{SSD(regression)}}{\text{SSD(total)}} \tag{18.9}$$

The computations are shown in Table 18.1 and give the following results for the linear model:

$$r^2 = \frac{5111}{7018} = 0.72$$

This means that 72% of the total variation is explained by the linear model.

18.4 MARITIME DISASTERS WITH MANY FATALITIES

Even in the twenty-first century, maritime losses with high number of human losses have taken place. According to Wikipedia, 74 accidents with human fatalities were reported in the period 2001–2020. Although the source lacks a more precise definition of a disaster, it lists cases with losses from 4 up to 1,864 fatalities. The median number of fatal accidents per year was 4 but with minimum 0 and maximum 11. A considerable part of the accidents was overloaded ferries and vessels with refugees. Figure 18.4 shows that there is strong variation in the number of disasters from one year to the next.

Table 18.1 World ship losses. Computation of squared sum of deviations (SSD)

Year	Observed	Estimate	SSD(total)	SSD(regr)	SSD(res)
2005	98	102	739.8	948.6	13
2006	90	97	368.6	679.1	47
2007	87	92	262.4	454.5	26
2008	71	87	0.0	274/9	268
2009	96	83	635.0	140.2	178
2010	110	78	1536.6	50.4	1030
2011	84	73	174.2	5.6	118
2012	59	68	139.2	5.7	89
2013	60	64	116.6	50.7	14
2014	54	59	282.2	140.7	24
2015	56	54	219.0	275.6	3
2016	40	49	948.6	455.4	89
2017	48	45	519.8	680.2	11
2018	38	40	1075.8	949/9	4
Mean	70.8	SSD	7018.4	5111.4	1915
St. dev.	22.4	St. dev.	22.4	19.1	11.7

No. of fatal accidents

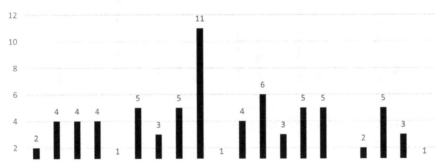

Figure 18.4 Fatal maritime accidents in 2001–2020.

Source: Maritime disasters, Wikipedia.

The CDF (cumulative distribution function) for the annual number of disasters is shown in Figure 18.5 and indicates that the probability of having more than seven events per year is 5%.

Just as the number of accidents varied considerably from one year to another, the accumulated number of fatalities per year showed an even greater variation as shown in Figure 18.6. The contrast is considerable having no fatalities in 2016 to 2401 fatalities in 2002. Although observing the low numbers from 2016, it seems speculative to indicate any

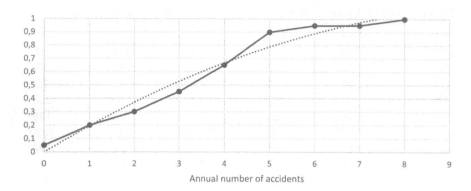

Figure 18.5 Cumulative density function (CDF) for the annual number of disasters.
Source: Wikipedia.

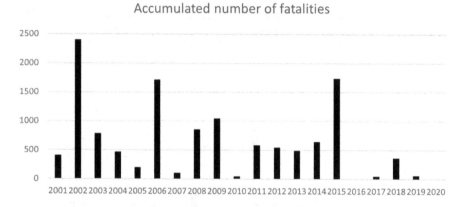

Figure 18.6 Total number of fatalities per year.
Source: Wikipedia.

downward trend noting the fact that the numbers were low also in 2007 and 2010.

The average number of fatalities per accident was 169 and varying from 20 to 1,864. A simple, empirical plot of the loss number is shown in Figure 18.7. Rather than being a true CDF, the plot is based on a numerical approach giving an ordered sequence of the values. The distribution indicates a probability of 5% of experiencing an accident with more than 690 fatalities. Despite the fact that both the annual number of accidents and the number of fatalities vary dramatically annually, it is still possible to establish pseudo-type statistical distribution for these parameters.

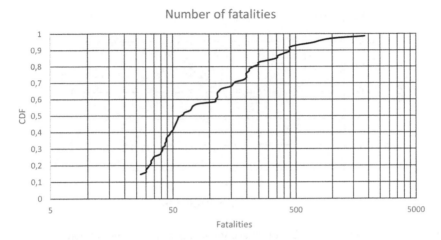

Figure 18.7 Cumulative distribution of fatalities for an accident.

Source: Wikipedia.

Table 18.2 Economic loss of maritime accident

Range of X (loss in 1000 NOK)	Point value X_i	Observed N_i	Accum. AN_i	CDF
1–100	20	48	48	0.324
100–200	120	35	83	0.561
200–500	300	24	107	0.723
500–1,000	600	16	123	0.831
1,000–2,000	1,200	10	133	0.899
2,000–5,000	3,000	7	140	0.946
5,000–10,000	6,000	4	144	0.973
10,000–20,000	12,000	2	146	0.986
20,000–50,000	30,000	1	147	0.993
	ΣN	147		
	$\Sigma N+1$	148		

18.5 FITTING A NON-PARAMETRIC DISTRIBUTION

Rather than estimating the parameters of a theoretical distribution, one may generate an empirical distribution directly based on the observed data. Let us take data for the economic loss as a result of selected ship accidents as a case to demonstrate the approach (Table 18.2). A non-parametric or empirical distribution is established as follows:

1. Select ranges for the loss variable (column 1).
2. Estimate average point value for each range (column 2).

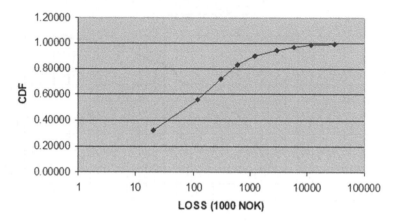

Figure 18.8 Cumulative distribution of economic loss of a maritime accident.

3. List the number of observations in each range N_i (column 3). The sum of observations is given below the column (ΣN).
4. Compute the accumulated number (AN) as follows:
   $AN_{i+1} = AN_i + N_{i+1}$
5. Compute the "artificial" CDF value in the following manner:
   $F(x) = AN_i / \Sigma N + 1$

The "trick" of adding 1 to ΣN is done so that CDF shall approach the value of 1.0 asymptotically.

The resulting CDF estimates are shown in the right-most column. The distribution is plotted in Figure 18.8 with a logarithmic scale for the abscissa.

18.6 THE LOGNORMAL DISTRIBUTION

18.6.1 Definitions

Certain consequence parameters, such as the number of lives lost or the size of an oil spill, seem to follow a skewed distribution. Stated simply this means that accidents with minor or limited consequences represent the majority of the total number of events. However, a small number of accidents lead to great or catastrophic consequences. The lifetime of a system or a component that degrades over time may be described by a lognormal distribution and the consequence of an accident also seems to be well described by this distribution.

A variable may in certain cases follow an exponential relationship like $x = \exp(w)$. A special case is when W has a normal distribution. Then the distribution of X is a lognormal distribution. It follows from the following transformation $\ln(X) = W$. The range of X is $(0,\infty)$. Let us assume that W is normally distributed with mean θ and variance w^2, and then the cumulative distribution function (CDF) for X is as follows:

$$F(x) = P[X \le x] = P[\exp(W) \le x] = P[W \le \ln(x)]. \qquad (18.10)$$

$$= P\left[Z \le \frac{\ln(x) - \theta}{w}\right] = \Phi\left[\frac{\ln(x) - \theta}{w}\right]$$

for $x > 0$ and Z is the standard normal variable. Notice that the distribution is defined only for positive values of X. The probability density function (PDF) is:

$$f(x) = \frac{1}{\sqrt{2\pi}} \frac{1}{x\,w} \exp\left[-\frac{(\ln(x) - \theta)^2}{2\,w^2}\right] \qquad (18.11)$$

where variance and mean are:

$$w^2 = \ln\left[1 + \left(\frac{\sigma}{\mu}\right)^2\right] \qquad (18.12)$$

$$\theta = \ln(\mu) - \frac{1}{2}\sigma^2 \qquad (18.13)$$

The mean and variance of the normal distributed W can be expressed as follows:

$$\mu = e^{\theta + \frac{w^2}{2}} \qquad (18.14)$$

$$\sigma^2 = \mu^2 \left[\exp w^2 - 1\right] \qquad (18.15)$$

Let us assume that the annual number of accidents shows an approximate normal distribution with mean value 6.5 and standard deviation 3.5. The normal distribution assumption would mean also having negative numbers which are meaningless. A better alternative might be to specify positive numbers only and decreasing probability for higher numbers. An alternative approach is therefore to say that the annual number is lognormal distributed. The problem of finding the probability of having more eight accidents is based on this distribution. By applying the transformations given above, the parameters can be found:

$$w^2 = \ln\left[1 + \left(\frac{\sigma}{\mu}\right)^2\right] = \ln\left[1 + \left(\frac{3.5}{6.5}\right)^2\right] = 0.25$$

$$\theta = \ln(\mu) - \frac{1}{2}\sigma^2 = \ln(6.5) - \frac{1}{2}3.5^2 = -4.25$$

$$P[X > 8] = 1 - \Phi\left[\frac{\ln(x) - \theta}{w}\right] = 1 - \Phi\left[\frac{\ln(8) - (-4.25)}{\sqrt{0.25}}\right]$$

$$= 1 - \Phi[1.266] = 1 - (0.3980 + 0.5) = 0.102$$

It is in other words a probability 10.2% of having more than eight accidents per year.

18.6.2 Fitting a parametric distribution to observed data

Vose (2000) has given some basic rules for deciding whether to apply a theoretical distribution when we are going to model a stochastic variable. Two key points are as follows:

- Does the theoretical range of the variable match that of the fitted distribution?
- Does the distribution reflect the characteristics of the observed variable?

In order to illustrate the practical approach, we will use a set of data for cargo oil outflow as a result of ship accident (see Table 18.3).
   It has been proposed that oil outflow volume may be described by a log-normal distribution because:

- The distribution range is positive numbers.
- It is highly skewed.
- The outflow may be seen as a product of a number of failures: accident event, vessel load condition, and penetration of hull barrier.

In the following, a stepwise description of the approach is given. The calculations were done using MS Excel and are summarized in Table 18.4.

Table 18.3 Oil outflow distribution based on 22 accidents

Spill size (tonnes)	Number of accidents
10–100	9
100–500	8
500–1,000	2
1,000–5,000	1
5,000–10,000	1
10,000–50,000	1

Table 18.4 Excel datasheet: Estimation of lognormal distribution

Range: X	X	Observations	Observed PDF	Observed CDF	Estimated CDF	Estimated PDF	Squared diff PDF
10–100	20	9	0.3913	0.3913	0.41628	0.41628	0.0006
100–500	200	8	0.3478	0.7391	0.69874	0.28246	0.0016
500–1,000	600	2	0.0870	0.8261	0.80789	0.10915	0.0003
1,000–5,000	2,000	1	0.0435	0.8696	0.89490	0.08701	0.0006
5000–10,000	6,000	1	0.0435	0.9130	0.94546	0.05056	0.0011
10,000–50,000	20,000	1	0.0435	0.9565	0.97644	0.03098	0.0004
							0.0047
Mean	4,803.3	Sum	22				
St. dev.	7,771.3	Sum+1	23				

The three first steps are as follows:

1. Specify the ranges for observed outflow amount in tonnes.
2. Select subjectively a point value X within each range.
3. List the number of observations N for each range.

Next, the observed PDF is computed:

$$f_i(x) = \frac{N_i}{\sum N_i + 1}$$

and from this, the CDF values can be calculated:

$$F_i(x) = f_i(x) + f_{i-1}(x) \quad \text{where } f_0(x) = 0$$

The theoretical distribution function is estimated by means of the Solver function in MS Excel. The objective is to estimate a lognormal distribution $LN(\mu, \sigma)$ that best fits the empirical data. The first step is to select a set of arbitrary values for the parameters μ_1 and σ_1 for a lognormal distribution. With these parameter values the estimated CDF is computed for the values of X.

The estimated PDF values are simply computed by applying the following formula:

$$f_i(x) = F_i(x) - F_{i-1}(x)$$

Next, the squared difference between the observed and estimated PDF is calculated and then added together. The Solver function in MS Excel can then be used to minimize the SSD by optimizing the values of μ_1 and σ_1. The solution found by Solver is:

$$\mu_1 = 3.66 \text{ and } \sigma_1 = 3.14$$

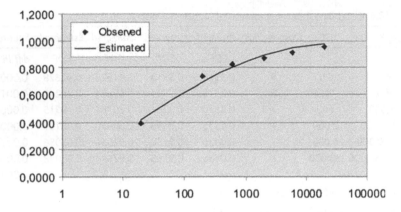

Figure 18.9 Oil outflow from ship accidents given by a lognormal distribution.

The results of the calculations are shown in Table 18.4. The theoretical distribution function is plotted in Figure 18.9.

18.7 ESTIMATING A WORST-CASE SCENARIO

18.7.1 A simple approach based on distribution function

In some cases, we may be more concerned about worst-case scenarios rather than the average consequences experienced from accidents. In Section 18.4, maritime disasters were discussed, and the average annual number of disasters was $\Theta = 3.7$. The empirical distribution of the number of fatalities is shown in Table 18.5. From this, we can find that the probability of having a disaster with more than 1,500 fatalities is:

$$P(N > 1,500) = 1 - P(N < 1,500) = 1 - \text{CDF}(1,500) = 1 - 0.9846 = 0.0154$$

The *return period* is defined as the *average time* between events of a certain magnitude, and may be written:

$$T = \frac{1}{\theta \cdot P(N > 1,500)} = \frac{1}{3.7 \cdot 0.0154} = 17.6 \text{ years}$$

It is interesting to notice that the return period is of the same order as the observation period which was 20 years. Secondly, it is worth noting that 90% of the observations have less than 500 fatalities which means that the basis for estimating the extremes was weak. This is very often the case, since extreme events fortunately are rare.

Table 18.5 Cumulative distribution of number of fatalities in maritime disasters

Fatalities range	Point value	Observations	Empirical PDF	Empirical CDF
20–39	30	10	0.1538	0.1538
40–49	45	10	0.1538	0.3077
50–99	75	13	0.2000	0.5077
100–199	150	11	0.1692	0.6769
200–249	225	6	0.0923	0.7692
250–399	325	5	0.0769	0.8462
400–499	450	4	0.0615	0.9077
501–999	750	3	0.0462	0.9538
1,000–2,000	1,500	2	0.0308	0.9846

Another way of stating the risk of a catastrophic scenario is to ask what is the probability of having this event in any given year? This may be answered by considering the annual number of accidents which was found to be 3.7 per year. This is rounded off to 4 per year. We can then calculate the probability that at least one of these four accidents leads to more than 1,500 fatalities. This can be regarded as binomial situation with $n=4$ and $p=0.0154$:

$$p(x) = \left\{ \frac{n!}{x!(n-x)!} \right\} p^x (1-p)^{n-x} \tag{18.16}$$

$$p(1) = \left\{ \frac{4!}{1! \cdot 3!} \right\} 0.0154^1 \cdot (1-0.0154)^3 = 0.0588$$

The corresponding return period is in other words $T = 1/0.0588 = 17.0$ years. This is close to the previous estimate.

18.7.2 The extreme value distributions

Another approach to finding extreme values is to use extreme value theory (EVT). EVT offers methods for dealing with events that happen rarely (Fisher & Tippett, 1928) and has shown great progress during the last decades. It has found application in fields like analysis of natural phenomena, insurance, and risk analysis (Fasen et al., 2014).

There are basically two main approaches in EVT:

1. *The Block Maxima Approach* is based on observation of the maxima within each observation period. It can be shown that these observations follow the *Generalized Extreme Value* (GEV) distribution:

$$G(x) = \exp \left\{ -\left[1 + \varepsilon \left(\frac{x-\mu}{\sigma} \right) \right]^{-\frac{1}{\varepsilon}} \right\} \tag{18.17}$$

where
    ε=shape parameter
    μ=location parameter
    σ=scale parameter

2. *The Threshold Model* is based on the observation and analysis of maxima above a specified limiting value. These maxima can be described with the so-called *Generalized Pareto* distribution:

$$H(y) = 1 - \left(1 + \varepsilon\frac{y}{\sigma}\right)^{-\frac{1}{\varepsilon}}$$

(18.18)

where
    ε=shape parameter
    σ=scale parameter

Figure 18.10 shows the largest disaster annually in the period of 2001–2020. See data in Section 18.4. The maxima vary from zero to around 1,000 fatalities with one exception: 1,864 fatalities in 2002 by the loss of *Le Joola* in Senegal.

The estimation of the parameters for the GEV distribution is illustrated in Table 18.6. The probability density function (GEV PDF) is estimated by minimizing the squared difference relative to the empirical PDF. The resulting distribution is shown in Figure 18.11. The highest maximum (1,864 fatalities) is evidently complicating the estimation of the tail of the distribution. The corresponding return period is plotted in Figure 18.12. By applying the EVT method, the return period for 1,900 fatalities is 6.4 years which is considerably less than the previous estimate. The findings should be taken with some caution, both with respect to having few observations in the range from 500 to 1,100 fatalities and the mentioned single observation of 1,864 fatalities.

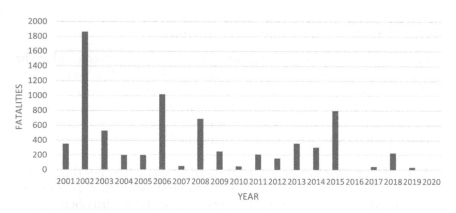

Figure 18.10 Annual largest disaster in terms of fatalities.

Source: Wikipedia.

Table 18.6 Estimation of the Generalized Pareto distribution for spill size

Fatalities range	Point value	Observations	Empirical PDF	Empirical CDF	GEV CDF	GEV PDF	SSD	GEV (I-CDF)
0–199	100	7	0.3333	0.3333	0.3298	0.3298	0.00001	0.6702
200–399	300	8	0.3810	0.7143	0.7162	0.3865	0.00003	0.2838
400–599	500	1	0.0476	0.7619	0.8270	0.1108	0.00399	0.1730
600–799	700	1	0.0476	0.8095	0.8773	0.0503	0.00001	0.1227
800–999	900	1	0.0476	0.8571	0.9056	0.0283	0.00037	0.0944
1000–1199	1,100	1	0.0476	0.9048	0.9237	0.0181	0.00087	0.0763
1200–1399	1,300	0	0.0000	0.9048	0.9362	0.0125	0.00016	0.0638
1400–1599	1,500	0	0.0000	0.9048	0.9453	0.0091	0.00008	0.0547
1600–1799	1,700	0	0.0000	0.9048	0.9522	0.0069	0.00005	0.0478
1800–1999	1,900	1	0.0476	0.9524	0.9576	0.0054	0.00178	0.0424
	3,000	0	0.0000		0.9743	0.0000		0.0257
	4,000	0	0.0000		0.9812	0.0000		0.0188
	Sum	20	0.9524			0.9576	0.00735	
	Location	110						
	Scale	101						
	Shape	0.9						

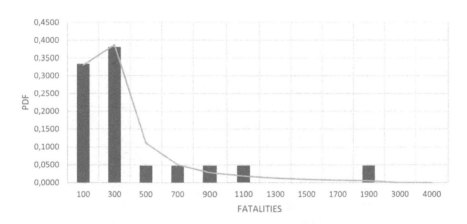

Figure 18.11 Disaster fatalities: probability density function of observed maxima and estimated extreme value distribution (GEV).

18.7.3 EVT estimation of tanker oil spills

Spill of tanker oil cargo from accidents has been a major safety and environmental problem since the late 1960s.

Table 18.7 lists all spills equal or greater than 10,000 tonnes since 1967 and adding up to 52 cases. Brief study of the table shows that both the

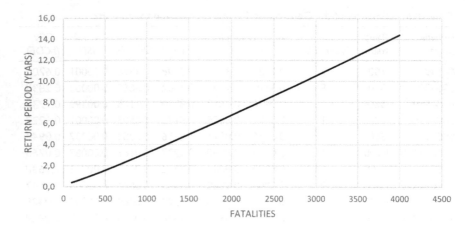

Figure 18.12 Return period for maximum fatalities.

Table 18.7 Tanker spills greater than 10,000 tonnes

Vessel	Year	Spill (tonnes)
Torrey Canyon	1967	119,000
R. C. Stoner	1967	20,000
World Glory	1968	46,000
Ocean Eagle	1968	12,500
Othello	1970	60,000
Arrow	1970	10,330
Texaco Danmark	1971	107,000
Texaco Oklahoma	1971	31,500
Wafra	1971	27,000
Sea Star	1972	115,000
Oswego-Guardian	1972	10,000
Napier	1973	30,000
Metula	1974	51,000
Jakob Maersk	1975	88,000
Epic Colocotronis	1975	61,000
Tarik Ibn Ziyad	1975	20,000
Corinthos	1975	36,000
Urquiola	1976	100,000
Argo Merchant	1976	28,000
St. Peter	1976	38,100
Hawaiian Patriot	1977	95,000
Verpet	1977	30,500
Borag	1977	34,000

(*continued*)

Table 18.7 (Continued)   Tanker spills greater than
10,000 tonnes

Vessel	Year	Spill (tonnes)
Amoco Cadiz	1978	223,000
Brzilian Marina	1978	10,039
Cabo Tamar	1978	12,500
Atlantic Empress	1979	287,000
Independenta	1979	94,000
Betelguese	1979	64,000
Irenes Serenade	1980	100,000
Tanio	1980	13,500
Globe Asimi	1981	17,000
Castillo de Bellver	1983	252,000
Nova	1985	70,000
Odyssey	1988	132,000
Khark 5	1989	70,000
Exxon Valdez	1989	37,000
Pacificos	1989	10,000
Mega Borg	1990	16,500
ABT Summer	1991	260,000
Haven	1991	144,000
Kirki	1991	17,230
Aegean Sea	1992	74,000
Katina P	1992	67,000
Braer	1993	85,000
Seki	1994	15,900
Sea Empress	1996	72,000
Erika	1999	25,000
Prestige	2002	63,000
Tasman Spirit	2003	30,000
Hebei Spirit	2007	11,000
Sanchi	2018	113,000

Source: ITOPF, Wikipedia.

frequency and spill size have decreased in later years. However, the focus here is to analyze the distributional characteristics of these large spills. The period from 1967 to 2018 represents 52 years. The average number of spills for the whole period is therefore 1 spill per year.

By applying the Threshold Model approach the extreme value distribution is estimated as shown in Table 18.8. The return period has been computed with two alternative assumptions: 1 spill per year for the whole period and 0.25 spills per year from year 2000 (conservatively 5 spills in the past two decades).

Table 18.8 Estimation of the generalized Pareto distribution for spill size

Spill size range	Point value	Spills	Empirical PDF	Gen. Pareto CDF	Gen. Pareto PDF	SSD	1-CDF
10–29,000	30,000	17	0.3208	0.3219	0.3219	0.00000	0.6781
30–59,000	60,000	10	0.1887	0.5395	0.2176	0.00084	0.4605
60–89,000	90,000	11	0.2075	0.6868	0.1473	0.00363	0.3132
90–119,000	120,000	8	0.1509	0.7867	0.0999	0.00261	0.2133
120–149,000	150,000	2	0.0377	0.8545	0.0678	0.00090	0.1455
150–179,000	180,000	0	0.0000	0.9006	0.0461	0.00213	0.0994
180–219,000	210,000	0	0.0000	0.9320	0.0314	0.00099	0.0680
210–239,000	240,000	1	0.0189	0.9534	0.0214	0.00001	0.0466
240–269,000	270,000	2	0.0377	0.9680	0.0146	0.00053	0.0320
270–300,000	300,000	1	0.0189	0.9780	0.0100	0.00008	0.0220
	400,000			0.9937			0.0063
	500,000			0.9981			0.0019
Sum		52	0.9811			0.01171	
Scale	77.066						
Shape	0.01						

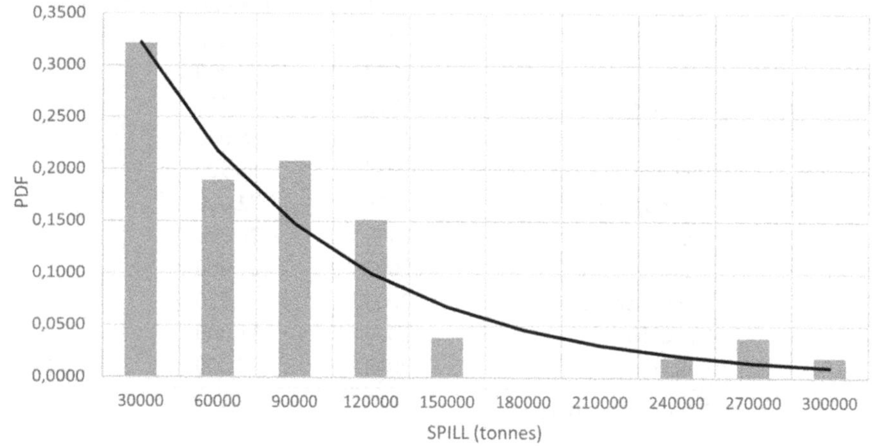

Figure 18.13 Spill size: observed and estimated probability distribution (Generalized Pareto).

Let us first recall the last accident in 2018, namely the spill of 113,000 tonnes from *Sanchi*. This corresponds to a return period of 5 years considering the whole period compared to 23 years looking at the last two decades (2000–2020). This means that the *Sanchi* spill was not a very improbable event. A more optimistic view is to state that there has only been one spill greater than 10,000 tonnes in the last decade (namely the *Sanchi* spill). A simple estimate for the spill frequency may then be set to 1/10=0.1 spill/year.

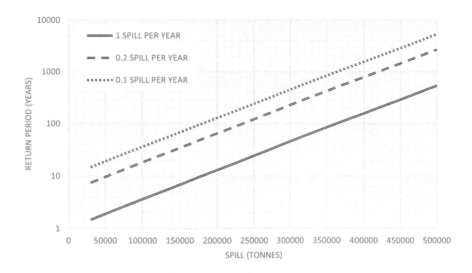

Figure 18.14 Return period for spills greater than 10,000 tonnes.

The corresponding return period would then increase to 47 years and still not an unlikely event. If a return period of 100 years is taken as target, this corresponds to a spill of 180,000 tonnes. This raises the question of whether the tanker spill risk has been brought to an acceptable level.

18.8 ANALYSIS OF COMPETENCE – CORRELATION COEFFICIENT

An important aspect of any safety program is to continuously assess attitudes and competence among the crew or employees. There are different sources that may be used for such an assessment:

- Examination scores
- Inspection and evaluation of work behavior
- Questionnaire study
- Assessment of personnel by their supervisors

In order to cross-check this kind of information, one may perform correlations on the data from such studies. Let us take the following situation: a company has invested in a safety awareness and training program and has later done an evaluation of the competence of the workforce. This leaves us with two sets of data:

1. Training program examination score (Score).
2. Safety rating by supervisor (Rating).

Table 18.9 Assessment of safety program

Crew member	Rating (X)	Score (Y)
1	4	5
2	9	8
3	7	9
4	9	8
5	3	4
6	7	8
7	8	8
8	5	7
9	10	8
10	6	5
11	8	9
12	8	7
13	6	7
14	9	10
15	8	10
Mean	7.1	7.5
St. dev.	2.0	1.8

The assessment data on the competence of the crew of a vessel are shown in Table 18.9. Both sets of data were based on a ranking scale from 1 (low) to 10 (high). It can be seen that the mean Score is 7.5, which is somewhat higher than the mean Rating which is 7.1.

The standard deviation for the exam scores (1.8) is slightly lower than for the ratings (2.0). One possible interpretation of these observations might be that:

- The exam scores are overestimating the safety competence.
- The rating approach is better at differentiating the competence among the individual crew members.

The results can be plotted in a scatter diagram as shown in Figure 18.15. Although a certain correlation between score and rating can be found, there is also a considerable scatter of the data. A more precise measure of the consistency between the two assessment parameters is the linear correlation coefficient:

$$r = \frac{\Sigma(X_i - \bar{X})(Y_i - \bar{Y})}{\sqrt{\Sigma(X_i - \bar{X})^2 \ \Sigma(Y_i - \bar{Y})^2}}$$

(18.19)

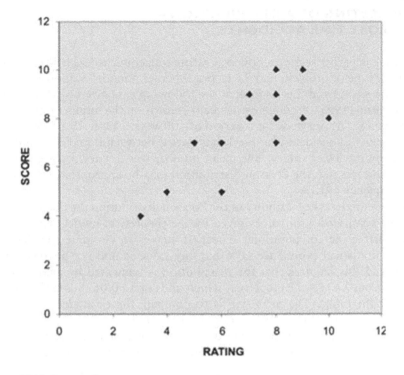

Figure 18.15 Scatter diagram: score versus rating.

or:

$$r = \frac{n \sum xy - \sum x \sum y}{\sqrt{n \sum x^2 - (\sum x)^2} \cdot \sqrt{n \sum y^2 - (\sum y)^2}}$$ (18.20)

which is more computation friendly expression.

With $n=15$ data sets the calculation is as follows:

$$\sum x = 107 \sum y = 113 \sum xy = 844 \sum x^2 = 819 \sum y^2 = 895$$

$$r = \frac{15 \cdot 844 - 107 \cdot 113}{\sqrt{15 \cdot 819 - 107^2} \cdot \sqrt{15 \cdot 895 - 113^2}} = 0.77$$

It can therefore be concluded that there is reasonable but not strong agreement between the exam score and rating. The safety management might therefore decide to use the examination method only, as this is more efficient and time-saving method than the rating. However, alternatively some effort should be directed toward improvement of the examination program in order to improve the differentiation between the candidates.

18.9 TESTING OF A DISTRIBUTION – LOST TIME ACCIDENTS

We have in earlier sections looked at estimating empirical distributions, but another type of problem may be to test whether a model is appropriate for the data set at hand. Let us look at the following case given by ReVelle and Stephenson (1995). A company has kept records on the number of lost-time accidents (LTA) per week for a period of 100 weeks. Table 18.10 shows that the number of accidents per week has varied between 0 and 3 with dominance on the lower values. The mean number was 0.9 accidents per week. This indicates that the Poisson distribution may be appropriate to describe the frequency of LTA.

In order to test the feasibility of the Poisson distribution the observed data will be compared with values given by the theoretical model. Table 18.11 summarizes the computations. The first step is to compute the PDF for $\lambda=0.9$. Statistical experience says that any value should not be lower than 0.05, and this requires that the distribution is truncated by grouping the values from $X=[3,4,5,6]$ together, which adds up to 0.0628 (see the shaded area in the table). The next step is to compute the estimated number of weeks on the basis of the 100 weeks observation period.

It can be proven that the squared sum of N relative differences between the number of observed weeks (O) and estimated weeks (E) is chi-square distributed:

$$\chi^2_{calc} = \sum_{i=1}^{N} \frac{(O_i - E_i)^2}{E_i} \qquad (18.21)$$

The distribution applies for positive values and has the parameter ν, which denotes the number of degrees of freedom and is given by:

$$\nu = (N-1)-1 \qquad (18.22)$$

Table 18.10 Number of lost-time accidents per week

LTA/Week X	Observed weeks N	X·N
0	45	0
1	29	29
2	17	34
3	9	27
4		
5		
6		
Sum:	100	90
Mean:		0.9

Table 18.11 Testing of Poisson distribution

LTA/week X	Poisson PDF	Poisson PDF corrected	Expected weeks	Observed weeks	$(O - E)^2/E$
0	0.4066	0.4066	40.66	45	0.4639
1	0.3659	0.3659	36.59	29	1.5749
2	0.1647	0.1647	16.47	17	0.0173
3	0.0494	0.0628	6.28	9	1.1767
4	0.0111				
5	0.0020				
6	0.0003				
Sum	0.0628	1.0000	100.00	100	3.2328

where N is the number of observations and where second term (-1) is the consequence of introducing an estimate for λ in the computations. Assuming our hypothesis that the number of observations is following a Poisson distribution is true (H_0), the calculated value of the chi-square criterion should be less than the critical value. We are then able to test the assumption of a Poisson distribution as follows:

From Table 18.11: $\chi^2_{calc} = 3.2328$
Degrees of freedom: $\nu = (4-1)-1 = 2$
Significance level: $\alpha = 0.95$
Tabulated value: $\chi^2_{\nu=2\,\alpha=0.95} = 5.991$

It can be concluded the postulated Poisson distribution is valid because the calculated chi-square value is less than the tabulated value.

18.10 CHOOSING AMONG ALTERNATIVE TRAINING PROGRAMS

The chi-square test can also be useful for testing other models. Let us take the following case described by ReVelle and Stephenson (1995). A company has tried out training programs of different duration: 1, 3, 5, and 10 days. The attending crew members were subject to a rating by their supervisors 6 months after the training session. The supervisor used the following ranking: excellent, good, or poor. The result of the assessment is shown in the upper section of Table 18.12. Inspection of the data may support the suspicion that there is no clear relationship between course duration and rating. It is interesting to note the low number of "excellent" ratings for the participants in the 10-day program.

Against this background, it may be interesting to test the following H_0 hypothesis:

H_0 = No correlation between Duration and Rating

Given this hypothesis, the distribution of the number of crew members would follow this computational rule:

$$\text{Cell}\left\{\text{Row}_i, \text{Column}_j\right\} = \frac{\text{Sum}(\text{Row}_i) * \text{Sum}(\text{Column}_j)}{\text{Sum}(\text{Rows \& Columns})} \tag{18.23}$$

which expresses the assumption of independence by the fact that the number in each cell is only determined by the column and row sums. Applying this rule to each cell in the table, we get the expected result shown in the middle part of Table 18.12.

Table 18.12 Analysis of training effectiveness

Observed rating

	Excellent	Good	Poor	Sum
I day	6	12	0	18
3 days	12	25	6	43
5 days	14	31	12	57
10 days	2	23	7	32
Sum	34	91	25	150

Estimated rating

	Excellent	Good	Poor	Sum
I day	4.08	10.92	3	18
3 days	9.75	26.09	7.17	43
5 days	12.92	34.58	9.50	57
10 days	7.25	19.41	5.33	32
Sum	34	91	25	150

Squared difference

	Excellent	Good	Poor	Sum
I day	0.9035	0.1068	3.0000	4.0103
3 days	0.5209	0.0453	0.1899	0.7561
5 days	0.0903	0.3706	0.6579	1.1188
10 days	3.8048	0.6626	0.5208	4.9883
				10.8736

Based on the data in the upper and middle parts of the table, we are now in the position to calculate the chi-square value as outlined in the previous section:

$$\chi^2_{calc} = \sum_{i=1}^{N} \frac{(O_i - E_i)^2}{E_i} \tag{18.24}$$

The result of for each cell is shown in the lower part of Table 18.12 and the sum is:

$$\chi^2_{calc} = 10.87$$

The number of degrees of freedom is given by the following formula:

$$v = (\text{Number of rows} - 1)(\text{Number of columns} - 1) = (4 - 1)(3 - 1) = 6$$

Assuming a significance level $\alpha = 0.99$ we get the tabulated value $\chi^2_{6,0.99} = 16.8$.

It can be concluded that the "no relationship" hypothesis holds as the calculated value is less than the tabulated value. This means that the variation in ratings is not more than would be expected under the null-hypothesis H_0. Or in other words, it is not possible to explain the variation in rating by the course duration.

It should, however, be pointed out that this conclusion is based on a very high value for the significance level: $\alpha = 0.99$. This reflects our concern of *not rejecting a true H_0*. If we decided to be more open to the alternative hypothesis that there is a relationship between course duration and rating, we might have set the significance level somewhat lower:

$$\alpha = 0.95$$

$$\chi^2_{6,0.95} = 12.6$$

The result did not, however, change our conclusion as the tabulated value still is higher than the one calculated.

18.11 THE EFFECT OF TIME – CONTROL CHARTS

The control chart (C-chart) approach was introduced as a technique in statistical quality control. Although different approaches have been advocated, it may basically be seen as a graphical method for checking whether a defined process is in a "state of statistical control" or out of control (Dhillon, 2005). Or more concrete, whether a measurement or observation falls outside the upper or lower control limits. This approach may also be applied in monitoring risk- or safety-related processes.

The number of accidents per time unit may be modeled with a Poisson process where the mean is:

$$\mu = \frac{N}{T} \tag{18.25}$$

where
  N = number of accidents
  T = time period

The standard deviation is:

$$\sigma = (\mu)^{1/2} \tag{18.26}$$

The upper and lower control limits may then be expressed as follows:

$$U_{CL} = \mu + 3\sigma \quad L_{CL} = \mu - 3\sigma$$

An unpublished investigation of 96 powered groundings for the period from 2001 through 2015 may be used as a case for studying the effect of hour and month on the accident frequency. The data was based on accident reports from the United Kingdom, Canada, and Australia.

Arranging the accidents to the six watch periods as shown in Table 18.13 gave the following mean frequency:

$$\mu = \frac{96}{6} = 16 \text{ per watch period}$$

And standard deviation:

$$\sigma = (16)^{1/2} = 4$$

Inspection of the table shows that there is considerable variation in the frequency. It can be seen that period 2 (early morning) is almost twice as high as period 5 (late afternoon).

Table 18.13 Powered grounding accidents per watch period

Period	Watch : Hours	No. of accidents	Normalized
1	00:00–03:59	17	106
2	04:00–07:59	21	131
3	08:00–11:59	14	88
4	12:00–15:59	17	106
5	16:00–19:59	11	69
6	20:00–23:59	16	100
	Total	96	

The corresponding control limits are:

$$U_{CL} = 16 + 3 \cdot 4 = 28 \quad L_{CL} = 16 - 3 \cdot 4 = 4$$

It can be concluded that none of the observations fall outside of the control limits. There is in other words no support for stating that any watch period is either more or less risky. A very simple support for this conclusion is the fact that a considerable number of powered groundings are neither related to extreme conditions nor stressful situations.

Some studies have also focused on the effect of the time of the year. In most parts of the world there are seasonal variations in weather and sea conditions. In Table 18.14, the monthly variation in number of accidents is shown based on the data set referenced above.

Mean frequency and standard deviation are:

$$\mu = \frac{96}{12} = 8$$

$$\sigma = \sqrt{8} = 2.8$$

The difference between the highest and the lowest months is even greater in this case as the number of accidents is three times as high in February and May compared to December. The control limits are:

$$U_{CL} = 8 + 3 \cdot 2.8 = 16.4 \quad L_{CL} = 8 - 3 \cdot 2.8 \approx 0$$

Based on this it can be concluded that the month of the year shows no statistically significant effect on the accident frequency.

Table 18.14 Powered groundings per month of the year

Month	No. of accidents	Normalized
January	6	75
February	12	150
March	6	75
April	8	100
May	12	150
June	8	100
July	11	138
August	6	75
September	6	75
October	9	113
November	8	100
December	4	50
Total	96	

REFERENCES

Dhillon, B. S. (2005). *Reliability, quality, and safety for engineers*. Boca Raton: CRC Press.

Fasen, V. et al. (2014). Quantifying extreme risks. In C. Klüppenberg, et al., (Eds.), *Risk: A multidisciplinary introduction* (pp. 151–181). Dordrecht: Springer.

Fisher, R. A., Tippett, L. H. C. (1928). Limiting forms of the frequency distribution of the largest and smallest member of a sample. *Proceedings of the Cambridge Philosophical Society, 24*, 180–190.

ReVelle, J. B., Stephenson, J. (1995). *Safety training methods* (2nd edition), New York: Wiley.

Vose, D. (2000). *Risk analysis: A quantitative guide*. Chichester: John Wiley.

Wikipedia, List of maritime disasters. https://en.wikipedia.org/wiki/List_of_maritime_disasters. (Accessed April 2022.)

Chapter 19

Learning from accidents and incidents

19.1 INTRODUCTION

In traditional risk management, before risk assessment was introduced as a tool, safety management was solely based on learning from accidents and incidents. When failures and errors occurred, attempts were made to correct the situation, to avoid that the same thing would happen again. In practice, a lot of our learning is made by trying and (sometimes) failing. Fortunately, this happens without serious consequences most of the time.

The introduction of risk-based safety management does not mean that we focus only on what may happen in the future and forget about learning from the past. This is still equally important and investigating serious accidents in detail to learn as much as possible can give valuable information for improving our systems and operations. All the key actors in the maritime industry, including IMO, national maritime authorities, classification societies, and shipowners and operators, need this information to improve safety.

In this book, focus is on learning from accidents and incidents, and the main purpose of investigating is to understand how and why accidents occurred, so we can learn and avoid similar accidents in the future. However, there may also be other purposes of accident investigations. The police will investigate with the purpose of determining if anyone is to be held accountable for what happened. Similarly, insurance companies and others may investigate to find out who should pay the costs associated with an accident.

To get the maximum benefit from accidents, we need systematic methods for investigating accidents, gathering the information, and distributing it in a format that can help stakeholders learn from it. There are many accident investigation methods described in literature and we do not intend to describe all of these in this book. Detailed descriptions are provided on three methods: STEP, M-SCAT, and MTO. They are covered in separate sections in this chapter. Sklet (2004) has made a comparison of some selected methods for accident investigation, and this can act as a good starting point for identifying and evaluating also other methods.

There are also numerous guidelines and other supporting documentation for conducting accident investigations. Within the marine sector, ABS has published guidance notes on investigation of marine incidents (ABS, 2005). ESREDA (2009) has also published guidelines that provide a good overview. "Investigating accidents and incidents" published by HSE UK (2014) is also a good source of information.

In this chapter, we will start by looking at the regulations governing accident investigations. This will be followed by some basic theory that is important, first on causes of accidents and second on accident theories. Then we move on to some methods for performing accident investigations and also an introduction to how an investigation can be performed in practice and how to write a report. At the end of the chapter, there is an overview of some sources of accident reports.

19.2 REGULATIONS

The requirement for maritime accident investigations to be performed is laid down in the United Nations Convention on the Law of the Sea (UNCLOS), article 94 on Duties of the flag State, paragraph 7, "Each State shall cause an inquiry to be held by or before a suitably qualified person or persons into every marine casualty or incident of navigation on the high seas involving a ship flying its flag and causing loss of life or serious injury to nationals of another State or serious damage to ships or installations of another State or to the marine environment." This paragraph requires inquiries to be held for all accidents of a certain consequence.

SOLAS and MARPOL further regulate this in relation to injury to people and damage to the environment respectively. SOLAS regulation i/21 states that "Each Administration undertakes to conduct an investigation of any casualty occurring to any of its ships subject to the provisions of the present Convention when it judges that such an investigation may assist in determining what changes in the present Regulations might be desirable". This regulation also specifies the purpose of the investigation, namely, to find out if changes to the regulations are desirable. This clearly points toward using investigations to learn from and to improve safety. The same regulation also requires information to be supplied to IMO.

Article 23 of the Load Lines Convention also contains more or less the same requirements as SOLAS.

MARPOL Article 12 states that investigations should be performed if the effect on the environment is major, and information should be provided to IMO. There is also a separate article laying down reporting requirements for incidents, article 8.

In 2008, IMO adopted the Code of International Standards and Recommended Practices for a Safety Investigation into a Marine Casualty or Marine Incident (Casualty Investigation Code) (IMO, 2008a). To supplement this, SOLAS Chapter XI-1 was also amended, making Parts I and II of the Casualty Investigation Code mandatory. SOLAS Chapter XI-1 further extends the requirements for investigation to include all accidents that are classified as a "very serious marine casualty", defined as a marine casualty involving the total loss of the ship or a death or severe damage to the environment. The Code also recommends an investigation into other marine casualties and incidents, by the flag State of a ship involved, if it is considered likely that it would provide information that could be used to prevent future accidents.

The overall objective of the Casualty Investigation Codes is to establish a common approach that can be used in investigations. It is underlined that the purpose of the investigations is not to apportion blame or determine liability, but to contribute to prevent future accidents.

A couple of important principles are laid down in the mandatory part of the code. First, investigations shall be impartial and objective, with no direction or interference from anyone who may be affected by the outcome of the investigation. Second, the code aims to protect seafarers against criminal prosecution, although this is in the non-mandatory part of the Code.

The code also specifies five principles of investigation in the non-mandatory part:

- Independent, with no interference
- Safety-focused, not blame or liability
- Cooperation between states (e.g., coastal state and flag state)
- Priority like any other investigation
- Scope should cover not just immediate causes, but also underlying causes

IMO have also published an "In-the-Field Job Aid for Investigators" (IMO, not dated) that provides practical advice on preparing, collecting evidence, and analyzing the evidence from an investigation.

It was mentioned that the results from investigations should be reported to IMO. IMO has established the *Correspondence and Working Groups on Casualty Analysis* that analyze the reports and seek to provide recommendations to IMO. The recommendations are then approved and forwarded to relevant bodies in IMO as appropriate. Information about accidents is also available from the Global Integrated Shipping Information System (GISIS). This includes a Maritime Casualties and Incidents module database, which includes data on Maritime Casualties and Incidents (MCI).

The ISM Code (IMO, 2018) also contains requirements relating to reporting and analysis, not only of accidents but also non-conformities and hazardous occurrences. The safety management system should ensure that these are reported, investigated, and analyzed by the company and the objective is to improve safety and prevent pollution in the future.

To supplement the ISM Code, IMO has also published guidance on near-miss reporting (IMO, 2008b). Therein, the term near-miss is defined as "A sequence of events and/or conditions that could have resulted in loss". The difference between accident and near-miss is thus whether losses have occurred or not.

It may also be mentioned that EU has introduced a common reporting system and fundamental principles (EU, 2009) and common methodology for accident investigation (EU, 2011).

19.3 CAUSES OF ACCIDENTS AND NEAR-MISS

"Cause" is a problematic term that is hard to define in a good manner, in particular when we start talking about root causes or underlying causes. Very many different terms are also being used, e.g., direct causes, immediate, root, underlying, background, determining, influencing, contributing, active failures, and latent conditions.

For this purpose, it may be useful to distinguish between deterministic and probabilistic causes.

Deterministic causes are causes that, if they are present, always will lead to a given effect. Alternatively, for a given accident, we can also say that a deterministic cause is a cause that is such that, if it had *not* been present, the accident would not have occurred. A simple example of the latter can be grounding due to loss of power on a ship. If power had not been lost, the accident had not occurred. However, we cannot turn the logic around in this case and say that when loss of power occurs, grounding will also occur. For grounding to occur, other conditions also must be present, e.g., that loss of power occurs close to shore.

This points to an important point in accident investigations – that accidents often are the result of several events and conditions being present at the same time. In the above example, we could list the following:

- Loss of power
- Shore in vicinity
- Drift toward shore

If either of these is missing, a grounding will not occur. One could thus argue that all three seen together is a deterministic cause that together causes the accident to occur.

Probabilistic causes are different because they influence the probability of an event occurring. A common example to use is lack of maintenance. Lack of maintenance will not automatically lead to loss of power on a ship, but it is more probable that this will occur if the crew has failed to maintain the ship. The problem with probabilistic causes is that we cannot say for certain that they were the "cause" of the accident, but we can say that they contributed to the accident and that if they had not been present, an accident would have been less likely to occur. Even if we lose power, we cannot necessarily point to poor maintenance and say that if the engine had been better maintained, this would not have occurred since random failures can occur even for the best maintained equipment.

In accident investigations, the deterministic causes are usually fairly easy to identify and understand, while the probabilistic causes will tend to be more difficult to find. Identifying probabilistic causes also require more understanding of the mechanisms that influence human behavior and also failure of technical systems. For example, if we do not understand how fatigue influences human performance, it will be difficult to identify how fatigue may have been a factor in an accident.

Probabilistic causes

In Figure 19.1, a possible sequence of causes is shown, from the accident that occurred at the bottom, working our way upwards to find the root cause. "D" next to a cause indicates that it is a deterministic cause, while "P" indicates a probabilistic cause.

In this case, a collision with another vessel has occurred and the investigation quickly shows that the navigator on the bridge had fallen asleep. It is highly likely that the collision would have been avoided if this had not occurred. However, to get a full understanding of causes, we need to keep on asking "why" an event occurs or a condition is present. If we ask why the navigator fell asleep, we may perhaps conclude that this was because he was very tired. Here we see that there is no deterministic connection between these two. Even if a navigator is tired, he will not necessarily fall asleep. Vice versa, we can say that falling asleep also can happen to those who are not fatigued, e.g., if the watch is very boring. This is therefore a probabilistic cause.

Further, a causal chain has been constructed where the cause of fatigue is increased workload and the cause of this again was cost-cutting by the company operating the ship. And we can also eventually point to demanding market conditions and argue that this was the reason for the cost-cutting.

This is a constructed example but is at the same time highly realistic. What we can also see from this is that the deterministic causes often can be regarded as symptoms of the underlying, probabilistic causes. Understanding these and fixing these will often have a better effect than directly trying to affect the direct cause (e.g., by punishing the sleeping navigator). Further, as we move backward through the causal chain, we can also see that the underlying causes often have an effect on many types of events. Fatigued crew can also make many other mistakes, not just falling asleep on watch. Similarly, cost-cutting can have an effect on many other aspects of safety, e.g., maintenance as mentioned.

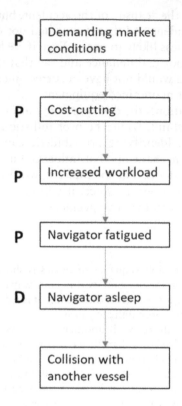

Figure 19.1 Possible sequence of causes.

19.4 ACCIDENT THEORIES

When we are investigating accidents, we will also have a more or less implicit understanding of what an accident is and how it occurs. This understanding will also influence what information we are looking for when collecting evidence and it will certainly influence our analysis of the evidence.

To illustrate this, we may start by considering some of the first "accident theories" that existed, namely that accidents were a result of "destiny" or were "Acts of God". In both cases, accidents are predetermined and occur whether we try to avoid them or not. With such an understanding of accidents, activities to manage safety are fairly meaningless.

Obviously, modern thinking about accidents is not based on an understanding like this, but they still tend to point to very different explanations.

One of the first scientific theories was the so-called "accident proneness theory". This postulated that certain people were more prone to have accidents than other, because they were more careless, clumsy, more willing to take risks, or for other reasons. This theory laid the ground for primarily blaming individuals when accidents occurred. The theory is not very much used today, although there are studies that indicate that there may be differences between individuals.

In broad terms, we can today talk about four main groups of models:

- Energy-barrier models
- Linear models
- Epidemiological model
- System models

Each of these is elaborated a bit more in the following subsections.

19.4.1 Energy-barrier models

The underlying idea behind energy-barrier models is that the source of accidents is energy in some form and that we use barriers (see more about barriers in Chapter 15) to protect various assets (people, the environment, economic values, etc.) (Haddon, 1973). The principle is illustrated in Figure 19.2.

Barriers can be physical, e.g., a fire-resistant wall that protects people from a fire, or it can also be immaterial barriers (e.g., maintenance to prevent equipment from failing, training to help people perform their job safely, and procedures describing a safe working practice).

Energy-barrier models have gained widespread use and are in practice underlying a lot of practical safety management work, although not necessarily explicitly recognized.

19.4.2 Sequential models

Sequential or Linear models are, as the name indicates, models where an accident is seen as a linear sequence of events. The first theory was the

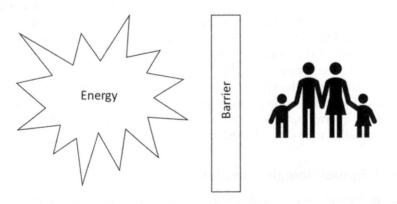

Figure 19.2 Energy-barrier model.

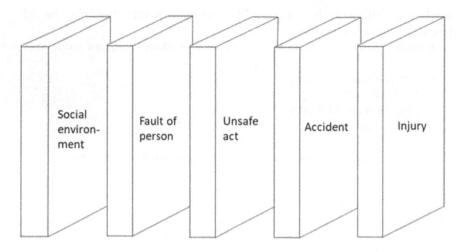

Figure 19.3 Domino model (based on Heinrich, 1931).

domino model (Heinrich, 1931), where each step in the sequence was described as a domino piece, and when one piece fell, all would fall and an accident would result (Figure 19.3).

If we look at the individual pieces, we can see that "Injury" corresponds to what we call the consequence of the accident. "Unsafe acts" are the direct causes, while "Fault of person" are the underlying personal characteristics of the individual who performs the unsafe act. Finally, "Social environment" is also termed "Ancestry" and relates to the background where this person is coming from. In this model, there is no room for the influence of the organization that an individual is working in.

Many developments of the domino model have been proposed, and in Section 19.7, the Loss Causation Model is described. This model also contains a generic set of "domino pieces", but they are described differently. In the Loss Causation Model, the sequence (from left to right) is Lack of Control by Management, Basic Causes, Immediate causes, Incident, and finally Loss. The two final steps are essentially the same as in the original domino model, but the other three pieces have been made more general and are focused on companies and their influence on accidents.

Of course, it may be argued that an unsafe act is sufficient to cause an accident. None of the other factors need to have failed and an accident may still occur. However, in accident investigations, we can use this model to look for explanations for why the unsafe act occurred.

19.4.3 Epidemiological models

Epidemiological models depart from linearity and imply a more complex view on what an accident is. The name comes from medicine, and the assumption

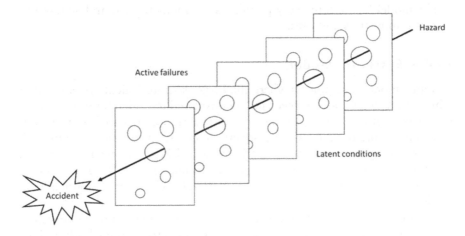

Figure 19.4 Swiss Cheese Model (adapted from Reason, 1997).

is that accidents occur due to a combination of latent factors (conditions that are present in the system) and active failures (errors, mistakes, equipment failure, etc.). When these are present at the same time, an accident occurs.

The most well-known epidemiological model is the Swiss Cheese Model, proposed by Reason (1997). He also introduced the terms latent conditions and active failures. The model is illustrated in Figure 19.4.

The Swiss Cheese Model is inspired by energy-barrier models and the name comes from the fact that the barriers are assumed to have holes in them (implying that they may fail) and can therefore be seen as slices of Swiss cheese. If the holes in the barriers are aligned, an accident occurs.

The holes in the barriers can be either latent conditions or active failures. Latent conditions are conditions that may be present for a long time, and which "weakens" the defenses against accidents, but which are not corrected. In the loss of *Herald of Free Enterprise*, the fact that the ship was designed with inadequate stability to tolerate flooding of the deck can be interpreted as a latent condition in operation. In addition, accidents are often triggered by active failures. Active failures are technical failures or errors in performing actions (human errors). In the *Herald of Free Enterprise* case, failure to close the bow door is an example of an active failure (or omission).

On the face of it, this looks as if it is very similar to the sequential models, with the barriers failing one by one. However, this should not be seen as a sequence, but rather as an illustration of how all barriers must fail at the same time. There is therefore not necessarily a link from failure of one to failure of the next. However, we see clearly the strong link to the energy-barrier model, with the additional feature that many barriers are illustrated.

The model has become very popular and is widely used and adapted to different industries and uses.

19.4.4 Systemic models

Systemic models are models where the larger sociotechnical system within which the accident occurs needs to be considered to understand the causes. By sociotechnical system, we mean a system that consists of both technical, human, organizational, and social elements. In practice, a ship with its crew and their organization can be regarded as a sociotechnical system and the same also applies for a company. Rasmussen (1997) was the first to propose a model with a sociotechnical hierarchy, shown in simplified form in Figure 19.5.

The hierarchy considers not just the technical systems and the operators immediately involved in the accident but takes into account both actions

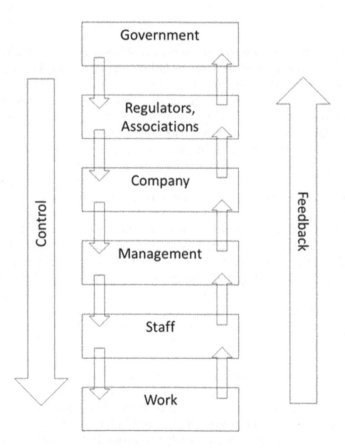

Figure 19.5 Sociotechnical hierarchy (based on Rasmussen, 1997).

and feedback from the levels above this, including supervisors, company management, various authorities, and up to the government level, where regulations that influence safety (and thus accidents) are passed. In investigations of major accidents, weaknesses in regulations are often pointed out. As was mentioned in the discussion of IMOs interest in investigations, one of the points is to identify recommendations for improving their regulations.

The hierarchy can also be seen as a number of "layers" of control loops, where the level above exercises control of the level below, and feedback is provided from the lower level to the level above. The feedback is used to modify the control actions, etc.

Flooding and capsize of Herald of Free Enterprise

WHAT HAPPENED?

The ro/ro passenger and freight ferry MV *Herald of Free Enterprise* (HFE), operated by Townsend Car Ferries Ltd., left the port of Zeebrugge bound for Dover at 18:00 hours on March 6, 1987. The vessel had a crew of 80 plus 81 cars, 47 freight vehicles, and approximately 460 passengers. Only a few minutes later when the vessel had turned and started to pick up forward speed, the water started to enter through the bow door onto the car deck which resulted in progressive heel and capsize. The vessel did not sink completely due to limited water depth. Due to the complicated evacuation and lack of rescue resources at least 150 passengers and 38 of the crew were lost (Figure 19.6).

Figure 19.6 Herald of Free Enterprise (Wikipedia, public domain photo).

CIRCUMSTANCES, CONTRIBUTING FACTORS

As a typical ro/ro vessel HFE had an enclosed superstructure above the main car deck. The doors were operated by hydraulic systems which were controlled manually on the car deck. The car deck was largely unrestricted as it had no sectioning or bulkheads. The deck was kept watertight with the closed bow and stern doors. Smaller leaks of water onto the deck could be removed by pumps (scuppers).

IMMEDIATE CAUSAL FACTORS

Due to high tide and mismatch between vessel and ramp design, the vessel had been trimmed by the bow to access the E deck. The vessel had not been trimmed to even keel before it left the port. Secondly, the bow doors were still open at the departure due to a series of failures of the crew. It was the job of the Assistant Bosun to close the doors, but he had left the watch and gone to bed. The Bosun observed the situation but did not see it as his task to intervene or notify the bridge. The Chief Officer was stationed at the bridge and could not inspect the closing of the doors himself, and even more seriously did not seek to get a positive confirmation of the closing. The Master was also passive in this respect.

As the vessel backed out and turned and started to pick up speed, the water started to flow onto the car deck through the bow door. The combination of trim nose down, possible overloading, increasing bow wave, and squat was sufficient to overcome the remaining freeboard or clearance to the deck at the bow. The fact that the vessel was in a turn may also have contributed to the sudden heel. As the car deck had no sectioning the water quickly started to accumulate along the deck side and thereby to build up a heeling moment due to the free surface effect.

BASIC CAUSAL FACTORS

During the investigation of the casualty, it was established that the management of the company had a critical role. The Master was under considerable pressure to keep the sailing schedule although the vessel had taken over the service on short notice. It was further clear that the Master of this vessel and a sister vessel requested installation of door indicators which would allow checking of the status of the doors from the bridge. This was denied by management on two occasions. It was also established that the vessels of this company regularly sailed over-loaded. The Masters had however no practical means of monitoring cargo intake and number of passengers. The policy onboard to accept "negative reporting" was fatal in this instance as nobody sought to positively confirm the closing of the doors. Apart from the inflowing water, the fact that the vessel was top-heavy may also have contributed to the sudden capsize.

Let us now identify some of the main factors that constitute this accident:

1. Causes
   - Vessel replaced another vessel on short notice
   - Vessel trimmed by the bow to match ramp
   - Pressure on Master to keep schedule
   - Policy on board to accept "negative" reporting
   - No monitor or indicator on bow door
   - Watch system in conflict with sailing schedule
2. Events
   - Assistant Bosun leaves watch and goes to bed
   - Bosun takes no action with respect to the open door
   - Bridge officers do not check closing of bow door
   - Vessel backs out from Zebrugge and turns to sea
   - Inflow of water through bow door
3. Consequences
   - Progressive heel
   - Capsize
   - Sinking in shallow water
   - Loss of 188 persons

The accident illustrated the vulnerability of ro-ro vessels with respect to flooding. Unlike traditional ship types they have large doors near the waterline (small freeboard) and few bulkheads to restrict water on deck. Experience has shown that even minor operational errors may have catastrophic consequences.

19.5 STEP

The first task in any accident investigation is to get an understanding of what has happened and who has been involved. Based on this, one can start to deduce why that accident happens. What we see in many accident reports is that the events are described in prose style. This may be acceptable to capture all pertinent information, but it has its obvious shortcomings as a starting point for causal analysis. Some accidents may develop gradually over a considerable time span and involve a number of actors in terms of persons and systems. It is therefore vital to place the individual events in a proper context and be given a certain structure and ordering.

A simple and illustrative method for presenting the actors involved and a timeline for the accident is a STEP diagram (Hendrick & Benner, 1986). An illustration of a diagram is shown in Figure 19.7.

The actors are shown to the left, listed vertically. An "actor" can be more or less anything that is relevant to understand how the accident developed, including persons, organizational units, vessels, systems, equipment, or objects. The timeline is used to illustrate events, conditions, and actions that are relevant to the development of the accident. Some of the advantages of using STEP is that it gives a good overview of the sequence of

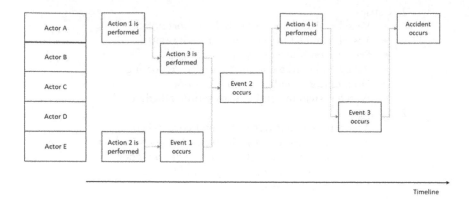

Figure 19.7 STEP diagram.

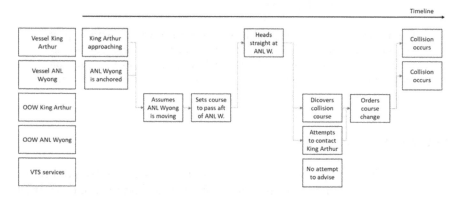

Figure 19.8 STEP Diagram for collision between King Arthur and ANL Wyong.

events at the same time as showing who/what is involved and the interactions between the events.

The number of actors illustrated and the level of detail in the timeline will depend on how detailed the investigation is, the complexity of the accident, and how deep into the background to the accident the investigation goes. In some cases, one may choose to go back to original design decisions or decisions relating to rules and regulations which may have been made years before the accident occurs.

STEP diagram for collision between two vessels

A concrete example of a STEP diagram is shown in Figure 19.8. This is loosely based on an actual collision that occurred between the gas carrier King Arthur and the container ship ANL Wyong in 2018 (MAIB, 2018).

From the diagram we can see that the starting point is that ANL Wyong is anchored, and that King Arthur is approaching the same area. The OOW (Officer On Watch) on King Arthur misunderstands the situation and believes that ANL Wyong is actually moving forward. He therefore sets the course of King Arthur so it will pass aft of ANL Wyong. However, since ANL Wyong is not moving, King Arthur ends up on a collision course. This is not discovered until too late and even if the OOW orders a course change, collision cannot be avoided. In the area where the collision occurred, there is also a vessel traffic service (VTS), but they did not provide any warnings. The OOW on ANL Wyong was also too late in contacting King Arthur to prevent the collision.

In this case, the actors are the two vessels, two individuals (the two OOWs), and an organization (the VTS service). Alternatively, a specific VTS operator could also have been the actor instead. If some sort of equipment failure had occurred (e.g., radar failure), the radar system could also have been an actor. If the accident report is consulted, we can also see that several other actors and also more events could have been included, but the diagram has been simplified for illustration purposes.

It may also be noted that in this particular case, inadequate situational awareness for the OOW on King Arthur was a key cause of the accident. He misinterpreted the information that he received regarding ANL Wyong and established a mental model that he used to predict the future position of ANL Wyong. Since this model was wrong as the vessel was stationary and not moving, his prediction of the future was also incorrect, and the wrong decision was made.

19.6 MTO METHOD

The MTO method (Man-Technology-Organization) is a method that combines three elements into one:

- A timeline, describing events leading up to the eventual consequence
- A description of deviations from what may be considered a "normal" situation
- An overview of barriers that have failed or worked during the accident sequence

In many respects, the method is very similar to what is described by US DoE in their workbook for conducting accident investigations (DOE, 1999).

In Figure 19.9, the elements of an MTO investigation are illustrated. Different variants are being used, but this is among others based on how the Petroleum Safety Authority Norway applies the method.

The timeline is illustrated in the middle, with the events leading up to the accident. Associated with the events are also causes, explaining why

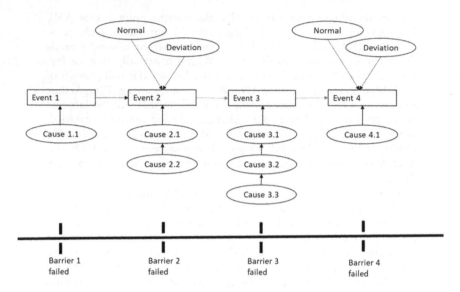

Figure 19.9 Illustration of the elements of the MTO method.

each event occurs. This can be necessary conditions or contributing factors. Each event can have several causes associated with it.

Above the timeline, the deviation analysis is shown. This can be done in different ways, but in the figure, the normal situation and the deviation is described. This can help to clarify precisely what the deviation was. A normal situation could, e.g., be described as follows "Procedure to be used" and the deviation is "Procedure was not used". Sometimes, only the deviation is described, without having to specify what the normal situation is. Deviations can be relevant for all events or just some of the events. There may also be more than one deviation for each event.

The third and final element, the barrier analysis, is shown at the bottom of the figure. In essence, this is just a listing of the barriers that have failed and also an illustration of where in the timeline the failure occurred. Barriers that have worked should also be included, although this is not always done in practice. The purpose is to illustrate clearly where failures occur, but at the same time it may also help us identify if there are missing barriers that should have been in place. The barrier analysis can thus be a help in identifying potential improvements.

Herald of free enterprise

An illustration of an MTO diagram for the capsizing of Herald of Free Enterprise is shown in Figure 19.10. This shows only a few of the events, including deviations from normal, causes of the events, and

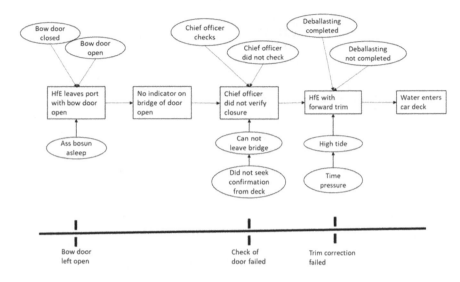

Figure 19.10 Extract from MTO diagram for capsizing of Herald of Free Enterprise.

barrier failures. It is noted that the diagram may be drawn in somewhat different ways, depending on the interpretation by the investigator and the focus of the investigation.

19.6.1 Flowcharting

It is clear that the STEP diagram is a suitable tool for outlining the main events in an accident. On the other side, it has also serious limitations in modeling the interaction or causality between the events. An alternative may be the use of a flowchart format where the events are described as nodes and interactions by connecting arrows. We may illustrate the approach with the *Exxon Valdez* grounding (Anonymous, 1990):

> The motor tanker Exxon Valdez left Valdez loaded with crude oil on the night of March 14, 1989. It soon ran into difficulties due to drifting ice in the restricted fairway and inadequate manning. The mate on watch deviated from the traffic separation scheme and took the vessel back too late to avoid grounding on Bligh Reef soon after midnight. Partly due to inadequate emergency preparedness the grounding resulted in a spill of 258,000 barrels of crude. The vessel itself had no technical shortcomings and was in fact the 'pride' of the Exxon fleet. The analysis revealed a number of management issues. The state of the vessel traffic service (VTS) in the area was inadequate and had its share of the responsibility for the disaster.

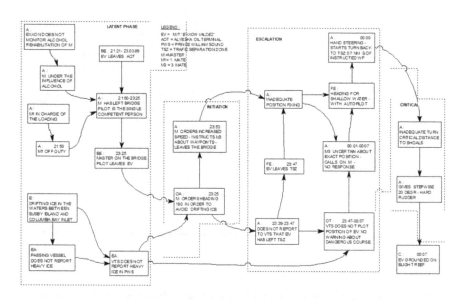

Figure 19.11 Exxon Valdez grounding – Flowchart of events.

The flowchart describing the main events is outlined in Figure 19.11. In order to provide some extra readability, the events are segmented in four phases:

1. Latency: Hazardous conditions and weaknesses
2. Initiation: Events that initiate the accident sequence
3. Escalation: Despite awareness of the situation and subsequent events it is difficult to handle the failures and errors
4. Critical: Last opportunity to avoid the accident

A summary of this accident is given in a tabular format in Table 19.1. This description summarizes key events under each phase, functional errors that explain why they took place, and potential preventive measures revealed as a result of the analysis process.

19.7 LOSS CAUSATION MODEL AND M-SCAT

19.7.1 Background

The Loss Causation Model (LCM) has been developed by the International Loss Control Institute Inc (ILCI) and is the core of the safety management approach promoted by DNV. It is primarily based on experience from land-based industry in the USA (Bird & Germain, 1992) and follows the

Table 19.1 Exxon Valdez grounding

Latent phase:	Functional error:	Preventive measures:
Inadequately manned relative to the operational requirements	Inadequate leadership Inadequate manning of the bridge Pilot did not report manning irregularity Inadequate planning of the voyage	Alcohol rehabilitation program Improve quality of management Leadership training Enforcement of port state control Modification of behavior

Summary of events:
Vessel leaves port fully loaded. Master under influence from alcohol and away from the bridge most of the time until pilot leaves the vessel. Voyage planning and adequate manning of the bridge have not been taken care of.

Initiating phase:	Functional error:	Preventive measures:
Inadequate bridge watchkeeping	Unqualified mate on watch Watch lacks a second officer required in Prince William Sound Inadequate voyage planning	Bridge management training Alcohol rehabilitation program Leadership training Behavior modification

Summary of events:
Vessel leaves Traffic Separation System (TSS) in order to avoid drifting ice. Watch is left to 3. Mate who is not checked out for this part of the voyage in Prince William Sound. Exxon requires 2 qualified officers on watch. Master gives an imprecise sailing order before he leaves the bridge.

Escalating phase:	Functional error:	Preventive measures:
Navigation without full control and consequent delayed turn back to safe waters (TSS)	Inadequate assessment and selection of waypoints No plotting in chart Increasing and too high speed Disturbed by drifting ice VTC did not monitor vessel	Training in coastal navigation Electronic chart display system (ECDIS) Upgrading of VTC facilities Training in VTC operations

Summary of events:
Vessel stays on a hazardous course for too long. Mate is unable to keep an updated position fixed relative to the shallow water at same time as he must observe the ice-infested waters. Starts to turn too late.

Critical phase:	Functional error:	Preventive measures:
Too late adjustment of turning radius (rate-of-turn).	Too late switching to hand steering Ordered to little rudder Maloperation of rudder control	Training in coastal navigation Training in ship handling Improved ergonomics of control systems

Summary of events:
The return to the TSS starts too late and the turning of the vessel is executed with too little rudder. There are some unclarified points with respect to whether the rudder is operated wrong or too little rudder is ordered. Vessel hits Bligh Reef.

Kristiansen (1995).

principles in the domino theory described earlier (Heinrich, 1950). In its present version, it constitutes a fairly complete system for management, planning, and control of industrial safety.

In this section, LCM and M-SCAT (Marine – Systematic Cause Analysis Technique) are described together since the model and the method are so closely connected.

19.7.2 The basics

The rationale of the LCM is that losses and safety problems can be tracked back to a lack of control in the organization. By losses are understood production problems, environmental pollution, property damage, personal injuries, employee health, etc.

The application of LCM may best be described according to the elements of the model outlined in Figure 19.12.

1. *Immediate causes*

    Immediate causes are defined as those circumstances that immediately precede the accident. They could also be labeled as unsafe acts or practices and unsafe conditions. Instead of "error", the word "substandard" is used, to avoid questions of blame.

2. *Basic causes*

    Basis causes are defined as those factors underlying the immediate causes. They are also referred to as root causes. Basic causes are split into *personal factors* (motivation, lack of skill, lack of knowledge, stress, etc.) and *job factors* (inadequate leadership or supervision, inadequate maintenance, unsuitable design or purchasing of equipment and tools, unclear procedures, etc.).

3. *Lack of control*

    Finally, lack of control is operationalized as inadequate and/or improper programs and standards and lack of monitoring of compliance with these standards. These are at the start of the event sequence and produce the conditions for basic causes. Control is traditionally seen as one of the basic management functions next to planning, organizing, and leading/directing. Lack of control may originate from:

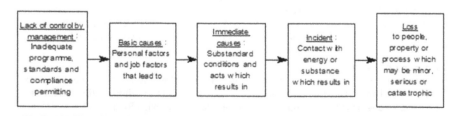

Figure 19.12 Loss Causation Model (Bird & Germain, 1992).

- Inadequate safety management system
- Inadequate performance standards
- Inadequate compliance with standards

Managers are encouraged to identify hazards using an appropriate technique such as HAZOP and Fault Tree Analysis. As well as hazards, means to control them should be identified. Once identified, a prioritization can be made. This implies that decisions must be made on the basis of risk and cost estimates. Again, several tools may be used such as Quantitative Risk Analysis and Reliability Analysis.

19.7.3 Taxonomy

The LCM approach to accident investigation and reporting puts emphasis on systematic coding. This allows the user to study trends and analyze correlation between accident factors. The risk manager may thereby monitor the risk picture of the operation. The approach assumes that the incident is reported both in terms of free text information and coded data. The main items are summarized in Table 19.2.

The Loss Causation Model in its original form was designed to deal with land-based industry and mainly focused on personal injury (work accidents) and economic loss. It can also be noticed that the database, apart from concrete findings, also contains an estimate of recurrence and potential seriousness of the kind of accident reported. Although this is merely subjective data it is useful for presentation of the risk picture for the system or operation. This may for example be done in a risk matrix showing probability versus severity.

Table 19.2 Items in the LCM investigation report

Section	Description
Identifying information	− Company, department, persons, date of incident − Kind of injury and nature of loss − Investigator identification
Risk	− Evaluation of loss potential (probability and seriousness)
Description	− Describe how the event occurred (free text)
Cause analysis	− Immediate and basic causes (free text)
Action plan	− What has and/or should be done to control the causes (free text)
Immediate causes	− Substandard actions and conditions (coded)
Basic causes	− Personal factors and job factors (coded)
Personal injury	− Type of contact and contact with (coded)
Review	− Assessment of the investigation and report
Site	− Sketch of site

A more detailed picture of the code structure is shown in Table 19.3. The different elements in the taxonomy will be commented upon.

Losses are used to denote the consequences of an accident and focus on personal injury and property/process damage. Both the object involved, and the degree of severity are indicated.

Incidents are only dealing with personal injury. It provides a classification of the kind of event or accident that directly affects the persons involved. It does not provide a description of the type of event that is related to the property or process involved. This weakness becomes evident if different accident types happen to be relevant for the system (property/process) and person. A fire in a building may for example lead to a fall accident to a worker involved. The fall accident may be coded (incident) but there is no coding option for the fire accident.

The explanation of why the incident took place is based on two sets of factors, namely "Immediate Causes" and "Basic Causes". The immediate causes are grouped in two:

Table 19.3 Incident analysis summary

Losses		Incidents		Inadequate control			
Type	No.	Type	No.	Program elements	P	S	C
Injury/illness:		Struck against		Leadership and			
First aid		Struck by		administration			
Medical treatment		Fall to lower level		Management training			
Lost workday Fatal		Fall on same level		Planned inspections			
		Caught in		Emergency			
Part of body harmed:		Caught between		preparedness			
		Overexertion		Organizational rules			
Head		Overstress		Accident/incident			
Eye		Contact with		analysis			
Hearing		Heat		Employee training			
Respiratory		Etc.		Personal protective			
Arm				equipment			
Etc.				Health control			
				Etc.			
Property or Process:							
Minor (less than $100)				*Legend:*			
Serious ($100–$999)							
Major ($1,000–$9,999)				P = *Inadequate*			
Catastrophic (over				*program*			
$10,000)				S = *Inadequate*			
				standards			
Type Property Damaged:				C = *Inadequate*			
				compliance			
Building							
Fixed equipment							
Motor vehicle							
Tools							
Etc.							

(Continued)

Table 19.3 (Continued) Incident analysis summary

Immediate causes		Basic causes	
Type	No.	Type	No.
Substandard Practices:		**Personal factors:**	
Operating without authority		Physical incapacity	
Failure to warn		Mental incapacity	
Failure to secure		Lack of knowledge	
Improper speed		Lack of skill	
Made safety device inoperable		Physical stress	
Used defective equipment		Psychological stress	
Etc.		Improper motivation	
Substandard conditions:		**Job factors:**	
Inadequate guards		Inadequate leadership/supervision	
Inadequate protection		Inadequate engineering	
Defective equipment		Inadequate purchasing	
Congestion		Inadequate maintenance	
Inadequate warning system		Inadequate tools/equipment/materials	
Fire Hazard		Inadequate work standards	
Explosion hazard		Wear and tear	
Etc.		Abuse and misuse	

- *Substandard practices*: Errors in terms of actions or omissions directly related to the process involved. It has 12 different categories of a fairly crude nature and does not indicate very precisely what kind of functions or tasks were affected.
- *Substandard conditions*: Denote inadequate work conditions, extreme environmental conditions, or acute events that affected a person or an operator.

The basic causes are also structured in two subsets as follows:

- *Personal factors*: Factors that make the operator less competent for the critical task or function. It may be a permanent or temporary disability or limitation. Another kind of explanation is lack of competence due to inadequate training. Wrong attitude may also be relevant.
- *Job factors*: Shortcomings or weaknesses related to work organization and management. This set of factors is mainly structured by management function, namely, leadership, engineering, purchasing, etc.

The last element in the causal analysis chain is to pinpoint what the authors have denoted "Inadequate Control". This was a fairly innovative idea in the sense that the method put more weight on the management part than had traditionally been the case. It is however somewhat difficult to see the distinction between Job Factors and Inadequate Control. The main difference is that the controls are more detailed and specific. Another interesting aspect is the option to make a distinction between different kinds of inadequacies: Inadequate program, standards, and compliance.

The method also offers a set of checklists for identifying Basic Causes termed SCAT (Systematic Causal Analysis Technique). Such factors are not readily visible in an accident investigation and require considerable knowledge and experience in work psychology and management. The checklist is seen as a help in identifying relevant basic causes.

19.7.4 M-SCAT

When DNV took over ILCI, and thereby the services related to the LCM method, they saw a need for adapting LCM to the marine and maritime business. As the critical part in accident investigation often is to identify causal factors it was decided to develop a taxonomy for marine operations which was called M-SCAT (marine version of the original SCAT scheme). The overall structure of the coding scheme was slightly modified. The new version is shown in Figure 19.13.

Apart from tailoring the framework to marine accidents M-SCAT also eliminated some of the shortcomings of the LCM taxonomy.

The M-SCAT taxonomy has the following basic classes:

- Type of Contact: Indicates the accident type and covers both Personal Injury and Property/Process Damage.
- Immediate/Direct Causes: Substandard Acts are directly related to maritime tasks. Substandard Conditions are directly related to ship systems and functions. Both categories are thereby less generic.

The rest of the taxonomy, Basic Causes, and Control Action Needs is more or less unaltered. This is to be expected, as they cover fundamental human and organizational behavior according to LCM.

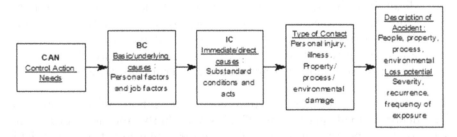

Figure 19.13 Marine – Systematic Cause Analysis Technique (M-SCAT).

19.8 ACCIDENT INVESTIGATION PROCESS

19.8.1 Overview of process

In Figure 19.14, an overview of the accident investigation process is shown. The steps can be briefly summarized as follows:

- Step 1: Initiation – includes deciding on whether to perform an investigation or not, establishing a mandate, and establishing the team that should perform the investigation.
- Step 2: Preparation – when the team has been established, they need to prepare for conducting the investigation, including various practical aspects relating to carrying out the investigation.
- Step 3: Gather evidence – this is an important part of the process and involves visiting the site, gathering various physical evidence, documents, and electronic evidence, as well as performing interviews.

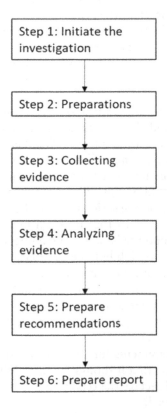

Figure 19.14 Overview of process.

- Step 4: Analyze evidence – in this step, the team establishes an understanding of what has happened and why. A timeline is established, barrier failures are identified, and any deviations are described.
- Step 5: Prepare recommendations.
- Step 6: Prepare draft and final report.

Steps 3 and 4 will typically flow over into each other, with iterations necessary as our understanding of what has happened increases. Need for further evidence gathering may then be necessary.

In the following subsection, each of the steps is described in greater detail. This is a description of a general investigation process. IMO have prepared a field guide for investigators that contains more details and can be used in a practical investigation (IMO, not dated). ABS (2005) also provides more advice on how to perform the investigation in practice.

19.8.2 Step 1: Initiate the investigation

As soon as practical after a report has been received about an accident or incident that requires investigation, the process should be initiated. Information will be lost as time goes by since the site may be changed, electronic evidence may be lost, and people will start to forget details about what happened.

Among the issues that need to be addressed before the investigation is started are:

- Determine the scope and terms of reference for the investigation. How wide-ranging should the investigation be and how deep into details and background causes should it go? What is the purpose of the investigation? In most cases, this will be to learn and provide recommendations. The scope and terms of reference should preferably be expressed in writing and should be signed by a manager at an appropriate level (usually depending on the seriousness of the accident).
- Who should take part in the investigation? An investigation team leader should be appointed and team members as necessary. The number should not be too high but need to match the scope and complexity of the investigation, the deadline set and the need for expertise in various areas (e.g., technical or operational expertise). Three to five persons is a typical team size.
- When should the investigation be finished and what is the deadline for submitting the report?
- What are the requirements for the report? Any particular format, or specific information that it should contain? Who is the intended audience? Who should receive the report in draft form for review before a final version is issued?
- What method should be used for the investigation? A common method and reporting format are advantageous as a familiar format will make

it easier for those who receive the report to quickly understand the report and the findings.

19.8.3 Step 2: Preparations

When the team has been established, they should convene for a kick-off meeting as quickly as possible. In this meeting, the facts that are known so far should be reviewed and what has already been done (e.g., in terms of securing evidence).

At this stage, all members of the team should make sure that they understand the objectives and the scope of the investigation and declare their competencies and qualifications. The team should also agree on relevant codes of conduct.

Further, the team should agree on an initial plan for who does what and a progress plan for the first steps in the investigation. Particular aspects that need to be investigated and in what order should be prioritized and the need for additional competence that is not already in the team must be identified.

At this stage, the team should also make sure that they have suitable facilities available for performing their work. Useful for the team is a conference room or similar where the team can work together and have discussions, suitable computer/internet access, whiteboard for sketching, e.g., a STEP diagram, space for drawings and other documents, etc.

The team may also need various other equipment for collecting evidence and for performing the investigation. This includes:

- Camera (photos and video)
- Recording equipment
- Paper and pens for sketches and note-taking
- Flashlight
- Tape measure
- Envelopes/boxes/bags/glasses for evidence and samples
- Stickers for marking samples
- Barrier tape for cordoning off the accident site
- Magnifying glass
- Flip-overs
- Tape
- Sticky notes
- etc.

19.8.4 Step 3: Collecting evidence

Evidence can come in many forms, and it may be necessary to prioritize what to focus on first. Evidence that may be lost or deteriorate with time is most important to secure first:

- The site of the accident should be secured and any evidence on the site should be collected. This may include photos and videos of the site, collecting equipment that has been used, noting position of people and equipment, etc.
- People directly involved and potential witnesses should be identified and also interviewed as early as possible. Those involved may leave the site and be difficult to get hold of (e.g., because they are leaving the ship to go off duty) and in general our memory of what has happened and what we have seen will quickly deteriorate. Interviewing relevant people as quickly as possible is therefore important to get as correct information as possible.
- Electronic evidence must be secured. This can take on different forms, from computer systems, control systems, navigation equipment, data recorders, video surveillance, etc. The status of many systems changes more or less continuously and ensuring that we collect information about the status when the accident occurred is important.
- Paper-based evidence such as temporary work documents, procedures, plans, organization charts, etc. are less likely to be modified and can therefore have lower priority in the data collection process.
- Similar with people that have not been directly involved or have witnessed the accident. They can also be interviewed later, since they will provide more general information and not so much specific to the accident. It is therefore less likely that they will forget important details.

Interviewing people is a skill that needs to be trained and that we will not go into a lot of detail on in this book. However, the IMO aid for investigators (IMO, not dated) provides useful advice.

When starting the interview, it is important to try to establish a personal rapport with the individual that is being interviewed. A "soft" start is helpful in this respect, by introducing yourself and also asking the person to present himself and his background. You should also ask permission to use a data recorder and explain why you want to use it and what the recording will be used for. Table 19.4 summarizes some important points to remember when conducting interviews.

19.8.5 Step 4: Analyzing evidence

The next step in the process is to analyze the evidence that has been collected. The first step is to ensure that the team has a common understanding of the sequence of events. This will typically be an iterative process until final agreement is reached. Additional evidence may also have to be collected before final conclusions can be drawn.

Next comes the equally important step of understanding why the accident occurred, i.e., the causes. Typically, hypotheses will be generated, discussed

Table 19.4 Some important points for interviews

Do	Don't
– Be polite – Behave in a natural manner – Keep interruption to minimum – Develop a friendly conversation – Display sincere interest – Use open questions – Specific questions can be used to obtain detailed information – Closed questions can be used to clarify points – Frequently summarize the information being given to ensure that you have understood everything correctly	– Do not use leading questions – Do not use hypothetical questions – Do not judge the interviewee – remain impartial – Do not interrogate the person – Do not ask for judgments or opinions – only facts – Do not voice your own opinions – Avoid discussions and arguments – Do not stress the person by asking them to speed up

among the team members and discarded, adjusted, or eventually accepted. Discussions between the team members are important in this phase, to test and discuss ideas and reach an agreement on the causes. This is also the stage where our accident model will be important.

At this stage, it is important not to have prejudices or preconceived beliefs or assumptions about the causes. If that is the case, it is far too easy to focus on the evidence that confirms our beliefs and disregard everything that points at other explanations. Considering all the evidence and being open for other explanations than we first assume is important.

The immediate causes are in many cases fairly easy to identify and describe, but the further back in the causal chain we go, to underlying causes, it becomes more difficult to tie the cause to the outcome in a definite manner. This is where the experience and judgment of the team members become important. Also, it is important to remember that the purpose of the investigation is not to prove connections between cause and outcome with 100% certainty. If the team has good reason to believe that something has contributed to the accident, this may in itself be sufficient to recommend that improvements should be made.

In the analysis of underlying causes, it is important to have human and organizational factors competence available in the investigation team. Methods exist that can help in identifying these causes, such as MORT, Tripod, and AcciMap (see, e.g., Sklet, 2004).

19.8.6 Step 5: Prepare recommendations

A crucial part of the results from an accident investigation is the recommendations. The EU common methodology for accident investigation (EU,

2011) among others describes a set of requirements to safety recommendations. These may be useful as a checklist when preparing recommendations. Generally, accident investigations can only provide recommendations and it will be up to authorities, companies, etc. to decide on what actually should be done. This will include both whether to implement the recommendations and/or to implement other actions than those that have been recommended.

To increase the likelihood of the recommendations being accepted and implemented, the checklist from EU may be a help. The list comprises the following points:

- Necessary – not very surprising, the recommendations should be necessary to avoid or reduce the risk associated with accidents in the future.
- Likely to be effective – the recommendation should be clearly linked to an identified weakness, and it should as far as possible be shown that it is likely to have an effect.
- Practicable – The recommendation must be practicable to implement, otherwise it is not likely to work in practice.
- Relevant – The recommendation should be relevant in relation to the investigation that has been performed and the findings from this.
- Stated in a clear, concise, and direct manner.
- Stated so that it can be the basis for corrective action plans, highlighting the safety gap that needs to be addressed.

19.8.7 Step 6: Prepare report

There is no standard format for accident investigation reports. However, it is recommended that each organization develops a standard format. In Section 19.9, a suggested format is described. Some general advice when writing the report (based on ABS, 2005):

- Start writing the report at the beginning of the investigation.
- Have the report reviewed for accuracy, clarity, language errors, and legal issues. A draft report should be issued for review before issuing the final version.
- Distinguish clearly between the facts that have been collected and any hypotheses, conclusions, and recommendations.
- Ensure that the report is written to the audience that it is intended for.
- Do not include information that is not necessary for the reader. Supplementary, detailed information can be placed in appendices or referenced to.
- Follow generally accepted guidelines for writing technical reports.

19.9 THE ACCIDENT REPORT

The accident report needs to be tailored to the audience that is going to read and use the report. This may influence the content of the report and perhaps in particular the style and level of details in the report. In a report intended to be read by fellow seafarers, much less detail about technical systems and way of operation is required compared to a report that is written for someone with no or little experience from the maritime industry.

A fixed format for accident reports does not exist. The format shown in Table 19.5 is partly based on ESREDA (2009) but with some modifications.

19.10 NEAR-MISS INVESTIGATIONS

Investigating near-misses is in principle not different from an accident investigation. Normally, the resources put into the investigation will however be more limited. The IMO guidance (IMO, 2008b) suggests that the following information should be gathered as a minimum:

1. Who and what was involved?
2. What happened, where, when, and in what sequence?
3. What were the potential losses and their potential severity?
4. What was the likelihood of a loss being realized?
5. What is the likelihood of a recurrence of the chain of events and/or conditions that led to the near-miss?

Near-misses that can be expected to happen again and that could have had serious consequences should be investigated in greater detail, using the accident investigation methods outlined in this previous section.

An important difference between accidents and near-misses is that it often is far more difficult to get reports on near-misses. There may be many reasons for this, e.g., fear of being blamed, too much hassle with reporting, seemingly no response on reports, and unsupportive company management. A good reporting culture is part of a good safety culture and some of the barriers against reporting can be overcome by:

- Supporting a "just culture", where reporting is encouraged and the person reporting does not risk blame, punishment, or any other negative reactions.
- Ensuring that reports can be made confidentially.
- Ensuring that adequate resources are available for investigations.
- Following up the near-miss report, providing feedback to the person reporting, and showing all involved that recommendations are taken seriously and implemented where relevant.

Table 19.5 Suggested content of an investigation report

Section	Outline
Summary	Short presentation of situation, key facts, events, causes, and lessons to be learned. Aimed at those who want to get a quick overview of the accident.
Background and purpose	Briefly explains why the investigation was initiated, includes the terms of reference, and describes the investigation team and the investigation process.
Analysis method used	This can include information about methods that have been used to collect evidence and to analyze the information that is gathered, and what accident theory is used as a basis for the investigation.
Factual information	This section summarizes the facts of the accident. This will include a description of the ship(s), relevant ship systems, crew, environmental conditions, and other relevant external factors that have influenced the accident. Further, the section should provide a narrative of the accident sequence and preferably also a graphical illustration of the events and actors (e.g., using a STEP diagram). It is important to distinguish facts clearly from analysis, presenting only facts in this section.
Analysis and results	This is an assessment of the facts and evidence provided in the previous section. Immediate and underlying causes should be presented and discussed, alternative explanations may be presented if this is relevant, or it is difficult to conclude based on available evidence. All failures, errors, and inadequacies both on a technical, crew, and organizational level should be identified and how they have contributed to the accident should be described.
Conclusions	Summarizes the investigation, briefly outlining the events, the consequences, and the causes.
Recommendations	Recommendations for improvements should be described and linked to the causes of the accident. They may address the shipowner, port, pilot company, maritime authority, or other relevant organizations.
Appendixes	Appendixes can vary significantly in number, content, and size. Examples of information that may be placed in appendixes are: - Vessel documentation, hull strength analysis, weather, and sea state data - Sea maps and track records - Background information, logs - Technical reports, analysis in detail of critical items or events - Lists of documents used as evidence - Lists of interviews (often not included in public versions of reports)

19.11 ACCIDENT INVESTIGATION REPORTS

Many maritime authorities and other organizations worldwide publish accident investigation reports on their websites. This is publicly available information that can be used for learning by anyone interested. In the following, some of these organizations are mentioned. We have focused on those that have reports available in English. Reports are also available from many other countries.

The web addresses provided below are correct as per May 2021 and the number of reports is also checked in the same period.

MAIB – Marine Accident Investigation Branch (UK)

This is the UK government agency that investigates marine accidents involving UK vessels worldwide and all vessels in UK territorial waters. On average, they receive between 1,500 and 1,800 reports of accidents each year and on average, about 30 investigations are performed. In addition to investigating accidents and publishing reports, they also issue safety bulletins to all mariners, with lessons learned from investigated accidents.

MAIB have close to 1,000 reports available on their website, mainly covering the period from 2000 and onwards.

Website: https://www.gov.uk/maib-reports

NTSB – National Transportation Safety Board (USA)

NTSB has a wide scope and performs investigations in all forms of transport in the USA, including aviation, railway, road, marine, and pipeline transportation. It is an independent agency that performs investigations and also provides recommendations to relevant parties.

As per May 2021, NTSB had close to 400 reports published on their website, going back to the early 1970s.

Website: https://www.ntsb.gov/investigations/AccidentReports/Pages/marine.aspx

Dutch Safety Board – about 100 reports

The Dutch Safety Board (Onderzoeksraad voor veiligheid) covers both transport and industry and publish their reports in both Dutch and English.

Website: https://www.onderzoeksraad.nl/

Norwegian Safety Investigation Authority – about 100 reports

The Norwegian Safety Investigation Authority (Statens havarikommisjon) covers transport in general, including maritime. In total, about 100 reports are published, although the majority are in Norwegian only. Some reports are however also available in English.

Website: https://havarikommisjonen.no/Marine/Published-reports

TSB - Transportation Safety Board of Canada

They also cover various modes of transport, but they have a large number of reports related to maritime, more than 500 since 1991. All are available in English.

Website: https://www.tsb.gc.ca/eng/rapports-reports/marine/index.html

ATSB – Australian Transport Safety Bureau

A final source that may be mentioned is the Australian Transport Safety Board that also has more than 350 reports published on their website.

https://www.atsb.gov.au/publications/safety-investigation-reports/?mode=Marine

REFERENCES

ABS. (2005). *Guidance notes on the investigation of marine incidents*. June 2005, Houston, TX: American Bureau of Shipping.

Anonymous. (1990). *Grounding of the US Tankship Exxon Valdez on Blight Reef, Prince William Sound near Valdez, Alaska*. National Transportation Safety Board, PB 90-916405. Washington, D.C.

Bird, F. E. & Germain, G. L. (1992). *Practical loss control leadership*. Loganville, Georgia: International Loss Control Institute Inc.

DOE. (1999). *Conducting accident investigations*. DOE Workbook, Revision 2, US Department of Energy, Washington, DC.

ESREDA. (2009). *Guidelines for safety investigations of accidents*. ESReDA Working Group on Accident Investigation, June 2009.

EU. (2009). *DIRECTIVE 2009/18/EC OF THE EUROPEAN PARLIAMENT AND OF THE COUNCIL of 23 April 2009 establishing the fundamental principles governing the investigation of accidents in the maritime transport sector and amending Council Directive 1999/35/EC and Directive 2002/59/EC of the European Parliament and of the Council*. https://www.legislation.gov.uk/eudr/2009/18/contents

EU. (2011). *COMMISSION REGULATION (EU) No 1286/2011 of 9 December 2011 adopting a common methodology for investigating marine casualties and incidents developed pursuant to Article 5(4) of Directive 2009/18/EC of the European Parliament and of the Council*. https://eur-lex.europa.eu/legal-content/EN/TXT/HTML/?uri=CELEX:32011R1286&from=EN

Haddon Jr, W. (1973). Energy damage and the ten countermeasure strategies. *Human Factors, 15*(4), 355–366.

Heinrich, H. W. (1950). *Industrial accident prevention*. New York: McGraw-Hill.

Hendrick, K., Benner Jr, L., & Benner, L. (1986). *Investigating accidents with STEP* (Vol. 13). Boca Raton: CRC Press.

HSE UK (2004). *Investigating accidents and incidents*, HSG245, HSE Books, Health and Safety Executive, London

IMO (2008a). *Code of international standards and recommended practices for a safety investigation into a marine casualty or marine incident (casualty investigation code)*. Resolution MSC.255(84).

IMO (2008b). *Guidance on near-miss reporting*. MSC-MEPC.7/Circ.7, October 2008

IMO (2018). *International safety management code*. London: International Maritime Organization.

IMO (not dated). *In-the-field job aid for investigators*. https://wwwcdn.imo.org/localresources/en/OurWork/IIIS/Documents/Casualty%20documents/In-the-field%20aide%20memoire.pdf

Kristiansen, S. (1995). *An Approach to Systematic Learning from Accidents*. IMAS'95: Management and Operation of Ships – Practical Techniques for Today and Tomorrow, May 24–25. Institute of Marine Engineers, London.

MAIB (2018). *Report on the investigation of the collision between the container vessel ANL Wyong and the gas carrier King Arthur in the approaches to Algeciras, Spain on 4 August 2018*. Southampton: Marine Accident Investigation Branch.

Rasmussen, J. (1997). Risk management in a dynamic society: A modelling problem. *Safety Science*, 27(2–3), 183–213.

Reason, J. (1997). *Managing the risks of organizational accidents*. Milton Park, Abingdon, Oxfordshire: Routledge.

Sklet, S. (2004). Comparison of some selected methods for accident investigation. *Journal of Hazardous Materials*, 111(1–3), 29–37.

Index

9 780367 518561